Information Decomposition of Target Effects from Multi-Source Interactions

Information Decomposition of Target Effects from Multi-Source Interactions

Special Issue Editors

Joseph Lizier
Nils Bertschinger
Juergen Jost
Michael Wibral

MDPI • Basel • Beijing • Wuhan • Barcelona • Belgrade

MDPI

Special Issue Editors

Joseph Lizier
The University of Sydney
Australia

Nils Bertschinger
Frankfurt Institute of Advanced Studies
(FIAS)
Germany

Juergen Jost
Max Planck Institute for Mathematics in the Sciences
Germany

Michael Wibral
Goethe University
Germany

Editorial Office
MDPI
St. Alban-Anlage 66 Basel, Switzerland

This is a reprint of articles from the Special Issue published online in the open access journal *Entropy* (ISSN 1099-4300) from 2017 to 2018 (available at: http://www.mdpi.com/journal/entropy/special_issues/Information_Decomposition_Target_Effects_Multi-Source_Interactions)

For citation purposes, cite each article independently as indicated on the article page online and as indicated below:

LastName, A.A.; LastName, B.B.; LastName, C.C. Article Title. *Journal Name* **Year**, *Article Number, Page Range.*

ISBN 978-3-03897-015-6 (Pbk)
ISBN 978-3-03897-016-3 (PDF)

Cover courtesy of Joseph Lizier.

Contents

About the Special Issue Editors

Joseph Lizier, Ph.D., is an ARC DECRA fellow and Senior Lecturer in Complex Systems in the Faculty of Engineering and IT at The University of Sydney (since 2015). His research focusses on studying the dynamics of information processing in biological and bio-inspired complex systems and networks, in particular for neural systems. He is a developer of the JIDT toolbox for measuring the dynamics of complex systems using information theory, and the related IDTxl toolbox for inferring effective network structure in neural data. His previous academic positions were at CSIRO ICT Centre (Sydney, 2012–2014) and Max Planck Institute for Mathematics in the Sciences (Leipzig, 2010–2012), with additional research experience in the telecommunications industry (Seeker Wireless, 2006–2010, and Telstra Research Laboratories, 2001–2006). He obtained a PhD in Computer Science (2010), and Bachelor degrees in Electrical Engineering (2001) and Science (1999), from The University of Sydney.

Nils Bertschinger, Dr. rer. nat., is Helmut O. Maucher-Stiftungsjuniorprofessor for systemic risk (since 2015) at the Frankfurt Institute for Advanced Studies and Goethe University (Frankfurt, Germany). He studied computer science at RWTH Aachen (diploma 2002) and received his PhD (2009) from the Max-Planck Institute for Mathematics in the Sciences (Leipzig, Germany) where he also worked as a scientist afterwards (2009–2015). His scientific interests are centered around information processing in complex systems. To this end, he pursues an inter-disciplinary approach applying methods from information theory and machine learning to different complex systems. Initially studying information processing in neural systems, he has considerably broadened his scope during his career, e.g. investigating social and economic systems. At FIAS he now investigates how systemic risks can develop and spread in financial systems.

Juergen Jost, studied mathematics, physics, economics, and philosophy at Bonn University from 1975–1980, received a PhD in Mathematics in 1980, was a Professor of Mathematics at Ruhr University Bochum from 1984–1996 and has been a Director of the Max Planck Institute for Mathematics in the Sciences, Leipzig, Germany, since 1996. He is also an honorary professor at the University of Leipzig and an external faculty member of the Santa Fe Institute for the Sciences of Complexity, USA. In 1993, he received the Gottfried Wilhelm Leibniz Award of the DFG (German Research Society) and in 2010, he obtained an ERC Advanced Grant. He is a member of the German National Academy Leopoldina, the Academy of Sciences and Literature at Mainz and the Saxonian Academy of Sciences. Jürgen Jost works on a general theory of complex systems and structure formation on the basis of information theory, dynamical systems, network theory and other mathematical ingredients. The potential applications, and the inspiration used, range from molecular biology (models of gene regulation and other molecular networks) to cognitive processes (theory of neural networks, invariant pattern recognition, modeling of basic phenomena in cognitive processing, learning theory) and social systems. Abstract topics include complexity measures, formal aspects of differentiation and integration and the emergence of higher level structures. He also works on various topics in pure mathematics (geometry, calculus of variations, dynamical systems) and the connections with high energy theoretical physics (quantum field theory, super string theory). He is the author of about 20 scientific monographs and advanced textbooks and of 300 publications in scientific journals and proceedings.

Michael Wibral, Ph.D., is Professor for Magnetoencephalography at the Brain Imaging Center of the Goethe University in Frankfurt am Main, Germany. His research focusses on studying information processing neural systems in general, and on uncovering the information theoretic signatures of predictive coding in particular. He has developed the MATLAB toolbox TRENTOOL for the analysis of information flows in neural data. Together with Joseph T. Lizier he intiated and contributed to the IDTxl toolbox for inferring effective network structure and partial information decompositions from data. He teaches information theory at the department of theoretical physics at the Goethe University. At present he is a guest scientist at the group of Theo Geisel at the Max Planck Institute for Dynamics and Self Organisation in Göttingen.

Editorial

Information Decomposition of Target Effects from Multi-Source Interactions: Perspectives on Previous, Current and Future Work

Joseph T. Lizier [1,*], Nils Bertschinger [2], Jürgen Jost [3,4] and Michael Wibral [5,6]

[1] Complex Systems Research Group and Centre for Complex Systems, Faculty of Engineering & IT, The University of Sydney, NSW 2006, Australia
[2] Frankfurt Institute of Advanced Studies (FIAS) and Goethe University, 60438 Frankfurt am Main, Germany; bertschinger@fias.uni-frankfurt.de
[3] Max Planck Institute for Mathematics in the Sciences, Inselstraße 22, 04103 Leipzig, Germany; jost@mis.mpg.de
[4] Santa Fe Institute, 1399 Hyde Park Road, Santa Fe, NM 87501, USA
[5] MEG Unit, Brain Imaging Center, Goethe University, 60528 Frankfurt, Germany; wibral@em.uni-frankfurt.de
[6] Max Planck Institute for Dynamics and Self-Organization, 37077 Göttingen, Germany
* Correspondence: joseph.lizier@sydney.edu.au; Tel.:+61-2-9351-3208

Received: 19 April 2018; Accepted: 19 April 2018; Published: 23 April 2018

Abstract: The formulation of the Partial Information Decomposition (PID) framework by Williams and Beer in 2010 attracted a significant amount of attention to the problem of defining redundant (or shared), unique and synergistic (or complementary) components of mutual information that a set of source variables provides about a target. This attention resulted in a number of measures proposed to capture these concepts, theoretical investigations into such measures, and applications to empirical data (in particular to datasets from neuroscience). In this Special Issue on "Information Decomposition of Target Effects from Multi-Source Interactions" at Entropy, we have gathered current work on such information decomposition approaches from many of the leading research groups in the field. We begin our editorial by providing the reader with a review of previous information decomposition research, including an overview of the variety of measures proposed, how they have been interpreted and applied to empirical investigations. We then introduce the articles included in the special issue one by one, providing a similar categorisation of these articles into: i. proposals of new measures; ii. theoretical investigations into properties and interpretations of such approaches, and iii. applications of these measures in empirical studies. We finish by providing an outlook on the future of the field.

Keywords: mutual information; information decomposition; unique information; redundant information; complementary information; redundancy; synergy

PACS: 89.70.Cf; 89.75.Fb; 05.65.+b; 87.19.lo

1. Background to Information Decomposition

Shannon information theory [1–3] has provided rigorous ways to capture our intuitive notions regarding uncertainty and information, and it has made an enormous impact in doing so. One of the fundamental measures here is mutual information $I(S; T)$, which captures the average information contained in samples s of a set of source variables S about samples t of another variable T, and vice versa. If we have two source variables S_1, S_2 and a target T, for example, we can measure:

1. the information held by one source about the target $I(S_1; T)$,

2. the information held by the other source about the target $I(S_2; T)$, and
3. the information jointly held by those sources together about the target $I(\{S_1, S_2\}; T)$.

Any other notion about the directed information relationship between these variables which can be captured by classical information-theoretic measures (e.g., conditional mutual information terms $I(S_1; T|S_2)$ and $I(S_2; T|S_1)$) is redundant with those three quantities.

However, intuitively, there is a strong desire to measure further notions of how this directed information interaction may be decomposed, e.g., for these two sources:

1. how much *redundant* or *shared* information $R(S_1, S_2 \rightarrow T)$ the two source variables hold about the target,
2. how much *unique* information $U(S_1 \backslash S_2 \rightarrow T)$ source variable S_1 holds about T that S_2 does not,
3. how much *unique* information $U(S_2 \backslash S_1 \rightarrow T)$ source variable S_2 holds about T that S_1 does not, and
4. how much *complementary* or *synergistic* information $C(S_1, S_2 \rightarrow T)$ can only be discerned by examining the two sources together.

These notions go beyond the traditional information-theoretic view of a channel serving the purpose of reliable communication, considering now the situation of multiple communication streams converging on a single target. This is a common situation in biology, and in particular in neuroscience, where, say, the ability of a target to synergistically fuse multiple information sources in a non-trivial fashion is likely to have its own intrinsic value, independently of reliability of communication.

The absence of (completely satisfactory) measures for such decompositions into redundant, unique and synergistic information has arguably been the most fundamental missing piece in classical information theory. Contemporary work on this problem was triggered by the formulation of the Partial Information Decomposition (PID) framework in a landmark paper by Williams and Beer [4] in 2010 (*note*: this paper was refined under an alternate title, and circulated privately only as [5]). This framework suggested that these quantities were related to the fundamental mutual information measures as follows and shown in Figure 1 for two source variables (with more complex relations for higher order interactions):

$$I(\{S_1, S_2\}; T) = R(S_1, S_2 \rightarrow T) + U(S_1 \backslash S_2 \rightarrow T) + U(S_2 \backslash S_1 \rightarrow T) + C(S_1, S_2 \rightarrow T), \tag{1}$$

$$I(S_1; T) = R(S_1, S_2 \rightarrow T) + U(S_1 \backslash S_2 \rightarrow T), \tag{2}$$

$$I(S_2; T) = R(S_1, S_2 \rightarrow T) + U(S_2 \backslash S_1 \rightarrow T). \tag{3}$$

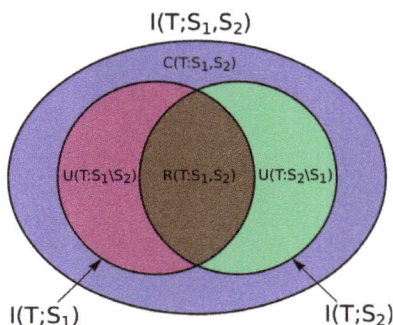

Figure 1. Partial information diagram for two sources to a target showing the relationship of the partial information quantities to the fundamental mutual information terms.

Crucially, the PID framework proposed that all these components coexist, subverting what had come to be the established interpretation [6] of the interaction information $II = I(S_1; T|S_2) - I(S_1; T)$, that $II > 0$ implied a synergistic interaction whilst $II < 0$ implied redundancy (and implying them to be mutually exclusive). Indeed, the PID framework revealed II as a net of synergy and redundancy terms (i.e., net synergy). Crucially, the PID framework proposed a set of axioms—*symmetry*, *self-redundancy* and *monotonicity*—that a measure of redundancy (for an arbitrary number of source variables to a target) should satisfy [5] (see summary e.g., in [7]). While these axioms were not sufficient to uniquely lock in a measure of redundancy, they do specify a partial ordering for redundancy terms across various joint collections of sources, and an algebra for how to compute *partial information atoms* attributed to such collections of sources (but no simpler collection) at nodes in a *partial information lattice* representing the hierarchy according to this ordering. This approach proved particularly appealing to the community.

In that paper, Williams and Beer [4] also proposed one measure of redundancy that satisfied the axioms they had laid out, known as $I_{\min}$. This measure found less favour in the community than the framework itself, encountering various criticisms such as that it did not distinguish "the *same* information or just the *same amount* of information" [8] (see also [7,9,10]), and did not satisfy a chain rule across multiple target variables [8]. However, perhaps the most controversy surrounded interpretation of the *Two-bit-copy* example (where a target is a copy of two IID input bits), which $I_{\min}$ suggested to be 1 bit redundant and 1 bit synergistic information, yet other authors felt should be 1 bit of unique information from each source because "the wires don't even touch" [10], p. 167. Indeed, the strong intuition some felt on this interpretation led Harder et al. [9] to suggest a 4th axiom (known as *identity*) requiring the redundancy in such copying situations to be equal to the mutual information between the two source variables.

Following these developments, the past few years witnessed a concentration of work by the community in proposing, contrasting, and investigating new measures to capture these notions of information decomposition. (See an earlier review by Wibral et al. [11], in Section 4 of that article). Primarily amongst these were the information-geometry based I_{red} from Harder et al. [9], and S_{VK} from Griffith and Koch [10] and $\widetilde{UI}$ from Bertschinger et al. [12], all of which were presented only for a pair of sources. The latter two approaches were later found to be equivalent, and attracted much attention due to being placed on a particularly rigorous mathematical footing, despite computational difficulties in solving the convex optimisation they require. For example, the derivation of the measure by Bertschinger et al. [12] followed directly (rather than being posed ad-hoc) from an assumption that existence of unique information depended only on the pairwise marginal distributions between the individual sources and the target (known as "Assumption (*)"). Furthermore, the measure was given an operational interpretation in terms of how unique information could be exploited in decision problems. Finally, many mathematical properties of the approach were proven by Bertschinger et al. [12] and in follow up papers by these authors [13,14].

Yet while many authors welcomed the new measures for satisfying the identity property, it was quickly realised that they did not completely solve the search for a redundancy measure for an arbitrary number of variables. This is because Rauh et al. [13] demonstrated that no redundancy measure can satisfy the identity property along with the original axioms of Williams and Beer [4] and still provide non-negative partial information atoms when we have more than two source variables.

As a consequence, the search for candidate redundancy measures continued, with various groups considering to drop either the identity property or one or more of the original Williams and Beer [4] axioms. Olbrich et al. [14] and Perrone and Ay [15] investigated the possibility of defining synergy via projections of probability distributions to those retaining only certain orders of interactions (in particular using exponential families), while Rosas et al. [16] sought similar decompositions for joint entropies. Some approaches sought to construct intermediate variables that could be used to represent components of the decomposition, e.g., the investigation of Gács-Körner common information by Griffith et al. [17], Griffith and Ho [18] and constructions of variables to contain synergy only by

Quax et al. [19]. Others investigated relatively simpler mechanisms such as the minimum mutual information (MMI) provided by any source by Barrett [20] (and the related approach by Chatterjee and Pal [21]).

Meanwhile, other theoretical developments were taking place in parallel. One line of work considered how these measures relate to concepts of distributed information processing in terms of information storage, transfer and modification [7,22–24]. Lizier et al. [7] made a case that information decomposition approaches should (at least) be interpretable on pointwise or event-wise realisations of the source and target variables, rather than only with their averages. Barrett [20] began considering continuous-valued variables, and indeed showed that the minimum mutual information was a unique form of the redundancy for linearly coupled Gaussian variables, for two sources, under the Williams and Beer [4] axioms and Bertschinger et al.'s [12] Assumption (*). Others provided detailed comparisons between the measures and catalogued results from various logic gates (e.g., [25]).

Despite the lingering issues surrounding a definitive measure of redundancy, the desire for using such measures has been intense, and applications have been made drawing on the variety of measures listed above. Computational neuroscience in particular emerged as a primary application area due to significant interest in questions surrounding how target neurons integrate information from large numbers of sources, as well as the availability of data sets on which to investigate these questions. For example, Timme et al. [25] contrasted I_{min} with several earlier candidates regarding the decomposition of information contributions between various electrode measurements from developing neural cultures, concentrating in particular on how redundancy and synergy generally increase during development. Later, Timme et al. [24] applied the PID view of information modification of Lizier et al. [7] to study dynamics of spiking activity of neural cultures incorporating history vectors of the target neuron, finding that neurons which modify "large amounts of information tended to receive connections from high out-degree neurons" in the effective network structure. Stramaglia et al. [26] use interaction information or net synergy interpretations to study interactions in electroencephalography (EEG) measurements in pre-seizure states for an epileptic patient. Further, Wibral et al. [27] applied PID to make various, theoretically proposed neural goal functions–such as infomax [28]–comparable, and were able to clarify whether the theories do indeed represent the information components that they had aimed at. Applications also began to emerge in examinations of biological data sets (e.g., [21,29]), and in gambling [30].

2. Contents of the Special Issue

In December 2016 we held an informal workshop on Partial Information Decomposition at the Frankfurt Institute for Advanced Studies and the Goethe University, bringing together some of the leading research groups in the field to discuss their latest developments. The workshop revealed a strong level of new activity in the area, and triggered deep discussions in particular regarding how further progress towards a measure may be made and which axiom(s) may need to be dropped/changed for this to occur. The attendees expressed a desire for publications of such new activity to be gathered in a common location, resulting in this Special Issue. The issue seeks to bring together the new efforts presented at the workshop, to capture a snapshot of current research, as well as to provide impetus for and focused scrutiny on newer work. We also seek to present progress to the wider community and attract further research in this area. In scope for the issue were research articles proposing new measures or pointing out future directions, review articles on existing approaches, commentary on properties and limitations of such approaches, philosophical contributions on how such measures may be used or interpreted, applications to empirical data (e.g., neural data), and more.

The contributions we have published can be classified under three key themes: new PID measures, theoretical investigations (including examinations of numerical estimators), and applications.

2.1. New Measures of Redundancy

Considering the first, perhaps not-so surprising theme, our Special Issue carries three papers proposing new measures of redundancy.

Rauh et al. [31] present the *extractable shared information* as a redundancy measure for the bivariate case. The key feature of this measure is that, in contrast to previous proposals, it satisfies the property of target or left monotonicity (i.e., that the redundancy is non-decreasing when more target variables are added [8], or restated here as redundancy being non-increasing when a new target variable is a function of the old target). This is achieved via a construction which translates any measure of shared information into one that satisfies this property. The authors then explore the properties of this measure, and show for example that it is not compatible with a Blackwell interpretation of unique information (see their other contribution, [32], discussed in Section 2.2).

Ince [33] constructs a measure I_{CCS} of redundancy by directly examining common values of pointwise mutual information (or change in surprisal) in each realisation of the variables. Interestingly, Ince [33] considers positive and negative pointwise information as fundamentally different and treats their occurrence separately, counting redundancy only from pointwise co-information terms when the signs of all relevant change in surprisal terms align. This necessitates considering redundant misinformation as well as redundant information (and related terms such as unique misinformation). The author argues for the justification of these new perspectives as well other properties of the measure, including replacing a requirement of monotonicity with subset equality (which had usually been considered only as part of monotonicity) and the use of a modified independent identity axiom introduced here. Ince [33] also provides a game-theoretic operational interpretation to argue for the approach presented, contrasting this with the decision-theoretic operational interpretation from Bertschinger et al. [12]. This line of work continues in a companion paper [34].

From a similar pointwise perspective, Finn and Lizier [35] build on earlier work to now directly identify positive and negative components of pointwise information from each source to the target as *specificity* and *ambiguity* [36], and argue that redundancies in these should be treated independently to avoid blurring them (in the same way that PID originally sought to avoid how interaction information blurs synergy and redundancy). The authors introduce a new example called "Pointwise Unique", where in any pointwise configuration only one source holds non-zero information about the target. They demonstrate that other existing measures do not identify unique information in this case, unlike their new approach. They also introduce a new operational interpretation of redundancy in terms of probability mass diagrams, and in allowing negative terms in net, show that their pointwise and component-wise approach is unique in satisfying a chain-rule over target variables. The latter feature also allows the approach to provide a consistent answer to *Two-bit-copy* of 1 bit redundant and 1 bit synergistic information, regardless of the order in which target bits are decomposed.

It is interesting to note that the latter two of these new approaches independently make similar departures from the status quo here: both taking a "bottom-up" pointwise information perspective, considering negative partial information terms, dropping the identity axiom, and being extendible to three or more source variables.

2.2. Theoretical Investigations

Next, the special issue contains a number of theoretical investigations into the properties of PID approaches in general and with regard to specific measures.

James and Crutchfield [37] make the case for measures of information decomposition beyond the standard Shannon measures by seeking to differentiate two examples of three variable systems: one constructed with dyadic dependencies and the other with triadic. Via a comprehensive analysis, they show that no standard Shannon measure can differentiate between the two examples, whilst various measures of information decomposition, e.g., Gács-Körner common information and the Bertschinger et al. [12] PID, are able to. Whilst these two PID approaches do provide such a differentiation, the authors express a general desire for the additional existence of a symmetric decomposition that does not partition variables into sources and targets.

Pica et al. [38] examine a two-source one-target PID from three perspectives in total, i.e., one perspective for each variable as the target, in order to examine commonalities between the perspectives. Assuming non-negativity but not any specific PID measure, they identify only seven non-negative information subatoms that are required to construct each of the three PIDs in full, subject to knowing the ordering of the three redundancy terms. The authors also suggest novel definitions for a split between source redundancy (arising from correlations between the source variables) and non-source redundancy. Indeed, the authors use their approach to provide further insights into the information structure of the dyadic-vs-triadic example of James and Crutchfield [37].

Rauh [39] identifies the cryptographic interpretation of secret sharing as a useful model to consider information decomposition, since secret sharing schemes incorporate specific understanding of which subsets of participants have information about the secret. The author establishes correspondence between secret sharing and PID, and then uses this approach as a model to explore the partial information lattice. Negative terms in the lattice are identified for more than two participants (analogous to the argument by Rauh et al. [13]), which leads the author to discuss whether and how such terms could or should be interpreted, and subsequently questions whether the lattice needs to be extended or improved in some fashion.

Rauh et al. [32] examine the decision-theoretic Blackwell partial order, which ranks information channels (with a common input) according to the utility that can be obtained when decisions are made on the channel outputs. The authors present the unexpected result that a coarse-graining of one channel output may actually result in improved utility. They go on to compare the Blackwell ordering to mutual information, and discuss implications of the result for information decomposition.

Faes et al. [40] utilise vector autoregressive Gaussian models and the MMI measure, coupled with the aforementioned perspective of information modification, to examine the decomposition of contributions from information sources to a target over various temporal scales. The method of investigating the decomposition of contributions across different scales is achieved by a combination of filtering and then downsampling, and synthetic examples in the first instance are used to demonstrate that the method can reveal quite different decompositions at different temporal scales due to contrasting fast and slow dynamics. The authors then apply the approach to intracranial EEG data obtained prior to and during epileptic seizures, revealing in particular how synergistic and unique information transfer components change with scale.

Makkeh et al. [41] consider the the convex optimisation problem that must be solved in order to evaluate the Bertschinger et al. [12] approach, continuing on from the original observations by Bertschinger et al. [12] that Mathematica could not directly solve these optimisation problems. The authors provide both theoretical and practical perspectives, discussing various algorithmic approaches to the problem and why some perform poorly, and empirically comparing the performance of a number of software packages. Importantly, the authors identify two software packages which perform satisfactorily, and make recommendations regarding their use here.

2.3. Applications of Information Decomposition

Applications of PID form a substantial class of papers in our special issue. As identified above, neural applications (in addition to the EEG analysis by Faes et al. [40] above) account for the largest portion of these.

Kay et al. [42] consider the PID between a neural receptive field input and the signal modulating (amplifying or suppressing) it, giving rise to an output signal. In particular they demonstrate that, contrary to intuition from some perspectives, a modulatory signal can affect the transmission of information about other inputs without being transmitted itself. The authors go on to apply the Ince [33] and Bertschinger et al. [12] PID measures, as well as a related decomposition of entropy by

Ince [34], to results from a visual contrast detection task in order to demonstrate that such forms of modulation may occur in real neural systems.

Wibral et al. [43] apply PID to decompose information storage, transfer and in particular information modification in developing neural cultures, following the perspective of Lizier et al. [7]. Utilising the Bertschinger et al. [12] PID measure via the publicly available IDT^{x}l toolkit [44], the authors identify the aforementioned components of information processing from pairs of input (multi-unit) spike train recordings to each output recording. They report that information modification initially rose during development with maturation of the culture (indicating intricate processing capabilities), followed by a decay when redundant information among neurons took over (possibly due to a lack of external inputs).

Moving on to artificial neural computation then, Tax et al. [45] also use PID to analyse neural development, but this time the development of a restricted Boltzmann machine during training. The authors focus on decomposing the information held by (sample pairs of) individual hidden neurons about the target variable to be classified, using I_{min} [4]. They observe a first phase where neurons appear to learn predominantly redundant information about the target, followed by a second phase where the neurons specialise to learn unique information about the target (also with a significant synergistic component). Further, the authors report that while larger networks appear to utilise higher order representations to a greater extent, individuals in smaller networks appear to learn more unique details, and conclude that perhaps network size pressure on learning can lead to disentangled representations.

Ghazi-Zahedi et al. [46] apply PID in order to further our understanding of morphological computation, "processes in the body that would otherwise have to be conducted by the brain". Examining the embodied concept of the sensorimotor loop model, the authors quantify morphological computation as synergistic information from the cognitive system's actuators and the current world state (incorporating both the system's morphology and the part of the environment that can be affected by and affects the system) to the next world state. The authors focus on the synergy measure of Perrone and Ay [15] for this purpose, comparing it to previous measures and finding it to be generally more reliably oriented with their intuition, though not in all cases.

As highlighted above, computational biology has also emerged as an interesting application area for PID, and here Maity et al. [47] use PID to examine cross-talk in biochemical networks between two mitogen-activated protein kinase (MAPK) pathways. The authors examine data from models of these pathways, using Gaussian model calculations of the information-theoretic terms and quantifying net synergy. They demonstrate differences in information decomposition between different pathway architectures, e.g., signal integration motifs and signal bifurcation motifs.

Sootla et al. [48] turn our attention to various canonical complex systems, demonstrating how PID can provide still new insights into these well-understood examples. Utilising the Bertschinger et al. [12] PID (building on work by some of the authors on estimators for this measure in another contribution to the special issue [41]), the authors begin by examining decomposition of information in triplets of spins in the 2D Ising model, while the temperature is varied. They report that redundant information is maximised at the critical point, whilst synergistic information peaks in the disordered phase. Next, the authors decompose information of cells in 1D elementary cellular automata (ECA) from the two neighbouring sources of those cells. They perform a dimensionality reduction on the PID atoms (as dimensions), identifying some (but not perfect) distinction in characteristics between Wolfram's rule classes.

3. Outlook

Information decomposition into redundant, unique and synergistic components has been recognised as a crucial theoretical problem which has proven far more difficult to solve than may have been expected. Thankfully, there is very strong activity in the community leading to progress on information decomposition approaches, which as outlined above is well reflected in this special issue.

We hope that our presentation of these papers will further the debate regarding which is the "right" measure of redundancy, which original assumptions or axioms may need to be dropped or changed (as per new measures and challenges to current thinking in Section 2.1), and how the approaches can and should be interpreted and/or extended (as per investigations in Section 2.2). Certainly there is a hunger for applications of information decomposition (as per Section 2.3), and again we hope that the special issue helps to disseminate and encourage these approaches.

Author Contributions: All authors edited multiple manuscripts for the special issue. J.T.L. wrote the first draft of this editorial, and all authors edited and approved the manuscript.

Acknowledgments: We thank all authors for their contributions, all participants of the workshop in Frankfurt, as well as the anonymous reviewers of the articles here, and editorial staff at Entropy. J.T.L. was supported through the Australian Research Council DECRA grant DE160100630. J.T.L. and M.W. were supported through a Universities Australia/German Academic Exchange Service (DAAD) Australia–Germany Joint Research Cooperation Scheme grant (2016–17): "Measuring neural information synthesis and its impairment" (PPP Australia Project-ID 57216857; IRMA ID: 180136). N.B. thanks Dr. h.c. Maucher for funding his position.

Conflicts of Interest: The authors declare no conflict of interest. The funding sponsors had no role in the design of the study; in the collection, analyses, or interpretation of data; in the writing of the manuscript, and in the decision to publish the results.

References

1. Shannon, C.E. A Mathematical Theory of Communication. *Bell Syst. Tech. J.* **1948**, *27*, 379–423, doi:10.1002/j.1538-7305.1948.tb01338.x.
2. Cover, T.M.; Thomas, J.A. *Elements of Information Theory*; John Wiley & Sons: Hoboken, NJ, USA, 2012.
3. MacKay, D. *Information Theory, Inference and Learning Algorithms*; Cambridge University Press: Cambridge, UK, 2003.
4. Williams, P.L.; Beer, R.D. Nonnegative decomposition of multivariate information. *arXiv* **2010**, arXiv:1004.2515. Available online: https://arxiv.org/abs/1004.2515 (accessed on 21 April 2018).
5. Williams, P.L.; Beer, R.D. Indiana University. Decomposing Multivariate Information, Privately communicated, 2010.
6. Schneidman, E.; Bialek, W.; Berry, M.J. Synergy, redundancy, and independence in population codes. *J. Neurosci.* **2003**, *23*, 11539–11553.
7. Lizier, J.T.; Flecker, B.; Williams, P.L. Towards a Synergy-Based Approach to Measuring Information Modification. In Proceedings of the 2013 IEEE Symposium on Artificial Life (IEEE ALIFE), Singapore, 16–19 April 2013; pp. 43–51.
8. Bertschinger, N.; Rauh, J.; Olbrich, E.; Jost, J. Shared Information—New Insights and Problems in Decomposing Information in Complex Systems. In Proceedings of the European Conference on Complex Systems 2012, Brussels, Belgium, 3–7 September 2012; Springer: Cham, Switzerland, 2013; pp. 251–269.
9. Harder, M.; Salge, C.; Polani, D. Bivariate measure of redundant information. *Phys. Rev. E* **2013**, *87*, 012130.
10. Griffith, V.; Koch, C. Quantifying Synergistic Mutual Information. In *Guided Self-Organization: Inception*; Prokopenko, M., Ed.; Springer: Berlin/Heidelberg, Germany, 2014; Volume 9, pp. 159–190, doi:10.1007/978-3-642-53734-9_6.
11. Wibral, M.; Lizier, J.T.; Priesemann, V. Bits from brains for biologically inspired computing. *Front. Robot. AI* **2015**, *2*, doi:10.3389/frobt.2015.00005.
12. Bertschinger, N.; Rauh, J.; Olbrich, E.; Jost, J.; Ay, N. Quantifying unique information. *Entropy* **2014**, *16*, 2161–2183.
13. Rauh, J.; Bertschinger, N.; Olbrich, E.; Jost, J. Reconsidering Unique Information: Towards a Multivariate Information Decomposition. In Proceedings of the 2014 IEEE International Symposium on Information Theory (ISIT), Honolulu, HI, USA, 29 June–4 July 2014; pp. 2232–2236.
14. Olbrich, E.; Bertschinger, N.; Rauh, J. Information decomposition and synergy. *Entropy* **2015**, *17*, 3501–3517.
15. Perrone, P.; Ay, N. Hierarchical Quantification of Synergy in Channels. *Front. Robot. AI* **2016**, *2*, 35.
16. Rosas, F.; Ntranos, V.; Ellison, C.J.; Pollin, S.; Verhelst, M. Understanding interdependency through complex information sharing. *Entropy* **2016**, *18*, 38.

17. Griffith, V.; Chong, E.K.; James, R.G.; Ellison, C.J.; Crutchfield, J.P. Intersection information based on common randomness. *Entropy* **2014**, *16*, 1985–2000.
18. Griffith, V.; Ho, T. Quantifying redundant information in predicting a target random variable. *Entropy* **2015**, *17*, 4644–4653.
19. Quax, R.; Har-Shemesh, O.; Sloot, P. Quantifying Synergistic Information Using Intermediate Stochastic Variables. *Entropy* **2017**, *19*, 85, doi:10.3390/e19020085.
20. Barrett, A.B. Exploration of synergistic and redundant information sharing in static and dynamical Gaussian systems. *Phys. Rev. E* **2015**, *91*, 052802.
21. Chatterjee, P.; Pal, N.R. Construction of synergy networks from gene expression data related to disease. *Gene* **2016**, *590*, 250–262, doi:10.1016/j.gene.2016.05.029.
22. Williams, P.L.; Beer, R.D. Generalized Measures of Information Transfer. *arXiv* **2011**, arXiv:1102.1507. Available online: https://arxiv.org/abs/1102.1507 (accessed on 21 April 2018).
23. Flecker, B.; Alford, W.; Beggs, J.M.; Williams, P.L.; Beer, R.D. Partial information decomposition as a spatiotemporal filter. *Chaos* **2011**, *21*, 037104, doi:10.1063/1.3638449.
24. Timme, N.M.; Ito, S.; Myroshnychenko, M.; Nigam, S.; Shimono, M.; Yeh, F.C.; Hottowy, P.; Litke, A.M.; Beggs, J.M. High-Degree Neurons Feed Cortical Computations. *PLoS Comput. Biol.* **2016**, *12*, 1–31.
25. Timme, N.; Alford, W.; Flecker, B.; Beggs, J.M. Synergy, redundancy, and multivariate information measures: an experimentalist's perspective. *J. Comput. Neurosci.* **2014**, *36*, 119–140, doi:10.1007/s10827-013-0458-4.
26. Stramaglia, S.; Cortes, J.M.; Marinazzo, D. Synergy and redundancy in the Granger causal analysis of dynamical networks. *New J. Phys.* **2014**, *16*, 105003, doi:10.1088/1367-2630/16/10/105003.
27. Wibral, M.; Priesemann, V.; Kay, J.W.; Lizier, J.T.; Phillips, W.A. Partial information decomposition as a unified approach to the specification of neural goal functions. *Brain Cogn.* **2017**, *112*, 25–38, doi:10.1016/j.bandc.2015.09.004.
28. Linsker, R. Self-organisation in a perceptual network. *IEEE Comput.* **1988**, *21*, 105–117.
29. Biswas, A.; Banik, S.K. Redundancy in information transmission in a two-step cascade. *Phys. Rev. E* **2016**, *93*, 052422, doi:10.1103/physreve.93.052422.
30. Frey, S.; Williams, P.L.; Albino, D.K. Information encryption in the expert management of strategic uncertainty. *arXiv* **2016**, arXiv:1605.04233. Available online: https://arxiv.org/abs/1605.04233 (accessed on 21 April 2018).
31. Rauh, J.; Banerjee, P.K.; Olbrich, E.; Jost, J.; Bertschinger, N. On Extractable Shared Information. *Entropy* **2017**, *19*, 328, doi:10.3390/e19070328.
32. Rauh, J.; Banerjee, P.K.; Olbrich, E.; Jost, J.; Bertschinger, N.; Wolpert, D. Coarse-Graining and the Blackwell Order. *Entropy* **2017**, *19*, 527, doi:10.3390/e19100527.
33. Ince, R. Measuring Multivariate Redundant Information with Pointwise Common Change in Surprisal. *Entropy* **2017**, *19*, 318, doi:10.3390/e19070318.
34. Ince, R.A.A. The Partial Entropy Decomposition: Decomposing multivariate entropy and mutual information via pointwise common surprisal. *arXiv* **2017**, arXiv:1702.01591. Available online: https://arxiv.org/abs/1702.01591 (accessed on 21 April 2018).
35. Finn, C.; Lizier, J.T. Pointwise Partial Information Decomposition Using the Specificity and Ambiguity Lattices. *Entropy* **2018**, *20*, 297, doi:10.3390/e20040297.
36. Finn, C.; Lizier, J.T. Probability Mass Exclusions and the Directed Components of Pointwise Mutual Information. *arXiv* **2018**, arXiv:1801.09223. Available online: https://arxiv.org/abs/1801.09223 (accessed on 21 April 2018).
37. James, R.G.; Crutchfield, J.P. Multivariate dependence beyond shannon information. *Entropy* **2017**, *19*, 531.
38. Pica, G.; Piasini, E.; Chicharro, D.; Panzeri, S. Invariant components of synergy, redundancy, and unique information among three variables. *Entropy* **2017**, *19*, 451.
39. Rauh, J. Secret sharing and shared information. *Entropy* **2017**, *19*, 601.
40. Faes, L.; Marinazzo, D.; Stramaglia, S. Multiscale information decomposition: exact computation for multivariate Gaussian processes. *Entropy* **2017**, *19*, 408.
41. Makkeh, A.; Theis, D.O.; Vicente, R. Bivariate Partial Information Decomposition: The Optimization Perspective. *Entropy* **2017**, *19*, 530.
42. Kay, J.W.; Ince, R.A.; Dering, B.; Phillips, W.A. Partial and Entropic Information Decompositions of a Neuronal Modulatory Interaction. *Entropy* **2017**, *19*, 560.

43. Wibral, M.; Finn, C.; Wollstadt, P.; Lizier, J.T.; Priesemann, V. Quantifying Information Modification in Developing Neural Networks via Partial Information Decomposition. *Entropy* **2017**, *19*, 494.
44. Wollstadt, P.; Lizier, J.T.; Finn, C.; Martinz-Zarzuela, M.; Vicente, R.; Lindner, M.; Martinez-Mediano, P.; Wibral, M. The Information Dynamics Toolkit, IDT^{xl}. Available online: https://github.com/pwollstadt/IDTxl (accessed on 25 August 2017).
45. Tax, T.; Mediano, P.A.; Shanahan, M. The partial information decomposition of generative neural network models. *Entropy* **2017**, *19*, 474.
46. Ghazi-Zahedi, K.; Langer, C.; Ay, N. Morphological computation: Synergy of body and brain. *Entropy* **2017**, *19*, 456.
47. Maity, A.K.; Chaudhury, P.; Banik, S.K. Information theoretical study of cross-talk mediated signal transduction in MAPK pathways. *Entropy* **2017**, *19*, 469.
48. Sootla, S.; Theis, D.; Vicente, R. Analyzing Information Distribution in Complex Systems. *Entropy* **2017**, *19*, 636, doi:10.3390/e19120636.

Article

On Extractable Shared Information

Johannes Rauh [1,*], **Pradeep Kr. Banerjee** [1], **Eckehard Olbrich** [1], **Jürgen Jost** [1] and
Nils Bertschinger [2]

[1] Max Planck Institute for Mathematics in the Sciences, 04103 Leipzig, Germany;
    pradeep@mis.mpg.de (P.K.B.); olbrich@mis.mpg.de (E.O.); jjost@mis.mpg.de (J.J.)
[2] Frankfurt Institute for Advanced Studies, 60438 Frankfurt, Germany; bertschinger@fias.uni-frankfurt.de
* Correspondence: jrauh@mis.mpg.de; Tel.: +49-341-9959-602

Received: 31 May 2017; Accepted: 22 June 2017; Published: 3 July 2017

Abstract: We consider the problem of quantifying the information shared by a pair of random variables X_1, X_2 about another variable S. We propose a new measure of shared information, called *extractable shared information*, that is left monotonic; that is, the information shared about S is bounded from below by the information shared about $f(S)$ for any function f. We show that our measure leads to a new nonnegative decomposition of the mutual information $I(S; X_1 X_2)$ into shared, complementary and unique components. We study properties of this decomposition and show that a left monotonic shared information is not compatible with a Blackwell interpretation of unique information. We also discuss whether it is possible to have a decomposition in which both shared and unique information are left monotonic.

Keywords: information decomposition; multivariate mutual information; left monotonicity; Blackwell order

MSC: 94A17

1. Introduction

A series of recent papers have focused on the bivariate information decomposition problem [1–6]. Consider three random variables S, X_1, X_2 with finite alphabets $\mathcal{S}$, $\mathcal{X}_1$ and $\mathcal{X}_2$, respectively. The total information that the pair (X_1, X_2) convey about the target S can have aspects of *shared* or *redundant* information (conveyed by both X_1 and X_2), of *unique* information (conveyed exclusively by either X_1 or X_2), and of *complementary* or *synergistic* information (retrievable only from the the joint variable (X_1, X_2)). In general, all three kinds of information may be present concurrently. One would like to express this by decomposing the mutual information $I(S; X_1 X_2)$ into a sum of nonnegative components with a well-defined operational interpretation. One possible application area is in the neurosciences. In [7], it is argued that such a decomposition can provide a framework to analyze neural information processing using information theory that can integrate and go beyond previous attempts.

For the general case of k finite source variables $(X_1, \ldots, X_k)$, Williams and Beer [3] proposed the partial information lattice framework that specifies how the total information about the target S is shared across the singleton sources and their disjoint or overlapping coalitions. The lattice is a consequence of certain natural properties of shared information (sometimes called the *Williams–Beer axioms*). In the bivariate case ($k = 2$), the decomposition has the form

$$I(S; X_1 X_2) = \underbrace{SI(S; X_1, X_2)}_{\text{shared}} + \underbrace{CI(S; X_1, X_2)}_{\text{complementary}} + \underbrace{UI(S; X_1 \backslash X_2)}_{\text{unique } (X_1 \text{ wrt } X_2)} + \underbrace{UI(S; X_2 \backslash X_1)}_{\text{unique } (X_2 \text{ wrt } X_1)}, \tag{1}$$

$$I(S; X_1) = SI(S; X_1, X_2) + UI(S; X_1 \backslash X_2), \tag{2}$$

$$I(S; X_2) = SI(S; X_1, X_2) + UI(S; X_2 \backslash X_1), \tag{3}$$

where $SI(S; X_1, X_2)$, $UI(S; X_1 \backslash X_2)$, $UI(S; X_2 \backslash X_1)$, and $CI(S; X_1, X_2)$ are nonnegative functions that depend continuously on the joint distribution of (S, X_1, X_2). The difference between shared and complementary information is the familiar co-information [8] (or interaction information [9]), a symmetric generalization of the mutual information for three variables,

$$CoI(S; X_1, X_2) = I(S; X_1) - I(S; X_1 | X_2) = SI(S; X_1, X_2) - CI(S; X_1, X_2).$$

Equations (1) to (3) leave only a single degree of freedom, i.e., it suffices to specify either a measure for SI, for CI or for UI.

Williams and Beer not only introduced the general partial information framework, but also proposed a measure of SI to fill this framework. While their measure has subsequently been criticized for "not measuring the right thing" [4–6], there has been no successful attempt to find better measures, except for the bivariate case ($k = 2$) [1,4]. One problem seems to be the lack of a clear consensus on what an ideal measure of shared (or unique or complementary) information should look like and what properties it should satisfy. In particular, the Williams–Beer axioms only put crude bounds on the values of the functions SI, UI and CI. Therefore, additional axioms have been proposed by various authors [4–6]. Unfortunately, some of these properties contradict each other [5], and the question for the right axiomatic characterization is still open.

The Williams–Beer axioms do not say anything about what should happen when the target variable S undergoes a local transformation. In this context, the following *left monotonicity* property was proposed in [5]:

(LM) $SI(S; X_1, X_2) \geq SI(f(S); X_1, X_2)$ for any function f. *(left monotonicity)*

Left monotonicity for unique or complementary information can be defined similarly. The property captures the intuition that shared information should only decrease if the target performs some *local* operation (e.g., coarse graining) on her variable S. As argued in [2], left monotonicity of shared and unique information are indeed desirable properties. Unfortunately, none of the measures of shared information proposed so far satisfy left monotonicity.

In this contribution, we study a construction that enforces left monotonicity. Namely, given a measure of shared information SI, define

$$\overline{SI}(S; X_1, X_2) := \sup_{f:S \to S'} SI(f(S); X_1, X_2), \tag{4}$$

where the supremum runs over all functions $f : S \to S'$ from the domain of S to an arbitrary finite set S'. By construction, $\overline{SI}$ satisfies left monotonicity, and $\overline{SI}$ is the smallest function bounded from below by SI that satisfies left monotonicity.

Changing the definition of shared information in the information decomposition framework Equations (1)–(3) leads to new definitions of unique and complementary information:

$$\overline{UI}^*(S; X_1 \backslash X_2) := I(S; X_1) - \overline{SI}(S; X_1, X_2),$$
$$\overline{UI}^*(S; X_2 \backslash X_1) := I(S; X_2) - \overline{SI}(S; X_1, X_2),$$
$$\overline{CI}^*(S; X_1, X_2) := I(S; X_1 X_2) - \overline{SI}(S; X_1, X_2) - \overline{UI}^*(S; X_1 \backslash X_2) - \overline{UI}^*(S; X_2 \backslash X_1).$$

In general, $\overline{UI}^*(S; X_1 \backslash X_2) \neq \overline{UI}(S; X_1 \backslash X_2) := \sup_{f:S \to S'} UI(f(S); X_1 \backslash X_2)$. Thus, our construction cannot enforce left monotonicity for both UI and SI in parallel.

Lemma 2 shows that $\overline{SI}$, $\overline{UI}^*$ and $\overline{CI}^*$ are nonnegative and thus define a nonnegative bivariate decomposition. We study this decomposition in Section 4. In Theorem 1, we show that our construction is not compatible with a decision-theoretic interpretation of unique information proposed in [1]. In Section 5, we ask whether it is possible to find an information decomposition in which both shared and unique information measures are left monotonic. Our construction cannot directly be generalized to ensure left monotonicity of two functions simultaneously. Nevertheless, it is possible that such a decomposition exists, and in Proposition 1, we prove bounds on the corresponding shared information measure.

Our original motivation for the definition of $\overline{SI}$ was to find a bivariate decomposition in which the shared information satisfies left monotonicity. However, one could also ask whether left monotonicity is a required property of shared information, as put forward in [2]. In contrast, Harder et al. [4] argue that redundancy can also arise by means of a mechanism. Applying a function to S corresponds to such a mechanism that singles out a certain aspect from S. Even if all the X_i share nothing about the whole S, they might still share information about this aspect of S, which means that the shared information will increase. With this intuition, we can interpret $\overline{SI}$ not as an improved measure of shared information, but as a measure of *extractable shared information*, because it asks for the maximal amount of shared information that can be extracted from S by further processing S by a local mechanism. More generally, one can apply a similar construction to arbitrary information measures. We explore this idea in Section 3 and discuss probabilistic generalizations and relations to other information measures. In Section 6, we apply our construction to existing measures of shared information.

2. Properties of Information Decompositions

2.1. The Williams–Beer Axioms

Although we are mostly concerned with the case $k = 2$, let us first recall the three axioms that Williams and Beer [3] proposed for a measure of shared information for arbitrarily many arguments:

(S) $SI(S; X_1, \ldots, X_k)$ is symmetric under permutations of $X_1, \ldots, X_k$,   *(Symmetry)*
(SR) $SI(S; X_1) = I(S; X_1)$,   *(Self-redundancy)*
(M) $SI(S; X_1, \ldots, X_{k-1}, X_k) \leq SI(S; X_1, \ldots, X_{k-1})$,
   with equality if $X_i = f(X_k)$ for some $i < k$ and some function f.   *(Monotonicity)*

Any measure of SI satisfying these axioms is nonnegative. Moreover, the axioms imply the following:

(RM) $SI(S; X_1, \ldots, X_k) \geq SI(S; f_1(X_1), \ldots, f_k(X_k))$ for all functions $f_1, \ldots, f_k$.   *(right monotonicity)*

Williams and Beer also defined a function

$$I_{\min}(S; X_1, \ldots, X_k) = \sum_s P_S(s) \min_i \left\{ \sum_{x_i} P_{X_i|S}(x_i|s) \log \frac{P_{S|X_i}(s|x_i)}{P_S(s)} \right\} \tag{5}$$

and showed that $I_{\min}$ satisfies their axioms.

2.2. The COPY example and the Identity Axiom

Let X_1, X_2 be independent uniformly distributed binary random variables, and consider the copy function $\text{COPY}(X_1, X_2) := (X_1, X_2)$. One point of criticism of $I_{\min}$ is the fact that X_1 and X_2 share $I_{\min}(\text{COPY}(X_1, X_2); X_1, X_2) = 1$ bit about $\text{COPY}(X_1, X_2)$ according to $I_{\min}$, even though they are independent. Harder et al. [4] argue that the shared information about the copied pair should equal the mutual information:

(Id) $SI(\text{COPY}(X_1, X_2); X_1, X_2) = I(X_1; X_2)$.   *(Identity)*

Ref. [4] also proposed a bivariate measure of shared information that satisfies **(Id)**. Similarly, the measures of bivariate shared information proposed in [1] satisfies **(Id)**. However, **(Id)** is incompatible with a nonnegative information decomposition according to the Williams–Beer axioms for $k \geq 3$ [2].

On the other hand, Ref. [5] uses an example from game theory to give an intuitive explanation how even independent variables X_1 and X_2 can have nontrivial shared information. However, in any case the value of 1 bit assigned by $I_{\min}$ is deemed to be too large.

Entropy **2017**, *19*, 328

2.3. The Blackwell Property and Property (∗)

One of the reasons that it is so difficult to find good definitions of shared, unique or synergistic information is that a clear operational idea behind these notions is missing. Starting from an operational idea about decision problems, Ref. [1] proposed the following property for the unique information, which we now propose to call *Blackwell property*:

(BP) For a given joint distribution $P_{SX_1X_2}$, $UI(S; X_1 \backslash X_2)$ vanishes if and only if there exists a random variable X_1' such that $S - X_2 - X_1'$ is a Markov chain and $P_{SX_1'} = P_{SX_1}$.   (Blackwell property)

In other words, the channel $S \rightarrow X_1$ is a *garbling* or *degradation* of the channel $S \rightarrow X_2$. Blackwell's theorem [10] implies that this garbling property is equivalent to the fact that any decision problem in which the task is to predict S can be solved just as well with the knowledge of X_2 as with the knowledge of X_1. We refer to Section 2 in [1] for the details.

Ref. [1] also proposed the following property:

(∗) SI and UI depend only on the marginal distributions P_{SX_1} and P_{SX_2} of the pairs (S, X_1) and (S, X_2).

This property was in part motivated by **(BP)**, which also depends only on the channels $S \rightarrow X_1$ and $S \rightarrow X_2$ and thus on P_{SX_1} and P_{SX_2}. Most information decompositions proposed so far satisfy property (∗).

3. Extractable Information Measures

One can interpret $\overline{SI}$ as a measure of *extractable shared information*. We explain this idea in a more general setting.

For fixed k, let $IM(S; X_1, \ldots, X_k)$ be an arbitrary information measure that measures one aspect of the information that $X_1, \ldots, X_k$ contain about S. At this point, we do not specify what precisely an information measure is, except that it is a function that assigns a real number to any joint distributions of $S, X_1, \ldots, X_k$. The notation is, of course, suggestive of the fact that we mostly think about one of the measures SI, UI or CI, in which the first argument plays a special role. However, IM could also be the mutual information $I(S; X_1)$, the entropy $H(S)$, or the coinformation $CoI(S; X_1, X_2)$. We define the corresponding *extractable* information measure as

$$\overline{IM}(S; X_1, \ldots, X_k) := \sup_f \ IM(f(S); X_1, \ldots, X_k), \tag{6}$$

where the supremum runs over all functions $f : \mathcal{S} \mapsto \mathcal{S}'$ from the domain of S to an arbitrary finite set $\mathcal{S}'$. The intuition is that $\overline{IM}$ is the maximal possible amount of IM one can "extract" from $(X_1, \ldots, X_k)$ by transforming S. Clearly, the precise interpretation depends on the interpretation of IM.

This construction has the following general properties:

1. Most information measures satisfy $IM(O; X_1, \ldots, X_k) = 0$ when O is a constant random variable. Thus, in this case, $\overline{IM}(S; X_1, \ldots, X_k) \geq 0$. Thus, for example, even though the coinformation can be negative, the *extractable coinformation* is never negative.
2. Suppose that IM satisfies left monotonicity. Then, $\overline{IM} = IM$. For example, entropy H and mutual information I satisfy left monotonicity, and so $\overline{H} = H$ and $\overline{I} = I$. Similarly, as shown in [2], the measure of unique information $\widetilde{UI}$ defined in [1] satisfies left monotonicity, and so $\overline{\widetilde{UI}} = \widetilde{UI}$.
3. In fact, $\overline{IM}$ is the smallest left monotonic information measure that is at least as large as IM.

The next result shows that our construction preserves monotonicity properties of the other arguments of IM. It follows that, by iterating this construction, one can construct an information measure that is monotonic in all arguments.

Lemma 1. *Let $f_1, \ldots, f_k$ be fixed functions. If IM satisfies $IM(S; f_1(X_1), \ldots, f_k(X_k)) \leq IM(S; X_1, \ldots, X_k)$ for all S, then $\overline{IM}(S; f_1(X_1), \ldots, f_k(X_k)) \leq \overline{IM}(S; X_1, \ldots, X_k)$ for all S.*

Proof. Let $f^* = \arg\max_f \{IM(f(S); f_1(X_1), \ldots, f_k(X_k))\}$. Then,

$$\overline{IM}(S; f_1(X_1), \ldots, f_k(X_k)) = IM(f^*(S); f_1(X_1), \ldots, f_k(X_k))$$
$$\overset{(a)}{\leq} IM(f^*(S); X_1, \ldots, X_k) \leq \sup_f IM(f(S); X_1, \ldots, X_k) = \overline{IM}(S; X_1, \ldots, X_k),$$

where (a) follows from the assumptions. □

As a generalization to the construction, instead of looking at "deterministic extractability," one can also look at "probabilistic extractability" and replace f by a stochastic matrix. This leads to the definition

$$\overline{\overline{IM}}(S; X_1, \ldots, X_k) := \sup_{P_{S'|S}} IM(S'; X_1, \ldots, X_k), \tag{7}$$

where the supremum now runs over all random variables S' that are independent of $X_1, \ldots, X_k$ given S. The function $\overline{\overline{IM}}$ is the smallest function bounded from below by IM that satisfies

(PLM) $IM(S; X_1, X_2) \geq IM(S'; X_1, X_2)$ whenever S' is independent of X_1, X_2 given S.

(probabilistic left monotonicity)

An example of this construction is the intrinsic conditional information $I(X; Y \downarrow Z) :=$ $\min_{P_{Z'|Z}} I(X; Y|Z')$, which was defined in [11] to study the secret-key rate, which is the maximal rate at which a secret can be generated by two agents knowing X or Y, respectively, such that a third agent who knows Z has arbitrarily small information about this key. The min instead of the max in the definition implies that $I(X; Y \downarrow Z)$ is "anti-monotone" in Z.

In this paper, we restrict ourselves to the deterministic notions, since many of the properties we want to discuss can already be explained using deterministic extractability. Moreover, the optimization problem (6) is a finite optimization problem and thus much easier to solve than Equation (7).

4. Extractable Shared Information

We now specialize to the case of shared information. The first result is that when we apply our construction to a measure of shared information that belongs to a bivariate information decomposition, we again obtain a bivariate information decomposition.

Lemma 2. *Suppose that SI is a measure of shared information, coming from a nonnegative bivariate information decomposition (satisfying Equations (1) to (3)). Then, $\overline{SI}$ defines a nonnegative information decomposition; that is, the derived functions*

$$\overline{UI}^*(S; X_1 \backslash X_2) := I(S; X_1) - \overline{SI}(S; X_1, X_2),$$
$$\overline{UI}^*(S; X_2 \backslash X_1) := I(S; X_2) - \overline{SI}(S; X_1, X_2),$$
$$and \quad \overline{CI}^*(S; X_1, X_2) := I(S; X_1 X_2) - \overline{SI}(S; X_1, X_2) - \overline{UI}^*(S; X_1 \backslash X_2) - \overline{UI}^*(S; X_2 \backslash X_1)$$

are nonnegative. These quantities relate to the original decomposition by

$$a) \ \overline{SI}(S; X_1, X_2) \geq SI(S; X_1, X_2),$$
$$b) \ \overline{CI}^*(S; X_1, X_2) \geq CI(S; X_1, X_2),$$
$$c) \ UI(f^*(S); X_1 \backslash X_2) \leq \overline{UI}^*(S; X_1 \backslash X_2) \leq UI(S; X_1 \backslash X_2),$$

where f^* is a function that achieves the supremum in Equation (4).

Proof.

a) $\overline{SI}(S; X_1, X_2) \geq SI(S; X_1, X_2) \geq 0,$

b) $\overline{CI}^*(S; X_1, X_2) = \overline{SI}(S; X_1, X_2) - CoI(S; X_1, X_2) \geq SI(S; X_1, X_2) - CoI(S; X_1, X_2)$
$\geq CI(S; X_1, X_2) \geq 0,$

c) $\overline{UI}^*(S; X_1\backslash X_2) = I(S; X_1) - \overline{SI}(S; X_1, X_2) \leq I(S; X_1) - SI(S; X_1, X_2) = UI(S; X_1\backslash X_2),$
$\overline{UI}^*(S; X_1\backslash X_2) = I(S; X_1) - \overline{SI}(S; X_1, X_2) \geq I(f^*(S); X_1) - SI(f^*(S); X_1, X_2)$
$= UI(f^*(S); X_1\backslash X_2) \geq 0,$

where we have used the data processing inequality. $\square$

Lemma 3. *1. If SI satisfies ($*$), then $\overline{SI}$ also satisfies ($*$).*
2. If SI is right monotonic, then $\overline{SI}$ is also right monotonic.

Proof. (1) is direct, and (2) follows from Lemma 1. $\square$

Without further assumptions on SI, we cannot say much about when $\overline{SI}$ vanishes. However, the condition that $\overline{UI}^*$ vanishes has strong consequences.

Lemma 4. *Suppose that $\overline{UI}^*(S; X_1\backslash X_2)$ vanishes, and let f^* be a function that achieves the supremum in Equation (4). Then, there is a Markov chain $X_1 - f^*(S) - S$. Moreover, $UI(f^*(S); X_1\backslash X_2) = 0$.*

Proof. Suppose that $\overline{UI}^*(S; X_1\backslash X_2) = 0$. Then, $I(S; X_1) = \overline{SI}(S; X_1, X_2) = SI(f^*(S); X_1, X_2) \leq I(f^*(S); X_1) \leq I(S; X_1)$. Thus, the data processing inequality holds with equality. This implies that $X_1 - f^*(S) - S$ is a Markov chain. The identity $UI(f^*(S); X_1\backslash X_2) = 0$ follows from the same chain of inequalities. $\square$

Theorem 1. *If UI has the Blackwell property, then $\overline{UI}^*$ does not have the Blackwell property.*

Proof. As shown in the example in the appendix, there exist random variables S, X_1, X_2 and a function f that satisfy

1. S and X_1 are independent given $f(S)$.
2. The channel $f(S) \rightarrow X_1$ is a garbling of the channel $f(S) \rightarrow X_2$.
3. The channel $S \rightarrow X_1$ is not a garbling of the channel $S \rightarrow X_2$.

We claim that f solves the optimization problem (4). Indeed, for an arbitrary function f',

$$SI(f'(S); X_1, X_2) \leq I(f'(S); X_1) \leq I(S; X_1) = I(f(S); X_1) = SI(f(S); X_1, X_2).$$

Thus, f solves the maximization problem (4).

If UI satisfies the Blackwell property, then (2) and (3) imply $UI(f(S); X_1\backslash X_2) = 0$ and $UI(S; X_1\backslash X_2) > 0$. On the other hand,

$$\overline{UI}^*(S; X_1 \backslash X_2) = I(S; X_1) - \overline{SI}(S; X_1, X_2) = I(S; X_1) - SI(f(S); X_1, X_2)$$
$$= I(S; X_1) - I(f(S); X_1) + UI(f(S); X_1\backslash X_2) = 0.$$

Thus, $\overline{UI}^*$ does not satisfy the Blackwell property. $\square$

Corollary 1. *There is no bivariate information decomposition in which UI satisfies the Blackwell property and SI satisfies left monotonicity.*

Proof. If SI satisfies left monotonicity, then $\overline{SI} = SI$. Thus, $UI = \overline{UI}^{*}$ cannot satisfy the Blackwell property by Theorem 1. □

5. Left Monotonic Information Decompositions

Is it possible to have an extractable information decomposition? More precisely, is it possible to have an information decomposition in which all information measures are left monotonic? The obvious strategy of starting with an arbitrary information decomposition and replacing each partial information measure by its extractable analogue does not work, since this would mean increasing all partial information measures (unless they are extractable already), but then their sum would also increase. For example, in the bivariate case, when SI is replaced by a larger function $\overline{SI}$, then UI needs to be replaced by a smaller function, due to the constraints (2) and (3).

As argued in [2], it is intuitive that UI be left monotonic. As argued above (and in [5]), it is also desirable that SI be left monotonic. The intuition for synergy is much less clear. In the following, we restrict our focus to the bivariate case and study the implications of requiring both SI and UI to be left monotonic. Proposition 1 gives bounds on the corresponding SI measure.

Proposition 1. *Suppose that SI, UI and CI define a bivariate information decomposition, and suppose that SI and UI are left monotonic. Then,*

$$SI(f(X_1, X_2); X_1, X_2) \leq I(X_1; X_2) \tag{8}$$

for any function f.

Before proving the proposition, let us make some remarks. Inequality (8) is related to the identity axiom. Indeed, it is easy to derive Inequality (8) from the identity axiom and from the assumption that SI is left monotonic. Although Inequality (8) may not seem counterintuitive at first sight, none of the information decompositions proposed so far satisfy this property (the function $I_{\wedge}$ from [12] satisfies left monotonicity and has been proposed as a measure of shared information, but it does not lead to a nonnegative information decomposition).

Proof. If SI is left monotonic, then

$$SI(f(X_1, X_2); X_1, X_2) \leq SI(\text{Copy}(X_1, X_2); X_1, X_2)$$
$$= I(\text{Copy}(X_1, X_2); X_1) - UI(\text{Copy}(X_1, X_2); X_1 \backslash X_2).$$

If UI is left monotonic, then

$$UI(\text{Copy}(X_1, X_2); X_1 \backslash X_2) \geq UI(X_1; X_1 \backslash X_2) = I(X_1; X_1) - SI(X_1; X_1, X_2).$$

Note that $I(X_1; X_1) = H(X_1) = I(\text{Copy}(X_1, X_2); X_1)$ and

$$SI(X_1; X_1, X_2) = I(X_1; X_2) - UI(X_1; X_2 \backslash X_1) = I(X_1; X_2).$$

Putting these inequalities together, we obtain $SI(f(X_1, X_2); X_1, X_2) \leq I(X_1; X_2)$. □

6. Examples

In this section, we apply our construction to Williams and Beer's measure, $I_{\min}$ [3], and to the bivariate measure of shared information, $\widetilde{SI}$, proposed in [1].

First, we make some remarks on how to compute the extractable information measure (under the assumption that one knows how to compute the underlying information measure itself). The optimization problem (4) is a discrete optimization problem. The search space is the set of functions from the support S of S to some finite set S'. For the information measures that we have in mind, we may restrict to surjective functions f, since the information measures only depend on events with positive probabilities. Thus, we may restrict to sets S' with $|S'| \leq |S|$. Moreover, the information measures are invariant under permutations of the alphabet S. Therefore, the only thing that matters about f is which elements from S are mapped to the same element in S'. Thus, any function $f : S \rightarrow S'$ corresponds to a partition of S, where $s, s' \in S$ belong to the same block if and only if $f(s) = f(s')$, and it suffices to look at all such partitions. The number of partitions of a finite set S is the *Bell number* $B_{|S|}$.

The Bell numbers increase super-exponentially, and for larger sets S, the search space of the optimization problem (4) becomes quite large. For smaller problems, enumerating all partitions in order to find the maximum is still feasible. For larger problems, one would need a better understanding about the optimization problem. For reference, some Bell numbers include:

n	3	4	6	10
B_n	5	15	203	115975

As always, symmetries may help, and so in the COPY example discussed below, where $|S| = 4$, it suffices to study six functions instead of $B_4 = 15$.

We now compare the measure $\overline{I}_{\min}$, an extractable version of Williams and Beer's measure $I_{\min}$ (see Equation (5) above), to the measure $\widetilde{SI}$, an extractable version of the measure $\widetilde{SI}$ proposed in [1]. For the latter, we briefly recall the definitions. Let Δ be the set of all joint distributions of random variables (S, X_1, X_2) with given state spaces $S, \mathcal{X}_1, \mathcal{X}_2$. Fix $P = P_{SX_1X_2} \in \Delta$. Define Δ_P as the set of all distributions $Q_{SX_1X_2}$ that preserves the marginals of the pairs (S, X_1) and (S, X_2), that is,

$$\Delta_P := \left\{ Q_{SX_1X_2} \in \Delta : Q_{SX_1} = P_{SX_1}, Q_{SX_2} = P_{SX_2}, \forall (S, X_1, X_2) \in \Delta \right\}.$$

Then, define the functions

$$\widetilde{UI}(S; X_1 \backslash X_2) := \min_{Q \in \Delta_P} I_Q(S; X_1 | X_2),$$

$$\widetilde{UI}(S; X_2 \backslash X_1) := \min_{Q \in \Delta_P} I_Q(S; X_2 | X_1),$$

$$\widetilde{SI}(S; X_1, X_2) := \max_{Q \in \Delta_P} CoI_Q(S; X_1, X_2),$$

$$\widetilde{CI}(S; X_1, X_2) := I(S; X_1 X_2) - \min_{Q \in \Delta_P} I_Q(S; X_1 X_2),$$

where the index Q in I_Q or CoI_Q indicates that the corresponding quantity is computed with respect to the joint distribution Q. The decomposition corresponding to $\widetilde{SI}$ satisfies the Blackwell property and the identity axiom [1]. $\widetilde{UI}$ is left monotonic, but $\widetilde{SI}$ is not [2]. In particular, $\overline{\widetilde{SI}} \neq \widetilde{SI}$. $\widetilde{SI}$ can be characterized as the smallest measure of shared information that satisfies property ($*$). Therefore, $\overline{\widetilde{SI}}$ is the smallest left monotonic measure of shared information that satisfies property ($*$).

Let $\mathcal{X}_1 = \mathcal{X}_2 = \{0, 1\}$ and let X_1, X_2 be independent uniformly distributed random variables. Table 1 collects values of shared information about $f(X_1, X_2)$ for various functions f (in bits).

Table 1. Shared information about $f(X_1, X_2)$ for various functions f (in bits).

f	$I_{\min}$	$\overline{I}_{\min}$	$\widetilde{SI}$	$\overline{\widetilde{SI}}$
COPY	1	1	0	1/2
AND/OR	$3/4 \log 4/3$	$3/4 \log 4/3$	$3/4 \log 4/3$	$3/4 \log 4/3$
XOR	0	0	0	0
SUM	1/2	1/2	1/2	1/2
X_1	0	0	0	0
f_1	1/2	1/2	0	0

The function $f_1 : \{00, 01, 10, 11\} \to \{0, 1, 2\}$ is defined as

$$f_1(X_1, X_2) := \begin{cases} X_1, & \text{if } X_2 = 1, \\ 2, & \text{if } X_2 = 0. \end{cases}$$

The SUM function is defined as $f(X_1, X_2) := X_1 + X_2$. Table 1 contains (up to symmetry) all possible non-trivial functions f. The values for the extractable measures are derived from the values of the corresponding non-extractable measures. Note that the values for the extractable versions differ only for COPY from the original ones. In these examples, $\overline{I}_{\min} = I_{\min}$, but as shown in [5], $I_{\min}$ is not left monotonic in general.

7. Conclusions

We introduced a new measure of shared information that satisfies the left monotonicity property with respect to local operations on the target variable. Left monotonicity corresponds to the idea that local processing will remove information in the target variable and thus should lead to lower values of measures which quantify information about the target variable. Our measure fits the bivariate information decomposition framework; that is, we also obtain corresponding measures of unique and synergistic information. However, we also have shown that left monotonicity for the shared information contradicts the Blackwell property of the unique information, which limits the value of a left monotonic measure of shared information for information decomposition.

We also presented an alternative interpretation of the construction used in this paper. Starting from an arbitrary measure of shared information SI (which need not be left monotonic), we interpret the left monotonic measure $\overline{SI}$ as the amount of shared information that can be extracted from S by local processing.

Our initial motivation for the construction of $\overline{SI}$ was the question to which extent shared information originates from the redundancy between the predictors X_1 and X_2 or is created by the mechanism that generated S. These two different flavors of redundancy were called *source redundancy* and *mechanistic redundancy*, respectively, in [4]. While $\overline{SI}$ cannot be used to completely disentangle source and mechanistic redundancy, it can be seen as a measure of the maximum amount of redundancy that can be created from S using a (deterministic) mechanism. In this sense, we believe that it is an important step forward towards a better understanding of this problem and related questions.

Acknowledgments: We thank the participants of the PID workshop at FIAS in Frankfurt in December 2016 for many stimulating discussions on this subject. Nils Bertschinger thanks H. C. Maucher for funding his position.

Author Contributions: The research was initiated by Jürgen Jost and carried out by all authors, with main contributions of Johannes Rauh and Pradeep Kr. Banerjee. The manuscript was written by Johannes Rauh, Pradeep Kr. Banerjee and Eckehard Olbrich. All authors have read and approved the final manuscript.

Conflicts of Interest: The authors declare no conflict of interest.

Appendix A. Counterexample in Theorem 1

Consider the joint distribution

$f(s)$	s	x_1	x_2	$P_{f(S)SX_1X_2}$
0	0	0	0	1/4
0	1	0	1	1/4
0	0	1	0	1/8
0	1	1	0	1/8
1	2	1	1	1/4

and the function $f : \{0, 1, 2\} \to \{0, 1\}$ with $f(0) = f(1) = 0$ and $f(2) = 1$. Then, X_1 and X_2 are independent uniform binary random variables, and $f(S) = \text{AND}(X_1, X_2)$. In addition, $S - f(S) - X_1$ is a Markov chain. By symmetry, the joint distributions of the pairs $(f(S), X_1)$ and $(f(S), X_2)$ are identical, and so the two channels $f(S) \rightarrow X_1$ and $f(S) \rightarrow X_2$ are identical, and, hence, trivially, one is a garbling of the other. However, one can check that the channel $S \rightarrow X_1$ is not a garbling of the channel $S \rightarrow X_2$.

This example is discussed in more detail in Rauh et al. [13].

References

1. Bertschinger, N.; Rauh, J.; Olbrich, E.; Jost, J.; Ay, N. Quantifying unique information. *Entropy* **2014**, *16*, 2161–2183.
2. Rauh, J.; Bertschinger, N.; Olbrich, E.; Jost, J. Reconsidering unique information: Towards a multivariate information decomposition. In Proceedings of 2014 IEEE International Symposium on Information Theory (ISIT), Honolulu, HI, USA, 29 June–4 July 2014; pp. 2232–2236.
3. Williams, P.; Beer, R. Nonnegative Decomposition of Multivariate Information. *arXiv* **2010**, arXiv:1004.2515v1.
4. Harder, M.; Salge, C.; Polani, D. A Bivariate measure of redundant information. *Phys. Rev. E* **2013**, *87*, 012130.
5. Bertschinger, N.; Rauh, J.; Olbrich, E.; Jost, J. Shared Information—New Insights and Problems in Decomposing Information in Complex Systems. In Proceedings of the European Conference on Complex Systems 2012, Brussels, Belgium, 2–7 September 2012; pp. 251–269.
6. Griffith, V.; Koch, C. Quantifying Synergistic Mutual Information. In *Guided Self-Organization: Inception*; Prokopenko, M., Ed.; Springer: Berlin/Heidelberg, Germany, 2014; Volume 9, pp. 159–190.
7. Wibral, M.; Priesemann, V.; Kay, J.W.; Lizier, J.T.; Phillips, W.A. Partial information decomposition as a unified approach to the specification of neural goal functions. *Brain Cogn.* **2017**, *112*, 25–38.
8. Bell, A.J. The Co-Information Lattice. In Proceedings of the Fourth International Workshop on Independent Component Analysis and Blind Signal Separation (ICA 03), Nara, Japan, 1–4 April 2003, pp. 921–926.
9. McGill, W. Multivariate information transmission. *IRE Trans. Inf. Theory* **1954**, *4*, 93–111.
10. Blackwell, D. Equivalent Comparisons of Experiments. *Ann. Math. Stat.* **1953**, *24*, 265–272.
11. Maurer, U.; Wolf, S. The intrinsic conditional mutual information and perfect secrecy. In Proceedings of 1997 IEEE International Symposium on Information Theory, Ulm, Germany, 29 June–4 July 1997.
12. Griffith, V.; Chong, E.K.P.; James, R.G.; Ellison, C.J.; Crutchfield, J.P. Intersection Information Based on Common Randomness. *Entropy* **2014**, *16*, 1985–2000.
13. Rauh, J.; Banerjee, P.K.; Olbrich, E.; Jost, J.; Bertschinger, N.; Wolpert, D. Coarse-graining and the Blackwell order. *arXiv* **2017**, arXiv:1701.07805.

Article

Measuring Multivariate Redundant Information with Pointwise Common Change in Surprisal

Robin A. A. Ince

Institute of Neuroscience and Psychology, University of Glasgow, Glasgow G12 8QB, UK;
robin.ince@glasgow.ac.uk

Received: 2 May 2017; Accepted: 27 June 2017; Published: 29 June 2017

Abstract: The problem of how to properly quantify redundant information is an open question that has been the subject of much recent research. Redundant information refers to information about a target variable S that is common to two or more predictor variables X_i. It can be thought of as quantifying overlapping information content or similarities in the representation of S between the X_i. We present a new measure of redundancy which measures the common change in surprisal shared between variables at the local or pointwise level. We provide a game-theoretic operational definition of unique information, and use this to derive constraints which are used to obtain a maximum entropy distribution. Redundancy is then calculated from this maximum entropy distribution by counting only those local co-information terms which admit an unambiguous interpretation as redundant information. We show how this redundancy measure can be used within the framework of the Partial Information Decomposition (PID) to give an intuitive decomposition of the multivariate mutual information into redundant, unique and synergistic contributions. We compare our new measure to existing approaches over a range of example systems, including continuous Gaussian variables. Matlab code for the measure is provided, including all considered examples.

Keywords: mutual information; redundancy; synergy; pointwise; local; surprisal; partial information decomposition; interaction information; co-information

---

1. Introduction

Information theory was originally developed as a formal approach to the study of man-made communication systems [1,2]. However, it also provides a comprehensive statistical framework for practical data analysis [3]. For example, mutual information is closely related to the log-likelihood ratio test of independence [4]. Mutual information quantifies the statistical dependence between two (possibly multi-dimensional) variables. When two variables (X and Y) both convey mutual information about a third, S, this indicates that some prediction about the value of S can be made after observing the values of X and Y. In other words, S is represented in some way in X and Y. In many cases, it is interesting to ask how these two representations are related—can the prediction of S be improved by simultaneous observation of X and Y (synergistic representation), or is one alone sufficient to extract all the knowledge about S which they convey together (redundant representation). A principled method to quantify the detailed structure of such representational interactions between multiple variables would be a useful tool for addressing many scientific questions across a range of fields [5–8]. Within the experimental sciences, a practical implementation of such a method would allow analyses that are difficult or impossible with existing statistical methods, but that could provide important insights into the underlying system.

Williams and Beer [6] present an elegant methodology to address this problem, with a non-negative decomposition of multivariate mutual information. Their approach, called the Partial Information Decomposition (PID), considers the mutual information within a set of variables. One variable is considered as a privileged *target* variable, here denoted S, which can be thought of as the independent variable in classical statistics. The PID then considers the mutual information conveyed about this target variable by the remaining *predictor* variables, denoted $\mathcal{X} = \{X_1, X_2, \ldots X_n\}$, which can be thought of as dependent variables. In practice the target variable S may be an experimental stimulus or parameter, while the predictor variables in $\mathcal{X}$ might be recorded neural responses or other experimental outcome measures. However, note that due to the symmetry of mutual information, the framework applies equally when considering a single (dependent) output in response to multiple inputs [7]. Williams and Beer [6] present a mathematical lattice structure to represent the set theoretic intersections of the mutual information of multiple variables [9]. They use this to decompose the mutual information $I(\mathcal{X}; S)$ into terms quantifying the unique, redundant and synergistic information about the independent variable carried by each combination of dependent variables. This gives a complete picture of the representational interactions in the system.

The foundation of the PID is a measure of redundancy between any collection of subsets of $\mathcal{X}$. Intuitively, this should measure the information shared between all the considered variables, or alternatively their common representational overlap. Williams and Beer [6] use a redundancy measure they term $I_{\min}$. However as noted by several authors this measure quantifies the minimum *amount* of information that all variables carry, but does not require that each variable is carrying the *same* information. It can therefore overstate the amount of redundancy in a particular set of variables. Several studies have noted this point and suggested alternative approaches [10–16].

In our view, the additivity of surprisal is the fundamental property of information theory that provides the possibility to meaningfully quantify redundancy, by allowing us to calculate overlapping information content. In the context of the well-known set-theoretical interpretation of information theoretic quantities as measures which quantify the area of sets and which can be visualised with Venn diagrams [9], co-information (often called interaction information) [17–20] is a quantity which measures the intersection of multiple mutual information values (Figure 1). However, as has been frequently noted, co-information conflates synergistic and redundant effects.

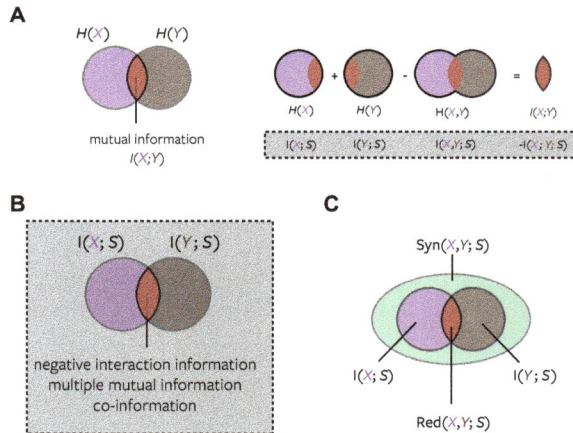

Figure 1. Venn diagrams of mutual information and interaction information. (**A**) Illustration of how mutual information is calculated as the overlap of two entropies; (**B**) The overlapping part of two mutual information values (negative interaction information) can be calculated in the same way—see dashed box in (**A**); (**C**) The full structure of mutual information conveyed by two variables about a third should separate redundant and synergistic regions.

We first review co-information and the PID before presenting I_{ccs}, a new measure of redundancy based on quantifying the common change in surprisal between variables at the local or pointwise level [21–25]. We provide a game-theoretic operational motivation for a set of constraints over which we calculate the maximum entropy distribution. This game-theoretic operational argument extends the decision theoretic operational argument of [12] but arrives at different conclusions about the fundamental nature of unique information. We demonstrate the PID based on this new measure with several examples that have been previously considered in the literature. Finally, we apply the new measure to continuous Gaussian variables [26].

2. Interaction Information (Co-Information)

2.1. Definitions

The foundational quantity of information theory is *entropy*, which is a measure of the variability or uncertainty of a probability distribution. The entropy of a discrete random variable X, with probability mass function $P(X)$ is defined as:

$$H(X) = \sum_{x \in X} p(x) \log_2 \frac{1}{p(x)} \tag{1}$$

This is the expectation over X of $h(x) = -\log_2 p(x)$, which is called the *surprisal* of a particular value x. If a value x has a low probability, it has high surprisal and vice versa. Many information theoretic quantities are similarly expressed as an expectation—in such cases, the specific values of the function over which the expectation is taken are called *pointwise* or *local* values [21–25]. We denote these local values with a lower case symbol. Following [7] we denote probability distributions with a capital latter, e.g., $P(X_1, X_2)$, but denote values of specific realisations, i.e., $P(X_1 = x_1, X_2 = x_2)$ with lower case shorthand $p(x_1, x_2)$.

Figure 1A shows a Venn diagram representing the entropy of two variables X and Y. One way to derive mutual information $I(X;Y)$ is as the intersection of the two entropies. This intersection can be calculated directly by summing the individual entropies (which counts the overlapping region twice) and subtracting the joint entropy (which counts the overlapping region once). This matches one of the standard forms of the definition of mutual information:

$$I(X;Y) = H(X) + H(Y) - H(X,Y) \tag{2}$$

$$= \sum_{x,y} p(x,y) \left[\log_2 \frac{1}{p(y)} - \log_2 \frac{1}{p(y|x)} \right] \tag{3}$$

Here $p(y|x)$ denotes the conditional probability of observing $Y = y$, given that $X = x$ has been observed: $p(y|x) = p(y,x)/p(x)$. Mutual information is the expectation of $i(x;y) = h(y) - h(y|x) = \log_2 \frac{p(y|x)}{p(y)}$, the difference in surprisal of value y when value x is observed. To emphasise this point we use a notation which makes explicit the fact that pointwise information measures a change in surprisal

$$i(x;y) = \Delta_y h(x) = h(x) - h(x|y) \tag{4}$$

$$= \Delta_x h(y) = h(y) - h(y|x) \tag{5}$$

Mutual information is non-negative, symmetric and equals zero if and only if the two variables are statistically independent (that is, $p(x,y) = p(x)p(y) \forall x \in X, y \in Y$) [2].

A similar approach can be taken when considering mutual information about a target variable S that is carried by two predictor variables X and Y (Figure 1B). Again the overlapping region can be calculated directly by summing the two separate mutual information values and subtracting the joint information. However, in this case the resulting quantity can be negative. Positive values of the intersection represent a net redundant representation: X and Y share the same information about S.

Negative values represent a net synergistic representation: X and Y provide more information about S together than they do individually.

In fact, this quantity was first defined as the negative of the intersection described above, and termed *interaction information* [17]:

$$\begin{aligned} I(X;Y;S) &= I(X,Y;S) - I(X;S) - I(Y;S) \\ &= I(S;X|Y) - I(S;X) \\ &= I(S;Y|X) - I(S;Y) \\ &= I(X;Y|S) - I(X;Y) \end{aligned} \tag{6}$$

The alternative equivalent formulations illustrate how the interaction information is symmetric in the three variables, and also represents for example, the information between S and X which is gained (synergy) or lost (redundancy) when Y is fixed (conditioned out).

This quantity has also been termed *multiple mutual information* [27], *co-information* [19], *higher-order mutual information* [20] and *synergy* [28–31]. Multiple mutual information and co-information use a different sign convention from interaction information. For odd numbers of variables (e.g., three X_1, X_2, S) co-information has the opposite sign to interaction information; positive values indicate net redundant overlap.

As for mutual information and conditional mutual information, the interaction information as defined above is an expectation over the joint probability distribution. Expanding the definitions of mutual information in Equation (6) gives:

$$I(X;Y;S) = \sum_{x,y,s} p(x,y,s) \log_2 \frac{p(x,y,s)p(x)p(y)p(s)}{p(x,y)p(x,s),p(y,s)} \tag{7}$$

$$I(X;Y;S) = \sum_{x,y,s} p(x,y,s) \left[\log_2 \frac{p(s|x,y)}{p(s)} - \log_2 \frac{p(s|x)}{p(s)} - \log_2 \frac{p(s|y)}{p(s)} \right] \tag{8}$$

As before we can consider the local or pointwise function

$$i(x;y;s) = \Delta_s h(x,y) - \Delta_s h(x) - \Delta_s h(y) \tag{9}$$

The negation of this value measures the overlap in the change of surprisal about s between values x and y (Figure 1A).

It can be seen directly from the definitions above that in the three variable case the interaction information is bounded:

$$\begin{aligned} I(X;Y;S) &\geq -\min\left[I(S;X), I(S;Y), I(X;Y)\right] \\ I(X;Y;S) &\leq \min\left[I(S;X|Y), I(S;Y|X), I(X;Y|S)\right] \end{aligned} \tag{10}$$

We have introduced interaction information for three variables, from a perspective where one variable is privileged (independent variable) and we study interactions in the representation of that variable by the other two. However, as noted interaction information is symmetric in the arguments, and so we get the same result whichever variable is chosen to provide the analysed information content.

Interaction information is defined similarly for larger numbers of variables. For example, with four variables, maintaining the perspective of one variable being privileged, the 3-way Venn diagram intersection of the mutual information terms again motivates the definition of interaction information:

$$\begin{aligned} I(W;X;Y;S) = &- I(W;S) - I(X;S) - I(Y;S) \\ &+ I(W,X;S) + I(W,Y;S) + I(Y,X;S) \\ &- I(W,X,Y;S) \end{aligned} \tag{11}$$

In the n-dimensional case the general expression for interaction information on a variable set $\mathcal{V} = \{\mathcal{X}, S\}$ where $\mathcal{X} = \{X_1, X_2, \ldots, X_n\}$ is:

$$I(\mathcal{V}) = -\sum_{\mathcal{T} \subseteq \mathcal{X}} (-1)^{|\mathcal{T}|} I(\mathcal{T}; S) \tag{12}$$

which is an alternating sum over all subsets $\mathcal{T} \subseteq \mathcal{X}$, where each $\mathcal{T}$ contains $|\mathcal{T}|$ elements of $\mathcal{X}$. The same expression applies at the local level, replacing I with the pointwise i. Dropping the privileged target S an equivalent formulation of interaction information on a set of n-variables $\mathcal{X} = \{X_1, X_2, \ldots, X_n\}$ in terms of entropy is given by [18,32]:

$$I(\mathcal{X}) = -\sum_{\mathcal{T} \subseteq \mathcal{X}} (-1)^{|\mathcal{X}| - |\mathcal{T}|} H(\mathcal{T}) \tag{13}$$

2.2. Interpretation

We consider as above a three variable system with a target variable S and two predictor variables X, Y, with both X and Y conveying information about S. The concept of redundancy is related to whether the information conveyed by X and that conveyed by Y is *the same* or *different*. Within a decoding (supervised classification) approach, the relationship between the variables is determined from predictive performance within a cross-validation framework [33,34]. If the performance when decoding X and Y together is the same as the performance when considering e.g., X alone, this indicates that the information in Y is completely redundant with that in X; adding observation of Y has no predictive benefit for an observer. In practice redundancy may not be complete as in this example; some part of the information in X and Y might be shared, while both variables also convey unique information not available in the other.

The concept of synergy is related to whether X and Y convey more information when observed together than they do when observed independently. Within the decoding framework this means higher performance is obtained by a decoder which predicts on a joint model of simultaneous X and Y observations, versus a decoder which combines independent predictions obtained from X and Y individually. The predictive decoding framework provides a useful intuition for the concepts, but has problems quantifying redundancy and synergy in a meaningful way because of the difficulty of quantitatively relating performance metrics (percent correct, area under ROC, etc.) between different sets of variables—i.e., X, Y and the joint variable (X, Y).

The first definition (Equation (6)) shows that interaction information is the natural information theoretic approach to this problem: it contrasts the information available in the joint response to the information available in each individual response (and similarly obtains the intersection of the multivariate mutual information in higher order cases). A negative value of interaction information quantifies the redundant overlap of Figure 1B, positive values indicate a net synergistic effect between the two variables. However, there is a major issue which complicates this interpretation: interaction information conflates synergy and redundancy in a single quantity (Figure 1B) and so does not provide a mechanism for separating synergistic and redundant information (Figure 1C) [6]. This problem arises for two reasons. First, local terms $i(x; y; s)$ can be positive for some values of x, y, s and negative for others. These opposite effects can then cancel in the overall expectation. Second, as we will see, the computation of interaction information can include terms which do not have a clear interpretation in terms of synergy or redundancy.

3. The Partial Information Decomposition

In order to address the problem of interaction information conflating synergistic and redundant effects, Williams and Beer [6] proposed a decomposition of mutual information conveyed by a set of predictor variables $\mathcal{X} = \{X_1, X_2, \ldots, X_n\}$, about a target variable S. They reduce the total multivariate mutual information, $I(\mathcal{X}; S)$, into a number of non-negative atoms representing the unique, redundant and synergistic information between all subsets of $\mathcal{X}$: in the two-variable case this corresponds to the four regions of Figure 1C. To do this they consider all subsets of $\mathcal{X}$, denoted $\mathbf{A_i}$, and termed *sources*. They show that the redundancy structure of the multivariate information is determined by the "collection of all sets of sources such that no source is a superset of any other"—formally the set of anti-chains on the lattice formed from the power set of $\mathcal{X}$ under set inclusion, denoted $\mathcal{A}(\mathcal{X})$. Together with a natural ordering, this defines a redundancy lattice [35]. Each node of the lattice represents a partial information atom, the value of which is given by a partial information (PI) function. Note there is a direct correspondence between the lattice structure and a Venn diagram representing multiple mutual information values. Each node on a lattice corresponds to a particular intersecting region in the Venn diagram. For two variables there are only four terms, but the advantage of the lattice representation becomes clearer for higher number of variables. The lattice view is much easier to interpret when there are a large number of intersecting regions that are hard to visualise in a Venn diagram. Figure 2 shows the structure of this lattice for $n = 2, 3$. The PI value for each node, denoted I_∂, can be determined via a recursive relationship (Möbius inverse) over the redundancy values of the lattice:

$$I_\partial(S; \alpha) = I_\cap(S; \alpha) - \sum_{\beta \prec \alpha} I_\partial(S; \beta) \tag{14}$$

where $\alpha \in \mathcal{A}(\mathcal{X})$ is a set of sources (each a set of input variables X_i) defining the node in question.

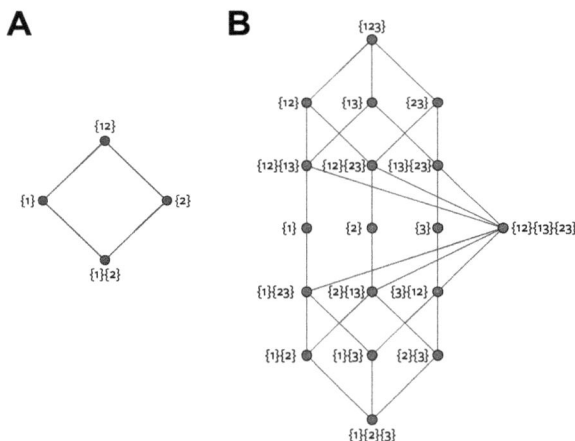

Figure 2. Redundancy lattice for (**A**) two variables; (**B**) three variables. Modified from [6].

The redundancy value of each node of the lattice, $I_\cap$, measures the total amount of redundant information shared between the sources included in that node. For example, $I_\cap(S; \{X_1\}\{X_2\})$ quantifies the redundant information content about S that is common to both X_1 and X_2. The partial information function, I_∂, measures the unique information contributed by only that node (redundant, synergistic or unique information within subsets of variables).

For the two variable case, if the redundancy function used for a set of sources is denoted $I_\cap(S; \mathbf{A_1}, \ldots, \mathbf{A_k})$ and following the notation in [6], the nodes of the lattice, their redundancy and their partial information values are given in Table 1.

Table 1. Full Partial Information Decomposition (PID) in the two-variable case. The four terms here correspond to the four regions in Figure 1C.

Node Label	Redundancy Function	Partial Information	Represented Atom
{12}	$I_\cap(S; \{X_1, X_2\})$	$I_\cap(S; \{X_1, X_2\})$ $- I_\cap(S; \{X_1\}) - I_\cap(S; \{X_2\})$ $+ I_\cap(S; \{X_1\}\{X_2\})$	unique information in X_1 and X_2 together (synergy)
{1}	$I_\cap(S; \{X_1\})$	$I_\cap(S; \{X_1\})$ $- I_\cap(S; \{X_1\}\{X_2\})$	unique information in X_1 only
{2}	$I_\cap(S; \{X_2\})$	$I_\cap(S; \{X_2\})$ $- I_\cap(S; \{X_1\}\{X_2\})$	unique information in X_2 only
{1}{2}	$I_\cap(S; \{X_1\}\{X_2\})$	$I_\cap(S; \{X_1\}\{X_2\})$	redundant information between X_1 and X_2

Note that we have not yet specified a redundancy function. A number of axioms have been proposed for any candidate redundancy measure [6,11]:

Symmetry:

$$I_\cap (S; \mathbf{A_1}, \dots, \mathbf{A_k}) \text{ is symmetric with respect to the } \mathbf{A_i}\text{'s.} \tag{15}$$

Self Redundancy:

$$I_\cap (S; \mathbf{A}) = I(S; \mathbf{A}) \tag{16}$$

Subset Equality:

$$I_\cap (S; \mathbf{A_1}, \dots, \mathbf{A_{k-1}}, \mathbf{A_k}) = I_\cap (S; \mathbf{A_1}, \dots, \mathbf{A_{k-1}}) \text{ if } \mathbf{A_{k-1}} \subseteq \mathbf{A_k} \tag{17}$$

Monotonicity:

$$I_\cap (S; \mathbf{A_1}, \dots, \mathbf{A_{k-1}}, \mathbf{A_k}) \leq I_\cap (S; \mathbf{A_1}, \dots, \mathbf{A_{k-1}}) \tag{18}$$

Note that previous presentations of these axioms have included subset equality as part of the monotonicity axiom; we separate them here for reasons that will become clear later. Subset equality allows the full power set of all combinations of sources to be reduced to only the anti-chains under set inclusion (the redundancy lattice). Self redundancy ensures that the top node of the redundancy lattice, which contains a single source $\mathbf{A} = \mathcal{X}$, is equal to the full multivariate mutual information and therefore the lattice structure can be used to decompose that quantity. Monotonicity ensures redundant information is increasing with the height of the lattice, and has been considered an important requirement that redundant information should satisfy.

Other authors have also proposed further properties and axioms for measures of redundancy [13,14]. In particular, Reference [11] propose an additional axiom regarding the redundancy between two sources about a variable constructed as a copy of those sources:

Identity Property (Harder et al.):

$$I_\cap ([\mathbf{A_1}, \mathbf{A_2}] ; \mathbf{A_1}, \mathbf{A_2}) = I(\mathbf{A_1}; \mathbf{A_2}) \tag{19}$$

In this manuscript we focus on redundant and synergistic mutual information. However, the concepts of redundancy and synergy can also be applied directly to entropy [36]. Redundant entropy is variation that is shared between two (or more) variables, synergistic entropy is additional uncertainty that arises when the variables are considered together, over and above what would be obtained if they were statistically independent. Note that since the global joint entropy quantity is maximised when the two variables are independent, redundant entropy is always greater than synergistic entropy [36]. However, local synergistic entropy can still occur: consider negative local information terms, which by definition quantify a synergistic local contribution to the joint entropy sum since $h(x,y) > h(x) + h(y)$. A crucial insight that results from this point of view is that mutual information itself quantifies both redundant and synergistic entropy effects—it is the difference between redundant and synergistic entropy across the two inputs [36]. With H_∂ denoting redundant or synergistic partial entropy analogous to partial information we have:

$$I(\mathbf{A_1}; \mathbf{A_2}) = H_\partial(\{\mathbf{A_1}\}\{\mathbf{A_2}\}) - H_\partial(\{\mathbf{A_1}, \mathbf{A_2}\}) \qquad (20)$$

This is particularly relevant for the definition of the identity axiom. We argue that the previously unrecognised contribution of synergistic entropy to mutual information (pointwise negative terms in the mutual information expectation sum) should not be included in an information redundancy measure.

Note that any information redundancy function can induce an entropy redundancy function by considering the information redundancy with the copy of the inputs. For example, for the bivariate case we can define:

$$H_\cap(\{\mathbf{A_1}\}\{\mathbf{A_2}\}) = I_\cap([\mathbf{A_1}, \mathbf{A_2}]; \mathbf{A_1}, \mathbf{A_2}) \qquad (21)$$

So any information redundancy measure that satisfies the identity property [12] cannot measure synergistic entropy [36], since for the induced entropy redundancy measure $H_\cap(\{\mathbf{A_1}\}\{\mathbf{A_2}\}) = I(\mathbf{A_1}; \mathbf{A_2})$ so from Equation (20) $H_\partial(\{\mathbf{A_1A_2}\}) = 0$. To address this without requiring introducing in detail the partial entropy decomposition [36], we propose a modified version of the identity axiom, which still addresses the two-bit copy problem but avoids the problem of including synergistic mutual information contributions in the redundancy measure. When $I(\mathbf{A_1}; \mathbf{A_2}) = 0$ there are no synergistic entropy effects because $i(a_1, a_2) = 0 \ \forall a_1, a_2$ so there are no misinformation terms and no synergistic entropy between the two inputs.

Independent Identity Property:

$$I(\mathbf{A_1}; \mathbf{A_2}) = 0 \implies I_\cap([\mathbf{A_1}, \mathbf{A_2}]; \mathbf{A_1}, \mathbf{A_2}) = 0 \qquad (22)$$

Please note that while this section primarily reviews existing work on the partial information decomposition, two novel contributions here are the explicit consideration of subset equality separate to monotonicity, and the definition of the independent identity property.

3.1. An Example PID: RDNUNQXOR

Before considering specific measures of redundant information that have been proposed for use with the PID, we first illustrate the relationship between the redundancy and the partial information lattice values with an example. We consider a system called RDNUNQXOR [10]. The structure of this system is shown in Figure 3A [37]. It consists of two three bit predictors, X_1 and X_2, and a four bit target S. This example is noteworthy, because an intuitive PID is obvious from the definition of the system, and it includes by construction 1 bit of each type of information decomposable with the PID.

All three variables share a bit (labelled b in Figure 3A). This means there should be 1 bit of redundant information. Bit b is shared between each predictor and the target so forms part of $I(X_i; S)$, and is also shared between the predictors, therefore it is shared or redundant information. All variables have one bit that is distributed according to a XOR configuration across the three variables (labelled a). This provides 1 bit of synergy within the system, because the value of bit a of S can only be predicted when X_1 and X_2 are observed together simultaneously [10]. Bits c and d are shared between S and each of X_1 and X_2 individually. So each of these contributes to $I(X_i; S)$, but as unique information.

We illustrate the calculation of the PID for this system (Figure 3B,C, Table 2). From the self-redundancy axiom, the three single-source terms can all be calculated directly from the classical mutual information values. The single predictors each have 2 bits of mutual information (the two bits shared with S). Both predictors together have four bits of mutual information with S, since the values of all four bits of S are all fully determined when both X_1 and X_2 are observed. Since by construction there is 1 bit shared redundantly between the predictors, we claim $I_\cap(S; \{1\}\{2\}) = 1$ bit and we have all the redundancy values on the lattice. Then from the summation procedure illustrated in Table 1 we can calculate the partial information values. For example, $I_\partial(S; \{1\}) = 2 - 1 = 1$, and $I_\partial(S; \{12\}) = 4 - 1 - 1 - 1 = 1$.

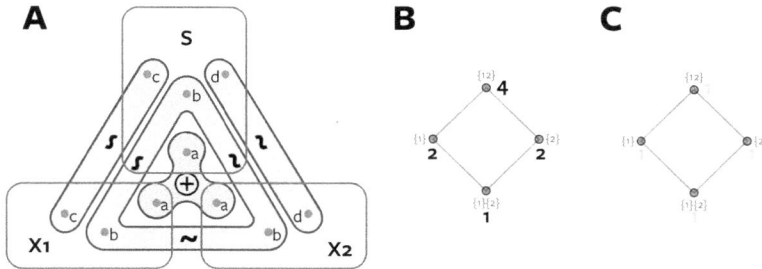

Figure 3. Partial Information Decomposition for RDNUNQXOR (**A**) The structure of the RDNUNQXOR system borrowing the graphical representation from [37]. S is a variable containing 4 bits (labelled a, b, c, d). X_1 and X_2 each contain 3 bits. $\sim$ indicates bits which are coupled (distributed identically) and $\oplus$ indicates the enclosed variables form the XOR relation; (**B**) Redundant information values on the lattice (black); (**C**) Partial information values on the lattice (green).

Table 2. PID for RDNUNQXOR (Figure 3).

Node	$I_\cap$	I_∂
$\{1\}\{2\}$	1	1
$\{1\}$	2	1
$\{2\}$	2	1
$\{12\}$	4	1

3.2. Measuring Redundancy With Minimal Specific Information: I_{min}

The redundancy measure proposed by Williams and Beer [6] is denoted I_{min} and derived as the average (over values s of S) minimum specific information [38,39] over the considered input sources. The information provided by a source $\mathbf{A}$ (as above a subset of dependent variables X_i) can be written:

$$I(S; \mathbf{A}) = \sum_s p(s) I(S = s; \mathbf{A}) \tag{23}$$

where $I(S = s; \mathbf{A})$ is the *specific information*:

$$I(S = s; \mathbf{A}) = \sum_a p(a|s) \left[\log_2 \frac{1}{p(s)} - \log_2 \frac{1}{p(s|a)} \right] \tag{24}$$

which quantifies the average reduction in surprisal of s given knowledge of $\mathbf{A}$. This splits the overall mutual information into the reduction in uncertainty about each individual target value. I_{min} is then defined as:

$$I_{min}(S; \mathbf{A_1}, \ldots, \mathbf{A_k}) = \sum_s p(s) \min_{\mathbf{A_i}} I(S = s; \mathbf{A_i}) \tag{25}$$

This quantity is the expectation (over S) of the minimum amount of information about each specific target value s conveyed by any considered source. I_{min} is non-negative and satisfies the axioms of symmetry, self redundancy and monotonicity, but not the identity property (neither Harder et al. or independent forms). The crucial conceptual problem with I_{min} is that it indicates the variables share a common *amount* of information, but not that they actually share the *same* information content [5,10,11].

The most direct example of this is the "two-bit copy problem", which motivated the identity axiom [5,10,11]. We consider two independent uniform binary variables X_1 and X_2 and define S as a direct copy of these two variables $S = (X_1, X_2)$. In this case $I_{min}(S; \{1\}\{2\}) = 1$ bit; for every s both X_1 and X_2 each provide 1 bit of specific information. However, both variables give different information about each value of s: X_1 specifies the first component, X_2 the second. Since X_1 and X_2 are independent by construction there should be no overlap. This illustrates that I_{min} can overestimate redundancy with respect to an intuitive notion of overlapping information content.

3.3. Measuring Redundancy With Maximised Co-Information: I_{broja}

A number of alternative redundancy measures have been proposed for use with the PID in order to address the problems with I_{min} (reviewed by [26]). Two groups have proposed an equivalent approach, based on the idea that redundancy should arise only from the marginal distributions $P(X_1, S)$ and $P(X_2, S)$ ([12], their Assumption *) and that synergy should arise from structure not present in those two marginals, but only in the full joint distribution $P(X_1, X_2, S)$. Please note that we follow their terminology and refer to this concept as Assumption * throughout. Griffith and Koch [10] frame this view as a minimisation problem for the multivariate information $I(S; X_1, X_2)$ over the class of distributions which preserve the individual source-target marginal distributions. Bertschinger et al. [12] seek to minimise $I(S; X_1 | X_2)$ over the same class of distributions, but as noted both approaches result in the same PID. In both cases the redundancy, $I_\cap(S; \{X_1\}\{X_2\})$, is obtained as the maximum of the co-information (negative interaction information) over all distributions that preserve the source-target marginals:

$$I_{\text{max-nii}}(S; \{X_1\}\{X_2\}) = \max_{Q \in \Delta_P} -I_Q(S; X_1; X_2) \tag{26}$$

$$\Delta_P = \{Q \in \Delta : Q(X_1, S) = P(X_1, S), Q(X_2, S) = P(X_2, S)\} \tag{27}$$

We briefly highlight here a number of conceptual problems with this approach. First, this measure satisfies the Harder et al. identity property (Equation (19)) [11,12] and is therefore incompatible with the notion of synergistic entropy [36]. Second, this measure optimises co-information, a quantity which conflates synergy and redundancy [6]. Given ([12], Assumption *) which states that unique and redundant information are constant on the optimisation space, this is equivalent to minimizing synergy [7].

$$I_{\text{max-nii}}(S; \{X_1\}\{X_2\}) = I_{\text{red}}(S; \{X_1\}\{X_2\}) - I_{\text{syn-min}}(S; \{X_1\}\{X_2\}) \tag{28}$$

where $I_{\text{syn-min}}(S; \{X_1\}\{X_2\})$ is the smallest possible synergy given the target-predictor marginal constraints, but is not necessarily zero. Therefore, the measure provides a bound on redundancy (under Assumption * [12]) but cannot measure the true value. Third, Bertschinger et al. [12] motivate the constraints for the optimisation from an operational definition of unique information based on decision theory. It is this argument which suggests that the unique information is constant on the optimisation space Δ_P, and which motivates a foundational axiom for the measure that equal target-predictor marginal distributions imply zero unique information. However, we do not agree that unique information is invariant to the predictor-predictor marginal distributions, or necessarily equals zero when target-predictor marginals are equal. We revisit the operational definition in Section 4.3 by considering a game theoretic extension which provides a different perspective. We use this to provide a counter-example that proves the decision theoretic argument is not a necessary condition for the existence of unique information, and therefore the I_{broja} procedure is invalid since redundancy is not fixed on Δ_P. We also demonstrate with several examples (Section 5) how the I_{broja} optimisation results in coupled predictor variables, suggesting the co-information optimisation is indeed maximising the source redundancy between them.

3.4. Other Redundancy Measures

Harder et al. [11] define a redundancy measure based on a geometric projection argument, which involves an optimisation over a scalar parameter λ, and is defined only for two sources, so can be used only for systems with two predictor variables. Griffith et al. [13] suggest an alternative measure motivated by zero-error information, which again formulates an optimisation problem (here maximisation of mutual information) over a family of distributions (here distributions Q which are a function of each predictor so that $H(Q|X_i) = 0$). Griffith and Ho [16] extend this approach by modifying the optimisation constraint to be $H(Q|X_i) = H(Q|X_i, Y)$.

4. Measuring Redundancy With Pointwise Common Change in Surprisal: I_{ccs}

We derive here from first principles a measure that we believe encapsulates the intuitive meaning of redundancy between sets of variables. We argue that the crucial feature which allows us to directly relate information content between sources is the additivity of surprisal. Since mutual information measures the expected change in pointwise surprisal of s when x is known, we propose measuring redundancy as the expected pointwise change in surprisal of s which is common to x and y. We term this *common change in surprisal* and denote the resulting measure $I_{ccs}(S; \alpha)$.

4.1. Derivation

As for entropy and mutual information we can consider a Venn diagram (Figure 1) for the change in surprisal of a specific value s for specific values x and y and calculate the overlap directly using local co-information (negative local interaction information). However, as noted before the interaction information can confuse synergistic and redundant effects, even at the pointwise level. Recall that mutual information $I(S; X)$ is the expectation of a local function which measures the pointwise change in surprisal $i(s; x) = \Delta_s h(x)$ of value s when value x is observed. Although mutual information itself is always non-negative, the pointwise function can take both positive and negative values. Positive values correspond to a reduction in the surprisal of s when x is observed, negative values to an increase in surprisal. Negative local information values are sometimes referred to as *misinformation* [23] and can be interpreted as representing synergistic entropy between S and X [36]. Mutual information is then the expectation of both positive (information) terms and negative (misinformation) terms. Table 3 shows how the possibility of local misinformation terms complicates pointwise interpretation of the local negative interaction information (co-information).

Note that the fourth column represents the local co-information which quantifies the set-theoretic overlap of the two univariate local information values. By considering the signs of all four terms, the two univariate local informations, the local joint information and their overlap, we can determine terms which correspond to redundancy and terms which correspond to synergy. We make an assumption that a decrease in surprisal of s (positive local information term) is a fundamentally different event to an increase in surprisal of s (negative local information). Therefore, we can only interpret the local co-information as a set-theoretic overlap in the case where all three local information terms have the same sign. If the joint information has a different sign to the individual informations (rows 5 and 6) the two variables together represent a fundamentally different change in surprisal than either do alone. While a full interpretation of what these terms might represent is difficult, we argue it is clear they cannot represent a common change in surprisal. Similarly, if the two univariate local informations have opposite sign, they cannot have any common overlap.

Table 3. Different interpretations of local interaction information terms. ? indicates that combination of terms does not admit a clear interpretation in terms of redundancy or synergy.

$\Delta_s h(x)$	$\Delta_s h(y)$	$\Delta_s h(x,y)$	$-i(x;y;s)$	Interpretation
+	+	+	+	redundant information
+	+	+	−	synergistic information
−	−	−	−	redundant misinformation
−	−	−	+	synergistic misinformation
+	+	−	...	?
−	−	+	...	?
+/−	−/+	...	...	?

The table shows that interaction information combines redundant information with synergistic misinformation, and redundant misinformation with synergistic information. As discussed, it also includes terms which do not admit a clear interpretation. We argue that a principled measure of redundancy should consider only redundant information and redundant misinformation. We therefore consider the pointwise negative interaction information (overlap in surprisal), but only for symbols corresponding to the first and third rows of Table 3. That is, terms where the sign of the change in surprisal for all the considered sources is equal, and equal also to the sign of overlap (measured with local co-information). In this way, we count the contributions to the overall mutual information (both positive and negative) which are genuinely shared between the input sources, while ignoring other (synergistic and ambiguous) interaction effects. We assert that conceptually this is exactly what a redundancy function should measure.

We denote the local co-information (negative interaction information if n is odd) with respect to a joint distribution Q as $c_q(a_1, \ldots, a_n)$, which is defined as [20]:

$$c_q(a_1, \ldots, a_n) = \sum_{k=1}^{n} (-1)^{k+1} \sum_{i_1 < \cdots < i_k} h_q\left(a_{i_1}, \ldots, a_{i_k}\right) \tag{29}$$

where $h_q(a_1, \ldots, a_n) = -\log q(a_1, \ldots, a_n)$ is pointwise entropy (surprisal). Then we define I_{ccs}, the common change in surprisal, as:

Definition 1.

$$I_{ccs}(S; A_1, \ldots, A_n) = \sum_{a_1, \ldots, a_n} \tilde{p}(a_1, \ldots, a_n) \Delta_s h^{\text{com}}(a_1, \ldots, a_n)$$

$$\Delta_s h^{\text{com}}(a_1, \ldots, a_n) = \begin{cases} c_{\tilde{p}}(a_1, \ldots, a_n, s) & \text{if } \operatorname{sgn} \Delta_s h(a_1) = \ldots = \operatorname{sgn} \Delta_s h(a_n) \\ & \quad = \operatorname{sgn} \Delta_s h(a_1, \ldots, a_n) = \operatorname{sgn} c(a_1, \ldots, a_n, s) \\ 0 & \text{otherwise} \end{cases} \tag{30}$$

where $\Delta_s h^{\text{com}}(a_1, \ldots, a_n)$ represents the common change in surprisal (which can be positive or negative) between the input source values, and $\tilde{P}$ is a joint distribution obtained from the observed joint distribution P (see below). I_{ccs} measures overlapping information content with co-information, by separating contributions which correspond unambiguously to redundant mutual information at the pointwise level, and taking the expectation over these local redundancy values.

Unlike $I_{\min}$ which considered each input source individually, the pointwise overlap computed with local co-information requires a joint distribution over the input sources, $\tilde{P}$ in order to obtain the local surprisal values $h_{\tilde{p}}(a_1, \ldots, a_n, s)$. We use the maximum entropy distribution subject to the constraints of equal bivariate source-target marginals, together with the equality of the n-variate joint target marginal distribution:

Definition 2.

$$\hat{P}(A_1,\ldots,A_n,S) = \arg\max_{Q\in\Delta_P} \sum_{a_1,\ldots,a_n,s} -q(a_1,\ldots,a_n,s)\log q(a_1,\ldots,a_n,s)$$

$$\Delta_P = \left\{ Q \in \Delta : \begin{array}{l} Q(A_i,S) = P(A_i,S) \text{ for } i = 1,\ldots,n \\ Q(A_1,\ldots,A_n) = P(A_1,\ldots,A_n) \end{array} \right\} \tag{31}$$

where $P(A_1,\ldots,A_n,S)$ is the probability distribution defining the system under study and here Δ is the set of all possible joint distributions on $A_1,\ldots,A_n,S$. We develop the motivation for the constraints in Section 4.3.1, and for using the distribution with maximum entropy subject to these constraints in in Section 4.3.2.

In a previous version of this manuscript we used constraints obtained from the decision theoretic operational definition of unique information [12]. We used the maximum entropy distribution subject to the constraints of pairwise target-predictor marginal equality:

Definition 3.

$$\hat{P}_{\text{ind}}(A_1,\ldots,A_n,S) = \arg\max_{Q\in\Delta_P} \sum_{a_1,\ldots,a_n,s} -q(a_1,\ldots,a_n,s)\log q(a_1,\ldots,a_n,s)$$

$$\Delta_P = \left\{ Q \in \Delta : \ Q(A_i,S) = P(A_i,S) \text{ for } i = 1,\ldots,n \right\} \tag{32}$$

This illustrates I_{ccs} can be defined in a way compatible with either operational perspective, depending on whether it is calculated using $\hat{P}$ or $\hat{P}_{\text{ind}}$. We suggest that if a reader favours the decision theoretic definition of unique information [12] over the new game-theoretic definition proposed here (Section 4.3.1) I_{ccs} can be defined in a way consistent with that, and still provides advantages over I_{broja}, which maximises co-information without separating redundant from synergistic contributions (Sections 3.3 and 4.3.2). We include Definition 3 here for continuity with the earlier version of this manuscript, but note that for all the examples considered here we use $\hat{P}$, following the game theoretic operational definition of unique information (Section 4.3.1).

Note that the definition of I_{min} in terms of minimum specific information [39] (Equation (25)) suggests as a possible extension the use of a form of *specific co-information*. In order to separate redundant from synergistic components this should be thresholded with zero to only count positive (redundant) contributions. This can be defined both in terms of target-specific co-information following I_{min} (for clarity these definitions are shown only for two variable inputs):

$$I_{\text{target specific coI}}(S; A_1, A_2) = \sum_s p(s) \max\left[I(S = s; A_1) + I(S = s; A_2) - I(S = s; A_1, A_2), 0\right] \tag{33}$$

or alternatively in terms of source-specific co-information:

$$I_{\text{source specific coI}}(S; A_1, A_2) = \sum_{a_1,a_2} p(a_1,a_2) \max\left[I(S; A_1 = a_1) + I(S; A_2 = a_2)\right. \tag{34}$$

$$\left. - I(S; A_1 = a_1, A_2 = a_2), 0\right] \tag{35}$$

I_{ccs} can be seen as a fully local approach within this family of measures. The first key ingredient of this family is to exploit the additivity of surprisal and hence use the co-information to quantify the overlapping information content (Figure 1); the second ingredient is to break down in some way the expectation summation in the calculation of co-information, to separate redundant and synergistic effects that are otherwise conflated. We argue the fully local view of I_{ccs} is required to fully separate redundant from synergistic effects. In either specific co-information calculation, when summing the contributions within the expectation over the non-specific variable any combination of terms listed in Table 3 could occur. Therefore, these specific co-information values could still conflate redundant information with synergistic misinformation.

4.2. Calculating I_{ccs}

We provide here worked examples of calculating I_{ccs} for two simple example systems. The simplest example of redundancy is when the system consists of a single coupled bit [10] (Example RDN), defined by the following distribution $P(X_1, X_2, S)$:

$$p(0,0,0) = p(1,1,1) = 0.5 \tag{36}$$

In this example $\hat{P} = P$; the maximum entropy optimisation results in the original distribution. Table 4 shows the pointwise terms of the co-information calculation. In this system for both possible configurations the change in surprisal from each predictor is 1 bit and overlaps completely. The signs of all changes in surprisal and the local co-information are positive, indicating that both these events correspond to redundant local information. In this case I_{ccs} is equal to the co-information.

Table 4. Pointwise values from $I_{ccs}(S; \{1\}\{2\})$ for RDN.

(x_1, x_2, s)	$\Delta_s h(x_1)$	$\Delta_s h(x_2)$	$\Delta_s h(x_1, x_2)$	$c(x_1; x_2; s)$	$\Delta_s h^{\text{com}}(x_1, x_2)$
$(0,0,0)$	1	1	1	1	1
$(1,1,1)$	1	1	1	1	1

The second example we consider is binary addition (see also Section 5.2.2), $S = X_1 + X_2$, with distribution $P(X_1, X_2, S)$ given by

$$p(0,0,0) = p(0,1,1) = p(1,0,1) = p(1,1,2) = 1/4 \tag{37}$$

In this example, again $\hat{P} = P$. The pointwise terms are shown in Table 5. For the events with $x_1 = x_2$, both predictors provide 1 bit local change in surprisal of s, but they do so independently since the change in surprisal when observing both together is 2 bits. Therefore, the local co-information is 0; there is no overlap. For the terms where $x_1 \neq x_2$, neither predictor alone provides any local information about s. However, together they provide a 1 bit change in surprisal. This is therefore a purely synergistic contribution, providing -1 bits of local co-information. However, since this is synergistic, it is not included in $\Delta_s h^{\text{com}}$. $I_{ccs}(S; \{1\}\{2\}) = 0$, although the co-information for this system is -0.5 bits. This example illustrates how interpreting the pointwise co-information terms allows us to select only those representing redundancy.

Table 5. Pointwise values from $I_{ccs}(S; \{1\}\{2\})$ for SUM.

(x_1, x_2, s)	$\Delta_s h(x_1)$	$\Delta_s h(x_2)$	$\Delta_s h(x_1, x_2)$	$c(x_1; x_2; s)$	$\Delta_s h^{\text{com}}(x_1, x_2)$
$(0,0,0)$	1	1	2	0	0
$(0,1,1)$	0	0	1	-1	0
$(1,0,1)$	0	0	1	-1	0
$(1,1,2)$	1	1	2	0	0

4.3. Operational Motivation for Choice of Joint Distribution

4.3.1. A Game-Theoretic Operational Definition of Unique Information

Bertschinger et al. [12] introduce an operational interpretation of unique information based on decision theory, and use that to argue the "unique and shared information should only depend on the marginal [source-target] distributions" $P(A_i, S)$ (their Assumption (*) and Lemma 2). Under the assumption that those marginals alone should specify redundancy they find I_{broja} via maximisation of co-information. Here we review and extend their operational argument and arrive at a different conclusion.

Bertschinger et al. [12] operationalise unique information based on the idea that if an agent, Alice, has access to unique information that is not available to a second agent, Bob, there should be some situations in which Alice can exploit this information to gain a systematic advantage over Bob ([7], Appendix B therein). They formalise this as a decision problem, with the systematic advantage corresponding to a higher expected reward for Alice than Bob. They define a decision problem as a tuple $(p, \mathcal{A}, u)$ where $p(S)$ is the marginal distribution of the target, S, $\mathcal{A}$ is a set of possible actions the agent can take, and $u(s, a)$ is the reward function specifying the reward for each $s \in S$ $a \in \mathcal{A}$. They assert that unique information exists if and only if there exists a decision problem in which there is higher expected reward for an agent making optimal decisions based on observation of X_1, versus an agent making optimal decisions on observations of X_2. This motivates their fundamental assumption that unique information depends only on the pairwise target-predictor marginals $P(X_1, S)$, $P(X_2, S)$ ([12] Assumption *), and their assertion that $P(X_1, S) = P(X_2, S)$ implies no unique information in either predictor.

We argue that the decision problem they consider is too restrictive, and therefore the conclusions they draw about the properties of unique and redundant information are incorrect. Those properties come directly from the structure of the decision problem; the reward function u is the same for both agents, and the agents play independently from one other. The expected reward is calculated separately for each agent, ignoring by design any trial by trial covariation in their observed evidence $P(X_1, X_2)$, and resulting actions.

While it is certainly true that if their decision problem criterion is met, then there is unique information, we argue that the decision problem advantage is not a necessary condition for the existence of unique information. We prove this by presenting below a counter-example, in which we demonstrate unique information without a decision theoretic advantage. To construct this example, we extend their argument to a game-theoretic setting, where we explicitly consider two agents playing against each other. Decision theory is usually defined as the study of individual agents, while situations with multiple interacting agents are the purview of game theory. Since the unique information setup includes two agents, it seems more natural to use a game theoretic approach. Apart from switching from a decision theoretic to a game theoretic perspective, we make exactly the same argument. It is possible to operationalise unique information so that unique information exists if and only if there exists a game (with certain properties described below) where one agent obtains a higher expected reward when both agents are playing optimally under the same utility function.

We consider two agents interacting in a game, specifically a non-cooperative, simultaneous, one-shot game [40] where both agents have the same utility function. Non-cooperative means the players cannot form alliances or agreements. Simultaneous (as opposed to sequential) means the players move simultaneously; if not actually simultaneous in implementation such games can be effectively simultaneous as long as each player is not aware of the other players actions. This is a crucial requirement for a setup to operationalise unique information because if the game was sequential, it would be possible for information to "leak" from the first players evidence, via the first players action, to the second. Restricting to simultaneous games prevents this, and ensures each game provides a fair test for unique information in each players individual predictor evidence. One-shot (as opposed to repeated) means the game is played only once as a one off, or at least each play is completely independent of any other. Players have no knowledge of previous iterations, or opportunity to learn from or adapt to the actions of the other player. The fact that the utility function is the same for the actions of each player makes it a fair test for any advantage given by unique information—both players are playing by the same rules. These requirements ensure that, as for the decision theoretic argument of [12], each player must chose an action to maximise their reward based only the evidence they observe from the predictor variable. If a player is able to obtain a systematic advantage, in the form of a higher expected reward for some specific game, given the game is fair and they are acting only on the information in the predictor they observe, then this must correspond to unique information in that predictor. This is the same as the claim made in [12] that higher expected reward in a specific decision problem implies unique information in the predictor.

In fact, if in addition to the above properties the considered game is also symmetric and non-zero-sum then this is exactly equivalent to the decision theoretic formulation. Symmetric means the utility function is invariant to changes of player identity (i.e., it is the same if the players swap places). Alternatively, an asymmetric game is one in which the reward is not necessarily unchanged if the identity of the players is switched. A zero-sum game is one in which there is a fixed reward that is distributed between the players while in a non-zero-sum game the reward is not fixed. The decision problem setup is non-zero-sum, since the action of one agent does not affect the reward obtained by the other agent. Both players consider the game as a decision problem and so play as they would in the decision theoretic framework (i.e., to choose an action based only on their observed evidence in such a way as to maximise their expected reward). This is because since the game is non-cooperative, simultaneous and one-shot they have no knowledge of or exposure to the other players actions.

We argue unique information should also be operationalised in asymmetric and zero-sum games, since these also satisfy the core requirements outlined above for a fair test of unique information. In a zero-sum game, the reward of each agent now also depends on the action of the other agent, therefore unique information is not invariant to changes in $P(X_1, X_2)$, because this can change the balance of rewards on individual realisations. Note that this does not require either player is aware of the others actions (because the game is simultaneous), they still chose an action based only on their own predictor evidence, but their reward depends also on the action of the other agent (although those actions themselves are invisible). The stochastic nature of the reward from the perspective of each individual agent is not an issue since, as for the decision theoretic approach, we consider only one-shot games. Alternatively, if an asymmetry is introduced to the game, for example by allowing one agent to set the stake in a gambling task, then again $P(X_1, X_2)$ affects the unique information. We provide a specific example for this second case, and specify an actual game which meets the above requirements and provides a systematic advantage to one player, demonstrating the presence of unique information. However, this system does not admit a decision problem which provides an advantage. This counter-example therefore proves that the decision theoretic operationalisation of [12] is not a necessary condition for the existence of unique information.

Borrowing notation from [12] we consider two agents, which each observe values from X_1 and X_2 respectively, and take actions $a_1, a_2 \in \mathcal{A}$. Both are subject the same core utility function $v(s, a)$, but we break the symmetry in the game by allowing one agent to perform a second action—setting the stake on each hand (realisation). This results in utility functions $u_i(s, a_i, x_1) = c(x_1)v(s, a_i)$, where c is a stake weighting chosen by agent 1 on the basis of their evidence. This stake weighting is not related to their guess on the value s (their action a_i), but serves here as a way to break the symmetry of the game while maintaining equal utility functions for each player. That is, although the reward here is a function also of x_1, it is the same function for both players, so $a_1 = a_2 \implies u_1(s, a_1, x_1) = u_2(s, a_2, x_1) \forall s, x_1$. In general, in the game theoretic setting the utility function can depend on the entire state of the world, $u(s, a_i, x_1, x_2)$, but here we introduce only an asymmetric dependence on x_1. Both agents have the same utility function as required for a fair test of unique information, but that utility function is asymmetric—it is not invariant to switching the players. The second agent is not aware of the stake weighting applied to the game when they choose their action. The tuple $(p, \mathcal{A}, \mathbf{u})$ defines the game with $\mathbf{u}(s, a_1, a_2, x_1) = [u_1(s, a_1, x_1), u_2(s, a_2, x_1)]$. In this case the reward of agent 2 depends on x_1, introducing again a dependence on $P(X_1, X_2)$. However, because both agents have the same asymmetric utility function, this game meets the intuitive requirements for an operational test of unique information. If there is no unique information, agent 1 should not be able to profit simply by changing the stakes on different trials. If they can profit systematically by changing the stakes on trials that are favourable to them based on the evidence they observe, that is surely an operationalisation of unique information. We emphasise again that we are considering here a non-cooperative, simultaneous, one-shot, non-zero-sum, asymmetric game. So agent 2 does not have any information about the stake weight on individual games, and cannot learn anything about the stake weight from repeated plays. Therefore, there is no way for unique information in X_1 to affect the action of agent 2 via the stake weight setting. The only difference from the decision theoretic framework is that here we consider an asymmetric utility function.

To demonstrate this, and provide a concrete counter-example to the decision theoretic argument [12] we consider a system termed REDUCEDOR (Joseph Lizier, *personal communication*). Figure 4A shows the probability distribution which defines this binary system. Table 6 shows the PIDs for this system. Figure 4B shows the distribution resulting from the I_{broja} optimisation procedure. Both systems have the same target-predictor marginals $P(X_i, S)$, but have different predictor-predictor marginals $P(X_1, X_2)$. I_{broja} reports zero unique information. I_{ccs} reports zero redundancy, but unique information present in both predictors.

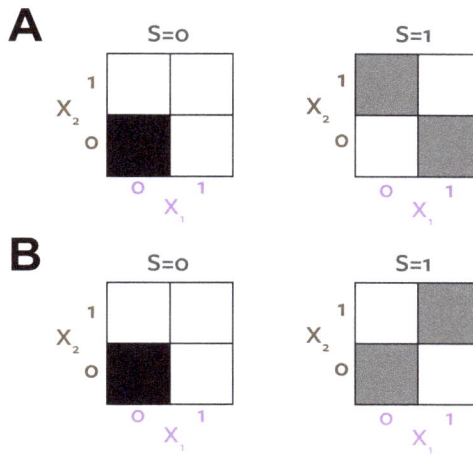

Figure 4. REDUCEDOR. (**A**) Probability distribution of REDUCEDOR system; (**B**) Distribution resulting from I_{broja} optimisation. Black tiles represent outcomes with $p = 0.5$. Grey tiles represent outcomes with $p = 0.25$. White tiles are zero-probability outcomes.

Table 6. Partial Information Decompositions (PIDs) for REDUCEDOR (Figure 4A).

Node	$I_\partial[I_{min}]$	$I_\partial[I_{broja}]$	$I_\partial[I_{ccs}]$
{1}{2}	0.31	0.31	0
{1}	0	0	0.31
{2}	0	0	0.31
{12}	0.69	0.69	0.38

In the I_{broja} optimised distribution (Figure 4B) the two predictors are directly coupled, $P(X_1 = 0, X_2 = 1) = P(X_1 = 1, X_2 = 0) = 0$. In this case there is clearly no unique information. The coupled marginals mean both agents see the same evidence on each realisation, make the same choice and therefore obtain the same reward, regardless of the stake weighting chosen by agent 1. However, in the actual system, the situation is different. Now the evidence is de-coupled, the agents never both see the evidence $x_i = 1$ on any particular realisation $P(X_1 = 1, X_2 = 1) = 0$. Assuming a utility function $v(s, a) = \delta_{sa}$ reflecting a guessing game task, the optimal strategy for both agents is to make a guess $a_i = 0$ when they observe $x_i = 0$, and guess $a_i = 1$ when they observe $x_i = 1$. If Alice (X_1) controls the stake weight she can choose $c(x_1) = 1 + x_1$ which results in a doubling of the reward when she observes $X_1 = 1$ versus when she observes $X_1 = 0$. Under the true distribution of the system for realisations where $x_1 = 1$, we know that $x_2 = 0$ and $s = 1$, so Bob will guess $a_2 = 0$ and be wrong (have zero reward). On an equal number of trials Bob will see $x_2 = 1$, guess correctly and Alice will win nothing, but those trials have half the utility of the trials that Alice wins due to the asymmetry resulting from her specifying the gambling stake. Therefore, on average, Alice will have a systematically higher reward as a result of exploiting her unique information, which is unique because on specific realisations it is available only to her. Similarly, the argument can be reversed, and if Bob gets to choose the stakes, corresponding to a utility weighting $c(x_2) = 1 + x_2$, he can exploit unique information available to him on a separate set of realisations.

Both games considered above would provide no advantage when applied to the I_{broja} distribution (Figure 4B). The information available to each agent when they observe $X_i = 1$ is not unique, because it always occurs together on the same realisations. There is no way to gain an advantage in any game since it will always be available simultaneously to the other agent. In both decompositions the information corresponding to prediction of the stimulus when $x_i = 1$ is quantified as 0.31 bits. I_{broja} quantifies this as redundancy because it ignores the structure of $P(X_1, X_2)$ and so does not consider the within trial relationships between the agents evidence. I_{broja} cannot distinguish between the two distributions illustrated in Figure 4. I_{ccs} quantifies the 0.31 bits as unique information in both predictors, because in the true system each agent sees the informative evidence on different trials, and so can exploit it to gain a higher reward in a certain game. I_{ccs} agrees with I_{broja} in the system in Figure 4B, because here the same evidence is always available to both agents, so is not unique.

We argue that this example directly illustrates the fact that unique information is not invariant to $P(X_1, X_2)$, and that the decision theoretic operational definition of [12] is too restrictive. The decision theory view says that unique information corresponds to an advantage which can be obtained only when two players go to different private rooms in a casino, play independently and then compare their winnings at the end of the session. The game theoretic view says that unique information corresponds to any obtainable advantage in a fair game (simultaneous and with equal utility functions), even when the players play each other directly, betting with a fixed pot, on the same hands at the same table. We have shown a specific example where there is an advantage in the second case, but not the first case. We suggest such an advantage cannot arise without unique information in the predictor and therefore claim this counter-example proves that the decision theoretic operationalisation is not a necessary condition for the existence of unique information. While this is a single specific system, we will see in the examples (Section 5) that the phenomenon of I_{broja} over-stating redundancy by neglecting unique information which is masked when the inputs are coupled occurs frequently. We argue this occurs because the I_{broja} optimisation maximises co-information. It therefore couples the predictors to maximise the contribution of source redundancy to the co-information, since the game theoretic operationalisation shows that redundancy is not invariant to the predictor-predictor marginal distribution.

4.3.2. Maximum Entropy Optimisation

For simplicity we consider first a two-predictor system. The game-theoretic operational definition of unique information provided in the previous section requires that the unique information (and hence redundancy) should depend only on the pairwise marginals $P(S, X_1)$, $P(S, X_2)$ and $P(X_1, X_2)$. Therefore, any measure of redundancy which is consistent with this operational definition should take a constant value over the family of distributions which satisfy those marginal constraints. This is the same argument applied in [12] but we consider here the game-theoretic extension to their decision theoretic operationalisation. Co-information itself is not constant over this family of distributions, because its value can be altered by third order interactions (i.e., those not specified by the pairwise marginals). Consider for example XOR. The co-information of this distribution is -1 bits, but the maximum entropy distribution preserving pairwise marginal constraints is the uniform distribution with a co-information of 0 bits. Therefore, if I_{ccs} were calculated using the full joint distribution it would not be consistent with the game-theoretic operational definition of unique information.

Since redundancy should be invariant given the specified marginals, our definition of I_{ccs} must be a function only of those marginals. However, we need a full joint distribution over the trivariate joint space to calculate the pointwise co-information terms. We use the maximum entropy distribution subject to the constraints specified by the game-theoretic operational definition (Equation (31)). The maximum entropy distribution is by definition the most parsimonious way to fill out a full trivariate distribution given only a set of bi-variate marginals [41]. It introduces no additional structure to the 3-way distribution over that which is specified by the constraints. Pairwise marginal constrained maximum entropy distributions have been widely used to study the effect of third and higher order interactions, because they provide a surrogate model which removes these effects [42–45]. Any distribution with lower entropy would by definition have some additional structure over that which is required to specify the unique and redundant information following the game-theoretic operationalisation.

Note that the definition of I_{broja} follows a similar argument. If redundancy was measured with co-information directly, it would not be consistent with the decision theoretic operationalisation [12]. Bertschinger et al. [12] address this by choosing the distribution which maximises co-information subject to the decision theoretic constraints. While we argue that maximizing entropy is in general a more principled approach than maximizing co-information, note that with the additional predictor marginal constraint introduced by the game-theoretic operational definition, both approaches are equivalent for two predictors (since maximizing co-information is equal to maximizing entropy given the constraints). However, once the distribution is obtained the other crucial difference is that I_{ccs} separates genuine redundant contributions at the local level, while I_{broja} computes the full co-information, which conflates redundant and synergistic effects (Table 3) [6].

We apply our game-theoretic operational definition in the same way to provide the constraints in Equation (31) for an arbitrary number of inputs. The action of each agent is determined by $P(A_i, S)$ (or equivalently $P(S|A_i)$) and the agent interaction effects (from zero-sum or asymmetric utility functions) are determined by $P(A_1, \ldots, A_n)$.

4.4. Properties

The measure I_{ccs} as defined above satisfies some of the proposed redundancy axioms (Section 3). The symmetry and self-redundancy axioms are satisfied from the properties of co-information [20]. For self-redundancy, consider that co-information for $n = 2$ is equal to mutual information at the pointwise level (Equation (29)):

$$
\begin{aligned}
c(s, a) &= h(s) + h(a) - h(s, a) \\
&= i(s; a) = \Delta_s h(a)
\end{aligned}
\tag{38}
$$

So $\operatorname{sgn} c(s, a) = \operatorname{sgn} \Delta_s h(a) \ \forall s, a$ and $I_{ccs}(S; A) = I(S; A)$. Subset equality is also satisfied. If $\mathbf{A_{l-1}} \subseteq \mathbf{A_l}$ then we consider values $a_{l-1} \in \mathbf{A_{l-1}}, a_l \in \mathbf{A_l}$ with $a_l = (a_l^{l-1}, a_l^+)$ and $a_l^{l-1} \in \mathbf{A_{l-1}} \cap \mathbf{A_l} = \mathbf{A_{l-1}}, a_l^+ \in \mathbf{A_l} \setminus \mathbf{A_{l-1}}$. Then

$$
p(a_{i_1}, \ldots, a_{i_j}, a_{l-1}, a_l^{l-1}, a_l^+) =
\begin{cases}
0 & \text{if } a_{l-1} \neq a_l^{l-1} \\
p(a_{i_1}, \ldots, a_{i_j}, a_l) & \text{otherwise}
\end{cases}
\tag{39}
$$

for any $i_1 < \cdots < i_j \in \{1, \ldots, l-2\}$. So for non-zero terms in Equation (30):

$$
h(a_{i_1}, \ldots, a_{i_j}, a_{l-1}, a_l) = h(a_{i_1}, \ldots, a_{i_j}, a_l)
\tag{40}
$$

Therefore all terms for $k \geq 2$ in Equation (29) which include a_{l-1}, a_l cancel with a corresponding $k - 1$ order term including a_l, so

$$
c(a_1, \ldots, a_{l-1}, a_l) = c(a_1, \ldots, a_{l-1})
\tag{41}
$$

and subset equality holds.

I_{ccs} does not satisfy the Harder et al. identity axiom [11] (Equation (19)); any distribution with negative local information terms serves as a counter example. These negative terms represent synergistic entropy which is included the standard mutual information quantity [36]. Therefore their omission in the calculation of I_{ccs} seems appealing; since they result from a synergistic interaction they should not be included in a measure quantifying redundant information. I_{ccs} does satisfy the modified independent identity axiom (Equation (22)), and so correctly quantifies redundancy in the two-bit copy problem (Section 3.2).

However, I_{ccs} does not satisfy monotonicity. To demonstrate this, consider the following example (Table 7, modified from [13], Figure 3).

Table 7. Example system with unique misinformation.

x_1	x_2	s	$p(x_1, x_2, s)$
0	0	0	0.4
0	1	0	0.1
1	1	1	0.5

For this system,

$$I(S; X_1) = I(S; X_1, X_2) = 1 \text{ bit}$$
$$I(S; X_2) = 0.61 \text{ bits}$$

Because of the self redundancy property, these values specify $I_\cap$ for the upper 3 values of the redundancy lattice (Figure 2A). The value of the bottom node is given by

$$I_\partial = I_\cap = I_{ccs}(S; \{1\}\{2\}) = 0.77 \text{ bits}$$

This value arises from two positive pointwise terms:

$$x_1 = x_2 = s = 0 \text{ (contributes 0.4 bits)}$$
$$x_1 = x_2 = s = 1 \text{ (contributes 0.37 bits)}$$

So $I_{ccs}(S; \{1\}\{2\}) > I_{ccs}(S; \{2\})$ which violates monotonicity on the lattice. How is it possible for two variables to share more information than one of them carries alone?

Consider the pointwise mutual information values for $I_{ccs}(S; \{2\}) = I(S; X_2)$. There are the same two positive information terms that contribute to the redundancy (since both are common with X_1). However, there is also a third misinformation term of -0.16 bits when $s = 0, x_2 = 1$. In our view, this demonstrates that the monotonicity axiom is incorrect for a measure of redundant information content. As this example shows a node can have *unique misinformation*.

For this example I_{ccs} yields the PID:

$$I_\partial(\{1\}\{2\}) = 0.77$$
$$I_\partial(\{1\}) = 0.23$$
$$I_\partial(\{2\}) = -0.16$$
$$I_\partial(\{12\}) = 0.16$$

While monotonicity has been considered a crucial axiom with the PID framework, we argue that subset equality, usually considered as part of the axiom of monotonicity, is the essential property that permits the use of the redundancy lattice. We have seen this lack of monotonicity means the PID obtained with I_{ccs} is not non-negative. We agree that while "negative ... atoms can *subjectively* be seen as flaw" [37], we argue here that in fact they are a necessary consequence of a redundancy measure that genuinely quantifies overlapping information content. Please note that in an earlier version of this manuscript we proposed thresholding with 0 to remove negative values. We no longer do so.

Mutual information is the expectation of a local quantity that can take both positive (local information) and negative (local misinformation) values, corresponding to redundant and synergistic entropy respectively [36]. Jensen's equality ensures that the final expectation value of mutual information is positive; or equivalently that redundant entropy is greater than synergistic entropy in any bivariate system. We argue that when breaking down the classical Shannon information into a partial information decomposition, there is no reason that those partial information values must be non-negative, since there is no way to apply Jensen's inequality to these partial values. We have illustrated this with a simple example where a negative unique information value is obtained, and inspection of the pointwise terms shows that this is indeed due to negative pointwise terms in the mutual information calculation for one predictor that are not present in the mutual information calculation for the other predictor: unique misinformation. Applying the redundancy lattice and the partial information decomposition directly to entropy can provide some further insights into the prevalence and effects of misinformation or synergistic entropy [36].

We conjecture that I_{ccs} is continuous in the underlying probability distribution [46] from the continuity of the logarithm and co-information, but not differentiable due to the thresholding with 0. Continuity requires that, at the local level,

$$c(s, a_1, a_2) < \min \left[i(s; a_1), i(s; a_2) \right] \tag{42}$$

when $\operatorname{sgn} i(s; a_1) = \operatorname{sgn} i(s; a_2) = \operatorname{sgn} i(s; a_1, a_2) = \operatorname{sgn} c(s, a_1, a_2)$. While this relationship holds for the full integrated quantities [20], it does not hold at the local level for all joint distributions. However, we conjecture that it holds when using the pairwise maximum entropy solution $\hat{P}$, with no higher order interactions. This is equivalent to saying that the overlap of the two local informations should not be larger than the smallest—an intuitive requirement for a set theoretic overlap. However, at this stage the claim of continuity remains a conjecture. In the Matlab implementation we explicitly test for violations of the condition in Equation (42), which do not occur in any of the examples we consider here. This shows that all the examples we consider here are at least locally continuous in the neighbourhood of the specific joint probability distribution considered.

In the next sections, we demonstrate with a range of example systems how the results obtained with this approach match intuitive expectations for a partial information decomposition.

4.5. Implementation

Matlab code is provided to accompany this article, which features simple functions for calculating the partial information decomposition for two and three variables [47].This includes implementation of I_{min} and the PID calculation of [6], as well as I_{ccs} and I_{broja}. Scripts are provided reproducing all the examples considered here. Implementations of I_{ccs} and I_{mmi} [26] for Gaussian systems are also included. To calculate I_{broja} and compute the maximum entropy distributions under marginal constraints we use the `dit` package [48–50]

5. Two Variable Examples

5.1. Examples from Williams and Beer (2010) [6]

We begin with the original examples of ([6], Figure 4), reproduced here in Figure 5.

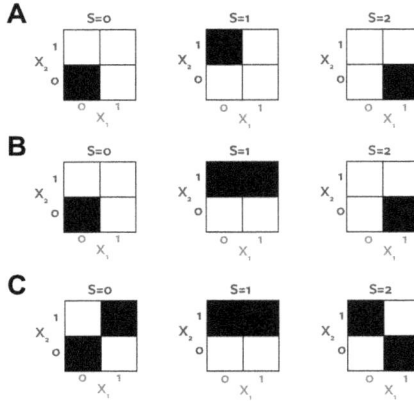

Figure 5. Probability distributions for three example systems (**A–C**). Black tiles represent equiprobable outcomes. White tiles are zero-probability outcomes. (**A,B**) modified from [6].

Table 8 shows the PIDs for the system shown in Figure 5A, obtained with I_{min}, I_{broja} and I_{ccs}. Note that this is equivalent to the system SUBTLE in ([13], Figure 4). I_{ccs} and I_{min} agree qualitatively here; both show both synergistic and redundant information. I_{broja} shows zero synergy. The pointwise computation of I_{ccs} includes two non-zero terms; when

$$x_1 = 0, x_2 = 1, s = 1 \text{ and when}$$
$$x_1 = 1, x_2 = 0, s = 2$$

For both of these local values, x_1 and x_2 are contributing the same reduction in surprisal of s (0.195 bits each for 0.39 bits overall redundancy). There are no other redundant local changes in surprisal (positive or negative). In this case, both the I_{broja} optimised distribution and the pairwise marginal maximum entropy distribution are equal to the original distribution. So here I_{broja} is measuring redundancy directly with co-information, whereas I_{ccs} breaks down the co-information to include only the two terms which directly represent redundancy. In the full co-information calculation of I_{broja} there is one additional contribution of -0.138 bits, which comes from the $x_1 = x_2 = s = 0$ event. In this case the local changes in surprisal of s from x_1 and x_2 are both positive (0.585), but the local co-information is negative (-0.415). This corresponds to the second row of Table 3—it is synergistic local information. Therefore this example clearly shows how the I_{broja} measure of redundancy erroneously includes synergistic effects.

Table 8. PIDs for example Figure 5A.

Node	$I_\partial[I_{min}]$	$I_\partial[I_{broja}]$	$I_\partial[I_{ccs}]$
{1}{2}	0.5850	0.2516	0.3900
{1}	0.3333	0.6667	0.5283
{2}	0.3333	0.6667	0.5283
{12}	0.3333	0	0.1383

Table 9 shows the PIDs for the system shown in Figure 5B. Here I_{broja} and I_{ccs} agree, but diverge qualitatively from I_{min}. I_{min} shows both synergy and redundancy, with no unique information carried by X_1 alone. I_{ccs} shows no synergy and redundancy, only unique information carried independently by X_1 and X_2. Reference [6] argue that "X_1 and X_2 provide 0.5 bits of redundant information corresponding to the fact that knowledge of either X_1 or X_2 reduces uncertainty about the outcomes

$S = 0, S = 2''$. However, while both variables reduce uncertainty about S, they do so in different ways—X_1 discriminates the possibilities $S = 0, 1$ vs. $S = 1, 2$ while X_2 allows discrimination between $S = 1$ vs. $S = 0, 2$. These discriminations represent different non-overlapping information content, and therefore should be allocated as unique information to each variable as in the I_{ccs} and I_{broja} PIDs. While the full outcome can only be determined with knowledge of both variables, there is no synergistic information because the discriminations described above are independent.

Table 9. PIDs for example Figure 5B.

Node	$I_\partial[I_{min}]$	$I_\partial[I_{broja}]$	$I_\partial[I_{ccs}]$
{1}{2}	0.5	0	0
{1}	0	0.5	0.5
{2}	0.5	1	1
{12}	0.5	0	0

To induce genuine synergy it is necessary to make the X_1 discrimination between $S = 0, 1$ and $S = 1, 2$ ambiguous without knowledge of X_2. Table 10 shows the PID for the system shown in Figure 5C, which includes such an ambiguity. Now there is no information in X_1 alone, but it contributes synergistic information when X_2 is known. Here, I_{min} correctly measures 0 bits redundancy, and all three PIDs agree (the other three terms have only one source, and therefore are the same for all measures from self-redundancy).

Table 10. PIDs for example Figure 5C.

Node	$I_\partial[I_{min}]$	$I_\partial[I_{broja}]$	$I_\partial[I_{ccs}]$
{1}{2}	0	0	0
{1}	0	0	0
{2}	0.25	0.25	0.25
{12}	0.67	0.67	0.67

5.2. Binary Logical Operators

The binary logical operators OR, XOR and AND are often used as example systems [10–12]. For XOR, the I_{ccs} PID agrees with both I_{min} and I_{broja} and quantifies the 1 bit of information as fully synergistic.

5.2.1. AND/OR

Figure 6 illustrates the probability distributions for AND and OR. This makes clear the equivalence between them; because of symmetry any PID should give the same result on both systems. Table 11 shows the PIDs. In this system I_{min} and I_{broja} agree, both showing no unique information. I_{ccs} shows less redundancy, and unique information in both predictors. The redundancy value with I_{ccs} falls within the bounds proposed in ([10], Figure 6.11).

To see where this unique information arises with I_{ccs} we can consider directly the individual pointwise contributions for the AND example (Table 12). $I_{ccs}(\{1\}\{2\})$ has a single pointwise contribution from the event $(0, 0, 0)$, only when both inputs are 0 is there redundant local information about the outcome. For the event $(0, 1, 0)$ (and symmetrically for $1, 0, 0$) x_1 conveys local information about s, while x_2 conveys local misinformation, therefore there is no redundancy, but a unique contribution for both x_1 and x_2. We can see in the $(1, 1, 1)$ event the change in surprisal of s from the two predictors is independent, so again contributes unique rather than redundant information. So the unique information in each predictor is a combination of unique information and misinformation terms.

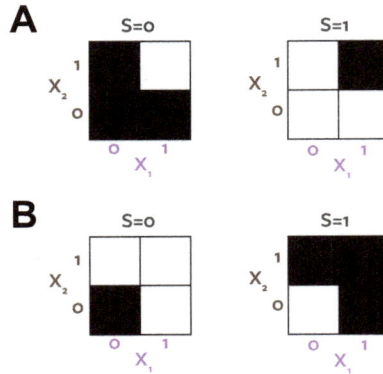

Figure 6. Binary logical operators. Probability distributions for (**A**) AND; (**B**): OR. Black tiles represent equiprobable outcomes. White tiles are zero-probability outcomes.

Table 11. PIDs for AND/OR.

Node	$I_\partial[I_{min}]$	$I_\partial[I_{broja}]$	$I_\partial[I_{ccs}]$
{1}{2}	0.31	0.31	0.10
{1}	0	0	0.21
{2}	0	0	0.21
{12}	0.5	0.5	0.29

Table 12. Pointwise values from $I_{ccs}(S; \{1\}\{2\})$ for AND.

(x_1, x_2, s)	$\Delta_s h(x_1)$	$\Delta_s h(x_2)$	$\Delta_s h(x_1, x_2)$	$c(x_1; x_2; s)$	$\Delta_s h^{com}(x_1, x_2)$
$(0,0,0)$	0.415	0.415	0.415	0.415	0.415
$(0,1,0)$	0.415	−0.585	0.415	−0.585	0
$(1,0,0)$	−0.585	0.415	0.415	−0.585	0
$(1,1,1)$	1	1	2	0	0

For I_{broja} the specific joint distribution that maximises the co-information in the AND example while preserving $P(X_i, S)$ ([12], Example 30, $\alpha = 1/4$) has an entropy of 1.5 bits. $\hat{P}(X_1, X_2, S)$ used in the calculation of I_{ccs} is equal to the original distribution and has an entropy of 2 bits. Therefore, the distribution used in I_{broja} has some additional structure above that specified by the individual joint target marginals and which is chosen to maximise the co-information (negative interaction information). As discussed above, interaction information can conflate redundant information with synergistic misinformation, as well as having other ambiguous terms when the signs of the individual changes of surprisal are not equal. As shown in Table 12, the AND system includes such ambiguous terms (rows 2 and 3, which contribute synergy to the interaction information). Any system of the form considered in ([12], Example 30) will have similar contributing terms. This illustrates the problem with using co-information directly as a redundancy measure, regardless of how the underlying distribution is obtained. The distribution selected to maximise co-information will be affected by these ambiguous and synergistic terms. In fact, it is interesting to note that for the I_{broja} distribution ($\alpha = 1/4$), $p(0, 1, 0) = p(1, 0, 0) = 0$ and the two ambiguous synergistic terms are removed from the interaction information. This indicates how the optimisation of the co-information might be driven by terms that cannot be interpreted as genuine redundancy. Further, the distribution used in I_{broja} has perfectly coupled marginals. This increases the source redundancy measured by the co-information. Under this distribution, the $(1, 1, 1)$ term now contributes 1 bit locally to the co-information. This is redundant because $x_1 = 1$ and $x_2 = 1$ always occur together. In the original distribution the $(1,1,1)$ term is independent because the predictors are independent.

We argue there is no fundamental conceptual problem with the presence of unique information in the AND example. Both variables share some information, have some synergistic information, but also have some unique information corresponding to the fact that knowledge of either variable taking the value 1 reduces the uncertainty of $s = 1$ independently (i.e., on different trials). If the joint target marginal distributions are equal, then by symmetry $I_\partial(\{1\}) = I_\partial(\{2\})$, but it is not necessary that $I_\partial(\{1\}) = I_\partial(\{2\}) = 0$ ([12], Corollary 8).

5.2.2. SUM

While not strictly a binary logic gate, we also consider the summation of two binary inputs. The AND gate can be thought of as a thresholded version of summation. Summation of two binary inputs is also equivalent to the system XORAND [10–12]. Table 13 shows the PIDs.

Table 13. PIDs for SUM.

Node	$I_\partial[I_{\min}]$	$I_\partial[I_{\text{broja}}]$	$I_\partial[I_{\text{ccs}}]$
{1}{2}	0.5	0.5	0
{1}	0	0	0.5
{2}	0	0	0.5
{12}	1	1	0.5

As with AND, $I_{\min}$ and I_{broja} agree, and both allocate 0 bits of unique information. Both of these methods always allocate zero unique information when the target-predictor marginals are equal. I_{ccs} differs in that it allocates 0 redundancy. This arises for a similar reason to the differences discussed earlier for REDUCEDOR (Section 4.3). The optimised distribution used in I_{broja} has directly coupled predictors:

$$P_{\text{broja}}(X_1 = 0, X_2 = 0) = P_{\text{broja}}(X_1 = 1, X_2 = 1) = 0.5$$
$$P_{\text{broja}}(X_1 = 0, X_2 = 1) = P_{\text{broja}}(X_1 = 1, X_2 = 0) = 0 \tag{43}$$

While the actual system has independent uniform marginal predictors ($P(i,j) = 0.25$). In the I_{broja} calculation of co-information the local events $(0,0,0)$ and $(1,1,2)$ both contribute redundant information, because X_1 and X_2 are coupled. However, the local co-information terms for the true distribution show that the contributions of $x_1 = 0$ and $x_2 = 0$ are independent when $s = 0$ (see Table 5). Therefore, with the true distribution these contributions are actually unique information. These differences arise because of the erroneous assumption within I_{broja} that the unique and redundant information should be invariant to the predictor-predictor marginal distribution (Section 4.3). Since they are not, the I_{broja} optimisation maximises redundancy by coupling the predictors.

The resulting I_{ccs} PID seems quite intuitive. Both X_1 and X_2 each tell whether the output sum is in $(0,1)$ or $(1,2)$, and they do this independently, since they are distributed independently (corresponding to 0.5 bits of unique information each). However, the final full discrimination of the output can only be obtained when both inputs are observed together, providing 0.5 bits of synergy. In contrast, I_{broja} measures 0.5 bits of redundancy. It is hard to see how summation of two independent variables should be redundant as it is not apparent how two independent summands can convey overlapping information about their sum. For AND, there is redundancy between two independent inputs. I_{ccs} shows that this arises from the fact that if $x_1 = 0$ then $y = 0$ and similarly if $x_2 = 0$ then $y = 0$. So when both x_1 and x_2 are zero they are both providing the same information content—that $y = 0$, so there is redundancy. In contrast, in SUM, $x_1 = 0$ tells that $y = 0$ or $y = 1$, but which of the two particular outputs is determined independently by the values of x_2. So the information each input conveys is independent (unique) and not redundant.

5.3. Griffith and Koch (2014) Examples

Griffith and Koch [10] present two other interesting examples: RDNXOR (their Figure 6.9) and RDNUNQXOR (their Figure 6.12).

RDNXOR consists of two two-bit (4 value) inputs X_1 and X_2 and a two-bit (4 value) output S. The first component of X_1 and X_2 redundantly specifies the first component of S. The second component of S is the XOR of the second components of X_1 and X_2. This system therefore contains 1 bit of redundant information and 1 bit of synergistic information; further every value $s \in S$ has both a redundant and synergistic contribution. I_{ccs} correctly quantifies the redundancy and synergy with the PID $(1, 0, 0, 1)$ (as do both I_{min} and I_{broja}).

RDNUNQXOR consists of two three-bit (8 value) inputs X_1 and X_2 and a four-bit (16 value) output S (Figure 3). The first component of S is specified redundantly by the first components of X_1 and X_2. The second component of S is specified uniquely by the second component of X_1 and the third component of S is specified uniquely by the second component of X_2. The fourth component of S is the XOR of the third components of X_1 and X_2. Again I_{ccs} correctly quantifies the properties of the system with the PID $(1, 1, 1, 1)$, identifying the separate redundant, unique and synergistic contributions (as does I_{broja} but not I_{min}).

Note that the PID with I_{ccs} also gives the expected results for examples RND and UNQ from [10] (see example scripts in the accompanying code [47]).

5.4. Dependence on Predictor-Predictor Correlation

To directly illustrate the fundamental conceptual difference between I_{ccs} and I_{broja} we construct a family of distributions with the same target-predictor marginals and investigate the resulting decomposition as we change the predictor-predictor correlation [51].

We restrict our attention to binary variables with uniformly distributed univariate marginal distributions. We consider pairwise marginals with a symmetric dependence of the form

$$p_c(0,0) = p_c(1,1) = (1+c)/4$$
$$p_c(0,1) = p_c(1,0) = (1-c)/4 \tag{44}$$

where the parameter c specified the correlation between the two variables. We fix $c = 0.1$ for the two target-predictor marginals:

$$P(X_1, S) = P_{0.1}(X_1, S)$$
$$P(X_2, S) = P_{0.1}(X_2, S) \tag{45}$$

Then with $P(X_1, X_2) = P_c(X_1, X_2)$ we can construct a trivariate joint distribution $P_c(S, X_1, X_2)$ which is consistent with these three pairwise marginals as follows [51]. This is a valid distribution for $-0.8 \le c \le 0.1$.

$$p_c(0,0,0) = c/4 + 1/4$$
$$p_c(0,0,1) = 1/40 - c/4$$
$$p_c(0,1,0) = 1/40 - c/4$$
$$p_c(0,1,1) = c/4 + 1/5$$
$$p_c(1,0,0) = 0 \tag{46}$$
$$p_c(1,0,1) = 9/40$$
$$p_c(1,1,0) = 9/40$$
$$p_c(1,1,1) = 1/20$$

Figure 7 shows I_{broja} and I_{ccs} PIDs for this system. By design the values of unique and redundant information obtained with I_{broja} do not change as a function of predictor-predictor correlation when the target-predictor marginals are fixed. With I_{ccs} the quantities change in an intuitive manner. When the predictors are positively correlated, they are redundant, when they are negatively correlated they convey unique information. When they are independent, there is an equal mix of unique and mechanistic redundancy in this system. This emphasises the different perspective also revealed in the REDUCEDOR example (Section 4.3) and the AND example (Section 5.2.1). I_{broja} reports the co-information for a distribution where the predictors are perfectly coupled. For all the values of c reported in Figure 7A, the I_{broja} optimised distribution has coupled predictor-predictor marginals:

$$P(X_1 = 0, X_2 = 1) = P(X_1 = 1, X_2 = 0) = 0$$
$$P(X_1 = 0, X_2 = 0) = P(X_1 = 1, X_2 = 1) = 0.5 \tag{47}$$

Therefore, I_{broja} is again insensitive to the sort of unique information that can be operationalised in a game-theoretic setting by exploiting the trial-by-trial relationships between predictors (Section 4.3).

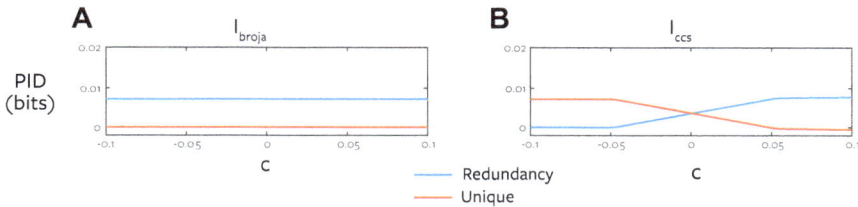

Figure 7. PIDs for binary systems with fixed target-predictor marginals as a function of predictor-predictor correlation. I_{broja} (**A**) and I_{ccs} (**B**) PIDs are shown for the system defined in Equation (46) as a function of the predictor-predictor correlation c.

6. Three Variable Examples

We now consider the PID of the information conveyed about S by three variables X_1, X_2, X_3. For three variables we do not compare to I_{broja}, since it is defined only for two input sources.

6.1. A Problem With the Three Variable Lattice?

Bertschinger et al. [14] identify a problem with the PID summation over the three-variable lattice (Figure 2B). They provide an example we term XORCOPY (described in Section 6.2.2) which demonstrates that any redundancy measure satisfying their redundancy axioms (particularly the Harder et al. identity axiom) cannot have only non-negative I_{∂} terms on the lattice. We provide here an alternative example of the same problem, and one that does not depend on the particular redundancy measure used. We argue it applies for any redundancy measure that attempts to measure overlapping information content.

We consider X_1, X_2, X_3 independent binary input variables. Y is a two-bit (4 value) output with the first component given by $X_1 \oplus X_2$ and the second by $X_2 \oplus X_3$. We refer to this example as DBLXOR. In this case the top four nodes have non-zero (redundant) information:

$$I_{\cap}(\{123\}) = I(\{123\}) = 2 \text{ bits}$$
$$I_{\cap}(\{12\}) = I_{\cap}(\{13\}) = I_{\cap}(\{23\}) = 1 \text{ bit}$$

We argue that all lower nodes on the lattice should have zero redundant (and partial) information. First, by design and from the properties of XOR no single variable conveys any information or can have any redundancy with any other source. Second, considering synergistic pairs, Figure 8A graphically illustrates the source-output joint distributions for the two-variable sources. Each value of the pairwise response (*x*-axes in Figure 8A) performs a different discrimination between the values of *Y* for each pair. Therefore, there is no way there can be redundant information between any of these synergistic pairs. Redundant information means the same information content. Since there are no discriminations (column patterns in the figure) that are common to more than one pair of sources, there can be no redundant information between them. Therefore, the information conveyed by the three two-variable sources is also independent and all lower nodes on the lattice are zero.

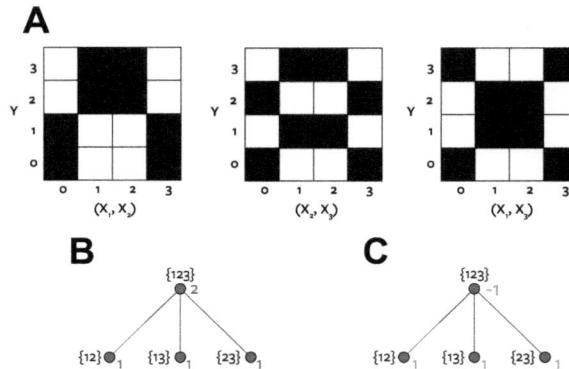

Figure 8. The DBLXOR example. (**A**) Pairwise variable joint distributions. Black tiles represent equiprobable outcomes. White tiles are zero-probability outcomes; (**B**) Non-zero nodes of the three variable redundancy lattice. Mutual information values for each node are shown in red; (**C**) PID. I_∂ values for each node are shown in green.

In this example, $I_\cap(\{123\}) = 2$ but there are three child nodes of $\{123\}$ each with $I_\partial = 1$ (Figure 8B). This leads to $I_\partial(\{123\}) = -1$. How can there be 3 bits of unique information in the lattice when there are only 2 bits of information in the system? In this case, we cannot appeal to the non-monotonicity of I_{ccs} since these values are monotonic on the lattice. There are also no negative pointwise terms in the calculation of $I(\{123\})$ so there is no synergistic misinformation that could explain a negative value.

In a previous version of this manuscript we argued that this problem arises because the three nodes in the penultimate level of the lattice are not disjoint, therefore not independent, and therefore mutual information is not additive over those nodes. We proposed a normalisation procedure to address such situations. However, we now propose instead to accept the negative values. As noted earlier (Section 4.4), negative values may subjectively be seen as a flaw [37], but given that mutual information itself is a summation of positive and negative terms, there is no a priori reason why a full decomposition must, or indeed can, be completely non-negative. In fact, in entropy terms, negative values are an essential consequence of the existence of mechanistic redundancy [36]. While in an information decomposition they can also arise from unique or synergistic misinformation, we propose that mechanistic redundancy is another explanation. In this particular example of DBLXOR, the negative $\{123\}$ term reflects a mechanistic redundancy between the three pairwise synergistic partial information terms that cannot be accounted for elsewhere on the lattice.

6.2. Other Three Variable Example Systems

6.2.1. Giant Bit and Parity

The most direct example of three-way information redundancy is the "giant bit" distribution [52]. This is the natural extension of example RDN (Section 4.2) with a single bit in common to all four variables, defined as:

$$P(0,0,0,0) = P(1,1,1,1) = 0.5 \tag{48}$$

Applying I_{ccs} results in a PID with $I_\partial(S; \{1\}\{2\}\{3\}) = 1$ bit, and all other terms zero.

A similarly classic example of synergy is the even parity distribution, a distribution in which an equal probability is assigned to all configurations with an even number of ones. The XOR distribution is the even parity distribution in the three variable (two predictor) case. Applying I_{ccs} results in a PID with $I_\partial(S; \{123\}) = 1$ bit, and all other terms zero.

Thus, the PID based on I_{ccs} correctly reflects the structure of these simple examples.

6.2.2. XORCOPY

This example was developed to illustrate the problem with the three variable lattice described above [14,53]. The system comprises three binary input variables X_1, X_2, X_3, with X_1, X_2 uniform independent and $X_3 = X_1 \oplus X_2$. The output Y is a three bit (8 value) system formed by copying the inputs $Y = (X_1, X_2, X_3)$. The PID with I_{min} gives:

$$I_\partial(\{1\}\{2\}\{3\}) = I_\partial(\{12\}\{13\}\{23\}) = 1 \text{ bit}$$

But since X_1 and X_2 are copied independently to the output it is hard to see how they can share information. Using common change in surprisal we obtain:

$$I_{ccs}(\{1\}\{23\}) = I_{ccs}(\{2\}\{13\}) = I_{ccs}(\{3\}\{12\}) = 1 \text{ bit}$$
$$I_{ccs}(\{12\}\{13\}\{23\}) = 2 \text{ bits}$$

The $I_{ccs}(\{i\}\{jk\})$ values correctly match the intuitive redundancy given the structure of the system, but result in a negative value similar to DBLXOR considered above. There are 3 bits of unique I_∂ among the nodes of the third level, but only 2 bits of information in the system. This results in the PID:

$$I_\partial(\{1\}\{23\}) = I_\partial(\{2\}\{13\}) = I_\partial(\{3\}\{12\}) = 1 \text{ bit}$$
$$I_\partial(\{12\}\{13\}\{23\}) = -1 \text{ bit}$$

As for DBLXOR we believe this provides a meaningful decomposition of the total mutual information, with the negative value here representing the presence of mechanistic redundancy between the nodes at the third level of the lattice. This mechanistic redundancy between synergistic pairs seems to be a signature property of an XOR mechanism.

6.2.3. Other Examples

Griffith and Koch [10] provide a number of other interesting three variable examples based on XOR operations, such as XORDUPLICATE (their Figure 6.6), XORLOSES (their Figure 6.7), XORMULTICOAL (their Figure 6.14). For all of these examples I_{ccs} provides a PID which matches what they suggest from the intuitive properties of the system (see `examples_3d.m` in accompanying code [47]). I_{ccs} also gives the correct PID for PARITYRDNRDN (which appeared in an earlier version of their manuscript).

We propose an additional example, XORUNQ, which consists of three independent input bits. The output consists of 2 bits (4 values), the first of which is given by $X_1 \oplus X_2$, and the second of which is a copy of X_3. In this case we obtain the correct PID:

$$I_\partial(\{3\}) = I_\partial(\{12\}) = 1 \text{ bit}$$

Another interesting example from [10] is ANDDUPLICATE (their Figure 6.13). In this example Y is a binary variable resulting from the binary AND of X_1 and X_2. X_3 is a duplicate of X_1. The PID we obtain for this system is shown in Figure 9.

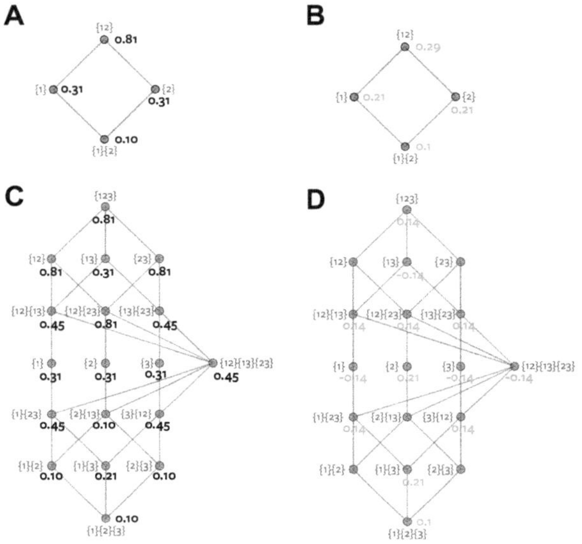

Figure 9. The ANDDUPLICATE example. (**A**) I_{ccs} values for AND; (**B**) Partial information values from the I_{ccs} PID for AND; (**C**) I_{ccs} values for ANDDUPLICATE; (**D**) Partial information values from the I_{ccs} PID for ANDDUPLICATE.

We can see that as suggested by [10],

$$I_\partial^{\text{ANDDUP}}(S; \{2\}) = I_\partial^{\text{AND}}(S; \{2\})$$
$$I_\partial^{\text{ANDDUP}}(S; \{1\}\{3\}) = I_\partial^{\text{AND}}(S; \{1\}) \tag{49}$$
$$I_\partial^{\text{ANDDUP}}(S; \{1\}\{2\}\{3\}) = I_\partial^{\text{AND}}(S; \{1\}\{2\})$$

The synergy relationship they propose, $I_\partial^{\text{ANDDUP}}(S; \{12\}\{23\}) = I_\partial^{\text{AND}}(S; \{12\})$ is not met, although the fundamental general consistency requirement relating 2 and 3 variable lattices is [36,54]:

$$
\begin{aligned}
I_\partial^{\text{AND}}(S; \{12\}) = &\ I_\partial^{\text{ANDDUP}}(S; \{12\}) \\
&+ I_\partial^{\text{ANDDUP}}(S; \{12\}\{13\}) + I_\partial^{\text{ANDDUP}}(S; \{12\}\{23\}) \\
&+ I_\partial^{\text{ANDDUP}}(S; \{12\}\{13\}\{23\}) \\
&+ I_\partial^{\text{ANDDUP}}(S; \{3\}\{12\})
\end{aligned}
\tag{50}
$$

Note that the preponderance of positive and negative terms with amplitude 0.14 bits is at first glance counter-intuitive, particularly the fact that $I_\partial^{\text{ANDDUP}}(S; \{1\}) = I_\partial^{\text{ANDDUP}}(S; \{3\}) = -0.146$ when X_3 is a copy of X_1. However, the 0.14 bits comes from a local misinformation term in the univariate predictor-target mutual information calculation for AND, which is not present in the joint mutual information calculation. This reflects the fact that, in entropy terms, $I(S; X_1)$ is not a proper subset of $I(S; X_1, X_2)$ [36]. A partial entropy decomposition of AND shows that $H_\partial(\{1\}\{23\}) = H_\partial(\{2\}\{13\}) = 0.14$. These are entropy terms that have an ambiguous interpretation and appear both in unique and synergistic partial information terms. It is likely that a higher-order entropy decomposition could shed more light on the structure of the ANDDUPLICATE PID.

7. Continuous Gaussian Variables

I_{ccs} can be applied directly to continuous variables. $\Delta_s h^{com}$ can be used locally in the same way, with numerical integration applied to obtain the expectation. Functions implementing this for Gaussian variables via Monte Carlo integration are included in the accompanying code [47]. Following Barrett [26] we consider the information conveyed by two Gaussian variables X_1, X_2 about a third Gaussian variable, S. We focus here on univariate Gaussians, but the accompanying implementation also supports multivariate normal distributions. Reference [26] show that for such Gaussian systems, all previous redundancy measures agree, and are equal to the minimum mutual information carried by the individual variables:

$$I_\cap(\{1\}\{2\}) = \min_{i=1,2} I(S; X_i) = I_{mmi}(\{1\}\{2\}) \tag{51}$$

Without loss of generality, we consider all variables to have unit variance, and the system is then completely specified by three parameters:

$$a = \text{Corr}(X_1, S)$$
$$c = \text{Corr}(X_2, S)$$
$$b = \text{Corr}(X_1, X_2)$$

Figure 10 shows the results for two families of Gaussian systems as a function of the correlation, b, between X_1 and X_2 ([26], Figure 3).

This illustrates again a key conceptual difference between I_{ccs} and existing measures. I_{ccs} is not invariant to the predictor-predictor marginal distributions (Section 5.4). When the two predictors have equal positive correlation with the target (Figure 10A,B), I_{mmi} reports zero unique information, and a constant level of redundancy regardless of the predictor-predictor correlation b. I_{ccs} transitions from having the univariate predictor information purely unique when the predictors are negatively correlated, to purely redundant when the predictors are strongly positively correlated. When the two predictors have unequal positive correlations with the target (Figure 10C,D), the same behaviour is seen. When the predictors are negatively correlated the univariate information is unique, as they become correlated both unique informations decrease as the redundancy between the predictors increases.

Having an implementation for continuous Gaussian variables is of practical importance, because for multivariate discrete systems sampling high dimensional spaces with experimental data becomes increasingly challenging. We recently developed a lower-bound approximate estimator of mutual information for continuous signals based on a Gaussian copula [3]. The Gaussian I_{ccs} measure therefore allows this approach to be used to obtain PIDs from experimental data.

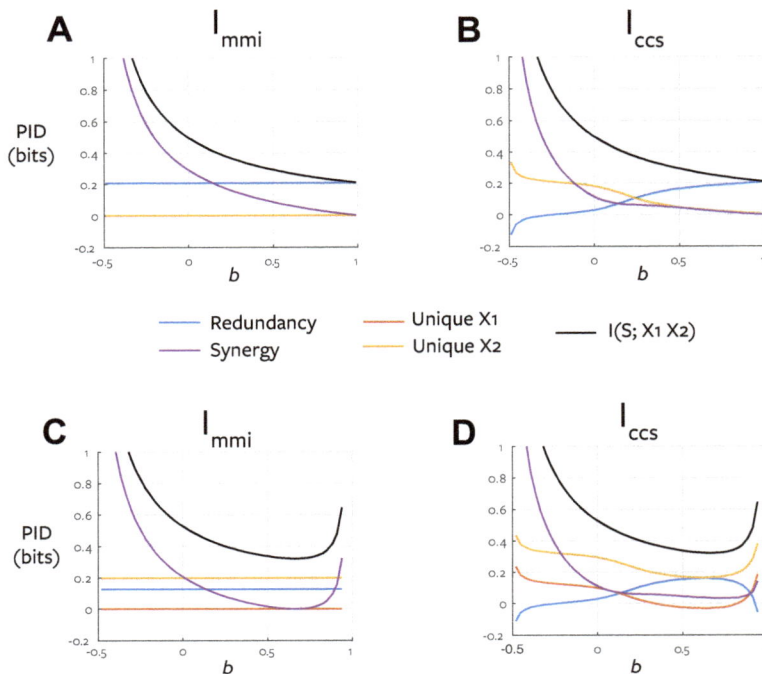

Figure 10. PIDs for Gaussian systems. (**A**) PID with I_{mmi} for $a = c = 0.5$ as a function of predictor-predictor correlation b; (**B**) PID with I_{ccs} for $a = c = 0.5$; (**C**) PID with I_{mmi} for $a = 0.4, c = 0.6$; (**D**) PID with I_{ccs} for $a = 0.4, c = 0.6$.

8. Discussion

We have presented I_{ccs}, a novel measure of redundant information based on the expected pointwise change in surprisal that is common to all input sources. Developing a meaningful quantification of redundant and synergistic information has proved challenging, with disagreement about even the basic axioms and properties such a measure should satisfy. Therefore, here we take a bottom-up approach, starting by defining what we think redundancy should measure at the pointwise level (common change in surprisal), and then exploring the consequences of this through a range of examples.

This new redundancy measure has several advantages over existing proposals. It is conceptually simple: it measures precisely the pointwise contributions to the mutual information which are shared unambiguously among the considered sources. This seems a close match to an intuitive definition of redundant information. I_{ccs} exploits the additivity of surprisal to directly measure the pointwise overlap as a set intersection, while removing the ambiguities that arise due to the conflation of pointwise information and misinformation effects by considering only terms with common sign (since a common sign is a prerequisite for there to be a common change in surprisal). I_{ccs} is defined for any number of input sources (implemented for 2 and 3 predictor systems), as well as any continuous system (implemented for multivariate Gaussian predictors and targets). Matlab code implementing the measure accompanies this article [47]. The code requires installation of Python and the dit toolbox [50]. The repository includes all the examples described herein, and it is straightforward for users to apply the method to any other systems or examples they would like.

To motivate the choice of joint distribution we use to calculate I_{ccs} we review and extend the decision theoretic operational argument of Bertschinger et al. [12]. We show how a game theoretic operationalisation provides a different perspective, and give a specific example where an exploitable game-theoretic advantage exists for each agent, but I_{broja} suggests there should be no unique information. We therefore conclude the decision theoretic formulation is too restrictive and that the balance of unique and redundant information is not invariant to changes in the predictor-predictor marginal distribution. This means that the optimisation in I_{broja} is not only minimising synergy, but could actually be increasing redundancy. Detailed consideration of several examples shows that the I_{broja} optimisation often results in distributions with coupled predictor variables, which maximises the source redundancy between them. For example, in the SUM system, the coupled predictors make the $(0,0,0)$ and $(1,1,2)$ events redundant, when in the true system the predictors are independent, so those events contribute unique information. However, we note that if required I_{ccs} can also be calculated following the decision theoretic perspective simply by using $\hat{P}_{ind}$.

I_{ccs} satisfies most of the core axioms for a redundancy measure, namely symmetry, self-redundancy and a modified identity property which reflects the fact that mutual information can itself include synergistic entropy effects [36]. Crucially, it also satisfies subset equality which has not previously been considered separately from monotonicity, but is the key axiom which allows the use of the reduced redundancy lattice. However, we have shown that I_{ccs} is not monotonic on the redundancy lattice because nodes can convey unique misinformation. This means the resulting PID is not non-negative. In fact, negative terms can occur even without non-monotonicity because for some systems (e.g., 3 predictor systems with XOR structures) mechanistic redundancy can result in negative terms [36]. We argue that while "negative ... atoms can subjectively be seen as flaw" [37] in fact, they are a necessary consequence of a redundancy measure that genuinely quantifies overlapping information content. We have shown that despite the negative values, I_{ccs} provides intuitive and consistent PIDs across a range of example systems drawn from the literature.

Mutual information itself is an expectation over positive and negative terms. While Jensen's inequality ensures that the overall expectation is non-negative, we argue there is no way to apply Jensen's inequality to decomposed partial information components of mutual information, whichever redundancy measure is used, and thus no reason to assume they must be non-negative. An alternative way to think about the negative values is to consider the positive and negative contributions to mutual information separately. The definition of I_{ccs} could easily be expanded to quantify redundant pointwise information separately from redundant pointwise misinformation (rows 1 and 3 of Table 3). One could then imagine two separate lattice decompositions, one for the pointwise information (positive terms) and one for the pointwise misinformation (negative terms). We conjecture that both of these lattices would be monotonic, and that the non-monotonicity of the I_{ccs} PID arises as a net effect from taking the difference between these. This suggests it may be possible to obtain zero unique information from a cancellation of redundant information with redundant misinformation, analogous to how zero co-information can result in the presence of balanced redundant and synergistic effects, and so exploring this approach is an interesting area for future work. It is also important to develop more formal analytical results proving further properties of the measure, and separate local information versus local misinformation lattices might help with this.

Rauh [55] recently explored an interesting link between the PID framework and the problem of cryptographic secret sharing. Intuitively, there should be a direct relationship between the two notions: an authorized set should have only synergistic information about the secret when all elements of the set are considered, and a shared secret scheme corresponds to redundant information about the secret between the authorized sets. Therefore, any shared secret scheme should yield a PID with a single non-negative partial information term equal to the entropy of the secret at the node representing the redundancy between the synergistic combinations of each authorised set within the inclusion-minimal access structure. Rauh [55] shows that if this intuitive relationship holds, then the PID cannot be non-negative. This finding further supports our suggestion that it may not be possible to obtain a non-negative PID from a redundancy measure that meaningfully quantifies overlapping information content; if such a measure satisfies the intuitive "secret sharing property" [55] it does not provide a non-negative PID. We note that I_{ccs} satisfies the secret sharing property for ([55], Example 1); whether it can be proved to do so in general is an interesting question for future research. These considerations suggest I_{ccs} might be useful in cryptographic applications.

Another important consideration for future research is how to address the practical problems of limited sampling bias [56] when estimating PID quantities from experimental data. Similarly, how best to perform statistical inference with non-parametric permutation methods is an open question. We suggest it is likely that different permutation schemes might be needed for the different PID terms, since trivariate conditional mutual information requires a different permutation scheme than bivariate joint mutual information [57].

How best to practically apply the PID to systems with more than three variables is also an important area for future research. The four variable redundancy lattice has 166 nodes, which already presents a significant challenge for interpretation if there are more than a handful of non-zero partial information values. We suggest that it might be useful to collapse together the sets of terms that have the same order structure. For example, for the three variable lattice the terms within the layers could be represented as shown in Table 14. While this obviously does not give the complete picture provided by the full PID, it gives considerably more detail than existing measures based on maximum entropy subject to different order marginal constraints, such as connected information [43]. We hope it might provide a more tractable practical tool that can still give important insight into the structure of interactions for systems with four or more variables.

Table 14. Order-structure terms for the three variable lattice. Resulting values for the example systems of a giant bit, even parity and DBLXOR (Section 6) are shown.

Level	Order-Structure Terms	Giant Bit	Parity	DBLXOR
7	(3)	0	1	-1
6	(2)	0	0	3
5	$(2,2)$	0	0	0
4	$(1), (2,2,2)$	0, 0	0, 0	0, 0
3	$(1,2)$	0	0	0
2	$(1,1)$	0	0	0
1	$(1,1,1)$	1	0	0

We have recently suggested that the concepts of redundancy and synergy apply just as naturally to entropy as to mutual information [36]. Therefore, the redundancy lattice and PID framework can be applied to entropy to obtain a partial entropy decomposition. A particular advantage of the entropy approach is that it provides a way to separately quantify source and mechanistic redundancy [11,36]. Just as mutual information is derived from differences in entropies, we suggest that partial information terms should be related to partial entropy terms. For any partial information decomposition, there should be a compatible partial entropy decomposition. We note that I_{ccs} is highly consistent with a PID based on a partial entropy decomposition obtained with a pointwise entropy redundancy measure which measures common surprisal [36]. More formal study of the relationships between the two approaches is an important area for future work. In contrast, it is hard to imagine an entropy decomposition compatible with I_{broja}. In fact, we have shown that I_{broja} is fundamentally incompatible with the notion of synergistic entropy. Since it satisfies the Harder et al. identity axiom, it induces a two variable entropy decomposition which always has zero synergistic entropy.

As well as providing the foundation for the PID, a conceptually well-founded and practically accessible measure of redundancy is a useful statistical tool in its own right. Even in the relatively simple case of two experimental dependent variables, a rigorous measure of redundancy can provide insights about the system that would not be possible to obtain with classical statistics. The presence of high redundancy could indicate a common mechanism is responsible for both sets of observations, whereas independence would suggest different mechanisms. To our knowledge the only established approaches that attempt to address such questions in practice are Representational Similarity Analysis [58] and cross-decoding methods such as the temporal generalisation method [59]. However, both these approaches can be complicated to implement, have restricted domains of applicability and cannot address synergistic interactions. We hope the methods presented here will provide a useful and accessible alternative allowing statistical analyses that provide novel interpretations across a range of fields.

Acknowledgments: I thank Jim Kay, Michael Wibral, Ryan James, Johannes Rauh and Joseph Lizier for useful discussions. I thank Ryan James for producing and maintaining the excellent dit package. I thank Daniel Chicharro for introducing me to the topic and providing many patient explanations and examples. I thank Eugenio Piasini, Christoph Daube and Philippe Schyns for useful comments on the manuscript. This work was supported by the Wellcome Trust [107802/Z/15/Z] and the Multidisciplinary University Research Initiative/Engineering and Physical Sciences Research Council [EP/N019261/1].

Conflicts of Interest: The author declares no conflict of interest.

References

1. Shannon, C. A mathematical theory of communication. *Bell Syst. Tech. J.* **1948**, *27*, 379–423.
2. Cover, T.; Thomas, J. *Elements of Information Theory*; Wiley: New York, NY, USA, 1991.
3. Ince, R.A.; Giordano, B.L.; Kayser, C.; Rousselet, G.A.; Gross, J.; Schyns, P.G. A statistical framework for neuroimaging data analysis based on mutual information estimated via a gaussian copula. *Hum. Brain Mapp.* **2017**, *38*, 1541–1573.
4. Sokal, R.R.; Rohlf, F.J. *Biometry*; WH Freeman and Company: New York, NY, USA, 1981.
5. Timme, N.; Alford, W.; Flecker, B.; Beggs, J.M. Synergy, redundancy, and multivariate information measures: An experimentalist's perspective. *J. Comput. Neurosci.* **2013**, *36*, 119–140.
6. Williams, P.L.; Beer, R.D. Nonnegative Decomposition of Multivariate Information. *Physics* **2010**, *1004*, 2515.
7. Wibral, M.; Priesemann, V.; Kay, J.W.; Lizier, J.T.; Phillips, W.A. Partial information decomposition as a unified approach to the specification of neural goal functions. *Brain Cogn.* **2017**, *112*, 25–38.
8. Lizier, J.T.; Prokopenko, M.; Zomaya, A.Y. A Framework for the Local Information Dynamics of Distributed Computation in Complex Systems. In *Guided Self-Organization: Inception*; Prokopenko, M., Ed.; Springer: Berlin/Heidelberg, Germany, 2014; pp. 115–158, doi:10.1007/978-3-642-53734-9_5.
9. Reza, F.M. *An Introduction to Information Theory*; McGraw-Hill: New York, NY, USA, 1961.
10. Griffith, V.; Koch, C. Quantifying Synergistic Mutual Information. In *Guided Self-Organization: Inception*; Prokopenko, M., Ed.; Springer: Berlin/Heidelberg, Germany, 2014; pp. 159–190.
11. Harder, M.; Salge, C.; Polani, D. Bivariate measure of redundant information. *Phys. Rev.* **2013**, *87*, 012130.
12. Bertschinger, N.; Rauh, J.; Olbrich, E.; Jost, J.; Ay, N. Quantifying Unique Information. *Entropy* **2014**, *16*, 2161–2183.
13. Griffith, V.; Chong, E.K.P.; James, R.G.; Ellison, C.J.; Crutchfield, J.P. Intersection Information Based on Common Randomness. *Entropy* **2014**, *16*, 1985–2000.
14. Bertschinger, N.; Rauh, J.; Olbrich, E.; Jost, J. Shared Information—New Insights and Problems in Decomposing Information in Complex Systems. In *Proceedings of the European Conference on Complex Systems 2012*; Gilbert, T., Kirkilionis, M., Nicolis, G., Eds.; Springer International Publishing: Berlin/Heidelberg, Germany, 2013; pp. 251–269, doi:10.1007/978-3-319-00395-5_35.
15. Olbrich, E.; Bertschinger, N.; Rauh, J. Information Decomposition and Synergy. *Entropy* **2015**, *17*, 3501–3517.
16. Griffith, V.; Ho, T. Quantifying Redundant Information in Predicting a Target Random Variable. *Entropy* **2015**, *17*, 4644–4653.
17. McGill, W.J. Multivariate information transmission. *Psychometrika* **1954**, *19*, 97–116.
18. Jakulin, A.; Bratko, I. Quantifying and Visualizing Attribute Interactions. *arXiv* **2003**, arXiv:cs/0308002.
19. Bell, A.J. The co-information lattice. In Proceedings of the 4th International Symposium on Independent Component Analysis and Blind Signal Separation (ICA2003), Nara, Japan, 1–4 April 2003; pp. 921–926.
20. Matsuda, H. Physical nature of higher-order mutual information: Intrinsic correlations and frustration. *Phys. Rev.* **2000**, *62*, 3096–3102.

21. Wibral, M.; Lizier, J.; Vögler, S.; Priesemann, V.; Galuske, R. Local active information storage as a tool to understand distributed neural information processing. *Front. Neuroinf.* **2014**, *8*, 1.
22. Lizier, J.T.; Prokopenko, M.; Zomaya, A.Y. Local information transfer as a spatiotemporal filter for complex systems. *Phys. Rev.* **2008**, *77*, 026110.
23. Wibral, M.; Lizier, J.T.; Priesemann, V. Bits from Biology for Computational Intelligence. *Quant. Biol.* **2014**, *185*, 1115–1117.
24. Van de Cruys, T. Two Multivariate Generalizations of Pointwise Mutual Information. In *Proceedings of the Workshop on Distributional Semantics and Compositionality*; Association for Computational Linguistics: Stroudsburg, PA, USA, 2011; pp. 16–20.
25. Church, K.W.; Hanks, P. Word Association Norms, Mutual Information, and Lexicography. *Comput. Linguist.* **1990**, *16*, 22–29.
26. Barrett, A.B. Exploration of synergistic and redundant information sharing in static and dynamical Gaussian systems. *Phys. Rev.* **2015**, *91*, 052802.
27. Han, T.S. Multiple mutual informations and multiple interactions in frequency data. *Inf. Control* **1980**, *46*, 26–45.
28. Gawne, T.; Richmond, B. How independent are the messages carried by adjacent inferior temporal cortical neurons? *J. Neurosci.* **1993**, *13*, 2758–2771.
29. Panzeri, S.; Schultz, S.; Treves, A.; Rolls, E. Correlations and the encoding of information in the nervous system. *Proc. Biol. Sci.* **1999**, *266*, 1001–1012.
30. Brenner, N.; Strong, S.; Koberle, R.; Bialek, W.; Steveninck, R. Synergy in a neural code. *Neural Comput.* **2000**, *12*, 1531–1552.
31. Schneidman, E.; Bialek, W.; Berry, M. Synergy, Redundancy, and Independence in Population Codes. *J. Neurosci.* **2003**, *23*, 11539–11553.
32. Ting, H. On the Amount of Information. *Theory Prob. Appl.* **1962**, *7*, 439–447.
33. Quian Quiroga, R.; Panzeri, S. Extracting information from neuronal populations: Information theory and decoding approaches. *Nat. Rev. Neurosci.* **2009**, *10*, 173–185.
34. Hastie, T.; Tibshirani, R.; Friedman, J. *The Elements of Statistical Learning*; Springer Series in Statistics: Berlin/Heidelberg, Germany, 2001; Volume 1.
35. Crampton, J.; Loizou, G. The completion of a poset in a lattice of antichains. *Int. Math. J.* **2001**, *1*, 223–238.
36. Ince, R.A.A. The Partial Entropy Decomposition: Decomposing multivariate entropy and mutual information via pointwise common surprisal. *arXiv* **2017**, arXiv:1702.01591.
37. James, R.G.; Crutchfield, J.P. Multivariate Dependence Beyond Shannon Information. *arXiv* **2016**, arXiv:1609.01233.
38. DeWeese, M.R.; Meister, M. How to measure the information gained from one symbol. *Netw. Comput. Neural Syst.* **1999**, *10*, 325–340.
39. Butts, D.A. How much information is associated with a particular stimulus? *Netw. Comput. Neural Syst.* **2003**, *14*, 177–187.
40. Osborne, M.J.; Rubinstein, A. *A Course in Game Theory*; MIT Press: Cambridge, MA, USA, 1994.
41. Jaynes, E. Information Theory and Statistical Mechanics. *Phys. Rev.* **1957**, *106*, 620–630.
42. Amari, S. Information Geometry of Multiple Spike Trains. In *Analysis of Parallel Spike Trains*; Grün, S., Rotter, S., Eds.; Springer: Berlin/Heidelberg, Germany, 2010; pp. 221–252.
43. Schneidman, E.; Still, S.; Berry, M., II; Bialek, W. Network Information and Connected Correlations. *Phys. Rev. Lett.* **2003**, *91*, 238701.
44. Ince, R.; Montani, F.; Arabzadeh, E.; Diamond, M.; Panzeri, S. On the presence of high-order interactions among somatosensory neurons and their effect on information transmission. *J. Phys. Conf. Ser.* **2009**, *197*, 012013.
45. Roudi, Y.; Nirenberg, S.; Latham, P. Pairwise Maximum Entropy Models for Studying Large Biological Systems: When They Can Work and When They Can't. *PLoS Comput. Biol.* **2009**, *5*, e1000380.
46. Lizier, J.T.; Flecker, B.; Williams, P.L. Towards a synergy-based approach to measuring information modification. In Proceedings of the 2013 IEEE Symposium on Artificial Life (ALIFE), Singapore, 16–19 April 2013; pp. 43–51.
47. Robince/partial-info-decomp. Available online: https://github.com/robince/partial-info-decomp (accessed on 29 June 2017).
48. Dit. Available online: https://github.com/dit/dit (accessed on 29 June 2017).

49. Dit: Discrete Information Theory. Available online: http://docs.dit.io/ (accessed on 29 June 2017).

50. James, R.G. cheebee7i. Zenodo. dit/dit v1.0.0.dev0 [Data set]. Available online: https://zenodo.org/record/235071#.WVMJ9nuVmpo (accessed on 28 June 2017).

51. Kay, J.W. On finding trivariate binary distributions given bivariate marginal distributions. Personal Communication, 2017.

52. Abdallah, S.A.; Plumbley, M.D. A measure of statistical complexity based on predictive information with application to finite spin systems. *Phys. Lett.* **2012**, *376*, 275–281.

53. Rauh, J.; Bertschinger, N.; Olbrich, E.; Jost, J. Reconsidering unique information: Towards a multivariate information decomposition. In Proceedings of the 2014 IEEE International Symposium on Information Theory (ISIT), Honolulu, HI, USA, 29 June–4 July 2014.

54. Chicharro, D.; Panzeri, S. Synergy and Redundancy in Dual Decompositions of Mutual Information Gain and Information Loss. *Entropy* **2017**, *19*, 71.

55. Rauh, J. Secret Sharing and Shared Information. *arXiv* **2017**, arXiv:1706.06998.

56. Panzeri, S.; Senatore, R.; Montemurro, M.A.; Petersen, R.S. Correcting for the Sampling Bias Problem in Spike Train Information Measures. *J. Neurophys.* **2007**, *96*, 1064–1072.

57. Ince, R.A.A.; Mazzoni, A.; Bartels, A.; Logothetis, N.K.; Panzeri, S. A novel test to determine the significance of neural selectivity to single and multiple potentially correlated stimulus features. *J. Neurosci. Methods* **2012**, *210*, 49–65.

58. Kriegeskorte, N.; Mur, M.; Bandettini, P. Representational Similarity Analysis—Connecting the Branches of Systems Neuroscience. *Front. Syst. Neurosci.* **2008**, *2*, 4, doi:10.3389/neuro.06.004.2008.

59. King, J.R.; Dehaene, S. Characterizing the dynamics of mental representations: The temporal generalization method. *Trends Cogn. Sci.* **2014**, *18*, 203–210.

entropy

MDPI

Article

Pointwise Partial Information Decomposition Using the Specificity and Ambiguity Lattices

Conor Finn [1,2,]*and Joseph T. Lizier [1]

[1]   Complex Systems Research Group and Centre for Complex Systems, Faculty of Engineering & IT,
      The University of Sydney, NSW 2006, Australia; joseph.lizier@sydney.edu.au
[2]   CSIRO Data61, Marsfield NSW 2122, Australia
[*]   Correspondence: conor.finn@sydney.edu.au

Received: 10 July 2017; Accepted: 10 April 2018; Published: 18 April 2018

Abstract: What are the distinct ways in which a set of predictor variables can provide information about a target variable? When does a variable provide unique information, when do variables share redundant information, and when do variables combine synergistically to provide complementary information? The redundancy lattice from the partial information decomposition of Williams and Beer provided a promising glimpse at the answer to these questions. However, this structure was constructed using a much criticised measure of redundant information, and despite sustained research, no completely satisfactory replacement measure has been proposed. In this paper, we take a different approach, applying the axiomatic derivation of the redundancy lattice to a single realisation from a set of discrete variables. To overcome the difficulty associated with signed pointwise mutual information, we apply this decomposition separately to the unsigned entropic components of pointwise mutual information which we refer to as the specificity and ambiguity. This yields a separate redundancy lattice for each component. Then based upon an operational interpretation of redundancy, we define measures of redundant specificity and ambiguity enabling us to evaluate the partial information atoms in each lattice. These atoms can be recombined to yield the sought-after multivariate information decomposition. We apply this framework to canonical examples from the literature and discuss the results and the various properties of the decomposition. In particular, the pointwise decomposition using specificity and ambiguity satisfies a chain rule over target variables, which provides new insights into the so-called two-bit-copy example.

Keywords: mutual information; pointwise information; information decomposition; unique information; redundant information; complementary information; redundancy; synergy

PACS: 89.70.Cf; 89.75.Fb; 05.65.+b; 87.19.lo

1. Introduction

   The aim of information decomposition is to divide the total amount of information provided by a set of predictor variables, about a target variable, into atoms of partial information contributed either individually or jointly by the various subsets of the predictors. Suppose that we are trying to predict a target variable T, with discrete state space $\mathcal{T}$, from a pair of predictor variables S_1 and S_2, with discrete state spaces $\mathcal{S}_1$ and $\mathcal{S}_2$. The mutual information $I(S_1; T)$ quantifies the information S_1 individually provides about T. Similarly, the mutual information $I(S_2; T)$ quantifies the information S_2 individually provides about T. Now consider the joint variable $S_{1,2}$ with the state space $\mathcal{S}_1 \times \mathcal{S}_2$. The (joint) mutual information $I(S_{1,2}; T)$ quantifies the total information S_1 and S_2 together provide about T. Although Shannon's information theory provides the prior three measures of information, there are four possible ways S_1 and S_2 could contribute information about T: the predictor S_1 could uniquely provide information about T; or the predictor S_2 could uniquely provide information about T;

both S_1 and S_2 could both individually, yet redundantly, provide the same information about T; or the predictors S_1 and S_2 could synergistically provide information about T which is not available in either predictor individually. Thus we have the following underdetermined set of equations,

$$
\begin{aligned}
I(S_{1,2};T) &= R(S_1,S_2\to T) + U(S_1\backslash S_2\to T) + U(S_2\backslash S_1\to T) + C(S_1,S_2\to T), \\
I(S_1;T) &= R(S_1,S_2\to T) + U(S_1\backslash S_2\to T), \\
I(S_2;T) &= R(S_1,S_2\to T) + U(S_2\backslash S_1\to T),
\end{aligned}
\tag{1}
$$

where $U(S_1\backslash S_2\to T)$ and $U(S_2\backslash S_1\to T)$ are the unique information provided by S_1 and S_2 respectively, $R(S_1,S_2\to T)$ is the redundant information, and $C(S_1,S_2\to T)$ is the synergistic or complementary information. (The directed notation is utilise here to emphasis the privileged role of the variable T.) Together, the equations in (1) form the bivariate information decomposition. The problem is to define one of the unique, redundant or complementary information—something not provided by Shannon's information theory—in order to uniquely evaluate the decomposition.

Now suppose that we are trying to predict a target variable T from a set of n finite state predictor variables $S = \{S_1, \ldots, S_n\}$. In this general case, the aim of information decomposition is to divide the total amount of information $I(S_1, \ldots, S_n; T)$ into atoms of partial information contributed either individually or jointly by the various subsets of S. But what are the distinct ways in which these subsets of predictors might contribute information about the target? Multivariate information decomposition is more involved than the bivariate information decomposition because it is not immediately obvious how many atoms of information one needs to consider, nor is it clear how these atoms should relate to each other. Thus the general problem of information decomposition is to provide both a structure for multivariate information which is consistent with the bivariate decomposition, and a way to uniquely evaluate the atoms in this general structure.

In the remainder of Section 1, we will introduce an intriguing framework called partial information decomposition (PID), which aims to address the general problem of information decomposition, and highlight some of the criticisms and weaknesses of this framework. In Section 2, we will consider the underappreciated pointwise nature of information and discuss the relevance of this to the problem of information decomposition. We will then propose a modified pointwise partial information decomposition (PPID), but then quickly repudiate this approach due to complications associated with decomposing the signed pointwise mutual information. In Section 3, we will discuss circumventing this issue by examining information on a more fundamental level, in terms of the unsigned entropic components of pointwise mutual information which we refer to as the specificity and the ambiguity. Then in Section 4—the main section of this paper—we will introduce the PPID using the specificity and ambiguity lattices and the measures of redundancy in Definitions 1 and 2. In Section 5, we will apply this framework to a number of canonical examples from the PID literature, discuss some of the key properties of the decomposition, and compare these to existing approaches to information decomposition. Section 6 will conclude the main body of the paper. Appendix A contains discussions regarding the so-called two-bit-copy problem in terms of Kelly gambling, Appendix B contains many of the technical details and proofs, while Appendix B contains some more examples.

1.1. Notation

The following notational conventions are observed throughout this article:

$T, \mathcal{T}, t, t^c$	denote the *target* variable, event space, event and complementary event respectively;
$S, \mathcal{S}, s, s^c$	denote the *predictor* variable, event space, event and complementary event respectively;
$\boldsymbol{S}, \boldsymbol{s}$	represent the *set* of n predictor variables $\{S_1, \ldots, S_n\}$ and events $\{s_1, \ldots, s_n\}$ respectively;
$\mathcal{T}^t, \mathcal{S}^s$	denote the *two-event partition* of the event space, i.e., $\mathcal{T}^t = \{t, t^c\}$ and $\mathcal{S}^s = \{s, s^c\}$;
$H(T), I(S;T)$	uppercase function names be used for *average* information-theoretic measures;
$h(t), i(s,t)$	lowercase function names be used for *pointwise* information-theoretic measures.

When required, the following index conventions are observed:

s^1, s^2, t^1, t^2 superscripts distinguish between different *different events* in a variable;

S_1, S_2, T_1, T_2 subscripts distinguish between *different variables*;

$S_{1,2}, s_{1,2}$ multiple superscripts represent *joint variables* and *joint events*.

Finally, to be discussed in more detail when appropriate, consider the following:

$A_1, \ldots, A_k$ sources are sets of predictor variables, i.e., $A_i \in \mathscr{P}_1(S)$ where $\mathscr{P}_1$ is the power set without $\varnothing$;

$a_1, \ldots, a_k$ source events are sets of predictor events, i.e., $a_i \in \mathscr{P}_1(s)$.

1.2. Partial Information Decomposition

The *partial information decomposition* (PID) of Williams and Beer [1,2] was introduced to address the problem of multivariate information decomposition. The approach taken is appealing as rather than speculating about the structure of multivariate information, Williams and Beer took a more principled, axiomatic approach. They start by considering potentially overlapping subsets of S called sources, denoted $A_1, \ldots, A_k$. To examine the various ways these sources might contain the same information, they introduce three axioms which "any reasonable measure for redundant information [$I_\cap$] should fulfil" ([3], p. 3502). Note that the axioms appear explicitly in [2] but are discussed in [1] as mere properties; a published version of the axioms can be found in [4].

W&B Axiom 1 (Commutativity). *Redundant information is invariant under any permutation σ of sources,*

$$I_\cap(A_1, \ldots, A_k \to T) = I_\cap(\sigma(A_1), \ldots, \sigma(A_k) \to T).$$

W&B Axiom 2 (Monotonicity). *Redundant information decreases monotonically as more sources are included,*

$$I_\cap(A_1, \ldots, A_{k-1} \to T) \leq I_\cap(A_1, \ldots, A_k \to T)$$

with equality if $A_k \supseteq A_i$ for any $A_i \in \{A_1, \ldots, A_{k-1}\}$.

W&B Axiom 3 (Self-redundancy). *Redundant information for a single source A_i equals the mutual information,*

$$I_\cap(A_i \to T) = I(A_i ; T).$$

These axioms are based upon the intuition that redundancy should be analogous to the set-theoretic notion of intersection (which is commutative, monotonically decreasing and idempotent). Crucially, Axiom 3 ties this notion of redundancy to Shannon's information theory. In addition to these three axioms, there is an (implicit) axiom assumed here known as *local positivity* [5], which is the requirement that all atoms be non-negative. Williams and Beer [1,2] then show how these axioms reduce the number of sources to the collection of sources such that no source is a superset of any other. These remaining sources are called *partial information atoms* (PI atoms). Each PI atom corresponds to a distinct way the set of predictors S can contribute information about the target T. Furthermore, Williams and Beer show that these PI atoms are partially ordered and hence form a lattice which they call the *redundancy lattice*. For the bivariate case, the redundancy lattice recovers the decomposition (1), while in the multivariate case it provides a meaningful structure for decomposition of the total information provided by an arbitrary number of predictor variables.

While the redundancy lattice of PID provides a structure for multivariate information decomposition, it does not uniquely determine the value of the PI atoms in the lattice. To do so requires a definition of a measure of redundant information which satisfies the above axioms. Hence, in order to complete the PID framework, Williams and Beer simultaneously introduced a measure of redundant information called I_{min} which quantifies redundancy as the minimum information that any source provides about a target event t, averaged over all possible events from T. However, not long after its introduction I_{min} was heavily criticised. Firstly, I_{min} does not distinguish between "whether different random variables carry the *same* information or just the *same amount* of information" ([5], p. 269; see also [6,7]). Secondly,

I_{min} does not possess the target chain rule introduced by Bertschinger et al. [5] (under the name left chain rule). This latter point is problematic as the target chain rule is a natural generalisation of the chain rule of mutual information—i.e., one of the fundamental, and indeed characterising, properties of information in Shannon's theory [8,9].

These issues with I_{min} prompted much research attempting to find a suitable replacement measure compatible with the PID framework. Using the methods of information geometry, Harder et al. [6] focused on a definition of redundant information called I_{red} (see also [10]). Bertschinger et al. [11] defined a measure of unique information $\widetilde{UI}$ based upon the notion that if one variable contains unique information then there must be some way to exploit that information in a decision problem. Griffith and Koch [12] used an entirely different motivation to define a measure of synergistic information S_{VK} whose decomposition transpired to be equivalent to that of $\widetilde{UI}$ [11]. Despite this effort, none of these proposed measures are entirely satisfactory. Firstly, just as for I_{min}, none of these proposed measures possess the target chain rule. Secondly, these measures are not compatible with the PID framework in general, but rather are only compatible with PID for the special case of bivariate predictors, i.e., the decomposition (1). This is because they all simultaneously satisfy the Williams and Beers axioms, local positivity, and the *identity property* introduced by Harder et al. [6]. In particular, Rauh et al. [13] proved that no measure satisfying the identity property and the Williams and Beer Axioms 1–3 can yield a non-negative information decomposition beyond the bivariate case of two predictor variables. In addition to these proposed replacements for I_{min}, there is also a substantial body of literature discussing either PID, similar attempts to decompose multivariate information, or the problem of information decomposition in general [3–5,7,10,13–28]. Furthermore, the current proposals have been applied to various problems in neuroscience [29–34]. Nevertheless (to date), there is no generally accepted measure of redundant information that is entirely compatible with PID framework, nor has any other well-accepted multivariate information decomposition emerged.

To summarise the problem, we are seeking a meaningful decomposition of the information provided an arbitrarily large set of predictor variables about a target variable, into atoms of partial information contributed either individually or jointly by the various subsets of the predictors. Crucially, the redundant information must capture when two predictor variables are carrying the same information about the target, not merely the same amount of information. Finally, any proposed measure of redundant information should satisfy the target chain rule so that net redundant information can be consistently computed for consistently for multiple target events.

2. Pointwise Information Theory

Both the entropy and mutual information can be derived from first principles as fundamentally *pointwise* quantities which measure the information content of individual events rather than entire variables. The pointwise entropy $h(t) = -\log p(t)$ quantifies the information content of a single event t, while the pointwise mutual information

$$i(s;t) = \log \frac{p(t|s)}{p(t)} = \log \frac{p(s,t)}{p(s)p(t)} = \log \frac{p(s|t)}{p(s)}, \tag{2}$$

quantifies the information provided by s about t, or vice versa. To our knowledge, these quantities were first considered by Woodward and Davies [35,36] who noted that the average form of Shannon's entropy "tempts one to enquire into other simpler methods of derivation [of the per state entropy]" ([35], p. 51). Indeed, they went on to show that the pointwise entropy and pointwise mutual information can be derived from two axioms concerning the addition of the information provided by the occurrence of individual events [36]. Fano [9] further formalised this pointwise approach by deriving both quantites from four postulates which "should be satisfied by a useful measure of information" ([9], p. 31). Taking the expectation of these pointwise quantities over all events recovers the average entropy $H(T) = \langle h(t) \rangle$ and average mutual information $I(S;T) = \langle i(s;t) \rangle$ first derived by Shannon [8]. Although both approaches arrive at the same average quantities, Shannon's treatment obfuscates the pointwise nature of the fundamental quantities. In contrast, the approach of Woodward, Davis and Fano makes this pointwise nature manifestly obvious.

It is important to note that, in contrast to the average mutual information, the pointwise mutual information is not non-negative. Positive pointwise information corresponds to the predictor event s raising the probability $p(t|s)$ relative to the prior probability $p(t)$. Hence when the event t occurs it can be said that the event s was *informative* about the event t. Conversely, negative pointwise information corresponds to the event s lowering the posterior probability $p(t|s)$ relative to the prior probability $p(t)$. Hence when the event t occurs we can say that the event s was *misinformative* about the event t. (Not to be confused with disinformation, i.e., intentionally misleading information.) Although a source event s may be misinformative about a particular target event t, a source event s is never misinformative about the target variable T since the pointwise mutual information averaged over all target realisations is non-negative [9]. The information provided by s is helpful for predicting T on average; however, in certain instances this (typically helpful) information is misleading in that it lowers $p(t|s)$ relative to $p(t)$—typically helpful information which subsequently turns out to be misleading is misinformation.

Finally, before continuing, there are two points to be made about the terminology used to describe pointwise information. Firstly, in certain literature (typically in the context of time-series analysis), the word *local* is used instead of pointwise, e.g., [4,18]. Secondly, in contemporary information theory, the word average is generally omitted while the pointwise quantities are explicitly prefixed; however, this was not always the accepted convention. Woodward [35] and Fano [9] both referred to pointwise mutual information as the *mutual information* and then explicitly prefixed the *average mutual information*. To avoid confusion, we will always prefix both pointwise and average quantities.

2.1. Pointwise Information Decomposition

Now that we are familiar with pointwise nature of information, suppose that we have a discrete realisation from the joint event space $\mathcal{T} \times S_1 \times S_2$ consisting of the target event t and predictor events s_1 and s_2. The pointwise mutual information $i(s_1;t)$ quantifies the information provided individually by s_1 about t, while the pointwise mutual information $i(s_2;t)$ quantifies the information provided individually by s_2 about t. The pointwise joint mutual information $i(s_{1,2};t)$ quantifies the total information provided jointly by s_1 and s_2 about t. In correspondence with the (average) bivariate decomposition (1), consider the pointwise bivariate decomposition, first suggested by Lizier et al. [4],

$$
\begin{aligned}
i(s_{1,2};t) &= r(s_1, s_2 \to t) + u(s_1 \backslash s_2 \to t) + u(s_2 \backslash s_1 \to t) + c(s_1, s_2 \to t),\\
i(s_1;t) &= r(s_1, s_2 \to t) + u(s_1 \backslash s_2 \to t),\\
i(s_2;t) &= r(s_1, s_2 \to t) + u(s_2 \backslash s_1 \to t).
\end{aligned}
\tag{3}
$$

Note that the lower case quantities denote the pointwise equivalent of the corresponding upper case quantities in (1). This decomposition could be considered for every discrete realisation on the support of the joint distribution $P(S_1, S_2, T)$. Hence, consider taking the expectation of these pointwise atoms over all discrete realisations,

$$
\begin{aligned}
U(S_1 \backslash S_2 \to T) &= \langle u(s_1 \backslash s_2 \to t) \rangle, & R(S_1, S_2 \to T) &= \langle r(s_1, s_2 \to t) \rangle,\\
U(S_2 \backslash S_1 \to T) &= \langle u(s_2 \backslash s_1 \to t) \rangle, & C(S_1, S_2 \to T) &= \langle c(s_1, s_2 \to t) \rangle.
\end{aligned}
\tag{4}
$$

Since the expectation is a linear operation, this will recover the (average) bivariate decomposition (1). Equation (3) for every discrete realisation, together with (1) and (4) form the bivariate pointwise information decomposition. Just as in (1), these equations are underdetermined requiring a separate definition of either the pointwise unique, redundant or complementary information for uniqueness. (Defining an average atom is sufficient for a unique bivariate decomposition (1), but still leaves the pointwise decomposition (3) within each realisation underdetermined).

2.2. Pointwise Unique

Now consider applying this pointwise information decomposition to the probability distribution *Pointwise Unique* (PWUNQ) in Table 1. In PWUNQ, observing 0 in either of S_1 or S_2 provides zero information about the target T, while complete information about the outcome of T is obtained by observing 1 or a 2 in either predictor. The probability distribution is structured such that in each of the four realisations, one predictor provides complete information while the other predictor provides zero information—the two predictors never provide the same information about the target which is justified by noting that one of the two predictors always provides zero pointwise information.

Given that redundancy is supposed to capture the same information, it seems reasonable to assume there must be zero pointwise redundant information for each realisation. This assumption is made without any measure of pointwise redundant information; however, no other possibility seems justifiable. This assertion is used to determine the pointwise redundant information terms in Table 1. Then using the pointwise information decomposition (3), we can then evaluate the other pointwise atoms of information in Table 1. Finally using (4), we get that there is zero (average) redundant information, and $1/2$ bit of (average) unique information from each predictor. From the pointwise perspective, the only reasonable conclusion seems to be that the predictors in PWUNQ must contain only unique information about the target.

Table 1. Example PWUNQ. For each realisation, the pointwise mutual information provided by each individual and joint predictor events, about the target event has been evaluated. Note that one predictor event always provides full information about the target while the other provides zero information. Based on the this, it is assumed that there must be zero redundant information. The pointwise partial information (PPI) atoms are then calculated via (3).

p	s_1	s_2	t	$i(s_1;t)$	$i(s_2;t)$	$i(s_{1,2};t)$	$u(s_1\setminus s_2 \to t)$	$u(s_2\setminus s_1 \to t)$	$r(s_1,s_2 \to t)$	$c(s_1,s_2 \to t)$
$1/4$	0	1	1	0	1	1	0	1	0	0
$1/4$	1	0	1	1	0	1	1	0	0	0
$1/4$	0	2	2	0	1	1	0	1	0	0
$1/4$	2	0	2	1	0	1	1	0	0	0
Expected values				$1/2$	$1/2$	1	$1/2$	$1/2$	0	0

However, in contrast to the above, $I_{\min}$, I_{red}, $\widetilde{UI}$, and $\mathcal{S}_{\mathrm{VK}}$ all say that the predictors in PWUNQ contain no unique information, rather only $1/2$ bit of redundant information plus $1/2$ bit of complementary information. This problem, which will be referred to as the *pointwise unique problem*, is a consequence of the fact that these measures all satisfy Assumption ($*$) of Bertschinger et al. [11], which (in effect) states that the unique and redundant information should only depend on the marginal distributions $P(S_1, T)$ and $P(S_2, T)$. In particular, any measure which satisfies Assumption ($*$) will yield zero unique information when $P(S_1, T)$ is isomorphic to $P(S_2, T)$, as is the case for PWUNQ. (Here, isomorphic should be taken to mean isomorphic probability spaces, e.g., [37], p. 27 or [38], p. 4.) It arises because Assumption ($*$) (and indeed the operational interpretation the led to its introduction) does not respect the pointwise nature of information. This operational view does not take into account the fact that individual events s_1 and s_2 may provide different information about the event t, even if the probability distributions $P(S_1, T)$ and $P(S_2, T)$ are the same. Hence, we contend that for any measure to capture the same information (not merely the same amount), it must respect the pointwise nature of information.

2.3. Pointwise Partial Information Decomposition

With the pointwise unique problem in mind, consider constructing an information decomposition with the pointwise nature of information as an inherent property. Let $a_1, \ldots, a_k$ be potentially intersecting subsets of the predictor events $s = \{s_1, \ldots, s_n\}$, called source events. Now consider rewriting the Williams and Beer axioms in terms of a measure of pointwise redundant information $i_\cap$ where the aim is to deriving a *pointwise partial information decomposition* (PPID).

PPID Axiom 1 (Symmetry). *Pointwise redundant information is invariant under any permutation σ of source events,*

$$i_\cap(a_1, \ldots, a_k \rightarrow t) = i_\cap(\sigma(a_1), \ldots, \sigma(a_k) \rightarrow T).$$

PPID Axiom 2 (Monotonicity). *Pointwise redundant information decreases monotonically as more source events are included,*

$$i_\cap(a_1, \ldots, a_{k-1} \rightarrow t) \leq i_\cap(a_1, \ldots, a_k \rightarrow t)$$

with equality if $a_k \supseteq a_i$ for any $a_i \in \{a_1, \ldots, a_{k-1}\}$.

PPID Axiom 3 (Self-redundancy). *Pointwise redundant information for a single source event a_i equals the pointwise mutual information,*

$$i_\cap(a_i \rightarrow t) = i(a_i ; t).$$

It seems that the next step should be to define some measure of pointwise redundant information which is compatible with these PPID axioms; however, there is a problem—the pointwise mutual information is not non-negative. While this would not be an issue for the examples like PWUNQ, where none of the source events provide negative pointwise information, it is an issue in general (e.g., see RDNERR in Section 5.4). The problem is that set-theoretic intuition behind Axiom 2 (monotonicity) makes little sense when considering signed measures like the pointwise mutual information.

Given the desire to address the pointwise unique problem, there is a need to overcome this issue. Ince [18] suggested that the set-theoretic intuition is only valid when all source events provide either positive or negative pointwise information. Ince contends that information and misinformation are "fundamentally different" ([18], p. 11) and that the set-theoretic intuition should be admitted in the difficult to interpret situations where both are present. We however, will take a different approach—one which aims to deal with these difficult to interpret situations whilst preserving the set-theoretic intuition that redundancy corresponds to overlapping information.

By way of a preview, we first consider precisely how an event s_1 provides information about an event t by the means of two distinct types of probability mass exclusion. We show how considering the process in this way naturally splits the pointwise mutual information into particular entropic components, and how one can consider redundancy on each of these components separately. Splitting the signed pointwise mutual information into these unsigned entropic components circumvents the above issue with Axiom 2 (monotonicity). Crucially, however, by deriving these entropic components from the probability mass exclusions, we retain the set-theoretic intuition of redundancy—redundant information will correspond to overlapping probability mass exclusions in the two-event partition $\mathcal{T}^t = \{t, t^c\}$.

3. Probability Mass Exclusions and the Directed Components of Pointwise Mutual Information

By definition, the pointwise information provided by s about t is associated with a change from the prior $p(t)$ to the posterior $p(t|s)$. As we explored from first principles in Finn and Lizier [39], this change is a consequence of the *exclusion* of probability mass in the target distribution $P(T)$ induced by the occurrence of the event s and inferred via the joint distribution $P(S, T)$. To be specific, when the event s occurs, one knows that the complementary event $s^c = \{\mathcal{S} \backslash s\}$ did not occur. Hence one can *exclude* the probability mass in the joint distribution $P(S, T)$ associated with the complementary event, i.e., exclude $P(s^c, T)$, leaving just the probability mass $P(s, T)$ remaining. The new target distribution $P(T|s)$ is evaluated by normalising this remaining probability mass. In [39] we introduced *probability mass diagrams* in order to visually explore the exclusion process. Figure 1 provides an example of such a diagram. Clearly, this process is merely a description of the definition of conditional probability. Nevertheless, we content that by viewing the change from the prior to the posterior in this way—by focusing explicitly on the exclusions rather than the resultant conditional probability—the vague intuition that redundancy corresponds to overlapping information becomes more apparent. This point will elaborated upon in Section 3.3. However, in order to do so, we need to first discuss the two distinct types of probability mass exclusion (which we do in Section 3.1) and then relate these to information-theoretic quantities (which we do in Section 3.2).

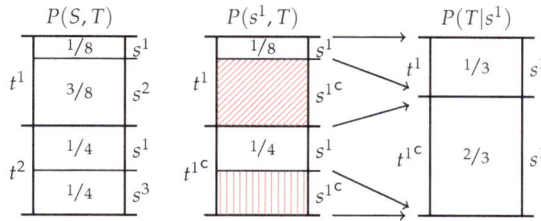

Figure 1. Sample probability mass diagrams, which use length to represent the probability mass of each joint event from $\mathcal{T} \times \mathcal{S}$. (**Left**) the joint distribution $P(S, T)$; (**Middle**) The occurrence of the event s^1 leads to exclusions of the complementary event s^{1^c} which consists of two elementary event, i.e., $s^{1^c} = \{s^2, s^3\}$. This leaves the probability mass $P(s^1, T)$ remaining. The exclusion of the probability mass $p(s^{1^c}, t^1)$ was misinformative since the event t^1 did occur. By convention, misinformative exclusions will be indicated with diagonal hatching. On the other hand, the exclusion of the probability mass $p(t^{1^c}, s^{1^c})$ was informative since the complementary event t^{1^c} did not occur. By convention, informative exclusions will be indicated with horizontal or vertical hatching; (**Right**) this remaining probability mass can be normalised yielding the conditional distribution $P(T|s^1)$.

3.1. Two Distinct Types of Probability Mass Exclusions

In [39] we examined the two distinct types of probability mass exclusions. The difference between the two depends on where the exclusion occurs in the target distribution $P(T)$ and the particular target event t which occurred. *Informative exclusions* are those which are confined to the probability mass associated with the set of elementary events in the target distribution which *did not occur*, i.e., exclusions confined to the probability mass of the complementary event $p(t^c)$. They are called such because the pointwise mutual information $i(s; t)$ is a monotonically increasing function of the total size of these exclusions $p(t^c)$. By convention, informative exclusions are represented on the probability mass diagrams by horizontal or vertical lines. On the other hand, the *misinformative exclusion* is confined to the probability mass associated with the elementary event in the target distribution which *did occur*, i.e., an exclusion confined to $p(t)$. It is referred to as such because the pointwise mutual information $i(s; t)$ is a monotonically decreasing function of the size of this type of exclusion $p(t)$. By convention, misinformative exclusions are represented on the probability mass diagrams by diagonal lines.

Although an event s may exclusively induce either type of exclusion, in general both types of exclusion are present simultaneously. The distinction between the two types of exclusions leads naturally to the following question—can one decompose the pointwise mutual information $i(s; t)$ into a positive informational component associated with the informative exclusions, and a negative informational component associated with the misinformative exclusions? This question is considered in detail in Section 3.2. However, before moving on, there is a crucial observation to be made about the pointwise mutual information which will have important implications for the measure of redundant information to be introduced later.

Remark 1. *The pointwise mutual information $i(s; t)$ depends only on the size of informative and misinformative exclusions. In particular, it does not depend on the apportionment of the informative exclusions across the set of elementary events contained in the complementary event t^c.*

In other words, whether the event s turns out to be net informative or misinformative about the event t—whether $i(s; t)$ is positive or negative—depends on the size of the two types of exclusions; but, to be explicit, does *not* depend on the distribution of the informative exclusion across the set of target events which did not occur. This remark will be crucially important when it comes to providing the operational interpretation of redundant information in Section 3.3. (It is also further discussed in terms of Kelly gambling [40] in Appendix A).

3.2. The Directed Components of Pointwise Information: Specificity and Ambiguity

We return now to the idea that one might be able to decompose the pointwise mutual information into a positive and negative component associated with the informative amd misinformative exclusions respectively. In [39] we proposed four postulates for such a decomposition. Before stating the postulates, it is important to note that although there is a "surprising symmetry" ([41], p. 23) between the information provided by s about t and the information provided by t about s, there is nothing to suggest that the components of the decomposition should be symmetric—indeed the intuition behind the decomposition only makes sense when considering the information is considered in a directed sense. As such, directed notation will be used to explicitly denote the information provided by s about t.

Postulate 1 (Decomposition). *The pointwise information provided by s about t can be decomposed into two non-negative components, such that $i(s;t) = i_+(s \to t) - i_-(s \to t)$.*

Postulate 2 (Monotonicity). *For all fixed $p(s,t)$ and $p(s^c,t)$, the function $i_+(s \to t)$ is a monotonically increasing, continuous function of $p(t^c, s^c)$. For all fixed $p(t^c, s)$ and $p(t^c, s^c)$, the function $i_-(s \to t)$ is a monotonically increasing continuous function of $p(s^c, t)$. For all fixed $p(s,t)$ and $p(t^c, s)$, the functions $i_+(s \to t)$ and $i_-(s \to t)$ are monotonically increasing and decreasing functions of $p(t^c, s^c)$, respectively.*

Postulate 3 (Self-Information). *An event cannot misinform about itself, $i_+(s \to s) = i(s;s) = -\log p(s)$.*

Postulate 4 (Chain Rule). *The functions $i_+(s_{1,2} \to t)$ and $i_-(s_{1,2} \to t)$ satisfy a chain rule, i.e.,*

$$
\begin{aligned}
i_+(s_{1,2} \to t) &= i_+(s_1 \to t) + i_+(s_2 \to t|s_1) \\
&= i_+(s_2 \to t) + i_+(s_1 \to t|s_2), \\
i_-(s_{1,2} \to t) &= i_-(s_1 \to t) + i_-(s_2 \to t|s_1) \\
&= i_-(s_2 \to t) + i_-(s_1 \to t|s_2)
\end{aligned}
$$

In Finn and Lizier [39], we proved that these postulates lead to the following forms which are unique up to the choice of the base of the logarithm in the mutual information in Postulates 1 and 3,

$$
\begin{aligned}
i^+(s_1 \to t) &= h(s_1) &&= -\log p(s_1), &&(5)\\
i^+(s_1 \to t|s_2) &= h(s_1|s_2) &&= -\log p(s_1|s_2), &&(6)\\
i^+(s_{1,2} \to t) &= h(s_{1,2}) &&= -\log p(s_{1,2}), &&(7)\\
i^-(s_1 \to t) &= h(s_1|t) &&= -\log p(s_1|t), &&(8)\\
i^-(s_1 \to t|s_2) &= h(s_1|t,s_2) &&= -\log p(s_1|t,s_2), &&(9)\\
i^-(s_{1,2} \to t) &= h(s_{1,2}|t) &&= -\log p(s_{1,2}|t). &&(10)
\end{aligned}
$$

That is, the Postulates 1–4 uniquely decompose the pointwise information provided by s about t into the following entropic components,

$$
\begin{aligned}
i(s;t) &= i^+(s \to t) - i^-(s \to t) \\
&= h(s) - h(s|t).
\end{aligned}
\tag{11}
$$

Although the decomposition of mutual information into entropic components is well-known, it is non-trivial that Postulates 1 and 3, based on the size of the two distinct types of probability mass exclusions, lead to this particular form, but not $i(s;t) = h(t) - h(t|s)$ or $i(s;t) = h(s) + h(t) - h(s,t)$.

It is important to note that although the original motivation was to decompose the pointwise mutual information into separate components associated with informative and misinformative exclusion, the decomposition (11) does not quite possess this direct correspondence:

- The positive informational component $i^+(s \to t)$ does not depend on t but rather only on s. This can be interpreted as follows: the less likely s is to occur, the more specific it is when it does occur, the greater the total amount of probability mass excluded $p(s^c)$, and the greater the potential for s to inform about t (or indeed any other target realisation).
- The negative informational component $i^-(s \to t)$ depends on both s and t, and can be interpreted as follows: the less likely s is to coincide with the event t, the more uncertainty in s given t, the greater size of the misinformative probability mass exclusion $p(s^c, t)$, and therefore the greater the potential for s to misinform about t.

In other words, although the negative informational component $i^-(s \to t)$ does correspond directly to the size of the misinformative exclusion $p(s^c, t)$, the positive informational component $i^+(s \to t)$ does not correspond directly to the size of the informative exclusion $p(t^c, s^c)$. Rather, the positive informational component $i^+(s \to t)$ corresponds to the *total* size of the probability mass exclusions $p(s^c)$, which is the sum of the sum of the informative and misinformative exclusions. For the sake of brevity, the positive informational component $i^+(s \to t)$ will be referred to as the *specificity*, while the negative informational component $i^-(s \to t)$ will be referred to as the *ambiguity*. The term ambiguity is due to Shannon: "[equivocation] measures the average ambiguity of the received signal" ([42], p. 67). Specificity is an antonym of ambiguity and the usage here is inline with the definition since the more specific an event s, the more information it could provide about t after the ambiguity is taken into account.

3.3. Operational Interpretation of Redundant Information

Arguing about whether one piece of information differs from another piece of information is nonsensical without some kind of unambiguous definition of what it means for two pieces of information to be the same. As such, Bertschinger et al. [11] advocate the need to provide an operational interpretation of what it means for information to be unique or redundant. This section provides our operational definition of what it means for information to be the same. This definition provides a concrete interpretation of what it means for information to be redundant in terms of overlapping probability mass exclusions.

The operational interpretation of redundancy adopted here is based upon the following idea: since the pointwise information is ultimately derived from probability mass exclusions, the *same information* must induce the *same exclusions*. More formally, the information provided by a set of predictor events $s_1, \ldots, s_k$ about a target event t must be the same information if each source event induces the same exclusions with respect to the two-event partition $\mathcal{T}^t = \{t, t^c\}$. While this statement makes the motivational intuition clear, it is not yet sufficient to serve as an operational interpretation of redundancy: there is no reference to the two distinct types of probability mass exclusions, the specific reference to the pointwise event space $\mathcal{T}^t$ has not been explained, and there is no reference to the fact the exclusions from each source may differ in size.

Informative exclusions are fundamentally different from misinformative exclusions and hence each type of exclusion should be compared separately: informative exclusions can overlap with informative exclusions, and misinformative exclusions can overlap with misinformative exclusions. In information-theoretic terms, this means comparing the specificity and the ambiguity of the sources separately—i.e., considering a measure of redundant specificity and a separate measure of redundant ambiguity. Crucially, these quantities (being pointwise entropies) are unsigned meaning that the difficulties associated with Axiom 2 (Monotonicity) and signed pointwise mutual information in Section 2.3 will not be an issue here.

The specific reference to the two-event partition $\mathcal{T}^t$ in the above statement is based upon Remark 1 and is crucially important. The pointwise mutual information does not depend on the apportionment of the informative exclusions across the set of events which did not occur, hence the pointwise redundant information should not depend on this apportionment either. In other words, it is immaterial if two predictor events s_1 and s_2 exclude different elementary events within the target complementary event t^c (assuming the probability mass excluded is equal) since with respect to the realised target event t the difference between the exclusions is only semantic. This has important implications for the comparison of exclusions from different predictor events. As the pointwise mutual information depends on, and only depends on, the size of the exclusions, then the only sensible comparison is a comparison of size. Hence, the common or overlapping exclusion must be the smallest exclusion. Thus, consider the following operational interpretation of redundancy:

Operational Interpretation (Redundant Specificity). *The redundant specificity between a set of predictor events $s_1, \dots, s_n$ is the specificity associated with the source event which induces the smallest total exclusions.*

Operational Interpretation (Redundant Ambiguity). *The redundant ambiguity between a set of predictor events $s_1, \dots, s_n$ is the ambiguity associated with the source event which induces the smallest misinformative exclusion.*

3.4. Motivational Example

To motivate the above operational interpretation, and in particular the need to treat the specificity separately to the ambiguity, consider Figure 2. In this pointwise example, two different predictor events provide the same amount of pointwise information since $P(T|s_1^1) = P(T|s_2^1)$, and yet the information provided by each event is in some way different since each excludes different sections of the target distribution $P(T)$. In particular, s_1^1 and s_2^1 both preclude the target event t^2, while s_2^1 additionally excludes probability mass associated with target events t^1 and t^3. From the perspective of the pointwise mutual information the events s_1^1 and s_2^1 seem to be providing the same information as

$$i(s_1^1 \rightarrow t^1) = i(s_2^1 \rightarrow t^1) = \log 4/3 \text{ bit.} \qquad (12)$$

However, from the perspective of the specificity and the ambiguity it can be seen that information is being provided in different ways since

$$\begin{aligned} i^+(s_1^1 \rightarrow t^1) &= \log 4/3 \text{ bit,} & i^-(s_1^1 \rightarrow t^1) &= 0 \text{ bit,} \\ i^+(s_2^1 \rightarrow t^1) &= \log 8/3 \text{ bit,} & i^-(s_2^1 \rightarrow t^1) &= 1 \text{ bit.} \end{aligned} \qquad (13)$$

Now consider the problem of decomposing information into its unique, redundant and complementary components. Figure 2 shows where exclusions induced by s_1^1 and s_2^1 overlap where they both exclude the target event t^2 which is an informative exclusion. This is the only exclusion induced by s_1^1 and hence all of the information associated with this exclusion must be redundantly provided by the event s_2^1. Without any formal framework, consider taking the redundant specificity and redundant ambiguity,

$$r^+(s_1^1, s_2^1 \rightarrow t^1) = i^+(s_1^1 \rightarrow t^1) = \log 4/3 \text{ bit,} \qquad (14)$$
$$r^-(s_1^1, s_2^1 \rightarrow t_1) = i^-(s_1^1 \rightarrow t^1) = \qquad 0 \text{ bit.} \qquad (15)$$

This would mean that the event s_2^1 provides the following unique specificity and unique ambiguity,

$$u^+(s_1^1 \backslash s_2^1 \rightarrow t^1) = i^+(s_1^1 \rightarrow t^1) - r^+(s_1^1, s_2^1 \rightarrow t^1) = 1 \text{ bit,} \qquad (16)$$
$$u^-(s_1^1 \backslash s_2^1 \rightarrow t^1) = i^-(s_1^1 \rightarrow t^1) - r^-(s_1^1, s_2^1 \rightarrow t^1) = 1 \text{ bit.} \qquad (17)$$

The redundant specificity $\log 4/3$ bit accounts for the overlapping informative exclusion of the event t^2. The unique specificity and unique ambiguity from s_2^1 are associated with its non-overlapping informative and misinformative exclusions; however, both of these 1 bit and hence, on net, s_2^1 is no more informative than s_1^1. Although obtained without a formal framework, this example highlights a need to consider the specificity and ambiguity rather than merely the pointwise mutual information.

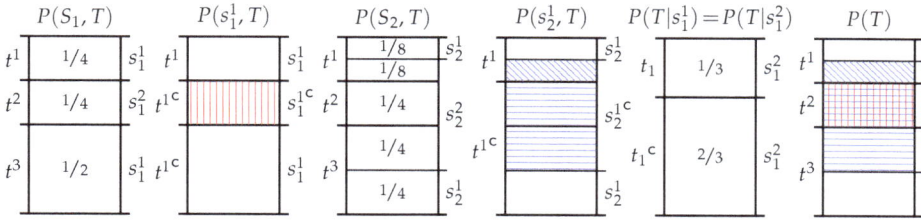

Figure 2. Sample probability mass diagrams for two predictors S_1 and S_2 to a given target T. Here events in the two different predictor spaces provide the same amount of pointwise information about the target event, $\log_2 4/3$ bits, since $P(T|s_1^1) = P(T|s_2^1)$, although each excludes different sections of the target distribution $P(T)$. Since they both provide the same amount of information, is there a way to characterise what information the additional unique exclusions from the event s_2^1 are providing?

4. Pointwise Partial Information Decomposition Using Specificity and Ambiguity

Based upon the argumentation of Section 3, consider the following axioms:

Axiom 1 (Symmetry). *Pointwise redundant specificity $i_\cap^+$ and pointwise redundant ambiguity $i_\cap^-$ are invariant under any permutation σ of source events,*

$$i_\cap^+ (a_1, \ldots, a_k \to t) = i_\cap^+ (\sigma(a_1), \ldots, \sigma(a_k) \to t),$$
$$i_\cap^- (a_1, \ldots, a_k \to t) = i_\cap^- (\sigma(a_1), \ldots, \sigma(a_k) \to t).$$

Axiom 2 (Monotonicity). *Pointwise redundant specificity $i_\cap^+$ and pointwise redundant ambiguity $i_\cap^-$ decreases monotonically as more source events are included,*

$$i_\cap^+ (a_1, \ldots, a_{k-1}, a_k \to t) \le i_\cap^+ (a_1, \ldots, a_{k-1} \to t),$$
$$i_\cap^- (a_1, \ldots, a_{k-1}, a_k \to t) \le i_\cap^- (a_1, \ldots, a_{k-1} \to t).$$

with equality if $a_k \supseteq a_i$ for any $a_i \in \{a_1, \ldots, a_{k-1}\}$.

Axiom 3 (Self-redundancy). *Pointwise redundant specificity $i_\cap^+$ and pointwise redundant ambiguity $i_\cap^-$ for a single source event a_i equals the specificity and ambiguity respectively,*

$$i_\cap^+ (a_i \to t) = i^+ (a_i \to t) = h(a_i),$$
$$i_\cap^- (a_i \to t) = i^- (a_i \to t) = h(a_i|t).$$

As shown in Appendix B.1, Axioms 1–3 induce two lattices—namely the *specificity lattice* and *ambiguity lattice*—which are depicted in Figure 3. Furthermore, each lattice is defined for every discrete realisation from $P(S_1, \ldots, S_n, T)$. The redundancy measures $i_\cap^+$ or $i_\cap^-$ can be thought of as a cumulative information functions which integrate the specificity or ambiguity uniquely contributed by each node as one moves up each lattice. Finally, just as in PID, performing a Möbius inversion over each lattice yielding the unique contributions of specificity and ambiguity from each sources event.

Similarly to PID, the specificity and ambiguity lattices provide a structure for information decomposition, but unique evaluation requires a separate definition of redundancy. However, unlike PID (or even PPID), this evaluation requires both a definition of pointwise redundant specificity and pointwise redundant ambiguity. Before providing these definitions, it is helpful to first see how the specificity and ambiguity lattices can be used to decompose multivariate information in the now familiar bivariate case.

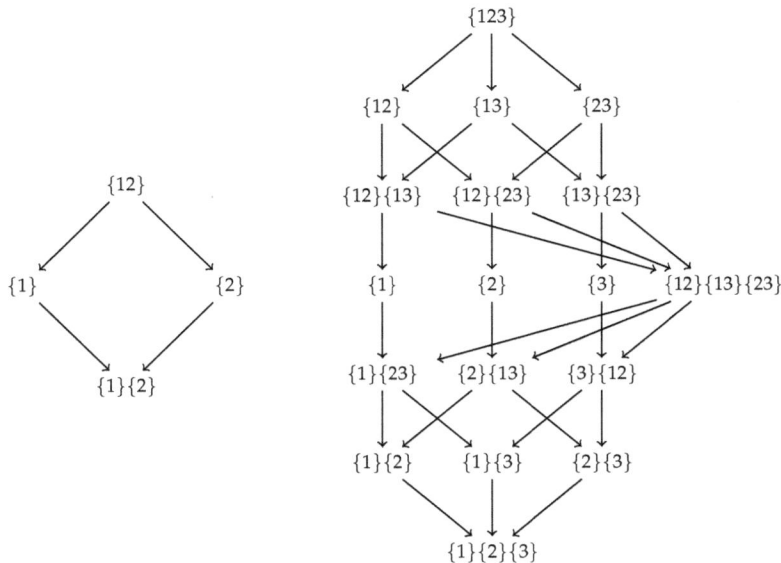

Figure 3. The lattice induced by the partial order $\preceq$ (A15) over the sources $\mathscr{A}(s)$ (A14). (**Left**) the lattice for $s = \{s_1, s_2\}$; (**Right**) the lattice for $s = \{s_1, s_2, s_3\}$. See Appendix B for further details. Each node corresponds to the self-redundancy (Axiom 3) of a source event, e.g., $\{1\}$ corresponds to the source event $\{\{s_1\}\}$, while $\{12, 13\}$ corresponds to the source event $\{\{s_{1,2}\}, \{s_{1,3}\}\}$. Note that the specificity and ambiguity lattices share the same structure as the redundancy lattice of partial information decomposition (PID) (cf. Figure 2 in [1]).

4.1. Bivariate PPID Using the Specificity and Ambiguity

Consider again the bivariate case where the aim is to decompose the information provided by s_1 and s_2 about t. The specificity lattice can be used to decompose the pointwise specificity,

$$
\begin{aligned}
i^+(s_{1,2} \to t) &= r^+(s_1, s_2 \to t) + u^+(s_1 \backslash s_2 \to t) + u^+(s_2 \backslash s_1 \to t) + c^+(s_1, s_2 \to t), \\
i^+(s_1 \to t) &= r^+(s_1, s_2 \to t) + u^+(s_1 \backslash s_2 \to t), \\
i^+(s_2 \to t) &= r^+(s_1, s_2 \to t) + u^+(s_2 \backslash s_1 \to t);
\end{aligned}
\tag{18}
$$

while the ambiguity lattice can be used to decompose the pointwise ambiguity,

$$
\begin{aligned}
i^-(s_{1,2} \to t) &= r^-(s_1, s_2 \to t) + u^-(s_1 \backslash s_2 \to t) + u^-(s_2 \backslash s_1 \to t) + c^-(s_1, s_2 \to t), \\
i^-(s_1 \to t) &= r^-(s_1, s_2 \to t) + u^-(s_1 \backslash s_2 \to t), \\
i^-(s_2 \to t) &= r^-(s_1, s_2 \to t) + u^-(s_2 \backslash s_1 \to t).
\end{aligned}
\tag{19}
$$

These equations share the same structural form as (3) only now decompose the specificity and the ambiguity rather than the pointwise mutual information, e.g., $r^+(s_1, s_2 \to t)$ denotes the redundant specificity while $u^-(s_1 \backslash s_2 \to t)$ denoted the unique ambiguity from s_1. Just as in for (3), this decomposition could be considered for every discrete realisation on the support of the joint distribution $P(S_1, S_2, T)$.

There are two ways one can be combine these values. Firstly, in a similar manner to (4), one could take the expectation of the atoms of specificity, or the atoms of ambiguity, over all discrete realisations yielding the average PI atoms of specificity and ambiguity,

$$
\begin{aligned}
U^+(S_1 \backslash S_2 \to T) &= \langle u^+(s_1 \backslash s_2 \to t) \rangle, & U^-(S_1 \backslash S_2 \to T) &= \langle u^-(s_1 \backslash s_2 \to t) \rangle, \\
U^+(S_2 \backslash S_1 \to T) &= \langle u^+(s_2 \backslash s_1 \to t) \rangle, & U^-(S_2 \backslash S_1 \to T) &= \langle u^-(s_2 \backslash s_1 \to t) \rangle, \\
R^+(S_1, S_2 \to T) &= \langle r^+(s_1, s_2 \to t) \rangle, & R^-(S_1, S_2 \to T) &= \langle r^-(s_1, s_2 \to t) \rangle, \\
C^+(S_1, S_2 \to T) &= \langle c^+(s_1, s_2 \to t) \rangle. & C^-(S_1, S_2 \to T) &= \langle c^-(s_1, s_2 \to t) \rangle.
\end{aligned}
\tag{20}
$$

Alternatively, one could subtract the pointwise unique, redundant and complementary ambiguity from the pointwise unique, redundant and complementary specificity yielding the pointwise unique, pointwise redundant and pointwise complementary information, i.e., recover the atoms from PPID,

$$
\begin{aligned}
r(s_1, s_2 \to t) &= r^+(s_1, s_2 \to t) - r^-(s_1, s_2 \to t), \\
u(s_1 \backslash s_2 \to t) &= u^+(s_1 \backslash s_2 \to t) - u^-(s_1 \backslash s_2 \to t), \\
u(s_2 \backslash s_1 \to t) &= u^+(s_2 \backslash s_1 \to t) - u^-(s_2 \backslash s_1 \to t), \\
c(s_1, s_2 \to t) &= c^+(s_1, s_2 \to t) - c^-(s_1, s_2 \to t).
\end{aligned}
\tag{21}
$$

Both (20) and (21) are linear operations, hence one could perform both of these operations (in either order) to obtain the average unique, average redundant and average complementary information, i.e., recover the atoms from PID,

$$
\begin{aligned}
R(S_1, S_2 \to T) &= R^+(S_1, S_2 \to T) - R^-(S_1, S_2 \to T), \\
U(S_1 \backslash S_2 \to T) &= U^+(S_1 \backslash S_2 \to T) - U^-(S_1 \backslash S_2 \to T), \\
U(S_2 \backslash S_1 \to T) &= U^+(S_2 \backslash S_1 \to T) - U^-(S_2 \backslash S_1 \to T), \\
C(S_1, S_2 \to T) &= C^+(S_1, S_2 \to T) - C^-(S_1, S_2 \to T).
\end{aligned}
\tag{22}
$$

4.2. Redundancy Measures on the Specificity and Ambiguity Lattices

Now that we have a structure for our information decomposition, there is a need to provide a definition of the pointwise redundant specificity and pointwise redundant ambiguity. However, before attempting to provide such a definition, there is a need to consider Remark 1 and the operational interpretation of in Section 3.3. In particular, the pointwise redundant specificity $i_\cap^+$ and pointwise redundant ambiguity $i_\cap^-$ should only depend on the size of informative and misinformative exclusions. They should not depend on the apportionment of the informative exclusions across the set of elementary events contained in the complementary event t^c. Formally, this requirement will be enshrined via the following axiom.

Axiom 4 (Two-event Partition). *The pointwise redundant specificity $i_\cap^+$ and pointwise redundant ambiguity $i_\cap^-$ are functions of the probability measures on the two-event partitions $\mathcal{A}_1^{a_1} \times \mathcal{T}^t, \ldots, \mathcal{A}_k^{a_k} \times \mathcal{T}^t$.*

Since the pointwise redundant specificity $i_\cap^+$ is specificity associated with the source event which induces the smallest total exclusions, and pointwise redundant ambiguity $i_\cap^-$ is the ambiguity associated with the source event which induces the smallest misinformative exclusion, consider the following definitions.

Definition 1. *The pointwise redundant specificity is given by*

$$r_{min}^+(a_1, \ldots, a_k \to t) = \min_{a_i} i^+(a_i \to t) = \min_{a_i} h(a_i). \tag{23}$$

Definition 2. *The pointwise redundant ambiguity is given by*

$$r_{min}^-(a_1, \ldots, a_k \to t) = \min_{a_i} i^-(a_i \to t) = \min_{a_j} h(a_j|t). \tag{24}$$

Theorem 1. *The definitions of r_{min}^+ and r_{min}^- satisfy Axioms 1–4.*

Theorem 2. *The redundancy measures r_{min}^+ and r_{min}^- increase monotonically on the $\langle \mathscr{A}(s), \preceq \rangle$.*

Theorem 3. *The atoms of partial specificity π^+ and partial ambiguity π^- evaluated using the measures r_{min}^+ and r_{min}^- on the specificity and ambiguity lattices (respectively), are non-negative.*

Appendix B.2 contains the proof of Theorems 1–3 and further relevant consideration of Defintions 1 and 2. As in (20), one can take the expectation of the either the pointwise redundant specificity r_{min}^+ or the pointwise redundant ambiguity r_{min}^- to get the average redundant specificity R_{min}^+ or the average redundant ambiguity R_{min}^-. Alternatively, just as in (21), one can recombine the pointwise redundant specificity r_{min}^+ and the pointwise redundant ambiguity r_{min}^- to get the pointwise redundant information r_{min}. Finally, as per (22), one could perform both of these (linear) operations in either order to obtain the average redundant information R_{min}. Note that while Theorem 3 proves that the atoms of partial specificity π^+ and partial ambiguity π^- are non-negative, it is trivial to see that r_{min} could be negative since when source events can redundantly provide misinformation about a target event. As shown in the following theorem, R_{min} can also be negative.

Theorem 4. *The atoms of partial average information Π evaluated by recombining and averaging $\pi^{\pm}$ are not non-negative.*

This means that the measure R_{min} does not satisfy local positivity. Nonetheless the negativity of R_{min} is readily explainable in terms of the operational interpretation of Section 3.3, as will be discussed further in Section 5.4. However, failing to satisfy local positivity does mean that r_{min} and R_{min} do not satisfy the *target monotonicity* property first discussed in Bertschinger et al. [5]. Despite this, as the following theorem shows, the measures do satisfy the target chain rule.

Theorem 5 (Pointwise Target Chain Rule). *Given the joint target realisation $t_{1,2}$, the pointwise redundant information r_{min} satisfies the following chain rule,*

$$\begin{aligned} r_{min}(a_1, \ldots, a_k \to t_{1,2}) &= r_{min}(a_1, \ldots, a_k \to t_1) + r_{min}(a_1, \ldots, a_k \to t_2|t_1), \\ &= r_{min}(a_1, \ldots, a_k \to t_2) + r_{min}(a_1, \ldots, a_k \to t_1|t_2). \end{aligned} \tag{25}$$

The proof of the last theorem is deferred to Appendix B.3. Note that since the expectation is a linear operation, Theorem 5 also holds for the average redundant information R_{min}. Furthermore, as these results apply to any of the source events, the target chain rule will hold for any of the PPI atoms, e.g., (21), and any of the PI atoms, e.g., (22). However, no such rule holds for the pointwise redundant specificity or ambiguity. The specificity depends only on the predictor event, i.e., does not depend on the target events. As such, when an increasing number of target events are considered, the specificity remains unchanged. Hence, a target chain rule cannot hold for the specificity, or the ambiguity alone.

5. Discussion

PPID using the specificity and ambiguity takes the ideas underpinning PID and applies them on a pointwise scale while circumventing the monotonicity issue associated with the signed pointwise mutual information. This section will explore the various properties of the decomposition in an example driven manner and compare the results to the most widely-used measures from the existing PID literature. (Further examples can be found in Appendix C.) The following shorthand notation will be utilised in the figures throughout this section:

$$i_1^+ = i^+(s_1 \to t), \qquad i_2^+ = i^+(s_2 \to t), \qquad i_{1,2}^+ = i^+(s_{1,2} \to t),$$
$$i_1^- = i^-(s_1 \to t), \qquad i_2^- = i^-(s_2 \to t), \qquad i_{1,2}^- = i^-(s_{1,2} \to t),$$

$$u_1^+ = u^+(s_1 \backslash s_2 \to t), \quad u_2^+ = u^+(s_2 \backslash s_1 \to t), \quad r^+ = r^+(s_1, s_2 \to t), \quad c^+ = c^+(s_1, s_2 \to t),$$
$$u_1^- = u^-(s_1 \backslash s_2 \to t), \quad u_2^- = u^-(s_2 \backslash s_1 \to t), \quad r^- = r^-(s_1, s_2 \to t), \quad c^- = c^-(s_1, s_2 \to t).$$

5.1. Comparison to Existing Measures

A similar approach to the decomposition presented in this paper is due to Ince [18], who also sought to define a pointwise information decomposition. Despite the similarity in this regard, the redundancy measure I_{CCS} presented in [18] approaches the pointwise monotonicity problem of Section 2.3 in a different way to the decomposition presented in this paper. Specifically, I_{CCS} aims to utilise the pointwise co-information as a measure of pointwise redundant information since it "quantifies the set-theoretic overlap of the two univariate [pointwise] information values" ([18], p. 14). There are, however, difficulties with this approach. Firstly (unlike the average mutual information and the Shannon inequalities), there are no inequalities which support this interpretation of pointwise co-information as the set-theoretic overlap of the univariate pointwise information terms—indeed, both the univariate pointwise information and the pointwise co-information are signed measures. Secondly, the pointwise co-information conflates the pointwise redundant information with the pointwise complementary information, since by (3) we have that

$$co\text{-}i(s_1; s_2; t) := i(s_1; t) + i(s_2; t) - i(s_{1,2}, t) = r(s_1, s_2 \to t) - c(s_1, s_2 \to t). \tag{26}$$

Aware of these difficulties, Ince defines I_{CCS} such that it only interprets the pointwise co-information as a measure of set-theoretic overlap in the case where all three pointwise information terms have the same sign, arguing that these are the only situations which admit a clear interpretation in terms of a common change in surprisal. In the other difficult to interpret situations, I_{CCS} defines the pointwise redundant information to be zero. This approach effectively assumes that $c(s_1, s_2 \to t) = 0$ in (26) when $i(s_1; t)$, $i(s_2; t)$ and $co\text{-}i(s_1; s_2; t)$ all have the same sign.

In a subsequent paper, Ince [19] also presented a partial entropy decomposition which aims to decompose multivariate entropy rather than multivariate information. As such, this decomposition is more similar to PPID using specificity and ambiguity than Ince's aforementioned decomposition. Although similar in this regard, the measure of pointwise redundant entropy H_{cs} presented in [19] takes a different approach to the one presented in this paper. Specifically, H_{cs} also uses the pointwise co-information as a measure of set-theoretic overlap and hence as a measure of pointwise redundant entropy. As the pointwise entropy is unsigned, the difficulties are reduced but remain present due to the signed pointwise co-information. In a manner similar to I_{CCS}, Ince defines H_{cs} such that it only interprets the pointwise co-information as a measure of set-theoretic overlap when it is positive. As per I_{CCS}, this effectively assumes that $c(s_1, s_2 \to t) = 0$ in (26) when all information terms have the same sign. When the pointwise co-information is negative, H_{cs} simply ignores the co-information by defining the pointwise redundant information to be zero. In contrast to both of Ince's approaches, PPID using specificity and ambiguity does not dispose of the set-theoretic intuition in these difficult to interpret situations. Rather, our approach considers the notion of redundancy in terms of overlapping exclusions—i.e., in terms of the underlying, unsigned measures which are amenable to a set-theoretic interpretation.

The measures of pointwise redundant specificity $r_{\min}^+$ and pointwise redundant ambiguity $r_{\min}^-$, from Definitions 1 and 2 are also similar to both the minimum mutual information I_{MMI} [17] and the original PID redundancy measure $I_{\min}$ [1]. Specifically, all three of these approaches consider the redundant information to be the minimum information provided about a target event t. The difference is that $I_{\min}$ applies this idea to the sources $A_1, \ldots, A_k$, i.e., to collections of entire predictor variables from S, while $r_{\min}^\pm$ apply this notion to the source events $a_1, \ldots, a_k$, i.e., to collections of predictor events from s. In other words, while the measure $I_{\min}$ can be regarded as being semi-pointwise (since it considers the information provided by the variables $S_1, \ldots, S_n$ about an event t), the measures $r_{\min}^\pm$ are fully pointwise (since they consider the information provided by the events $s_1, \ldots, s_n$ about an event t). This difference in approach is most apparent in the probability distribution PWUNQ—unlike PID, PPID using the specificity and ambiguity respects the pointwise nature of information, as we will see in Section 5.3.

PPID using specificity and ambiguity also share certain similarities with the bivariate PID induced by the measure $\widetilde{UI}$ of Bertschinger et al. [11]. Firstly, Axiom 4 can be considered to be a pointwise adaptation of their Assumption (∗), i.e., the measures $r_{\min}^\pm$ depend only on the marginal distributions $P(S_1, T)$ and $P(S_2, T)$ with respect to the two-event partitions $\mathcal{S}_1^{s_1} \times \mathcal{T}^t$ and $\mathcal{S}_2^{s_2} \times \mathcal{T}^t$. Secondly, in PPID using specificity and ambiguity, the only way one can only decide if there is complementary information $c(s_1, s_2 \to t)$ is by knowing the joint distribution $P(S_1, S_2, T)$ with respect to the joint two-event partitions $\mathcal{S}_1^{s_1} \times \mathcal{S}_2^{s_2} \times \mathcal{T}^t$. This is (in effect) a pointwise form of their Assumption (∗∗). Thirdly, by definition $r_{\min}^\pm$ are given by the minimum value that any one source event provides. This is the largest possible value that one could take for these quantities whilst still requiring that the unique specificity and ambiguity be non-negative. Hence, within each discrete realisation, $r_{\min}^\pm$ minimise the unique specificity and ambiguity whilst maximising the redundant specificity and ambiguity. This is similar to $\widetilde{UI}$ which minimises the (average) unique information while still satisfying Assumption (∗). Finally, note that since the measure $\mathcal{S}_{\text{VK}}$ produces a bivariate decomposition which is equivalent to that of $\widetilde{UI}$ [11], the same similarities apply between PPID using specificity and ambiguity and the decomposition induced by $\mathcal{S}_{\text{VK}}$ from Griffith and Koch [12].

5.2. Probability Distribution XOR

Figure 4 shows the canonical example of synergy, *exclusive-or* (XOR) which considers two independently distributed binary predictor variables S_1 and S_2 and a target variable $T = S_1$ XOR S_2. There are several important points to note about the decomposition of XOR. Firstly, despite providing zero pointwise information, an individual predictor event does indeed induce exclusions. However, the informative and misinformative exclusions are perfectly balanced such that the posterior (conditional) distribution is equal to the prior distribution, e.g., see the red coloured exclusions induced by $S_1 = 0$ in Figure 4. In information-theoretic terms, for each realisation, the pointwise specificity equals 1 bit since half of the total probability mass remains while the pointwise ambiguity also equals 1 bit since half of the probability mass associated with the event which subsequently occurs (i.e., $T = 0$), remains. These are perfectly balanced such that when recombined, as per (11), the pointwise mutual information is equal to 0 bit, as one would expect.

Secondly, $S_1 = 0$ and $S_2 = 0$ both induce the same exclusions with respect to the target pointwise event space $\mathcal{T}^{T=0}$. Hence, as per the operational interpretation of redundancy adopted in Section 3.3, there is 1 bit of pointwise redundant specificity and 1 bit of pointwise redundant ambiguity in each realisation. The presence of (a form of) redundancy in XOR is novel amongst the existing measures in the PID literature. (Ince [19] also identifies a form of redundancy in XOR.) Thirdly, despite the presence of this redundancy, recombining the atoms of pointwise specificity and ambiguity for each realisation, as per (21), leaves only one non-zero PPI atom: namely the pointwise complementary information $c(s_1, s_2 \to t) = 1$ bit. Furthermore, this is true for every pointwise realisation and hence, by (22), the only non-zero PI atom is the average complementary information $C(S_1, S_2 \to T) = 1$ bit.

$P(S_{1,2}, T)$ $S_1 = 0$ $S_2 = 0$ $S_{1,2} = 00$

	$P(S_{1,2},T)$			$S_1=0$			$S_2=0$			$S_{1,2}=00$
0	1/4	00	0		00	0		00	0	00
1	1/4	01	1		01	1		01	1	01
	1/4	10			10			10		10
0	1/4	11	0		11	0		11	0	11

p	s_1	s_2	t	i_1^+	i_1^-	i_2^+	i_2^-	i_{12}^+	i_{12}^-	r^+	u_1^+	u_2^+	c^+	r^-	u_1^-	u_2^-	c^-
1/4	0	0	0	1	1	1	1	2	1	1	0	0	1	1	0	0	0
1/4	0	1	1	1	1	1	1	2	1	1	0	0	1	1	0	0	0
1/4	1	0	1	1	1	1	1	2	1	1	0	0	1	1	0	0	0
1/4	1	1	0	1	1	1	1	2	1	1	0	0	1	1	0	0	0
Expected values				1	1	1	1	2	1	1	0	0	1	1	0	0	0

$R(S_1, S_2 \to T) = 0 \text{ bit}$ $U(S_1 \backslash S_2 \to T) = 0 \text{ bit}$ $U(S_2 \backslash S_1 \to T) = 0 \text{ bit}$ $C(S_1, S_2 \to T) = 1 \text{ bit}$

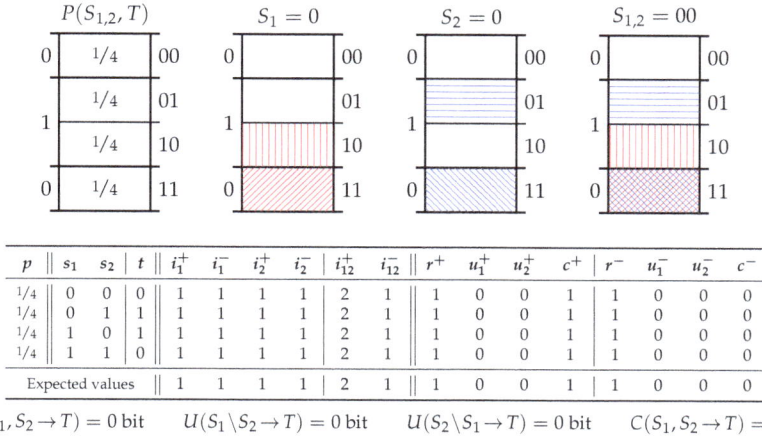

Figure 4. Example XOR. (**Top**) probability mass diagrams for the realisation $(S_1 = 0, S_2 = 0, T = 0)$; (**Middle**) For each realisation, the pointwise specificity and pointwise ambiguity has been evaluated using (5) and (8) respectively. The pointwise redundant specificity and pointwise redundant ambiguity are then determined using (23) and (24). The decomposition is calculated using (18) and (19). The expected specificity and ambiguity are calculated with (20); (**Bottom**) The average information is given by (22). As expected, XOR yields 1 bit of complementary information.

5.3. Probability Distribution PwUnq

Figure 5 shows the probability distribution PwUNQ introduced in Section 2.2. Recombining the decomposition via (21) yields the pointwise information decomposition proposed in Table 1—unsurprisingly, the explicitly pointwise approach results in a decomposition which does not suffer from the pointwise unique problem of Section 2.2.

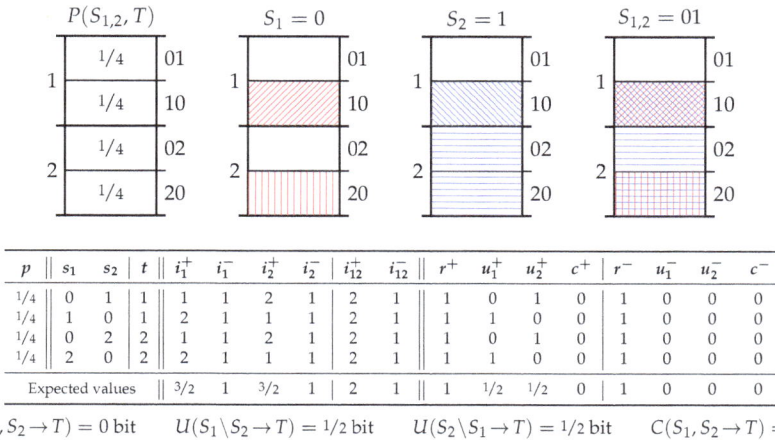

	$P(S_{1,2},T)$			$S_1=0$			$S_2=1$			$S_{1,2}=01$
1	1/4	01	1		01	1		01	1	01
	1/4	10			10			10		10
2	1/4	02	2		02	2		02	2	02
	1/4	20			20			20		20

p	s_1	s_2	t	i_1^+	i_1^-	i_2^+	i_2^-	i_{12}^+	i_{12}^-	r^+	u_1^+	u_2^+	c^+	r^-	u_1^-	u_2^-	c^-
1/4	0	1	1	1	1	2	1	2	1	1	0	1	0	1	0	0	0
1/4	1	0	1	2	1	1	1	2	1	1	1	0	0	1	0	0	0
1/4	0	2	2	1	1	2	1	2	1	1	0	1	0	1	0	0	0
1/4	2	0	2	2	1	1	1	2	1	1	1	0	0	1	0	0	0
Expected values				3/2	1	3/2	1	2	1	1	1/2	1/2	0	1	0	0	0

$R(S_1, S_2 \to T) = 0 \text{ bit}$ $U(S_1 \backslash S_2 \to T) = 1/2 \text{ bit}$ $U(S_2 \backslash S_1 \to T) = 1/2 \text{ bit}$ $C(S_1, S_2 \to T) = 0 \text{ bit}$

Figure 5. Example PwUNQ. (**Top**) probability mass diagrams for the realisation $(S_1 = 0, S_2 = 1, T = 1)$; (**Middle**) For each realisation, the pointwise partial information decomposition (PPID) using specificity and ambiguity is evaluated (see Figure 4 for details). Upon recombination as per (21), the PPI decomposition from Table 1 is attained; (**Bottom**) as does the average information—the decomposition does not have the pointwise unique problem.

In each realisation, observing a 0 in either source provides the same balanced informative and misinformative exclusions as in XOR. Observing either a 1 or 2 provides the same misinformative exclusion as observing the 0, but provides a larger informative exclusion than 0. This leaves only the probability mass associated with the event which subsequently occurs remaining (hence why observing a 1 and 2 is fully informative about the target). Information theoretically, in each realisation the predictor events provide 1 bit of redundant pointwise specificity and 1 bit of redundant pointwise ambiguity while the fully informative event additionally provides 1 bit of unique specificity.

5.4. Probability Distribution RDNERR

Figure 6 shows the probability distribution *redundant-error* (RDNERR) which considers two predictors which are nominally redundant and fully informative about the target, but where one predictor occasionally makes an erroneous prediction. Specifically, Figure 6 shows the decomposition of RDNERR where S_2 makes an error with a probability $\varepsilon = 1/4$. The important feature to note about this probability distribution is that upon recombining the specificity and ambiguity and taking the expectation over every realisation, the resultant average unique information from S_2 is $U(S_2 \backslash S_1 \rightarrow T) = -0.811$ bit.

On first inspection, the result that the average unique information can be negative may seem problematic; however, it is readily explainable in terms of the operational interpretation of Section 3.3. In RDNERR, a source event always excludes exactly $1/2$ of the total probability mass, thus every realisation contains 1 bit of redundant pointwise specificity. The events of the error-free S_1 induce only informative exclusions and as such provide 0 bit of pointwise ambiguity in each realisation. In contrast, the events in the error-prone S_2 always induce a misinformative exclusion, meaning that S_2 provides unique pointwise ambiguity in every realisation. Since S_2 never provides unique specificity, the average unique information is negative on average.

Despite the negativity of the average unique information, in is important to observe that S_2 provides 0.189 bit of information since S_2 also provides 1 bit of average redundant information. It is not that S_2 provides negative information on average (as this is not possible); rather it is that not all of the information provided by S_2 (i.e., the specificity) is "useful" ([42], p. 21). This is in contrast to S_1 which only provides useful specificity. To summarise, it is the unique ambiguity which distinguishes the information provided by variable S_2 from S_1, and hence why S_2 is deemed to provide negative average unique information. This form of uniqueness can only be distinguished by allowing the average unique information to be negative. This of course, requires abandoning the local positivity as a required property, as per Theorem 4. Few of the existing measures in the PID literature consider dropping this requirement as negative information quantities are typically regarded as being "unfortunate" ([43], p. 49). However, in the context of the pointwise mutual information, negative information values are readily interpretable as being misinformative values. Despite this, the average information from each predictor must be non-negative; however, it may be that what distinguishes one predictor from another are precisely the misinformative predictor events, meaning that the unique information is in actual fact, unique misinformation. Forgoing local positivity makes the PPID using specificity and ambiguity novel (the other exception in this regard is Ince [18] who was first to consider allowing negative average unique information.)

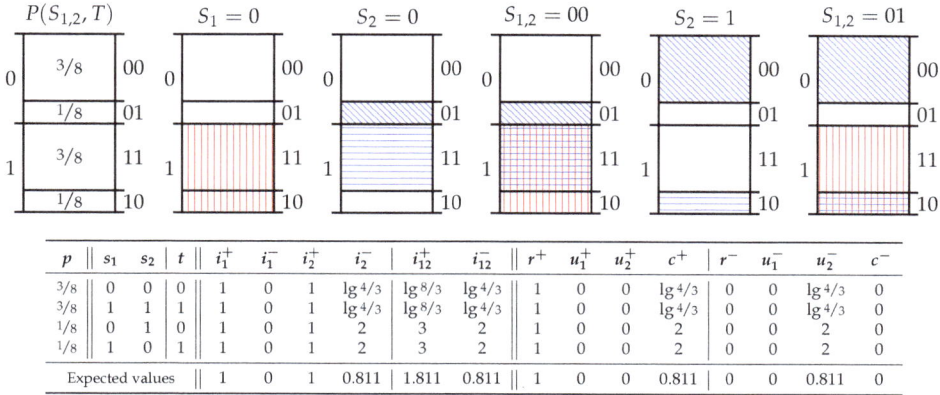

$P(S_{1,2}, T)$ $S_1 = 0$ $S_2 = 0$ $S_{1,2} = 00$ $S_2 = 1$ $S_{1,2} = 01$

	00	01	11	10
0	3/8			
		1/8		
1			3/8	
				1/8

p	s_1	s_2	t	i_1^+	i_1^-	i_2^+	i_2^-	i_{12}^+	i_{12}^-	r^+	u_1^+	u_2^+	c^+	r^-	u_1^-	u_2^-	c^-
3/8	0	0	0	1	0	1	lg 4/3	lg 8/3	lg 4/3	1	0	0	lg 4/3	0	0	lg 4/3	0
3/8	1	1	1	1	0	1	lg 4/3	lg 8/3	lg 4/3	1	0	0	lg 4/3	0	0	lg 4/3	0
1/8	0	1	0	1	0	1	2	3	2	1	0	0	2	0	0	2	0
1/8	1	0	1	1	0	1	2	3	2	1	0	0	2	0	0	2	0
Expected values			1	0	1	0.811	1.811	0.811	1	0	0	0.811	0	0	0.811	0	

$R(S_1, S_2 \to T) = 1$ bit $U(S_1 \setminus S_2 \to T) = 0$ bit $U(S_2 \setminus S_1 \to T) = -0.811$ bit $C(S_1, S_2 \to T) = 0.811$ bit

Figure 6. Example RDNERR. (**Top**) probability mass diagrams for the realisations $(S_1 = 0, S_2 = 0, T = 0)$ and $(S_1 = 0, S_2 = 1, T = 0)$; (**Middle**) for each realisation, the PPID using specificity and ambiguity is evaluated (see Figure 4 for details); (**Bottom**) the average PI atoms may be negative as the decomposition does not satisfy local positivity.

5.5. Probability Distribution TBC

Figure 7 shows the probability distribution *two-bit-copy* (TBC) which considers two independently distributed binary predictor variables S_1 and S_2, and a target variable T consisting of a separate elementary event for each joint event $S_{1,2}$. There are several important points to note about the decomposition of TBC. Firstly, due to the symmetry in the probability distribution, each realisation will have the same pointwise decomposition. Secondly, due to the construction of the target, there is an isomorphism (Again, isomorphism should be taken to mean isomorphic probability spaces, e.g., [37], p. 27 or [38], p. 4) between $P(T)$ and $P(S_1, S_2)$, and hence the pointwise ambiguity provided by any (individual or joint) predictor event is 0 bit (since given t, one knows s_1 and s_2). Thirdly, the individual predictor events s_1 and s_2 each exclude 1/2 of the total probability mass in $P(T)$ and so each provide 1 bit of pointwise specificity; thus, by (23), there is 1 bit of redundant pointwise specificity in each realisation. Fourthly, the joint predictor event $s_{1,2}$ excludes 3/4 of the total probability mass, providing 2 bit of pointwise specificity; hence, by (18), each joint realisation provides 1 bit of pointwise complementary specificity in addition to the 1 bit of redundant pointwise specificity. Finally, putting this together via (22), TBC consists of 1 bit of average redundant information and 1 bit of average complementary information.

Although "surprising" ([5], p. 268), according to the operational interpretation adopted in Section 3.3, two independently distributed predictor variables can share redundant information. That is, since the exclusions induced by s_1 and s_2 are the same with respect to the two-event partition $\mathcal{T}^t$, the information associated with these exclusions is regarded as being the same. Indeed, this probability distribution highlights the significance of specific reference to the two-event partition in Section 3.3 and Axiom 4. (This can be seen in the probability mass diagram in Figure 7, where the events $S_1 = 0$ and $S_2 = 0$ exclude different elementary target events within the complementary event 0^c and yet are considered to be the same exclusion with respect to the two-event partition $\mathcal{T}^0$.) That these exclusions should be regarded as being the same is discussed further in Appendix A. Now however, there is a need to discuss TBC in terms of Theorem 5 (Target Chain Rule).

$P(S_{1,2}, T)$			$S_1 = 0$	$S_1 = 0$	$S_2 = 0$	$S_2 = 0$	$S_{1,2} = 00$
0	1/4	00 0	0 0	0 0	0 0	0 0	0 00
1	1/4	01 1	0	0	1 1	1	1 01
2	1/4	10 2	1 0^c	1 0^c	2 0 0^c	2 0^c	2 10
3	1/4	11 3	1	1	3 1	3	3 11

p	s_1	s_2	t	$t_{1,2}$	$t_{1,3}$	$t_{2,3}$	i_1^+	i_1^-	i_2^+	i_2^-	i_{12}^+	i_{12}^-	r^+	u_1^+	u_2^+	c^+	r^-	u_1^-	u_2^-	c^-
1/4	0	0	0	00	00	00	1	0	1	0	2	0	1	0	0	1	0	0	0	0
1/4	0	1	1	01	01	11	1	0	1	0	2	0	1	0	0	1	0	0	0	0
1/4	1	0	2	10	11	01	1	0	1	0	2	0	1	0	0	1	0	0	0	0
1/4	1	1	3	11	10	10	1	0	1	0	2	0	1	0	0	1	0	0	0	0
Expected values							1	0	1	0	2	0	1	0	0	1	0	0	0	0

$$R(S_1, S_2 \to T) = 1 \text{ bit} \qquad U(S_1 \backslash S_2 \to T) = 0 \text{ bit} \qquad U(S_2 \backslash S_1 \to T) = 0 \text{ bit} \qquad C(S_1, S_2 \to T) = 1 \text{ bit}$$

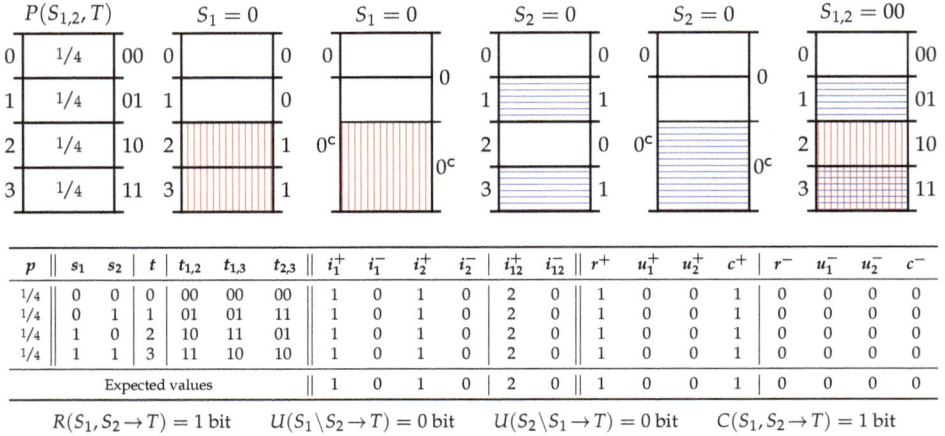

Figure 7. Example TBC. **(Top)** the probability mass diagrams for the realisation $(S_1 = 0, S_2 = 0, T = 00)$; **(Middle)** for each realisation, the PPID using specificity and ambiguity is evaluated (see Figure 4); **(Bottom)** the decomposition of XOR yields the same result as $I_{\min}$.

TBC was first considered as a "mechanism" ([6], p. 3) where "the wires don't even touch" ([12], p. 167), which merely copies or concatenates S_1 and S_2 into a composite target variable $T_{1,2} = (T_1, T_2)$ where $T_1 = S_1$ and $T_2 = S_2$. However, using causal mechanisms as a guiding intuition is dubious since different mechanisms can yield isomorphic probability distributions ([44], and references therein). In particular, consider two mechanisms which generate the composite target variables $T_{1,3} = (T_1, T_3)$ and $T_{2,3} = (T_2, T_3)$ where $T_3 = S_1$ XOR S_2. As can be seen in Figure 7, both of these mechanisms generate the same (isomorphic) probability distribution $P(S_1, S_2, T)$ as the mechanism generating $T_{1,2}$. If an information decomposition is to depend only on the probability distribution $P(S_1, S_2, T)$, and no other semantic details such as labelling, then all three mechanisms must yield the same information decomposition—this is not clear from the mechanistic intuition.

Although the decomposition of the various composite target variables must be the same, there is no requirement that the three systems must yield the same decomposition when analysed in terms of the individual components of the composite target variables. Nonetheless, there ought to be a consistency between the decomposition of the composite target variables and the decomposition of the component target variables—i.e., there should be a target chain rule. As shown in Theorem 5, the measures $r_{\min}$ and $R_{\min}$ satisfy the target chain rule, whereas $I_{\min}$, $\widetilde{UI}$, I_{red} and S_{VK} do not [5,7]. Failing to satisfy the target chain rule can lead to inconsistencies between the composite and component decompositions, depending on the order in which one considers decomposing the information (this is discussed further in Appendix A.3). In particular, Table 2 shows how $\widetilde{UI}$, I_{red} and S_{VK} all provide the same inconsistent decomposition for TBC when considered in terms of the composite target variable $T_{1,3}$. In contrast, $R_{\min}$ produces a consistent decomposition of $T_{1,3}$. Finally, based on the above isomorphism, consider the following (the proof is deferred to Appendix B.3).

Theorem 6. *The target chain rule, identity property and local positivity, cannot be simultaneously satisfied.*

Table 2. Shows the decomposition of the quantities in the first row induced by the measures in the first column. For consistency, the decomposition of $I(S_{1,2}; T_{1,3})$ should equal both the sum of the decomposition of $I(S_{1,2}; T_1)$ and $I(S_{1,2}; T_3 | T_1)$, and the sum of the decomposition of $I(S_{1,2}; T_3)$ and $I(S_{1,2}; T_1 | 3)$. Note that the decomposition induced by $\widetilde{UI}$, I_{red} and S_{VK} are not consistent. In contrast, $R_{\min}$ is consistent due to Theorem 5.

	$I(S_{1,2}; T_{1,3})$	$I(S_{1,2}; T_1)$	$I(S_{1,2}; T_3 \| T_1)$	$I(S_{1,2}; T_3)$	$I(S_{1,2}; T_1 \| T_3)$
$\widetilde{UI}$, I_{red}, S_{VK}	$U(S_1 \backslash S_2 \rightarrow T_{1,3}) = 1$ $U(S_2 \backslash S_1 \rightarrow T_{1,3}) = 1$	$U(S_1 \backslash S_2 \rightarrow T_1) = 1$	$U(S_2 \backslash S_1 \rightarrow T_3 \| T_1) = 1$	$C(S_1, S_2 \rightarrow T_3) = 1$	$R(S_1, S_2 \rightarrow T_1 \| T_3) = 1$
$R_{\min}$	$R(S_1, S_2 \rightarrow T_{1,3}) = 1$ $C(S_1, S_2 \rightarrow T_{1,3}) = 1$	$U(S_2 \backslash S_1 \rightarrow T_1) = -1$ $R(S_1, S_2 \rightarrow T_1) = 1$ $C(S_1, S_2 \rightarrow T_1) = 1$	$U(S_2 \backslash S_1 \rightarrow T_3 \| T_1) = 1$	$C(S_1, S_2 \rightarrow T_3) = 1$	$R(S_1, S_2 \rightarrow T_1 \| T_3) = 1$

5.6. Summary of Key Properties

The following are the key properties of the PPID using the specificity and ambiguity. Property 1 follows directly from the Definitions 1 and 2. Property 2 follows from Theorems 3 and 4. Property 3 follows from the probability distribution Tʙᴄ in Section 5.5. Property 4 was discussed in Section 4.2. Property 5 is proved in Theorem 5.

Property 1. *When considering the redundancy between the source events $a_1, \ldots, a_k$, at least one source event a_i will provide zero unique specificity, and at least one source event a_j will provide zero unique ambiguity. The events a_i and a_j are not necessarily the same source event.*

Property 2. *The atoms of partial specificity and partial ambiguity satisfy local positivity, $\pi^\pm \geq 0$. However, upon recombination and averaging, the atoms of partial information do not satisfy local positivity, $\Pi \geq 0$.*

Property 3. *The decomposition does not satisfy the identity property.*

Property 4. *The decomposition does not satisfy the target monotonicity property.*

Property 5. *The decomposition satisfies the target chain rule.*

6. Conclusions

The partial information decomposition of Williams and Beer [12] provided an intriguing framework for the decomposition of multivariate information. However, it was not long before "serious flaws" ([11], p. 2163) were identified. Firstly, the measure of redundant information $I_{\min}$ failed to distinguish between whether predictor variables provide the same information or merely the same amount of information. Secondly, $I_{\min}$ fails to satisfy the target chain rule, despite this addativity being one of the defining characteristics of information. Notwithstanding the problems, the axiomatic derivation of the redundancy lattice was too elegant to be abandoned and hence several alternate measures were proposed, i.e., I_{red}, $\widetilde{UI}$ and S_{VK} [6,11,12]. Nevertheless, as these measures all satisfy the identity property, they cannot produce a non-negative decomposition for an arbitrary number of variables [13]. Furthermore, none of these measures satisfy the target chain rule meaning they produce inconsistent decompositions for multiple target variables. Finally, in spite of satisfying the identity property (which many consider to be desirable), these measures still fail to identify when variables provide the same information, as exemplified by the pointwise unique problem presented in Section 2.

This paper took the axiomatic derivation of the redundancy lattice from PID and applied it to the unsigned entropic components of the pointwise mutual information. This yielded two separate redundancy lattices—the specificity and the ambiguity lattices. Then based upon an operational interpretation of redundancy, measures of pointwise redundant specificity $r^+_{\min}$ and pointwise redundant ambiguity r_{min}^- were defined. Together with specificity and ambiguity lattices, these measures were used to decompose multivariate information for an arbitrary number of variables. Crucially, upon recombination, the measure r_{min} satisfies the target chain rule. Furthermore, when applied to PWUNQ, these measures do not result in the pointwise unique problem. In our opinion, this demonstrates that the decomposition is indeed correctly identifying redundant information. However, others will likely disagree with this point given that the measure of redundancy does not satisfy the identity property. According to the identity property, independent variables can never provide the same information. In contrast, according to the operational interpretation adopted in this paper, independent variables can provide the same information if they happen to provide the same exclusions with respect to the two-event target distribution. In any case, the proof of Theorem 6 and the subsequent discussion in Appendix B.3, highlights the difficulties that the identity property introduces when considering the information provided about events in separate target variables. (See further discussion in Appendix A.3).

Our future work with this decomposition will be both theoretical and empirical. Regarding future theoretical work, given that the aim of information decomposition is to derive measures pertaining to sets of random variables, it would be worthwhile to derive the information decomposition from first principles in terms of measure theory. Indeed, such an approach would surely eliminate the semantic arguments (about what it means for information to unique, redundant or complementary), which currently plague the problem domain. Furthermore, this would certainly be a worthwhile exercise before attempting to generalise the information decomposition to continuous random variables. Regarding future empirical work, there are many rich data sets which could be decomposed using this decomposition including financial time-series and neural recordings, e.g., [28,33,34].

Acknowledgments: Joseph T. Lizier was supported through the Australian Research Council DECRA grant DE160100630. We thank Mikhail Prokopenko, Richard Spinney, Michael Wibral, Nathan Harding, Robin Ince, Nils Bertschinger, and Nihat Ay for helpful discussions relating to this manuscript. We also thank the anonymous reviewers for their particularly detailed and helpful feedback.

Author Contributions: C.F. and J.L. conceived the idea; C.F. designed, wrote and analyzed the computational examples; C.F. and J.L. wrote the manuscript.

Conflicts of Interest: The authors declare no conflict of interest. The founding sponsors had no role in the design of the study; in the collection, analyses, or interpretation of data; in the writing of the manuscript, and in the decision to publish the results.

Appendix A. Kelly Gambling, Axiom 4, and TBC

In Section 3.3, it was argued that the information provided by a set of predictor events $s_1, \ldots, s_k$ about a target event t is the same information if each source event induces the same exclusions with respect to the two-event partition $\mathcal{T}^t = \{t, t^c\}$. This was based on the fact that pointwise mutual information does not depend on the apportionment of the exclusions across the set of events which did not occur t^c. It was argued that since the pointwise mutual information is independent of these differences, the redundant mutual information should also be independent of these differences. This requirement was then integrated into the operational interpretation of Section 3.3 and was later enshrined in the form of Axiom 4. This appendix aims to justify this operational interpretation and argue why redundant information in TBC is not "unreasonably large" ([5], p. 269).

Appendix A.1. Pointwise Side Information and the Kelly Criterion

Consider a set of horses $\mathcal{T}$ running in a race which can be considered a random variable T with distribution $P(T)$. Say that for each $t \in \mathcal{T}$ a bookmaker offers odds of $o(t)$-for-1, i.e., the bookmaker will pay out $o(t)$ dollars on a \$1 bet if the horse t wins. Furthermore, say that there is no track take as $\sum_{t \in \mathcal{T}} 1/o(t) = 1$, and these odds are *fair*, i.e., $o(t) = 1/p(t)$ for all $t \in \mathcal{T}$ [40]. Let $\boldsymbol{b}(T)$ be the fraction of a gambler's capital bet on each horse $t \in \mathcal{T}$ and assume that the gambler stakes all of their capital on the race, i.e., $\sum_{t \in \mathcal{T}} b(t) = 1$.

Now consider an i.i.d. series of these races $T_1, T_2, \ldots$ such that $P(T_k) = P(T)$ for all $k \in \mathbb{N}$ and let $t_k \in \mathcal{T}$ represent the winner of the k-th race. Say that the bookmaker offers the same odds on each race and the gambler bets their entire capital on each race. The gambler's capital after m races D_m is a random variable which depends on two factors per race: the amount the gambler staked on each race winner t_k, and the odds offered on each winner t_k. That is,

$$D_m = \prod_{k=1}^{m} b(t_k)\, o(t_k), \tag{A1}$$

where monetary units $ have been chosen such that $D_0 = \$1$. The gambler's wealth grows (or shrinks) exponentially, i.e.,

$$D_m = 2^{m\, W(b,T)} \tag{A2}$$

where

$$W(b, T) = \frac{1}{m} \log D_m = \frac{1}{m} \sum_{k=1}^{m} \log b(t_k)\, o(t_k) = \mathrm{E}\big[\log b(t_k)\, o(t_k)\big] \tag{A3}$$

is the doubling rate of the gambler's wealth using a betting strategy $b(T)$. Here, the last equality is by the weak law of large numbers for large m.

Any reasonable gambler would aim to use an optimal strategy $b^*(T)$ which maximises the doubling rate $W(b, T)$. Kelly [40,43] proved that the optimal doubling rate is given by

$$W^*(T) = \max_b W(b, T) = \mathrm{E}\big[\log b^*(t_k)\, o(t_k)\big] \tag{A4}$$

and is achieved by using the proportional gambling scheme $b^*(T) = P(T)$. When the race T_k occurs and the horse t_k wins, the gambler will receive a payout of $b^*(t^k)\, o(t^k) = \$1$, i.e., the gambler receives their stake back regardless of the outcome. In the face of fair odds, the proportional Kelly betting scheme is the optimal strategy—non-terminating repeated betting with any other strategy will result in losses.

Now consider a gambler with access to a private wire S which provides (potentially useful) side information about the upcoming race. Say that these messages are selected from the set $\mathcal{S}$, and that the gambler receives the message s_k before the race T_k. Kelly [40,43] showed that the optimal doubling rate in the presence of this side information is given by

$$W^*(T|S) = \max_b W(b, T|S) = \mathrm{E}\big[\log b^*(t_k|s_k)\, o(t_k)\big], \tag{A5}$$

and is achieved by using the conditional proportional gambling scheme $b^*(T|s_k) = P(T|s_k)$. Both the proportional gambling scheme $b^*(T)$ and the conditional proportional gambling scheme $b^*(T|S)$ are based upon the *Kelly criterion* whereby bets are apportioned according to the best estimation of the outcome available. The financial value of the private wire to a gambler can be ascertained by comparing their doubling rate of the gambler with access to the side wire to that of a gambler with no side information, i.e.,

$$\Delta W = W^*(T|S) - W^*(T) = \mathrm{E}\big[\log b^*(t_k|s_k)\, o(t_k)\big] - \mathrm{E}\big[\log b^*(t_k)\, o(t_k)\big]$$
$$= \mathrm{E}\big[i(s_k; t_k)\big] = I(S; T). \tag{A6}$$

This important result due to Kelly [40] equates the increase in the doubling rate ΔW due to the presence of side information, with the mutual information between the private wire S and the horse race T. If on average, the gambler receives 1 bit of information from their private wire, then on average the gambler can expect to double their money per race. Furthermore, as one would expect, independent side information does not increase the doubling rate.

With no side information, the Kelly gambler always received their original stake back from the bookmaker. However, this is not true for the Kelly gambler with side information. Although their doubling rate is greater than or equal to that of the gambler with no side information, this is only true *on average*. Before the race T_k, the gambler receives the private wire message s_k and then, the horse t_k wins the race. From (A6), one can see that the return Δw_k for the k-th race is given by the pointwise mutual information,

$$\Delta w = i(s_k; t_k). \tag{A7}$$

Hence, just like the pointwise mutual information, the per race return can be positive or negative: if it is positive, the gambler will make a profit; if it is negative, the gambler will sustain a loss. Despite the potential for pointwise loses, the average return (i.e., the doubling rate) is, just like the average mutual information, non-negative—and indeed, is optimal. Furthermore, while a Kelly gambler with side information can lose money on any single race, they can never actually go bust. The Kelly gambler with side information s still hedges their risk by placing bets on all horses with a non-zero probability of winning according to their side information, i.e., according to $P(T|s_k)$. The only reason they would fail to place a bet on a horse is if their side information completely precludes any possibility of that horse winning. That is, a Kelly gambler with side information will never fall foul of gambler's ruin.

Appendix A.2. Justification of Axiom 4 and Redundant Information in TBC

Consider TBC semantically described in terms of a horse race. That is, consider a four horse race T where each horse has an equiprobable chance of winning, and consider the binary variables T_1, T_2, and T_3 which represent the following, respectively: the colour of the horse, black 0 or white 1; the sex of the jockey, female 0 or male 1; and the colour of the jockey's jersey, red 0 or green 1. Say that the four horses have the following attributes:

Horse 0 is a black horse $T_1 = 0$, ridden by a female jockey $T_2 = 0$, who is wearing a red jersey $T_3 = 0$.
Horse 1 is a black horse $T_1 = 0$, ridden by a male jockey $T_2 = 1$, who is wearing a green jersey $T_3 = 1$.
Horse 2 is a white horse $T_1 = 1$, ridden by a female jockey $T_2 = 0$, who is wearing a green jersey $T_3 = 1$.
Horse 3 is a white horse $T_1 = 1$, ridden by a male jockey $T_2 = 1$, who is wearing a red jersey $T_3 = 0$.

There are two important points to note. Firstly, the horses in the race T could also be uniquely described in terms of the composite binary variables $T_{1,2}$, $T_{1,3}$ or $T_{2,3}$. Secondly, if one knows T_1 and T_2 then one knows T_3 (which can be represented by the relationship $T_3 = T_1$ XOR T_2). Finally, consider private wires S_1 and S_2 which *independently* provide the colour of the horse and the colour of the jockey's jersey (respectively) before the upcoming race, i.e., $S_1 = T_1$ and $S_2 = T_2$.

Now say a bookmaker offers fair odds of 4-for-1 on each horse in the race T. Consider two gamblers who each have access to one of S_1 and S_2. Before each race, the two gamblers receive their respective private wire messages and place their bets according to the Kelly strategy. This means that each gambler lays half of their, say $1, stake on each of their two respective non-excluded horses: unknowingly, both of the gamblers have placed a bet on the soon-to-be race winner, and each gambler has placed a distinct bet on one of the two soon-to-be losers. The only horse neither has bet upon is also a soon-to-be loser. (See [5] for a related description of TBC in term of the game-theoretic notions of shared and common knowledge). After the race, the bookmaker pays out $2 dollars to each gambler: both have doubled their money. This happens because both of the gamblers had one bit of 1 bit of information about the race, i.e., pointwise mutual information. In particular, both gamblers improved their probability of predicting the eventual race winner. It did not matter, in any way, that the gamblers had each laid distinct bets on one of the three eventual race losers. The fact that they laid different bets on the horses which did not win, made no difference to their winnings. The apportionment of the exclusions across the set of events which did not occur, makes no difference to the pointwise mutual information. With respect to what occurred (i.e., with respect to which horse won), the fact the that they excluded different losers is only semantic. When it came to predicting the would-be-winner, both gamblers had the same predictive power; they both had the same freedom of choice with regards to selecting what would turn out to be the eventual race winner—they had the same information. It is for this reason that this information should be regarded as redundant information, regardless of the independence of the information sources. Hence, the introduction of both the operational interpretation of redundancy in Section 3.3 and Axiom 4 in Section 4.2.

Now consider a third gambler who has access to both private wires S_1 and S_2, i.e., $S_{1,2}$. Before the race, this gambler receives both private wire messages which, in total, precludes three of the horses from winning. This gambler then places the entirety of their $1 stake on the remaining horse which is sure to win. After the race, the bookmaker pays out $4: this gambler has quadrupled their money as they had 2 bit of pointwise mutual information about the race. Having both private wire messages simultaneously gave this gambler a 1 bit informational edge over the two gamblers with access to a single side wire. While each of the singleton gamblers had 1 bit of independent information, the only way one could profit from the independence of this information is by having both pieces of information simultaneously—this makes this 1 bit of information complementary. Although this may seem "palpably strange" ([12], p. 167), it is not so strange when from the following perspective: the only way to exploit two pieces of independent information is by having both pieces together simultaneously.

Appendix A.3. Accumulator Betting and the Target Chain Rule

Say that in addition to the 4-for-1 odds offered on the race T, the bookmaker also offers fair odds of 2-for-1 on each of the binary variables T_1, T_2 and T_3. Now, in addition to being able to directly gamble on the race T, one could indirectly gamble on T by placing a so-called *accumulator* bet on any pair of T_1, T_2 and T_3. An accumulator is a series of chained bets whereby any return from one bet is automatically staked on the next bet; if any bet in the chain is lost then the entire chain is lost. For example, a gambler could place 4-for-1 bet on horse 0 by placing the following accumulator bet: a 2-for1 bet on a black horse winning that chains into a 2-for-1 bet on the winning jockey being female (or equivalently, vice versa). In effect, these accumulators enable a gambler to bet on T by instead placing a chained bet on the independent component variables within the (equivalent) joint variables $T_{1,2}$, $T_{1,3}$ and $T_{2,3}$. Now consider again the three gamblers from the prior section, i.e., the two gamblers who each have a private wire S_1 and S_2, and the third gamble who has access to $S_{1,2}$. Say that they must each place a, say $1, accumulator bet on $T_{1,3}$—what should each gambler do according to the Kelly criterion?

For the sake of clarity, consider only the realisation where the horse $T = 0$ subsequently wins (due to the symmetry, the analysis is equivalent for all realisations). First consider the accumulator whereby the gamblers first bet on the colour of the winning horse T_1, which chains into a bet on the colour of the winning jockey's jersey T_3. Suppose that the private wire S_1 communicates that the winning horse will be black, while the private wire S_2 communicates that the winning horse will be ridden by a female jockey, i.e., $S_1 = 0$ and $S_2 = 0$. Following to the Kelly strategy, the gambler with access to $S_1 = 0$ takes out two $0.5 accumulator bets. Both of these accumulators feature the same initial bet on the winning horse being black since $T_1 = S_1 = 0$. Hence both bets return $1 each which become the stake on the next bet in each accumulator. This gambler knows nothing about the colour of the jockey's jersey T_3. As such, one accumulator chains into a bet on the winning jersey being red $T_3 = 0$, while the other chains into a bet on it being green $T_3 = 1$. When the horse $T = 0$ wins, the stake bet on the green jersey is lost while bet on red jersey pays out $2. This gambler had 1 bit of side information and so doubled their money. Now consider the gambler with private wire S_2, who knows nothing about T_1 or T_3 individually. Nonetheless, this gambler knows that the winner must be a female jockey $T_2 = 0$. As such, this gambler knows that if a black horse $T_1 = 0$ wins then its jockey must be wearing a red jersey $T_3 = 0$, or if a white horse $T_1 = 0$ wins then its jockey must be wearing a green jersey $T_3 = 1$ (since $T_3 = T_1$ XOR T_2). Thus this gambler can also utilise the Kelly strategy to place the following two $0.5 accumulator bets: the first accumulator bets on the winning horse being black $T_1 = 0$ and then chains into a bet on the winner's jersey being red $T_3 = 0$, while the second accumulator bets on the winning horse being white $T_1 = 1$ and then chains into a bet on the winner's jersey being green

$T_3 = 1$. When the horse $T = 0$ wins, the first accumulator pays out \$2, while the second accumulator is be lost. Hence, this gambler also doubles their money and so also had 1 bit of side information. Finally, consider the gambler with access to both private wires $S_{1,3}$, who can place an accumulator on the black horse $T_1 = 0$ winning chaining into a bet on the winning jockey wearing red $T_3 = 0$. This gambler can quadruple their stake, and so must possess 2 bit of side information.

Each of the three gamblers have the same final return regardless of whether the gamblers are betting on the variable T, or placing accumulator bets on the variables $T_{1,2}$, $T_{1,3}$ or $T_{2,3}$. However, the paths to the final result differs between the gamblers, reflecting the difference between the information the each gambler had about the sub-variables T_1, T_2 or T_3. Given the result of Kelly [40], the proposed information decomposition should reflect these differences, but yet still arrive at the same result—in other words, the information decomposition should satisfy a target chain rule. This is clear if the Kelly interpretation of information is to remain as a "duality" ([43], p. 159) in information theory.

Appendix B. Supporting Proofs and Further Details

This appendix contains many of the important theorems and proofs relating to PPID using specificity and ambiguity.

Appendix B.1. Deriving the Specificity and Ambiguity Lattices from Axioms 1–4

The following section is based directly on the original work of Williams and Beer [1,2]. The difference is that we now consider sources events a_i rather than sources A_i.

Proposition A1. *Both $i_\cap^+$ and $i_\cap^-$ are non-negative.*

Proof. Since $\emptyset \subseteq a_i$ for any a_i, Axioms 2 and 3 imply

$$i_\cap^+(a_1,\ldots,a_k \to t) \geq i_\cap^+(a_1,\ldots,a_k,\emptyset \to t) = i_\cap^+(\emptyset \to t) = h(\emptyset) \quad = 0, \tag{A8}$$

$$i_\cap^-(a_1,\ldots,a_k \to t) \geq i_\cap^-(a_1,\ldots,a_k,\emptyset \to t) = i_\cap^-(\emptyset \to t) = h(\emptyset|t) = 0. \tag{A9}$$

Hence, both $i_\cap^+(a_1,\ldots,a_k \to t)$ and $i_\cap^-(a_1,\ldots,a_k \to t)$ are non-negative. □

Proposition A2. *Both $i_\cap^+$ and $i_\cap^-$ are bounded from above by the specificity and the ambiguity from any single source event, respectively.*

Proof. For any single source a_i, Axioms 2 and 3 yield

$$h(a_i)= i_\cap^+(a_i \to t)= i_\cap^+(a_i, a_i \to t)\geq i_\cap^+(a_i,\ldots \to t), \tag{A10}$$

$$h(a_i|t)= i_\cap^-(a_i \to t)= i_\cap^-(a_i, a_i \to t)\geq i_\cap^+(a_i,\ldots \to t), \tag{A11}$$

as required. □

In keeping with Williams and Beer's approach [1,2], consider all of the distinct ways in which a collection of source events $a = \{a_1,\ldots,a_k\}$ could contribute redundant information. Thus far we have assumed that the redundancy measure can be applied to any collection of source events, i.e., $\mathscr{P}_1(a)$ where $\mathscr{P}_1$ denotes the power set with the empty set removed. Recall that the sources events are themselves collections of predictor events, i.e., $\mathscr{P}_1(s)$. That is, we can apply both $i_\cap^+$ and $i_\cap^-$ to elements of $\mathscr{P}_1(\mathscr{P}_1(s))$. However, this can be greatly reduced using Axiom 2 which states that if $a_i \subseteq a_j$, then

$$i_\cap^+(a_j, a_i,\ldots \to t) = i_\cap^+(a_i,\ldots \to t), \tag{A12}$$

$$i_\cap^-(a_j, a_i,\ldots \to t) = i_\cap^-(a_i,\ldots \to t). \tag{A13}$$

Hence, one need only consider the collection of source events such that no source event is a superset of any other in order,

$$\mathscr{A}(s) = \left\{ \alpha \in \mathscr{P}_1(\mathscr{P}_1(s)) \mid \forall\, a_i,\, a_j \in \alpha,\, a_i \not\subset a_j \right\}. \tag{A14}$$

This collection $\mathscr{A}(s)$ captures all the distinct ways in the source events could provide redundant information.

As per Williams and Beer's PID, this set of source events $\mathscr{A}(s)$ is structured. Consider two sets of source events $\alpha, \beta \in \mathscr{A}(s)$. If for every source event $b \in \beta$ there exists a source event $a \in \alpha$ such that $a \subseteq b$, then all of the redundant specificity and ambiguity shared by $b \in \beta$ must include any redundant specificity and ambiguity shared by $a \in \alpha$. Hence, a partial order $\preceq$ can be defined over the elements of the domain $\mathscr{A}(s)$ such that any collection of predictors event coalitions precedes another if and only if the latter provides any information the former provides,

$$\forall \alpha, \beta \in \mathscr{A}(s),\, (\alpha \preceq \beta \iff \forall b \in \beta,\, \exists a \in \alpha \mid a \subseteq b). \tag{A15}$$

Applying this partial ordering to the elements of the domain $\mathscr{A}(s)$ produces a lattice which has the same structure as the redundancy lattice from PID, i.e., the structure of the sources events here is the same as the structure of the sources in PID. (Figure 3 depicts this structure for the case of 2 and 3 predictor variables.) Applying $i_\cap^+$ to these sources events yields a *specificity lattice* while applying $i_\cap^-$ yields an *ambiguity lattice*.

Similar to $I_\cap$ in PID, the redundancy measures $i_\cap^+$ or $i_\cap^-$ can be thought of as a cumulative information functions which integrate the specificity or ambiguity uniquely contributed by each node as one moves up each lattice. In order in evaluate the unique contribution of specificity and ambiguity from each node in the lattice, consider the Möbius inverse [45,46] of $i_\cap^+$ and $i_\cap^-$. That is, the specificity and ambiguity of a node α is given by

$$i_\cap^\pm(\alpha \to t) = \sum_{\beta \preceq \alpha} i_\partial^\pm(\beta \to t) \qquad \forall\, \alpha,\, \beta \in \mathscr{A}(s). \tag{A16}$$

Thus the unique contributions of *partial specificity* i_∂^+ and *partial ambiguity* i_∂^- from each node can be calculated recursively from the bottom-up, i.e.,

$$i_\partial^\pm(\alpha \to t) = i_\cap^\pm(\alpha \to t) - \sum_{\beta \prec \alpha} i_\partial^\pm(\beta \to t). \tag{A17}$$

Theorem A1. *Based on the principle of inclusion-exclusion, we have the following closed-from expression for the partial specificity and partial ambiguity,*

$$i_\partial^\pm(\alpha \to t) = i_\cap^\pm(\alpha \to t) - \sum_{\varnothing \neq \gamma \subseteq \alpha^-} (-1)^{|\gamma|-1}\, i_\cap^\pm(\bigwedge \gamma \to t) \tag{A18}$$

Proof. For $\mathscr{B} \subseteq \mathscr{A}(s)$, define the sub-additive function $f^\pm(\mathscr{B}) = \sum_{\beta \in \mathscr{B}} = i^\pm(\beta \to t)$. From (A16), we get that $i_\cap^\pm(\alpha \to t) = f^\pm(\downarrow \alpha)$ and

$$i_\partial^\pm(\alpha \to t) = f^\pm(\downarrow \alpha) - f^\pm(\dot{\downarrow} \alpha) = f^\pm(\downarrow \alpha) - f^\pm(\bigcup_{\beta \in \alpha^-} \downarrow \beta). \tag{A19}$$

By the principle of inclusion-exclusion (e.g., see [46], p. 195) we get that

$$i_\partial^\pm(\alpha \to t) = f^\pm(\downarrow \alpha) - \sum_{\varnothing \neq \gamma \subseteq \alpha^-} (-1)^{|\gamma|-1}\, f^\pm(\bigcap_{\beta \in \gamma} \beta) \tag{A20}$$

For any lattice L and $A \subseteq L$, we have that $\bigcap_{a \in A} \downarrow a = \downarrow (\bigwedge A)$ (see [47], p. 57) thus

$$i_{\partial}^{\pm}(\alpha \rightarrow t) = f^{\pm}(\downarrow \alpha) - \sum_{\varnothing \neq \gamma \subseteq \alpha^-} (-1)^{|\gamma|-1} f^{\pm}(\bigwedge \gamma)$$

$$= f^{\pm}(\downarrow \alpha) - \sum_{\varnothing \neq \gamma \subseteq \alpha^-} (-1)^{|\gamma|-1} i^{\pm}(\bigwedge \gamma \rightarrow t) \tag{A21}$$

as required. $\square$

Similarly to PID, the specificity and ambiguity lattices provide a structure for information decomposition—unique evaluation requires a separate definition of redundancy. However, unlike PID (or even PPID), this evaluation requires both a definition of pointwise redundant specificity and pointwise redundant ambiguity.

Appendix B.2. Redundancy Measures on the Lattices

In Section 4.2, Definitions 1 and 2 provided the require measures. This section will prove some of the key properties of these measures when they are applies to the lattices derived in the previous section. The correspondence with the approach taken by Williams and Beer [1,2] continues in this section. However, sources events a_i are used in place of sources A_i and the measures $r_{min}^{\pm}$ are used in place of I_{min}. Note that the basic concepts from lattice theory and the notion used here are the same as found in ([1], Appendix B).

Theorem 1. *The definitions of r_{min}^{+} and r_{min}^{-} satisfy Axioms 1–4.*

Proof. Axioms 1, 3 and 4 follow trivially from the basic properties of the minimum. The main statement of Axiom 2 also immediately follows from the properties of the minimum; however, there is a need to verify the equality condition. As such, consider a_k such that $a_k \supseteq a_i$ for some $a_i \in \{a_1, \ldots, a_{k-1}\}$. From Postulate 4, we have that $h(a_k) \geq h(a_i)$ and hence that $\min_{a_j \in \{a_1, \ldots, a_k\}} h(a_j) = \min_{a_j \in \{a_1, \ldots, a_{k-1}\}} h(a_j)$, as required for r_{min}^{+}. *Mutatis mutandis*, similar follows for r_{min}^{-}. $\square$

Theorem 2. *The redundancy measures r_{min}^{+} and r_{min}^{-} increase monotonically on the $\langle \mathscr{A}(s), \preceq \rangle$.*

The proof of this theorem will require the following lemma.

Lemma A1. *The specificity and ambiguity $i^{\pm}(a \rightarrow t)$ are increasing functions on the lattice $\langle \mathscr{P}_1(s), \subseteq \rangle$*

Proof. Follows trivially from Postulate 4. $\square$

Proof of Theorem 2. Assume there exists $\alpha, \beta \in \mathscr{A}(s)$ such that $\alpha \prec \beta$ and $r_{min}^{\pm}(\beta \rightarrow t) < r_{min}^{\pm}(\alpha \rightarrow t)$. By definition, i.e., (23) and (24), there exists $b \in \beta$ such that $i^{\pm}(b \rightarrow t) < i^{\pm}(a \rightarrow t)$ for all $a \in \alpha$. Hence, by Lemma A1, there does not exist $a \in \alpha$ such that $a \subseteq b$. However, by assumption $\alpha \prec \beta$ and hence there exists $a \in \alpha$ such that $a \subseteq b$, which is a contradiction. $\square$

Theorem A2. *When using $r_{min}^{\pm}$ in place of the general redundancy measures $i_{\cap}^{\pm}$, we have the following closed-from expression for the partial specificity π^{+} and partial ambiguity π^{-},*

$$\pi^{\pm}(\alpha \rightarrow t) = r_{min}^{\pm}(\alpha \rightarrow t) - \max_{\beta \in \alpha^-} \min_{b \in \beta} i^{\pm}(b \rightarrow t). \tag{A22}$$

Proof. Let $i_{\cap}^{+} = r_{min}^{+}$ and $i_{\cap}^{-} = r_{min}^{-}$ in the general closed form expression for $i_{\partial}^{\pm}$ in Theorem A1,

$$\pi^{\pm}(\alpha \rightarrow t) = r_{min}^{\pm}(\alpha \rightarrow t) - \sum_{\varnothing \neq \gamma \subseteq \alpha^-} (-1)^{|\gamma|-1} \min_{b \in \bigwedge \gamma} i^{\pm}(b \rightarrow t). \tag{A23}$$

Since $\alpha \wedge \beta = \alpha \cup \beta$ (see [1], Equation (23)), and by Postulate 4, we have that

$$\pi^{\pm}(\alpha \to t) = r_{min}^{\pm}(\alpha \to t) - \sum_{\emptyset \neq \gamma \subseteq \alpha^-} (-1)^{|\gamma|-1} \min_{\beta \in \gamma} \min_{b \in \beta} i^{\pm}(b \to t). \tag{A24}$$

By the maximum-minimums identity (see [48]), we have that, $\max \alpha^- = \sum_{\emptyset \neq \gamma \subseteq \alpha^-} (-1)^{|\gamma|-1} \min \gamma$, and hence

$$\pi^{\pm}(\alpha \to t) = r_{min}^{\pm}(\alpha \to t) - \max_{\beta \in \alpha^-} \min_{b \in \beta} i^{\pm}(\alpha \to t). \tag{A25}$$

as required. □

Theorem 3. *The atoms of partial specificity π^+ and partial ambiguity π^- evaluated using the measures r_{min}^+ and r_{min}^- on the specificity and ambiguity lattices (respectively), are non-negative.*

Proof. It $\alpha = \perp$, the $\pi^{\pm}(\alpha \to t) = r_{min}^{\pm} \geq 0$ by the non-negativity of entropy. If $\alpha \neq \perp$, assume there exists $\alpha \in \mathscr{A}(s) \backslash \{\perp\}$ such that $\pi^{\pm}(\alpha \to t) < 0$. By Theorem A2,

$$\pi^{\pm}(\alpha \to t) = \min_{a \in \alpha} i^{\pm}(a \to t) - \max_{\beta \in \alpha^-} \min_{b \in \beta} i^{\pm}(b \to t). \tag{A26}$$

From this it can be seen that there must exist $\beta \in \alpha^-$ such that for all $b \in \beta$, we have that $i^{\pm}(a \to t) < i^{\pm}(b \to t)$ for some $a \in \alpha$. By Postulate 4 there does not exist $b \in \beta$ such that $b \subset a$. However, since by definition, $\beta \prec \alpha$ there exists $b \in \beta$ such that $b \subset a$, which is a contradiction. □

Theorem 4. *The atoms of partial average information Π evaluated by recombining and averaging $\pi^{\pm}$ are not non-negative.*

Proof. The proof is by the counter-example using RDNERR. □

Appendix B.3. Target Chain Rule

By using the appropriate conditional probabilities in Definitions 1 and 2, one can easily obtain the conditional pointwise redundant specificity,

$$r_{min}^+(a_1, \ldots, a_k \to t_1 | t_2) = \min_{a_i} h(a_i | t_2), \tag{A27}$$

or the conditional pointwise redundant ambiguity,

$$r_{min}^-(a_1, \ldots, a_k \to t_1 | t_2) = \min_{a_j} h(a_j | t_{1,2}). \tag{A28}$$

As per (21) these could be recombined, e.g., via (21), to obtain the conditional redundant information,

$$r_{min}(a_1, \ldots, a_k \to t_1 | t_2) = r_{min}^+(a_1, \ldots, a_k \to t_1 | t_2) - r_{min}^-(a_1, \ldots, a_k \to t_1 | t_2). \tag{A29}$$

The relationship between the regular forms and the conditional forms of the redundant specificity and redundant ambiguity has some important consequences.

Proposition A3. *The conditional pointwise redundant specificity provided by $a_1, \ldots, a_k$ about t_1 given t_2 is equal to pointwise redundant ambiguity provided by $a_1, \ldots, a_k$ about t_2 with the conditioned variable,*

$$r_{min}^+(a_1, \ldots, a_k \to t_1 | t_2) = r_{min}^-(a_1, \ldots, a_k \to t_2). \tag{A30}$$

Proof. By (24) and (A27). □

Proposition A4. *The pointwise redundant specificity provided by* $a_1, \ldots, a_k$ *is independent of the target event and even the target variable itself,*

$$r_{min}^+ (a_1, \ldots, a_k \to t_1) = r_{min}^+ (a_1, \ldots, a_k \to t_2) \qquad \forall\, t_1, t_2, T_1, T_2. \tag{A31}$$

Proof. By inspection of (23). □

Proposition A5. *The conditional pointwise redundant ambiguity provided by* $a_1, \ldots, a_k$ *about* t_1 *given* t_2 *is equal to the pointwise redundant ambiguity provided by* $a_1, \ldots, a_k$ *about* $t_{1,2}$,

$$r_{min}^- (a_1, \ldots, a_k \to t_1 | t_2) = r_{min}^- (a_1, \ldots, a_k \to t_{1,2}). \tag{A32}$$

Proof. By (24) and (A28). □

Note that specificity itself is not a function of the target event or variable. Hence, all of the target dependency is bound up in the ambiguity. Now consider the following.

Theorem 5 (Pointwise Target Chain Rule). *Given the joint target realisation* $t_{1,2}$, *the pointwise redundant information* r_{min} *satisfies the following chain rule,*

$$\begin{aligned} r_{min} (a_1, \ldots, a_k \to t_{1,2}) &= r_{min} (a_1, \ldots, a_k \to t_1) + r_{min} (a_1, \ldots, a_k \to t_2 | t_1), \\ &= r_{min} (a_1, \ldots, a_k \to t_2) + r_{min} (a_1, \ldots, a_k \to t_1 | t_2). \end{aligned} \tag{25}$$

Proof. Starting from r_{min}, by Corollary A4 and Corollary A5 we get that

$$\begin{aligned} r_{min} (a_1, \ldots, a_k \to t_{1,2}) &= r_{min}^+ (a_1, \ldots, a_k \to t_{1,2}) - r_{min}^- (a_1, \ldots, a_k \to t_{1,2}), \\ &= r_{min}^+ (a_1, \ldots, a_k \to t_1) - r_{min}^- (a_1, \ldots, a_k \to t_2 | t_1), \end{aligned} \tag{A33}$$

Then, by Corollary A3 we get that

$$\begin{aligned} r_{min} (a_1, \ldots, a_k \to t_{1,2}) &= r_{min}^+ (a_1, \ldots, a_k \to t_1) - r_{min}^- (a_1, \ldots, a_k \to t_1) \\ &\quad + r_{min}^- (a_1, \ldots, a_k \to t_1) - r_{min}^- (a_1, \ldots, a_k \to t_2 | t_1), \\ &= r_{min}^+ (a_1, \ldots, a_k \to t_1) - r_{min}^- (a_1, \ldots, a_k \to t_1) \\ &\quad + r_{min}^+ (a_1, \ldots, a_k \to t_2 | t_1) - r_{min}^- (a_1, \ldots, a_k \to t_2 | t_1), \\ &= r_{min} (a_1, \ldots, a_k \to t_1) + r_{min} (a_1, \ldots, a_k \to t_2 | t_1), \end{aligned} \tag{A34}$$

as required for the first equality in (25). *Mutatis mutandis*, we obtain the second equality in (25). □

Theorem 6. *The target chain rule, identity property and local positivity, cannot be simultaneously satisfied.*

Proof. Consider the probability distribution TBC, and in particular, the isomorphic probability distributions $P(T_{1,2})$ and $P(T_{1,3})$. By the identity property,

$$U(S_1 \backslash S_2 \to T_{1,2}) = 1 \text{ bit}, \qquad U(S_2 \backslash S_1 \to T_{1,2}) = 1 \text{ bit}, \tag{A35}$$

and hence, $R(S_1, S_2 \to T_{1,2}) = 0$ bit. On the other hand, by local positivity,

$$C(S_1, S_2 \to T_3) = 1 \text{ bit}, \qquad R(S_1, S_2 \to T_1 | T_3) = 1 \text{ bit} \tag{A36}$$

Then by the target chain rule,

$$C(S_1, S_2 \to T_{1,3}) = 1 \text{ bit} \qquad R(S_1, S_2 \to T_{1,3}) = 1 \text{ bit}, \tag{A37}$$

Finally, since $P(T_{1,2})$ is isomorphic to $P(T_{1,3})$ we have that, $R(S_1, S_2 \rightarrow T_{1,3}) = R(S_1, S_2 \rightarrow T_{1,2})$, which is a contradiction. □

Theorem 6 can be informally generalised as follows: it is not possible to simultaneously satisfy the target chain rule, the identity property, and have only $C(S_1, S_2 \rightarrow T) = 1$ bit in the probability distribution XOR without having negative (average) PI atoms in probability distributions where there is no ambiguity from any source. To see this, again consider decomposing the isomorphic probability distributions $P(T_{1,2})$ and $P(T_{1,3})$. In line with (A35), decomposing $T_{1,2}$ via the identity property yields $C(S_1, S_2 \rightarrow T_{1,2}) = 0$ bit. On the other hand, decomposing $T_{1,3}$ yields $C(S_1, S_2 \rightarrow T_3) = 1$ bit. Since $P(T_{1,2})$ is isomorphic to $P(T_{1,3})$, the target chain rule requires that,

$$C(S_1, S_2 \rightarrow T_1 | T_3) = -1 \text{ bit}, \qquad U(S_1 \backslash S_2 \rightarrow T_1 | T_3) = 1 \text{ bit}, \qquad U(S_2 \backslash S_1 \rightarrow T_1 | T_3) = 1 \text{ bit}. \qquad (A38)$$

That is, one would have to accept the negative (average) PI atom $C(S_1, S_2 \rightarrow T_1 | T_3) = -1$ bit despite the fact that there are no non-zero pointwise ambiguity terms upon splitting any of $i(s_1; t_1 | t_3)$, $i(s_2; t_1 | t_3)$ and $i(s_{1,2}; t_1 | t_3)$ into specificity and ambiguity. Although this does not constitute a formal proof that the identity property is incompatible with the target chain rule, one would have to accept and find a way to justify $C(S_1, S_2 \rightarrow T_1 | T_3) = -1$ bit. Since there is no ambiguity in $i(s_1; t_1 | t_3)$, $i(s_2; t_1 | t_3)$ and $i(s_{1,2}; t_1 | t_3)$, this result is not reconcilable within the framework of specificity and ambiguity.

Appendix C. Additional Example Probability Distributions

Appendix C.1. Probability Distribution TBEP

Figure A1 shows the probability distribution three bit–even parity (TBEP) which considers binary predictors variables S_1, S_2 and S_3 which are constrained such that together their parity is even. The target variable T is simply a copy of the predictors, i.e., $T = T_{1,2,3} = (T_1, T_2, T_3)$ where $T_1 = S_1$, $T_2 = S_2$ and $T_3 = S_3$. (Equivalently, the target can be represented by any four state variable T.) It was introduced by Bertschinger et al. [5] and revisited by Rauh et al. [13] who (as mentioned in Section 5.5) used it to prove the following by counter-example: there is no measure of redundant average information for more than two predictor variables which simultaneously satisfies the Williams and Beer Axioms, the identity property, and local positivity. The measures I_{red}, $\widetilde{UI}$ and S_{VK} these properties. Hence, this probability distribution which has been used to demonstrate that these measures are not consistent with the PID framework in the general case of an arbitrary number of predictor variables.

This example is similar to TBC in the several ways. Firstly, due to the symmetry in the probability distribution, each realisation will have the same pointwise decomposition. Secondly, there is an isomorphism between the probability distributions $P(T)$ and $P(S_1, S_2, S_3)$, and hence the pointwise ambiguity provided by any (individual or joint) predictor event is 0 bit (since given t, one knows s_1, s_2 and s_3). Thirdly, the individual predictor events s_1, s_2 and s_3 each exclude 1/2 of the total probability mass in $P(T)$ and so each provide 1 bit of pointwise specificity. Thus, there is 1 bit of three-way redundant, pointwise specificity in each realisation. Fourthly, the joint predictor event $s_{1,2,3}$ excludes 3/4 of the total probability mass, providing 2 bit of pointwise specificity (which is similar to TBC). However, unlike TBC, one could consider the three joint predictor events $s_{1,2}$, $s_{1,3}$ and $s_{2,3}$. These joint pairs also exclude 3/4 of the total probability mass each, and hence also each provide 2 bit of pointwise specificity. As such, there is 1 bit of pointwise, three-way redundant, pairwise complementary specificity between these three joint pairs of source events, in addition to the 1 bit of three-way redundant, pointwise specificity. Finally, putting this together and averaging over all realisations, TBEP consists of 1 bit of three-way redundant information and 1 bit of three-way redundant, pairwise complementary information. The resultant average decomposition is the same as the decomposition induced by I_{min} [5].

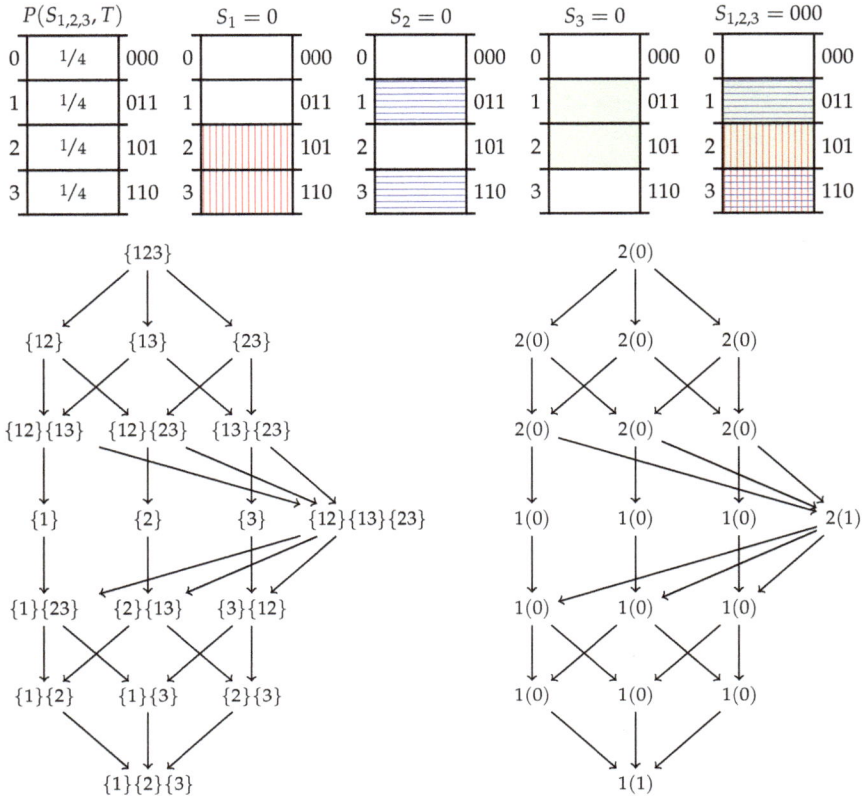

Figure A1. Example TBEP. (**Top**) probability mass diagram for realisation ($S_1=0, S_2=0, S_3=0, T=000$); (**Bottom left**) With three predictors, it is convenient to represent to decomposition diagrammatically. This is especially true TBEP as one only needs to consider the specificity lattice for one realisation; (**Bottom right**) The specificity lattice for the realisation ($S_1=0, S_2=0, S_3=0, T=000$). For each source event the left value corresponds to the value of $i_\cap^+$, evaluated using $r_{\min}^+$, while the right value (surrounded by parenthesis) corresponds to the partial information π^+.

Appendix C.2. Probability Distribution UNQ

Figure A2 shows the decomposition of the probability distribution *unique* (UNQ). Note that this probability distribution corresponds to RDNERR where the error probability $\varepsilon = 1/2$, and hence the similarity in the resultant distributions. The results may initially seem unusual, that the predictor S_1 is not uniquely informative since $U(S_1 \backslash S_2 \to T) = 0$ bit as one might intuitively expect. Rather it is deemed to be redundantly informative $RI = 1$ bit with the predictor S_2 which is also uniquely misinformative $U(S_2 \backslash S_1 \to T) = -1$ bit. This is because both S_1 and S_2 provide $I^+(S_1 \to T) = I^+(S_2 \to T) = 1$ bit of specificity; however the information provided by S_2 is unique in that the 1 bit provided is not "useful" ([42], p. 21) and hence $I(S_2 \to T) = 1$ bit while $I(S_2 \to T) = 1$ bit. Finally, the complementary information $C(S_1, S_2 \to T) = 1$ bit is required by the decomposition in order to balance this 1 bit of unique ambiguity. The results in this example partly explain our preference for term *complementary information* as opposed to *synergistic information*—while $C(S_1, S_2 \to T) = 1$ bit is readily explainable, it would be dubious to refer to this as *synergy* given that S_1 enables perfect predictions of T without any knowledge of S_2.

Entropy **2018**, *20*, 297

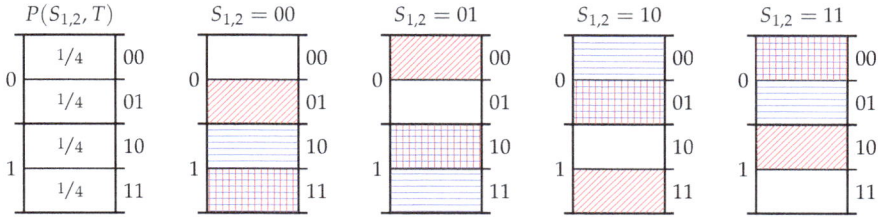

p	s_1	s_2	t	i_1^+	i_1^-	i_2^+	i_2^-	i_{12}^+	i_{12}^-	r^+	u_1^+	u_2^+	c^+	r^-	u_1^-	u_2^-	c^-
1/4	0	0	0	1	0	1	1	2	1	1	0	0	1	0	0	1	0
1/4	0	1	0	1	0	1	1	2	1	1	0	0	1	0	0	1	0
1/4	1	0	1	1	0	1	1	2	1	1	0	0	1	0	0	1	0
1/4	1	1	1	1	0	1	1	2	1	1	0	0	1	0	0	1	0
Expected values				1	0	1	1	2	1	1	0	0	1	0	0	1	0

$R(S_1, S_2 \rightarrow T) = 1$ bit $\quad U(S_1 \backslash S_2 \rightarrow T) = 0$ bit $\quad U(S_2 \backslash S_1 \rightarrow T) = -1$ bit $\quad C(S_1, S_2 \rightarrow T) = 1$ bit

Figure A2. Example UNQ. (**Top**) the probability mass diagrams for every single possible realisation; (**Middle**) for each realisation, the PPID using specificity and ambiguity is evaluated (see Figure 4); (**Bottom**) the atoms of (average) partial infromation obtained through recombination of the averages.

Appendix C.3. Probability Distribution AND

Figure A3 shows the decomposition of the probability distribution *and* (AND). Note that the probability distribution *or* (OR) has the same decomposition as the target distributions are isomorphic.

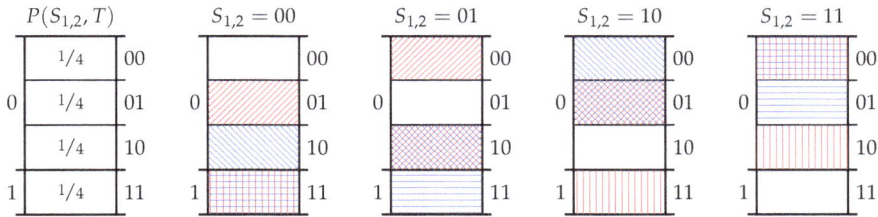

p	s_1	s_2	t	i_1^+	i_1^-	i_2^+	i_2^-	i_{12}^+	i_{12}^-	r^+	u_1^+	u_2^+	c^+	r^-	u_1^-	u_2^-	c^-
1/4	0	0	0	1	lg 3/2	1	lg 3/2	2	lg 3	1	0	0	1	lg 3/2	0	0	1
1/4	0	1	0	1	lg 3/2	1	lg 3	2	lg 3	1	0	0	1	lg 3/2	0	1	0
1/4	1	0	0	1	lg 3	1	lg 3/2	2	lg 3	1	0	0	1	lg 3/2	1	0	0
1/4	1	1	1	1	0	1	0	2	0	1	0	0	1	0	0	0	0
Expected values				1	0.689	1	0.689	2	1.189	1	0	0	1	0.439	0.250	0.250	0.25

$R(S_1, S_2 \rightarrow T) = 0.561$ bit $\quad U(S_1 \backslash S_2 \rightarrow T) = -0.25$ bit $\quad U(S_2 \backslash S_1 \rightarrow T) = -0.25$ bit $\quad C(S_1, S_2 \rightarrow T) = 0.75$ bit

Figure A3. Example AND. (**Top**) the probability mass diagrams for every single possible realisation; (**Middle**) for each realisation, the PPID using specificity and ambiguity is evaluated (see Figure 4); (**Bottom**) the atoms of (average) partial infromation obtained through recombination of the averages.

References and Note

1. Williams, P.L.; Beer, R.D. Information decomposition and synergy. Nonnegative decomposition of multivariate information. *arXiv* **2010**, arXiv:1004.2515.

2. Williams, P.L.; Beer, R.D. Indiana University. Decomposing Multivariate Information. Privately communicated, 2010. This unpublished paper is highly similar to [1]. Crucially, however, this paper derives the redundancy lattice from the W&B Axioms 1–3 of Section 1. In contrast, [1] derives the redundancy lattice as a property of the particular measure I_{min}.
3. Olbrich, E.; Bertschinger, N.; Rauh, J. Information decomposition and synergy. *Entropy* **2015**, *17*, 3501–3517. [CrossRef]
4. Lizier, J.T.; Flecker, B.; Williams, P.L. Towards a synergy-based approach to measuring information modification. In Proceedings of the IEEE Symposium on Artificial Life (ALife), Singapore, 16–19 April 2013; pp. 43–51.
5. Bertschinger, N.; Rauh, J.; Olbrich, E.; Jost, J. Shared information—New insights and problems in decomposing information in complex systems. In Proceedings of the European Conference on Complex Systems, Brussels, Belgium, 3–7 September 2012; Springer: Cham, The Netherland, 2013; pp. 251–269.
6. Harder, M.; Salge, C.; Polani, D. Bivariate measure of redundant information. *Phys. Rev. E* **2013**, *87*, 012130. [CrossRef]
7. Griffith, V.; Chong, E.K.; James, R.G.; Ellison, C.J.; Crutchfield, J.P. Intersection information based on common randomness. *Entropy* **2014**, *16*, 1985–2000. [CrossRef]
8. Shannon, C.E. A Mathematical Theory of Communication. *Bell Syst. Tech. J.* **1948**, *27*, 379–423. [CrossRef]
9. Fano, R. *Transmission of Information*; The MIT Press: Cambridge, MA, USA, 1961.
10. Harder, M. Information driven self-organization of agents and agent collectives. Ph.D. Thesis, University of Hertfordshire, Hertfordshire, UK, 2013.
11. Bertschinger, N.; Rauh, J.; Olbrich, E.; Jost, J.; Ay, N. Quantifying unique information. *Entropy* **2014**, *16*, 2161–2183. [CrossRef]
12. Griffith, V.; Koch, C. Quantifying Synergistic Mutual Information. In *Guided Self-Organization: Inception*; Prokopenko, M., Ed.; Springer: Berlin/Heidelberg, Germany, 2014; Volume 9, pp. 159–190.
13. Rauh, J.; Bertschinger, N.; Olbrich, E.; Jost, J. Reconsidering unique information: Towards a multivariate information decomposition. In Proceedings of the 2014 IEEE International Symposium on Information Theory, Honolulu, HI, USA, 29 June–4 July 2014; pp. 2232–2236.
14. Perrone, P.; Ay, N. Hierarchical Quantification of Synergy in Channels. *Front. Robot. AI* **2016**, *2*, 35. [CrossRef]
15. Griffith, V.; Ho, T. Quantifying redundant information in predicting a target random variable. *Entropy* **2015**, *17*, 4644–4653. [CrossRef]
16. Rosas, F.; Ntranos, V.; Ellison, C.J.; Pollin, S.; Verhelst, M. Understanding interdependency through complex information sharing. *Entropy* **2016**, *18*, 38. [CrossRef]
17. Barrett, A.B. Exploration of synergistic and redundant information sharing in static and dynamical Gaussian systems. *Phys. Rev. E* **2015**, *91*, 052802. [CrossRef]
18. Ince, R. Measuring Multivariate Redundant Information with Pointwise Common Change in Surprisal. *Entropy* **2017**, *19*, 318. [CrossRef]
19. Ince, R.A. The Partial Entropy Decomposition: Decomposing multivariate entropy and mutual information via pointwise common surprisal. *arXiv* **2017**, arXiv:1702.01591.
20. Chicharro, D.; Panzeri, S. Synergy and Redundancy in Dual Decompositions of Mutual Information Gain and Information Loss. *Entropy* **2017**, *19*, 71. [CrossRef]
21. Rauh, J.; Banerjee, P.K.; Olbrich, E.; Jost, J.; Bertschinger, N. On Extractable Shared Information. *Entropy* **2017**, *19*, 328. [CrossRef]
22. Rauh, J.; Banerjee, P.K.; Olbrich, E.; Jost, J.; Bertschinger, N.; Wolpert, D. Coarse-Graining and the Blackwell Order. *Entropy* **2017**, *19*, 527. [CrossRef]
23. Rauh, J. Secret sharing and shared information. *Entropy* **2017**, *19*, 601. [CrossRef]
24. Faes, L.; Marinazzo, D.; Stramaglia, S. Multiscale information decomposition: Exact computation for multivariate Gaussian processes. *Entropy* **2017**, *19*, 408. [CrossRef]
25. Pica, G.; Piasini, E.; Chicharro, D.; Panzeri, S. Invariant components of synergy, redundancy, and unique information among three variables. *Entropy* **2017**, *19*, 451. [CrossRef]
26. James, R.G.; Crutchfield, J.P. Multivariate dependence beyond shannon information. *Entropy* **2017**, *19*, 531. [CrossRef]
27. Makkeh, A.; Theis, D.O.; Vicente, R. Bivariate Partial Information Decomposition: The Optimization Perspective. *Entropy* **2017**, *19*, 530. [CrossRef]

28. Kay, J.W.; Ince, R.A.; Dering, B.; Phillips, W.A. Partial and Entropic Information Decompositions of a Neuronal Modulatory Interaction. *Entropy* **2017**, *19*, 560. [CrossRef]

29. Angelini, L.; de Tommaso, M.; Marinazzo, D.; Nitti, L.; Pellicoro, M.; Stramaglia, S. Redundant variables and Granger causality. *Phys. Rev. E* **2010**, *81*, 037201. [CrossRef]

30. Stramaglia, S.; Angelini, L.; Wu, G.; Cortes, J.M.; Faes, L.; Marinazzo, D. Synergetic and redundant information flow detected by unnormalized Granger causality: Application to resting state fMRI. *IEEE Trans. Biomed. Eng.* **2016**, *63*, 2518–2524. [CrossRef]

31. Ghazi-Zahedi, K.; Langer, C.; Ay, N. Morphological computation: Synergy of body and brain. *Entropy* **2017**, *19*, 456. [CrossRef]

32. Maity, A.K.; Chaudhury, P.; Banik, S.K. Information theoretical study of cross-talk mediated signal transduction in MAPK pathways. *Entropy* **2017**, *19*, 469. [CrossRef]

33. Tax, T.; Mediano, P.A.; Shanahan, M. The partial information decomposition of generative neural network models. *Entropy* **2017**, *19*, 474. [CrossRef]

34. Wibral, M.; Finn, C.; Wollstadt, P.; Lizier, J.T.; Priesemann, V. Quantifying Information Modification in Developing Neural Networks via Partial Information Decomposition. *Entropy* **2017**, *19*, 494. [CrossRef]

35. Woodward, P.M. *Probability and Information Theory: With Applications to Radar*; Pergamon Press: Oxford, UK, 1953.

36. Woodward, P.M.; Davies, I.L. Information theory and inverse probability in telecommunication. *Proc. IEE-Part III Radio Commun. Eng.* **1952**, *99*, 37–44. [CrossRef]

37. Gray, R.M. *Probability, Random Processes, and Ergodic Properties*; Springer: New York, NY, USA, 1988.

38. Martin, N.F.; England, J.W. *Mathematical Theory of Entropy*; Cambridge University Press: Cambridge, UK, 1984.

39. Finn, C.; Lizier, J.T. Probability Mass Exclusions and the Directed Components of Pointwise Mutual Information. *arXiv* **2018**, arXiv:1801.09223.

40. Kelly, J.L. A new interpretation of information rate. *Bell Labs Tech. J.* **1956**, *35*, 917–926. [CrossRef]

41. Ash, R. *Information Theory*; Interscience tracts in pure and applied mathematics; Interscience Publishers: Geneva, Switzerland, 1965.

42. Shannon, C.E.; Weaver, W. *The Mathematical Theory of Communication*; University of Illinois Press: Champaign, IL, USA, 1998.

43. Cover, T.M.; Thomas, J.A. *Elements of Information Theory*; John Wiley & Sons: Hoboken, NJ, USA, 2012.

44. Pearl, J. *Probabilistic Reasoning in Intelligent Systems: Networks of Plausible Inference*; Morgan Kaufmann Publishers Inc.: San Francisco, CA, USA, 1988.

45. Rota, G.C. On the foundations of combinatorial theory I. Theory of Möbius functions. *Probab. Theory Relat. Field* **1964**, *2*, 340–368.

46. Stanley, R.P. Enumerative Combinatorics. In *Cambridge Studies in Advanced Mathematics*, 2nd ed.; Cambridge University Press: Cambridge, UK, 2012; Volume 1.

47. Davey, B.A.; Priestley, H.A. *Introduction to Lattices and Order*, 2nd ed.; Cambridge University Press: Cambridge, UK, 2002.

48. Ross, S.M. *A First Course in Probability*, 8th ed.; Pearson Prentice Hall: Upper Saddle River, NJ, USA, 2009.

![entropy logo] *entropy*

MDPI

Article

Multivariate Dependence beyond Shannon Information

Ryan G. James *and James P. Crutchfield

Complexity Sciences Center, Physics Department, University of California at Davis, One Shields Avenue, Davis, CA 95616, USA; chaos@ucdavis.edu
*   Correspondence: rgjames@ucdavis.edu

Received: 20 June 2017; Accepted: 24 September 2017; Published: 7 October 2017

Abstract: Accurately determining dependency structure is critical to understanding a complex system's organization. We recently showed that the transfer entropy fails in a key aspect of this—measuring information flow—due to its conflation of dyadic and polyadic relationships. We extend this observation to demonstrate that Shannon information measures (entropy and mutual information, in their conditional and multivariate forms) can fail to accurately ascertain multivariate dependencies due to their conflation of qualitatively different relations among variables. This has broad implications, particularly when employing information to express the organization and mechanisms embedded in complex systems, including the burgeoning efforts to combine complex network theory with information theory. Here, we do not suggest that any aspect of information theory is wrong. Rather, the vast majority of its informational measures are simply inadequate for determining the meaningful relationships among variables within joint probability distributions. We close by demonstrating that such distributions exist across an arbitrary set of variables.

Keywords: stochastic process; transfer entropy; causation entropy; partial information decomposition; network science

PACS: 89.70.+c; 05.45.Tp; 02.50.Ey; 02.50.−r

1. Introduction

Information theory is a general, broadly applicable framework for understanding a system's statistical properties [1]. Due to its focus on probability distributions, it allows one to compare dissimilar systems (e.g., species abundance to ground state configurations of a spin system) and has found many successes in the physical, biological and social sciences [2–19] far outside its original domain of communication. Often, the issue on which it is brought to bear is discovering and quantifying dependencies [20–25]. Here, we define a dependency to be any deviation from statistical independence. It is possible for a single multivariate distribution to consist of many, potentially overlapping, dependencies. Consider the simple case of three variables X, Y, Z, where X and Y are coin flips and Z is their concatenation. We would say here that there are two dependencies: an XZ dependency and a YZ dependency. It is important to note that, though there are some similarities, this notion of a dependency is distinct from that used within the Bayesian network community.

The past two decades, however, produced a small, but important body of results detailing how standard Shannon information measures are unsatisfactory for determining some aspects of dependency and shared information. Within information-theoretic cryptography, the conditional mutual information has proven to be a poor bound on secret key agreement [26,27]. The conditional mutual information has also been shown to be unable to accurately measure information flow ([28] and references therein). Finally, the inability of standard methods of decomposing the joint entropy to provide any semantic understanding of how information is shared has motivated entirely new methods of decomposing information [29,30]. Common to all these is the fact that conditional mutual information conflates intrinsic dependence with conditional dependence. To be clear, the conditional mutual information between X and Y given Z cannot distinguish the case where X and Y are related ignoring Z (intrinsic dependence) from the case where X and Y are related due to the influence of Z (conditional dependence).

Here, we demonstrate a related, but deeper issue: Shannon information measures—entropy, mutual information and their conditional and multivariate versions—can fail to distinguish joint distributions with vastly differing internal dependencies.

Concretely, we start by constructing two joint distributions, one with dyadic sub-dependencies and the other with strictly triadic sub-dependencies. From there, we demonstrate that no standard Shannon-like information measure, and exceedingly few nonstandard methods, can distinguish the two. Stated plainly: when viewed through Shannon's lens, these two distributions are erroneously equivalent. While distinguishing these two (and their internal dependencies) may not be relevant to a mathematical theory of communication, it is absolutely critical to a mathematical theory of information storage, transfer and modification [31–34]. We then demonstrate two ways in which these failures generalize to the multivariate case. The first generalizes our two distributions to the multivariate and polyadic case via "dyadic camouflage". The second details a method of embedding an arbitrary distribution into a larger variable space using hierarchical dependencies, a technique we term "dependency diffusion". In this way, one sees that the initial concerns about information measures can arise in virtually any statistical multivariate analysis. In this short development, we assume a working knowledge of information theory, such as found in standard textbooks [35–38].

2. Development

We begin by considering the two joint distributions shown in Table 1. The first represents dyadic relationships among three random variables X, Y, and Z. Additionally, the second, the triadic relationships among them. (This distribution was first considered as RDNXOR in [39], though for other, but related reasons.) These appellations are used for reasons that will soon be apparent. How are these distributions structured? Are they structured identically or are they qualitatively distinct? It is clear from inspection that they are not identical, but a lack of isomorphism is less obvious.

We can develop a direct picture of underlying dependency structure by casting the random variables' four-symbol alphabet used in Table 1 into composite binary random variables, as displayed in Table 2. It can be readily verified that the dyadic distribution follows three simple rules: $X_0 = Y_1$, $Y_0 = Z_1$ and $Z_0 = X_1$; in particular, three dyadic rules. The triadic distribution similarly follows simple rules: $X_0 + Y_0 + Z_0 = 0 \mod 2$ (the XOR relation [40], or equivalently, any one of them is the XOR of the other two), and $X_1 = Y_1 = Z_1$; two triadic rules.

While this expansion to binary sub-variables is not unique, it is representative of the distributions. One could expand the dyadic distribution, for example, in such a way that some of the sub-variables would be related by XOR. However, those same sub-variables would necessarily be involved in other relationships, limiting their expression in a manner similar to that explored in [41]. This differs from our triadic distribution in that its two sub-dependencies are independent. That these binary expansions are, in fact, representative and that the triadic distribution cannot be written in a way that relies only on dyadic relationships can be seen in the connected information explored later in the section. For the dyadic distribution, there is no difference between the maximum entropy distribution constraining pairwise interactions from the distribution itself. However, the maximum entropy distribution obtained by constraining the pairwise interactions in the triadic distribution has a larger entropy than the triadic distribution itself, implying that there is structure that exists beyond the pairwise interactions.

Table 1. The (a) dyadic and (b)triadic probability distributions over the three random variables X, Y and Z that take values in the four-letter alphabet $\{0, 1, 2, 3\}$. Though not directly apparent from their tables of joint probabilities, the dyadic distribution is built from dyadic (pairwise) sub-dependencies, while the triadic from triadic (three-way) sub-dependencies.

	(a) Dyadic				(b) Triadic		
X	Y	Z	Pr	X	Y	Z	Pr
0	0	0	1/8	0	0	0	1/8
0	2	1	1/8	1	1	1	1/8
1	0	2	1/8	0	2	2	1/8
1	2	3	1/8	1	3	3	1/8
2	1	0	1/8	2	0	2	1/8
2	3	1	1/8	3	1	3	1/8
3	1	2	1/8	2	2	0	1/8
3	3	3	1/8	3	3	1	1/8

Table 2. Expansion of the (a) dyadic and (b) triadic distributions. In both cases, the variables from Table 1 were interpreted as two binary random variables, translating, e.g., $X = 3$ into $(X_0, X_1) = (1, 1)$. In this light, it becomes apparent that the dyadic distribution consists of the sub-dependencies $X_0 = Y_1$, $Y_0 = Z_1$ and $Z_0 = X_1$, while the triadic distribution consists of $X_0 + Y_0 + Z_0 = 0 \mod 2$ and $X_1 = Y_1 = Z_1$. These relationships are pictorially represented in Figure 1.

(a) Dyadic							(b) Triadic						
X		Y		Z			X		Y		Z		
X_0	X_1	Y_0	Y_1	Z_0	Z_1	Pr	X_0	X_1	Y_0	Y_1	Z_0	Z_1	Pr
0	0	0	0	0	0	1/8	0	0	0	0	0	0	1/8
0	0	1	0	0	1	1/8	0	1	0	1	0	1	1/8
0	1	0	0	1	0	1/8	0	0	1	0	1	0	1/8
0	1	1	0	1	1	1/8	0	1	1	1	1	1	1/8
1	0	0	1	0	0	1/8	1	0	0	0	1	0	1/8
1	0	1	1	0	1	1/8	1	1	0	1	1	1	1/8
1	1	0	1	1	0	1/8	1	0	1	0	0	0	1/8
1	1	1	1	1	1	1/8	1	1	1	1	0	1	1/8

These dependency structures are represented pictorially in Figure 1. Our development from this point on will not use any knowledge of these structures, but rather, it will attempt to distinguish the structures using only information measures.

What does an information-theoretic analysis say? Both the dyadic and triadic distributions describe events over three variables, each with an alphabet size of four. Each consists of eight joint events, each with a probability of 1/8. As such, each has a joint entropy of $H[X, Y, Z] = 3$ bit (The SI standard unit for time is the second, and its symbol is s; analogously, the standard (IEC 60027-2, ISO/IEC 80000-13) unit for information is the bit, and its symbol is bit. As such, it is inappropriate to write 3 bits, just as it would be inappropriate to write 3 ss). Our first observation is that *any* entropy—conditional or not—and any mutual information—conditional or not—will be identical for the two distributions. Specifically, the entropy of any variable conditioned on the other two vanishes: $H[X \mid Y, Z] = H[Y \mid X, Z] = H[Z \mid X, Y] = 0$ bit; the mutual information between any two variables conditioned on the third is unity: $I[X : Y \mid Z] = I[X : Z \mid Y] = I[Y : Z \mid X] = 1$ bit; and the three-way co-information also vanishes: $I[X : Y : Z] = 0$ bit. These conclusions are compactly summarized in the form of the information diagrams (I-diagrams) [42,43] shown in Figure 2. This diagrammatically represents all of the possible Shannon information measures (I-measures) [43] of the distribution: effectively, all the multivariate extensions of the standard Shannon measures, called atoms. It is important to note that the analogy between information theory and set theory should not be taken too far: while set cardinality is strictly nonnegative, information atoms need not be; see [38] for more details. The values of the information atoms are identical between the two distributions.

As a brief aside, it is of interest to note that it has been suggested (e.g., in [44,45], among others) that zero co-information implies that at least one variable is independent of the others; that is, in this case, a lack of three-way interactions. Krippendorff [46] early on demonstrated that this is not the case, though these examples more clearly exemplify this fact.

We now turn to the implications of the two information diagrams, Figure 2a,b, being identical. There are measures [20,22,44,47–53] and expansions [54–56] purporting to measure or otherwise extract the complexity, magnitude or structure of dependencies within a multivariate distribution. Many of these techniques, including those just cited, are sums and differences of atoms in these information diagrams. As such, they are unable to differentiate these distributions.

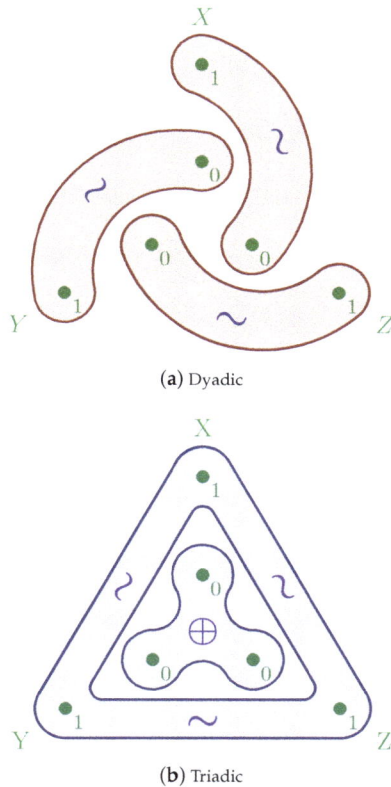

(a) Dyadic

(b) Triadic

Figure 1. Dependency structure for the (a) dyadic and (b) triadic distributions. Here, ~ denotes that two or more variables are distributed identically, and ⊕ denotes the enclosed variables form the XOR relation. Note that although these dependency structures are fundamentally distinct, their information diagrams (Figure 2) are identical.

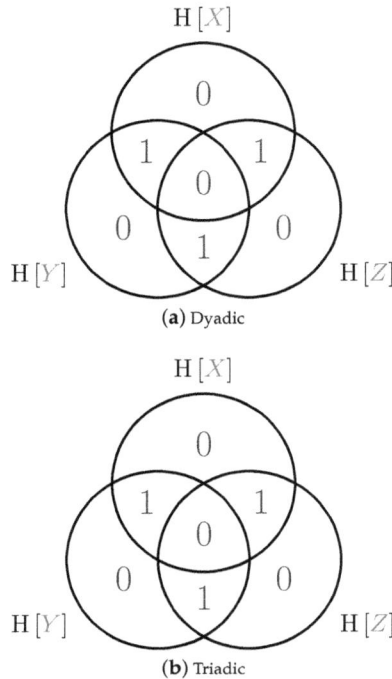

Figure 2. Information diagrams for the (**a**) dyadic and (**b**) triadic distributions. For the three-variable distributions depicted here, the diagram consists of seven atoms: three conditional entropies (each with value 0 bit), three conditional mutual information (each with value 1 bit) and one co-information (0 bit). Note that the two diagrams are identical, meaning that although the two distributions are fundamentally distinct, no standard information-theoretic measure can differentiate the two.

To drive home the point that the concerns raised here are very broad, Table 3 enumerates the result of applying a great many information measures to this pair of distributions. It is organized from top to bottom into four sections: entropies, mutual information, common information and other measures.

None of the entropies, dependent only on the probability mass function of the distribution, can distinguish the two distributions. Nor can any of the mutual information, as they are functions of the information atoms in the I-diagrams of Figure 2.

The common information, defined via auxiliary variables satisfying particular properties, can potentially isolate differences in the dependencies. Though only one of them—the Gács–Körner common information K [•] [57,58], involving the construction of the largest "subrandom variable" common to the variables—discerns that the two distributions are not equivalent because the triadic distribution contains the subrandom variable $X_1 = Y_1 = Z_1$ common to all three variables.

Finally, only two of the other measures identify any difference between the two. Some fail because they are functions of the probability mass function. Others, like the TSE complexity [59] and erasure entropy [60], fail since they are functions of the I-diagram atoms. Only the intrinsic mutual information I [• ↓ •] [26] and the reduced mutual information I [• ⇓ •] [27] distinguish the two since the dyadic distribution contains three dyadic sub-variables each of which is independent of the third variable, whereas in the triadic distribution, the conditional dependence of the XOR relation can be destroyed.

Table 3. Suite of information measures applied to the dyadic and triadic distributions, where: H [●] is the Shannon entropy [35], H_2 [●] is the order-2 Rényi entropy [61], S_q [●] is the Tsallis entropy [62], I [●] is the co-information [44], T [●] is the total correlation [47], B [●] is the dual total correlation [48,63], J [●] is the CAEKL mutual information [49], II [●] is the interaction information [64], K [●] is the Gács–Körner common information [57], C [●] is the Wyner common information [65,66], G [●] is the exact common information [67], F [●] is the functional common information,[a] M [●] is the MSS common information,[b] I [● ↓ ●] is the intrinsic mutual information [26],[c] I [● ⇓ ●] is the reduced intrinsic mutual information [27],[c ,d] X [●] is the extropy [68], R [●] is the residual entropy or erasure entropy [60,63], P [●] is the perplexity [69], D [●] is the disequilibrium [51], C_{LMRP} [●] is the LMRP complexity [51] and TSE [●] is the TSE complexity [59]. Only the Gács–Körner common information and the intrinsic mutual information, highlighted, are able to distinguish the two distributions; the Gács–Körner common information via the construction of a sub-variable ($X_1 = Y_1 = Z_1$) common to X, Y and Z and the intrinsic mutual information via the relationship $X_0 = Y_1$ being independent of Z.

Measures	Dyadic	Triadic
$H[X,Y,Z]$	3 bit	3 bit
$H_2[X,Y,Z]$	3 bit	3 bit
$S_2[X,Y,Z]$	0.875 bit	0.875 bit
$I[X:Y:Z]$	0 bit	0 bit
$T[X:Y:Z]$	3 bit	3 bit
$B[X:Y:Z]$	3 bit	3 bit
$J[X:Y:Z]$	1.5 bit	1.5 bit
$II[X:Y:Z]$	0 bit	0 bit
$K[X:Y:Z]$	0 bit	1 bit
$C[X:Y:Z]$	3 bit	3 bit
$G[X:Y:Z]$	3 bit	3 bit
$F[X:Y:Z]$ [a]	3 bit	3 bit
$M[X:Y:Z]$ [b]	3 bit	3 bit
$I[X:Y↓Z]$ [c]	1 bit	0 bit
$I[X:Y⇓Z]$ [c,d]	1 bit	0 bit
$X[X,Y,Z]$	1.349 bit	1.349 bit
$R[X:Y:Z]$	0 bit	0 bit
$P[X,Y,Z]$	8	8
$D[X,Y,Z]$	0.761 bit	0.761 bit
$C_{LMRP}[X,Y,Z]$	0.381 bit	0.381 bit
$TSE[X:Y:Z]$	2 bit	2 bit

[a] $F[\{X_i\}] = \min\limits_{\substack{\perp\!\!\!\perp X_i|V \\ V=f(\{X_i\})}} H[V]$, where $\perp\!\!\!\perp X_i|V$ means that the X_i are conditionally independent given V.

[b] $M[\{X_i\}] = H[V(X_i \searrow X_{\bar{i}})]$, where $X \searrow Y$ is the minimal sufficient statistic [35] of X about Y, and V denotes the informational union of variables. [c] Though this measure is generically dependent on which variable(s) is chosen to be conditioned on, due to the symmetry of the dyadic and triadic distributions, the values reported here are insensitive to permutations of the variables. [d] The original work [27] used the slightly more verbose notation I [● ↓↓ ●].

Figure 3 demonstrates three different information expansions—that, roughly speaking, group variables into subsets of difference sizes or "scales"—applied to our distributions of interest. The first is the complexity profile [55]. At scale k, the complexity profile is the sum of all I-diagram atoms consisting of at least k variables conditioned on the others. Here, since the I-diagrams are identical, so are the complexity profiles. The second profile is the marginal utility of information [56], which is a derivative of a linear programming problem whose constraints are given by the I-diagram, so here, again, they are identical. Finally, we have Schneidman et al.'s connected information [70], which comprise the differences in entropies of the maximum entropy distributions whose k- and $k-1$-way marginals are fixed to match those of the distribution of interest. Here, all dependencies are detected once pairwise marginals are fixed in the dyadic distribution, but it takes the full joint distribution to realize the XOR sub-dependency in the triadic distribution.

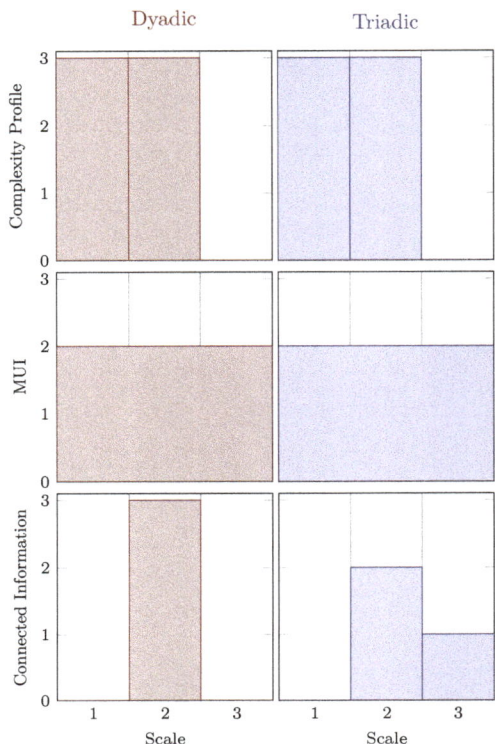

Figure 3. Suite of information expansions applied to the dyadic and triadic distributions: the complexity profile [55], the marginal utility of information [56] and the connected information [70]. The complexity profile and marginal utility of information profiles are identical for the two distributions as a consequence of the information diagrams (Figure 2) being identical. The connected information, quantifying the amount of dependence realized by fixing *k*-way marginals, is able to distinguish the two distributions. Note that although each of the *x*-axes is a scale, exactly what that means depends on the measure. Furthermore, while the scale for both the complexity profile and the connected information is discrete, the scale for the marginal utility of information is continuous.

While it is well known that causality cannot be determined from probability distributions alone [71], here we point out a related, though different issue. While causality is, in some sense, the determination of the precedence within a dependency, the results above demonstrate that many measures of Granger-like causality are insensitive to the order (dyadic, triadic, etc.) of a dependency (Note that here we do not demonstrate that the order of a dependency cannot be determined from the probability distribution, as Pearl has done for causality [71]. Rather, our demonstration is limited to Shannon-like information measures.). Neither the transfer entropy [20], the transinformation [53], the directed information [52], the causation entropy [22], nor any of their generalizations based on conditional mutual information differentiate between intrinsic relationships and those induced by the variables they condition on (As discussed there, the failure of these measures stems from the possibility of conditional dependence, whereas the aim for these directed measures is to quantify the information flow from one time series to another excluding the influences of the second. In this light, we propose $T'_{X \to Y} = I\left[X_0^t : Y_t \downarrow Y_0^t\right]$ [26] as an incremental improvement over the transfer entropy). This limitation underlies prior criticisms of these functions as measures of information flow [28]. Furthermore, separating out these contributions to the transfer entropy has been discussed in the context of the partial information decomposition in [72].

A promising approach to understanding informational dependencies is the partial information decomposition (PID) [29]. This framework seeks to decompose a mutual information of the form $I[(I_0, I_1) : O]$ into four nonnegative components: the information R that both inputs I_0 and I_1 redundantly provide the output O, the information U_0 that I_0 uniquely provides O, the information U_1 that I_1 uniquely provides O and, finally, the information S that both I_0 and I_1 synergistically or collectively provide O.

Under this decomposition, our two distributions take on very different characteristics (Here, we quantified the partial information lattice using the best-in-class technique of [73], though calculations using three other techniques [74–76] match. There is a recent debate suggesting that the measure of Bertschinger et al. is not in fact correct, but it is likely, due to the agreement among these measures, that any "true" measure of redundancy would result in the same decomposition. The original PID measure I_{min}, however, assigns both distributions: 1 bit of redundant information and 1 bit of synergistic information.). For both, the decomposition is invariant as far as which variables are selected as I_0, I_1 and O. For the dyadic distribution, PID identifies both bits in $I[(I_0, I_1) : O]$ as unique, one from each input I_i, corresponding to the dyadic sub-dependency shared by I_i and O. Orthogonally, for the triadic distribution PID identifies one of the bits as redundant, stemming from $X_1 = Y_1 = Z_1$, and the other as synergistic, resulting from the XOR relation among X_0, Y_0 and Z_0. These decompositions are displayed pictorially in Figure 4.

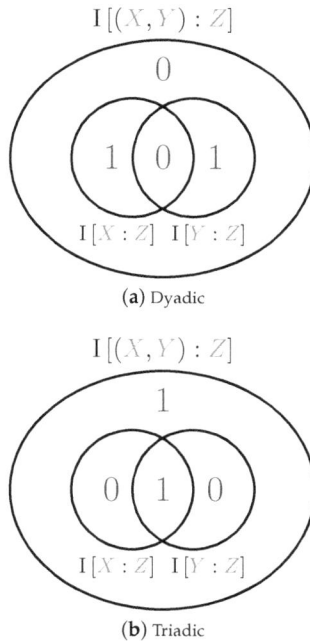

(**a**) Dyadic

(**b**) Triadic

Figure 4. Partial information decomposition diagrams for the (**a**) dyadic and (**b**) triadic distributions. Here, X and Y are treated as inputs and Z as output, but in both cases, the decomposition is invariant to permutations of the variables. In the dyadic case, the relationship is realized as 1 bit of unique information from X to Z and 1 bit of unique information from Y to Z. In the triadic case, the relationship is quantified as X and Y providing 1 bit of redundant information about Z while also supplying 1 bit of information synergistically about Z.

Another somewhat similar approach is that of integrated information theory [77]. However, this approach requires a known dynamic over the variables and is, in addition, highly sensitive to the dynamic. Here, in contrast, we considered only simple probability distributions without any assumptions as to how they might arise from the dynamics of interacting agents. That said, one might associate an integrated information measure with a distribution via the maximal information integration over all possible dynamics that give rise to the distribution. We leave this task for a later study.

3. Discussion

The broad failure of Shannon information measures to differentiate the dyadic and polyadic distributions has far-reaching consequences. Consider, for example, an experiment where a practitioner places three probes into a cluster of neurons, each probe touching two neurons and reporting zero when they are both quiescent, one when the first is excited but the second quiescent, two when the second is excited, but the first quiescent, and three when both are excited. Shannon-like measures—including the transfer entropy and related measures—would be unable to differentiate between the dyadic situation consisting of three pairs of synchronized neurons, the triadic situation consisting of a trio of synchronized neurons and a trio exhibiting the XOR relation, a relation requiring nontrivial sensory integration. Such a situation might arise when probing the circuitry of the *Drosophila melanogaster* connectome [78], for instance.

Furthermore, while partitioning each variable into sub-variables made the dependency structure clear, we do not believe that such a refinement should be a necessary step in discovering such a structure. Consider that we demonstrated that refinement is not strictly needed, since the partial information decomposition (as quantified using current techniques) was able to discover the distribution's internal structure without it.

These results, observations and the broad survey clearly highlight the need to extend Shannon's theory. In particular, the extension must introduce a fundamentally new measure, not merely sums and differences of the standard Shannon information measures. While the partial information decomposition was initially proposed to overcome the interpretational difficulty of the (potentially negative valued) co-information, we see here that it actually overcomes a vastly more fundamental weakness with Shannon information measures. While negative information atoms can subjectively be seen as a flaw, the inability to distinguish dyadic from polyadic relations is a much deeper and objective issue.

This may lead one to consider the partial information decomposition as the needed extension to Shannon theory. As it currently stands, we do not. The partial information decomposition depends on interpreting some random variables as "inputs" and others as "outputs". While this may be perfectly natural in some contexts, it is not satisfactory in general. It is possible that, were an agreeable multivariate partial information measure to be developed, the decomposition of, e.g., $I[(X_0, X_1, X_2) : X_0 X_1 X_2]$ could lead to a satisfactory symmetric decomposition. In any case, there has been longstanding interest in creating a symmetric decomposition analogous to the partial information decomposition [46] with some recent progress [79–81].

4. Dyadic Camouflage and Dependency Diffusion

The dyadic and triadic distributions we analyzed thus far were deliberately chosen to have small dimensionality in an effort to make them and the failure of Shannon information measures as comprehensible and intuitive as possible. Since a given dataset may have exponentially many different three-variable subsets, even just the two distributions examined here represent hurdles that information-based methods of dependency assessment must overcome. However, this is simply a starting point. We will now demonstrate that there exist distributions of arbitrary size whose k-way dependencies are masked, meaning the k-way co-information ($k \geq 3$) are all zero, and so, from the perspective of Shannon information theory, are indistinguishable from a distribution of the same size containing only dyadic relationships. Furthermore, we show how any such distribution may be obfuscated over any larger set of variables. This likely mandates a search over all partitions of all subsets of a system, making the problem of finding such distributions in the EXPTIME computational complexity class [82], meaning such a procedure will take time exponential in the size of the distribution.

Specifically, consider the four-variable parity distribution consisting of four binary variables such that $X_0 + X_1 + X_2 + X_3 = 0 \mod 2$. This is a straightforward generalization of the XOR distribution used in constructing the triadic distribution. We next need a generalization of the "giant bit" [63], which we call dyadic camouflage, to mix with the parity, informationally "canceling out" the higher-order mutual information even though dependencies of such orders exist in the distribution. An example dyadic camouflage distribution for four variables is given in Figure 5.

W	X	Y	Z	Pr
0	0	0	0	1/8
0	1	3	1	1/8
1	0	2	2	1/8
1	1	1	3	1/8
2	2	3	3	1/8
2	3	0	2	1/8
3	2	1	1	1/8
3	3	2	0	1/8

(a) Distribution

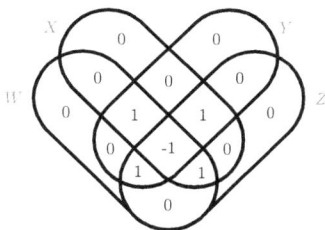

(b) I-diagram

Figure 5. Dyadic camouflage distribution: This distribution, when uniformly and independently mixed with the four-variable parity distribution (in which each variable is the parity of the other three), results in a distribution whose I-diagram incorrectly implies that the distribution contains only dyadic dependencies. The atoms of the camouflage distribution are constructed so that they cancel out the "interior" atoms of the parity distribution (whose $I[W : X : Y|Z] = I[W : X : Z|Y] = I[W : Y : Z|X] = I[X : Y : Z|W] = -1$ and $I[W : X : Y : Z] = 1$), leaving just the parity distribution's pairwise conditional atoms: $I[W : X|YZ]$, $I[W : Y|XZ]$, $I[W : Z|XY]$, $I[X : Y|WZ]$, $I[X : Z|WY]$, and $I[Y : Z|WX]$, all equal to one, while all others are zero.

Generically, consider an n-variable parity distribution, that is a distribution where $\sum X_i = 0 \mod 2$. It has an associated n-variable dyadic camouflage distribution with an alphabet size for each random variable of 2^{n-2}, and the entire joint distribution consists of $2^{\frac{(n-2)\cdot(n-1)}{2}}$ equally likely outcomes, both numbers determined due to entropy considerations. Specifically, in a parity distribution, each variable has 1 bit of entropy, and when mixed with its camouflage, it should have $n-1$ bits. Therefore, each variable in the camouflage distribution needs $n-2$ bits of entropy and needs, with uniform probability over those outcomes, 2^{n-2} characters in the alphabet. Furthermore, since the parity distribution itself has $n-1$ bits total, while its camouflaged form will have n choose two $(=(n-2)\cdot(n-1)/2)$ bits and, again with uniform probability, there must be $2^{(n-2)\cdot(n-1)/2}$ outcomes. The distribution is constrained such that any two variables are completely determined by the remaining $n - 2$. Moreover, each m-variable ($m < n$) sub-distribution consists of m mutually independent random variables.

The goal, then, is to construct such a distribution. One method of doing so is to begin by writing down one variable in increasing lexicographic order such that it has the correct number of outcomes; e.g., column W in Figure 5a. Then, find $n - 1$ permutations of this column such that any two columns are determined from the remaining $n - 2$. While such a search may be difficult, a distribution with these properties provably exists [83].

Finally, one can obfuscate any distribution by embedding it in a larger collection of random variables. Given a distribution D over n variables, associate each random variable i of D with a k-variable subset of a distribution D' in such a way that there is a mapping from the k-outcomes in the subset of D' to the outcome of the variable i in D. For example, one can embed the XOR distribution over X, Y, Z into six variables $X_0, X_1, Y_0, Y_1, Z_0, Z_1$ via $X_0 \oplus X_1 = X$, $Y_0 \oplus Y_1 = Y$ and $Z_0 \oplus Z_1 = Z$. In other words, the parity of (Z_0, Z_1) is equal to the XOR of the parities of (X_0, X_1) and (Y_0, Y_1). In this way, one must potentially search over all partitions of all subsets of D' in order to discover the distribution D hiding within. We refer to this method of obfuscation as dependency diffusion.

The first conclusion is that the challenges of conditional dependence can be found in joint distributions over arbitrarily large sets of random variables. The second conclusion, one that heightens the challenge to discovery, is that even finding which variables are implicated in polyadic dependencies can be exponentially difficult. Together, the camouflage and diffusion constructions demonstrate how challenging it is to discover, let alone work with, multivariate dependencies. This difficulty strongly implies that the current state of information-theoretic tools is vastly underpowered for the types of analyses required of our modern, data-rich sciences.

It is unlikely that the parity plus dyadic camouflage distribution discussed here is the only example of Shannon measures conflating the arity of dependencies and thus producing an information diagram identical to that of a qualitatively distinct distribution. This suggests an important challenge: find additional, perhaps simpler, joint distributions exhibiting this phenomenon.

5. Conclusions

To conclude, we constructed two distributions that cannot be distinguished using conventional (and many non-conventional) Shannon-like information measures. In fact, of the more than two dozen measures we surveyed, only five were able to separate the distributions: the Gács–Körner common information, the intrinsic mutual information, the reduced intrinsic mutual information, the connected information and the partial information decomposition.

The failure of the Shannon-type measures is perhaps not surprising: nothing in the standard mathematical theories of information and communication suggests that such measures *should* be able to distinguish these distributions [84]. However, distinguishing dependency structures such as dyadic from triadic relationships is of the utmost importance to the sciences; consider for example determining multi-drug interactions in medical treatments. Critically, since interpreting dependencies in random distributions is traditionally the domain of information theory, we propose that new extensions to information theory are needed.

These results may seem like a deal-breaking criticism of employing information theory to determine dependencies. Indeed, these results seem to indicate that much existing empirical work and many interpretations have simply been wrong and, worse even, that the associated methods are misleading while appearing quantitatively consistent. We think not, though. With the constructive and detailed problem diagnosis given here, at least we can see this issue. It is now a necessary step to address it. This leads us to close with a cautionary quote:

> "The tools we use have a profound (and devious!) influence on our thinking habits, and, therefore, on our thinking abilities" (Edsger W. Dijkstra [85]) .

Acknowledgments: We thank N. Barnett and D. P. Varn for helpful feedback. James P. Crutchfield thanks the Santa Fe Institute for its hospitality during visits as an External Faculty member. This material is based on work supported by, or in part by, the U. S. Army Research Laboratory and the U. S. Army Research Office under Contracts W911NF-13-1-0390 and W911NF-13-1-0340.

Author Contributions: Both authors contributed equally to the theoretical development of the work. R.G.J. carried out all numerical calculations. Both authors contributed equally to the writing of the manuscript.

Conflicts of Interest: The authors declare no conflict of interest.

Appendix A. A Python Discrete Information Package

Hand calculating the information quantities used in the main text, while profitably done for a few basic examples, soon becomes tedious and error prone. We provide a Jupyter notebook [86] making use of dit ("Discrete Information Theory") [87], an open source Python package that readily calculates these quantities.

References

1. Kullback, S. *Information Theory and Statistics*; Dover: New York, NY, USA, 1968.
2. Quastler, H. *Information Theory in Biology*; University of Illinois Press: Urbana-Champaign, IL, USA, 1953.
3. Quastler, H. The status of information theory in biology—A roundtable discussion. In *Symposium on Information Theory in Biology*; Yockey, H.P., Ed.; Pergamon Press: New York, NY, USA, 1958; 399p.
4. Kelly, J. A new interpretation of information rate. *IRE Trans. Inf. Theory* **1956**, *2*, 185–189.
5. Brillouin, L. *Science and Information Theory*, 2nd ed.; Academic Press: New York, NY, USA, 1962.
6. Bialek, W.; Rieke, F.; De Ruyter Van Steveninck, R.R.; Warland, D. Reading a neural code. *Science* **1991**, *252*, 1854–1857.
7. Strong, S.P.; Koberle, R.; de Ruyter van Steveninck, R.R.; Bialek, W. Entropy and information in neural spike trains. *Phys. Rev. Lett.* **1998**, *80*, 197.
8. Ulanowicz, R.E. The central role of information theory in ecology. In *Towards an Information Theory of Complex Networks*; Dehmer, M., Mehler, A., Emmert-Streib, F., Eds.; Springer: Berlin, Germany, 2011; pp. 153–167.
9. Grandy, W.T., Jr. *Entropy and the Time Evolution of Macroscopic Systems*; Oxford University Press: Oxford, UK, 2008; Volume 141.
10. Harte, J. *Maximum Entropy and Ecology: A Theory of Abundance, Distribution, and Energetics*. Oxford University Press: Oxford, UK, 2011.
11. Nalewajski, R.F. *Information Theory of Molecular Systems*; Elsevier: Amsterdam, The Netherlands, 2006.
12. Garland, J.; James, R.G.; Bradley, E. Model-free quantification of time-series predictability. *Phys. Rev. E* **2014**, *90*, 052910.
13. Kafri, O. Information theoretic approach to social networks. *J. Econ. Soc. Thought* **2017**, *4*, 77.
14. Varn, D.P.; Crutchfield, J.P. Chaotic crystallography: How the physics of information reveals structural order in materials. *Curr. Opin. Chem. Eng.* **2015**, *777*, 47–56.
15. Varn, D.P.; Crutchfield, J.P. What did Erwin mean? The physics of information from the materials genomics of aperiodic crystals and water to molecular information catalysts and life. *Phil. Trans. R. Soc. A* **2016**, *374*, doi:10.1098/rsta.2015.0067.
16. Zhou, X.-Y.; Rong, C.; Lu, T.; Zhou, P.; Liu, S. Information functional theory: Electronic properties as functionals of information for atoms and molecules. *J. Phys. Chem. A* **2016**, *120*, 3634–3642.
17. Kirst, C.; Timme, M.; Battaglia, D. Dynamic information routing in complex networks. *Nat. Commun.* **2016**, *7*, 11061
18. Izquierdo, E.J.; Williams, P.L.; Beer, R.D. Information flow through a model of the *C. elegans* klinotaxis circuit. *PLoS ONE* **2015**, *10*, e0140397.
19. James, R.G.; Burke, K.; Crutchfield, J.P. Chaos forgets and remembers: Measuring information creation, destruction, and storage. *Phys. Lett. A* **2014**, *378*, 2124–2127.
20. Schreiber, T. Measuring information transfer. *Phys. Rev. Lett.* **2000**, *85*, 461.
21. Fiedor, P. Partial mutual information analysis of financial networks. *Acta Phys. Pol. A* **2015**, *127*, 863–867.
22. Sun, J.; Bollt, E.M. Causation entropy identifies indirect influences, dominance of neighbors and anticipatory couplings. *Phys. D Nonlinear Phenom.* **2014**, *267*, 49–57.

23. Lizier, J.T.; Prokopenko, M.; Zomaya, A.Y. Local information transfer as a spatiotemporal filter for complex systems. *Phys. Rev. E* **2008**, *77*, doi:10.1103/PhysRevE.77.026110.
24. Walker, S.I.; Kim, H.; Davies, P.C.W. The informational architecture of the cell. *Phil. Trans. R. Soc. A* **2016**, *273*, doi:10.1098/rsta.2015.0057.
25. Lee, U.; Blain-Moraes, S.; Mashour, G.A. Assessing levels of consciousness with symbolic analysis. *Phil. Trans. R. Soc. Lond. A* **2015**, *373*, doi:10.1098/rsta.2014.0117.
26. Maurer, U.; Wolf, S. The intrinsic conditional mutual information and perfect secrecy. In Proceedings of the 1997 IEEE International Symposium on Information Theory, Ulm, Germany, 29 June–4 July 1997; p. 8.
27. Renner, R.; Skripsky, J.; Wolf, S. A new measure for conditional mutual information and its properties. In Proceedings of the 2003 IEEE International Symposium on Information Theory, Yokohama, Japan, 29 June–4 July 2003; p. 259.
28. James, R.G.; Barnett, N.; Crutchfield, J.P. Information flows? A critique of transfer entropies. *Phys. Rev. Lett.* **2016**, *116*, 238701.
29. Williams, P.L.; Beer, R.D. Nonnegative decomposition of multivariate information. *arXiv* **2010**, arXiv:1004.2515.
30. Bertschinger, N.; Rauh, J.; Olbrich, E.; Jost, J. Shared information: New insights and problems in decomposing information in complex systems. In *Proceedings of the European Conference on Complex Systems 2012*; Springer: Berlin, Germany, 2013; pp. 251–269.
31. Lizier, J.T. The Local Information Dynamics of Distributed Computation in Complex Systems. Ph.D. Thesis, University of Sydney, Sydney, Austrilia, 2010.
32. Ay, N.; Polani, D. Information flows in causal networks. *Adv. Complex Syst.* **2008**, *11*, 17–41.
33. Chicharro, D.; Ledberg, A. When two become one: The limits of causality analysis of brain dynamics. *PLoS ONE* **2012**, *7*, e32466.
34. Lizier, J.T.; Prokopenko, M. Differentiating information transfer and causal effect. *Eur. Phys. J. B Condens. Matter Complex Syst.* **2010**, *73*, 605–615.
35. Cover, T.M.; Thomas, J.A. *Elements of Information Theory*; John Wiley & Sons: New York, NY, USA, 2012.
36. Yeung, R.W. *A First Course in Information Theory*; Springer Science & Business Media: Berlin, Germany, 2012.
37. Csiszar, I.; Körner, J. *Information Theory: Coding Theorems for Discrete Memoryless Systems*; Cambridge University Press: Cambridge, UK, 2011.
38. MacKay, D.J.C. *Information Theory, Inference and Learning Algorithms*; Cambridge University Press: Cambridge, UK, 2003.
39. Griffith, V.; Koch, C. Quantifying synergistic mutual information. In *Guided Self-Organization: Inception*; Springer: Berlin, Germany, 2014; pp. 159–190.
40. Cook, M. Networks of Relations. Ph.D. Thesis, California Institute of Technology, Pasadena, CA, USA, 2005.
41. Merchan, L.; Nemenman, I. On the sufficiency of pairwise interactions in maximum entropy models of networks. *J. Stat. Phys.* **2016**, *162*, 1294–1308.
42. Reza, F.M. *An Introduction to Information Theory*; Courier Corporation: North Chelmsford, MA, USA, 1961.
43. Yeung, R.W. A new outlook on Shannon's information measures. *IEEE Trans. Inf. Theory* **1991**, *37*, 466–474.
44. Bell, A.J. The co-information lattice. In Proceedings of the 4th International Workshop on Independent Component Analysis and Blind Signal Separation, Nara, Japan, 1–4 April 2003; Amari, S.M.S., Cichocki, A., Murata, N., Eds.; Springer: New York, NY, USA, 2003; Volume ICA 2003, pp. 921–926.
45. Bettencourt, L.M.A.; Stephens, G.J.; Ham, M.I.; Gross, G.W. Functional structure of cortical neuronal networks grown in vitro. *Phys. Rev. E* **2007**, *75*, 021915.
46. Krippendorff, K. Information of interactions in complex systems. *Int. J. Gen. Syst.* **2009**, *38*, 669–680.
47. Watanabe, S. Information theoretical analysis of multivariate correlation. *IBM J. Res. Dev.* **1960**, *4*, 66–82.
48. Han, T.S. Linear dependence structure of the entropy space. *Inf. Control* **1975**, *29*, 337–368.
49. Chan, C.; Al-Bashabsheh, A.; Ebrahimi, J.B.; Kaced, T.; Liu, T. Multivariate mutual information inspired by secret-key agreement. *Proc. IEEE* **2015**, *103*, 1883–1913.
50. James, R.G.; Ellison, C.J.; Crutchfield, J.P. Anatomy of a bit: Information in a time series observation. *Chaos Interdiscip. J. Nonlinear Sci.* **2011**, *21*, 037109.
51. Lamberti, P.W.; Martin, M.T.; Plastino, A.; Rosso, O.A. Intensive entropic non-triviality measure. *Physica A* **2004**, *334*, 119–131.

52. Massey, J. Causality, feedback and directed information. In Proceedings of the International Symposium on Information Theory and Its Applications, Waikiki, HI, USA, 27–30 November 1990; Volume ISITA-90, pp. 303–305.
53. Marko, H. The bidirectional communication theory: A generalization of information theory. *IEEE Trans. Commun.* **1973**, *21*, 1345–1351.
54. Bettencourt, L.M.A.; Gintautas, V.; Ham, M.I. Identification of functional information subgraphs in complex networks. *Phys. Rev. Lett.* **2008**, *100*, 238701.
55. Bar-Yam, Y. Multiscale complexity/entropy. *Adv. Complex Syst.* **2004**, *7*, 47–63.
56. Allen, B.; Stacey, B.C.; Bar-Yam, Y. Multiscale Information Theory and the Marginal Utility of Information. *Entropy* **2017**, *19*, 273.
57. Gács, P.; Körner, J. Common information is far less than mutual information. *Probl. Control Inf.* **1973**, *2*, 149–162.
58. Tyagi, H.; Narayan, P.; Gupta, P. When is a function securely computable? *IEEE Trans. Inf. Theory* **2011**, *57*, 6337–6350.
59. Ay, N.; Olbrich, E.; Bertschinger, N.; Jost, J. A unifying framework for complexity measures of finite systems. In *Proceedings of the European Conference on Complex Systems 2006 (ECCS06)*; European Complex Systems Society (ECSS): Paris, France, 2006.
60. Verdu, S.; Weissman, T. The information lost in erasures. *IEEE Trans. Inf. Theory* **2008**, *54*, 5030–5058.
61. Rényi, A. On measures of entropy and information. In Proceedings of the Fourth Berkeley Symposium on Mathematical Statistics and Probability, Oakland, CA, USA, 20 June–30 July 1960; pp. 547–561.
62. Tsallis, C. Possible generalization of Boltzmann-Gibbs statistics. *J. Stat. Phys.* **1988**, *52*, 479–487.
63. Abdallah, S.A.; Plumbley, M.D. A measure of statistical complexity based on predictive information with application to finite spin systems. *Phys. Lett. A* **2012**, *376*, 275–281.
64. McGill, W.J. Multivariate information transmission. *Psychometrika* **1954**, *19*, 97–116.
65. Wyner, A.D. The common information of two dependent random variables. *IEEE Trans. Inf. Theory* **1975**, *21*, 163–179.
66. Liu, W.; Xu, G.; Chen, B. The common information of n dependent random variables. In Proceedings of the 2010 48th Annual Allerton Conference on Communication, Control, and Computing (Allerton), Monticello, IL, USA, 29 September–1 October 2010; pp. 836–843.
67. Kumar, G.R.; Li, C.T.; El Gamal, A. Exact common information. In Proceedings of the 2014 IEEE International Symposium on Information Theory (ISIT), Honolulu, HI, USA, 29 June–4 July 2014; pp. 161–165.
68. Lad, F.; Sanfilippo, G.; Agrò, G. Extropy: Complementary dual of entropy. *Stat. Sci.* **2015**, *30*, 40–58.
69. Jelinek, F.; Mercer, R.L.; Bahl, L.R.; Baker, J.K. Perplexity—A measure of the difficulty of speech recognition tasks. *J. Acoust. Soc. Am.* **1977**, *62*, S63.
70. Schneidman, E.; Still, S.; Berry, M.J.; Bialek, W. Network information and connected correlations. *Phys. Rev. Lett.* **2003**, *91*, 238701.
71. Pearl, J. *Causality*; Cambridge University Press: Cambridge, UK, 2009.
72. Williams, P.L.; Beer, R.D. Generalized measures of information transfer. *arXiv* **2011**, arXiv:1102.1507.
73. Bertschinger, N.; Rauh, J.; Olbrich, E.; Jost, J.; Ay, N. Quantifying unique information. *Entropy* **2014**, *16*, 2161–2183.
74. Harder, M.; Salge, C.; Polani, D. Bivariate measure of redundant information. *Phys. Rev. E* **2013**, *87*, 012130.
75. Griffith, V.; Chong, E.K.P.; James, R.G.; Ellison, C.J.; Crutchfield, J.P. Intersection information based on common randomness. *Entropy* **2014**, *16*, 1985–2000.
76. Ince, R.A.A. Measuring multivariate redundant information with pointwise common change in surprisal. *arXiv* **2016**, arXiv:1602.05063.
77. Albantakis, L.; Oizumi, M.; Tononi, G. From the phenomenology to the mechanisms of consciousness: Integrated information theory 3.0. *PLoS Comput. Biol.* **2014**, *10*, e1003588.
78. Takemura, S.; Bharioke, A.; Lu, Z.; Nern, A.; Vitaladevuni, S.; Rivlin, P.K.; Katz, W.T.; Olbris, D.J.; Plaza, S.M.; Winston, P.; et al. A visual motion detection circuit suggested by Drosophila connectomics. *Nature* **2013**, *500*, 175–181.
79. Rosas, F.; Ntranos, V.; Ellison, C.J.; Pollin, S.; Verhelst, M. Understanding interdependency through complex information sharing. *Entropy* **2016**, *18*, 38.

80. Ince, R.A. The Partial Entropy Decomposition: Decomposing multivariate entropy and mutual information via pointwise common surprisal. *Entropy* **2017**, *19*, 318.

81. Pica, G.; Piasini, E.; Chicharro, D.; Panzeri, S. Invariant components of synergy, redundancy, and unique information among three variables. *Entropy* **2017**, *19*, 451.

82. Garey, M.R.; Johnson, D.S. *Computers and Intractability: A Guide to the Theory of NP-Completeness;* W. H. Freeman: New York, NY, USA, 1979.

83. Chen, Q.; Cheng, F.; Lie, T.; Yeung, R.W. A marginal characterization of entropy functions for conditional mutually independent random variables (with application to Wyner's common information). In Proceedings of the 2015 IEEE International Symposium on Information Theory (ISIT), Hong Kong, China, 14–19 June 2015; pp. 974–978.

84. Shannon, C.E. The bandwagon. *IEEE Trans. Inf. Theory* **1956**, *2*, 3.

85. Dijkstra, E.W. How do we tell truths that might hurt? In *Selected Writings on Computing: A Personal Perspective;* Springer: Berlin, Germany, 1982; pp. 129–131.

86. Jupyter. Available online: https://github.com/jupyter/notebook (accessed on 7 October 2017).

87. James, R.G.; Ellison, C.J.; Crutchfield, J.P. Dit: Discrete Information Theory in Python. Available online: https://github.com/dit/dit (accessed on 7 October 2017).

entropy

MDPI

Article

Invariant Components of Synergy, Redundancy, and Unique Information among Three Variables

Giuseppe Pica [1,*], **Eugenio Piasini** [1], **Daniel Chicharro** [1,2] **and Stefano Panzeri** [1,*]

[1] Neural Computation Laboratory, Center for Neuroscience and Cognitive Systems @UniTn, Istituto Italiano di Tecnologia, 38068 Rovereto (TN), Italy; eugenio.piasini@iit.it (E.P.); daniel.chicharro@iit.it (D.C.)

[2] Department of Neurobiology, Harvard Medical School, Boston, MA 02115, USA

*   Correspondence: giuseppe.pica@iit.it (G.P.); stefano.panzeri@iit.it (S.P.)

Received: 27 June 2017; Accepted: 25 August 2017; Published: 28 August 2017

Abstract: In a system of three stochastic variables, the Partial Information Decomposition (PID) of Williams and Beer dissects the information that two variables (sources) carry about a third variable (target) into nonnegative information atoms that describe redundant, unique, and synergistic modes of dependencies among the variables. However, the classification of the three variables into two sources and one target limits the dependency modes that can be quantitatively resolved, and does not naturally suit all systems. Here, we extend the PID to describe trivariate modes of dependencies in full generality, without introducing additional decomposition axioms or making assumptions about the target/source nature of the variables. By comparing different PID lattices of the same system, we unveil a finer PID structure made of seven nonnegative information subatoms that are invariant to different target/source classifications and that are sufficient to describe the relationships among all PID lattices. This finer structure naturally splits redundant information into two nonnegative components: the source redundancy, which arises from the pairwise correlations between the source variables, and the non-source redundancy, which does not, and relates to the synergistic information the sources carry about the target. The invariant structure is also sufficient to construct the system's entropy, hence it characterizes completely all the interdependencies in the system.

Keywords: information theory; information decomposition; redundancy; synergy; multivariate dependencies

---

1. Introduction

Shannon's mutual information [1] provides a well established, widely applicable tool to characterize the statistical relationship between two stochastic variables. Larger values of mutual information correspond to a stronger relationship between the instantiations of the two variables in each single trial. Whenever we study a system with more than two variables, the mutual information between any two subsets of the variables still quantifies the statistical dependencies between these two subsets; however, many scientific questions in the analysis of complex systems require a finer characterization of how all variables simultaneously interact [2–6]. For example, two of the variables, A and B, may carry either redundant or synergistic information about a third variable C [7–9], but considering the value of the mutual information $I((A, B) : C)$ alone is not enough to distinguish these qualitatively different information-carrying modes. To achieve this finer level of understanding, recent theoretical efforts have focused on decomposing the mutual information between two subsets of variables into more specific information components (see e.g., [6,10–12]). Nonetheless, a complete framework for the information-theoretic analysis of multivariate systems is still lacking.

Characterizing the fine structure of the interactions among three stochastic variables can improve the understanding of many interesting problems across different disciplines [13–16]. For instance, this is the case for many important questions in the study of neural information processing. Determining quantitatively how two neurons encode information about an external sensory stimulus [7–9] requires describing the dependencies between the stimulus and the activity of the two neurons. Determining how the stimulus information carried by a neural response relates to the animal's behaviour [17–19] requires the analysis of the simultaneous three-wise dependencies among the stimulus, the neural activity and the subject's behavioral report. More generally, a thorough understanding of even the simplest information-processing systems would require the quantitative description of all different ways two inputs carry information about one output [20].

In systems where legitimate assumptions can be made about which variables act as sources of information and which variable acts as the target of information transmission, the partial information decomposition (PID) [10] provides an elegant framework to decompose the mutual information that one or two (source) variables carry about the third (target) variable into a finer lattice of redundant, unique and synergistic information atoms. However, in many systems the *a priori* classification of variables into sources and target is arbitrary, and limits the description of the distribution of information within the system [21]. Furthermore, even when one classification is adopted, the PID atoms do not characterize completely all the possible modes of information sharing between the sources and the target. For example, two sources can carry redundant information about the target irrespective of the strength of the correlations between them and, as a consequence, the PID redundancy atom can be larger than zero even if the sources have no mutual information [3,22,23]. Hence, the value of the PID redundancy measure cannot distinguish how the correlations between two variables contribute to the information that they share about a third variable.

In this paper, we address these limitations by extending the PID framework without introducing further axioms or assumptions about the three-variable (or *trivariate*) structure to analyze. We compare the atoms from the three possible PID lattices that are induced by the three possible choices for the target variable in the system. By tracking how the PID information modes change across different lattices, we move beyond the partial perspective intrinsic to a single PID lattice and unveil the finer structure common to all PID lattices. We find that this structure can be fully described in terms of a unique minimal set of seven information-theoretic quantities, which is invariant to different classifications of the variables. These quantities are derived from the PID atoms based on the relationships between different PID lattices.

The first result of this approach is the identification of two nonnegative subatomic components of the redundant information that any pair of variables carries about the third variable. The first component, that we name source redundancy (SR), quantifies the part of the redundancy which arises from the correlations of the sources. The second component, that we name non-source redundancy (NSR), quantifies the part of the redundancy which is not related to the source correlations. Interestingly, we find that whenever the non-source redundancy is larger than zero then also the synergy is larger than zero. The second result is that the minimal set induces a unique nonnegative decomposition of the full joint entropy $H(X, Y, Z)$ of the system. This allows us to dissect completely the distribution of information of any trivariate system in a general way that is invariant with respect to the source/target classification of the variables. To illustrate the additional insights of this new approach, we finally apply our framework to paradigmatic examples, including discrete and continuous probability distributions. These applications confirm our intuitions and clarify the practical usefulness of the finer PID structure. We also briefly discuss how our methods might be extended to the analysis of systems with more than three variables.

2. Preliminaries and State of the Art

Williams and Beer proposed an influential axiomatic construction that, in the general multivariate case, allows decomposing the mutual information that a set of sources has about a target into a series of redundant, synergistic, and unique contributions to the information. In the bivariate source case, i.e., for a system with two sources, this decomposition can be used to break down the mutual information

$I(X : (Y, Z))$ that two stochastic variables Y, Z (the sources) carry about a third variable X (the target) into the sum of four nonnegative atoms [10]:

- the Shared Information $SI(X : \{Y; Z\})$, which is the information about the target that is shared between the two sources (the redundancy);
- the Unique Informations $UI(X : \{Y \backslash Z\})$ and $UI(X : \{Z \backslash Y\})$, which are the separate pieces of information about the target that can be extracted from one of the sources, but not from the other;
- the Complementary Information $CI(X : \{Y; Z\})$, which is the information about the target that is only available when both of the sources are jointly observed (the synergy).

This construction is commonly known as the Partial Information Decomposition (PID). Sums of subsets of the four PID atoms provide the classical mutual information quantities between each of the sources and the target, $I(X : Y)$ and $I(X : Z)$, and the conditional mutual information quantities whereby one of the sources is the conditioned variable, $I(X : Z|Y)$ and $I(X : Y|Z)$. Such relationships are displayed with a color code in Figure 1.

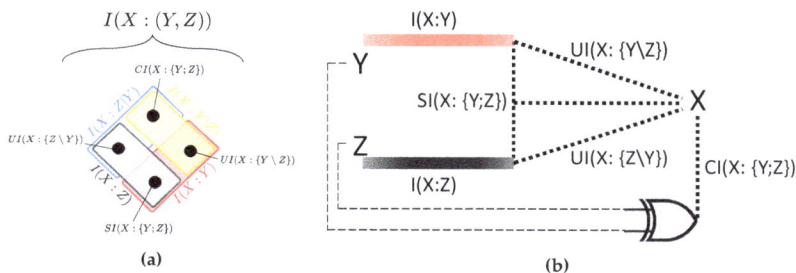

Figure 1. The Partial Information Decomposition as defined by Williams and Beer's axioms [10]. (a) The mutual information of the sources Y, Z about the target X is decomposed into four atoms: the redundancy $SI(X : \{Y; Z\})$, the unique informations $UI(X : \{Y \backslash Z\})$, $UI(X : \{Z \backslash Y\})$, and the synergy $CI(X : \{Y; Z\})$. The colored rectangles represent the linear equations that relate the four PID atoms to four Shannon information quantities; (b) An exploded view of the allotment of information between the sources Y, Z and the target X: each PID atom of panel (a) corresponds to a thick dotted line, while the colored stripes represent the two pairwise mutual informations between each of the sources and the target (with the same color code as in (a)). Each of the mutual informations splits into the sum of the redundancy with its corresponding unique information. The circuit-diagram symbol for the XOR operation is associated to the synergistic component $CI(X\{Y; Z\})$ only for illustration, as XOR is often taken as a paradigmatic example of synergistic interaction between variables.

The PID decomposition of Ref. [10] is based upon a number of axioms that formalize some properties that a measure of redundancy should have. These axioms are expressed in simple terms as follows: (1) the redundancy should be symmetric under any permutation of the sources (weak symmetry); (2) for a single source, the redundancy should equal the mutual information between the source and the target (self-redundancy); (3) the redundancy should not increase if a new source is added (monotonicity) [3,24]. However, these axioms do not determine univocally the value of the four PID atoms. For example, some definitions of redundancy imply that all PID atoms are nonnegative (global positivity) [10,11,22], while some authors have questioned that this property should be always satisfied [25]. Further, the specific redundancy measure proposed in Ref. [10] has been questioned as it can lead to unintuitive results [22], and thus many attempts have been devoted to finding alternative measures [22,25–30] compatibly with an extended number of axioms, such as the *identity axiom* proposed in [22]. Other work has studied in more detail the lattice structure that underpins the PID, indicating the duality between information gain and information loss lattices [12]. Even though there is no consensus on how to build partial information decompositions in systems with more than two sources, for trivariate systems the measures of redundancy, synergy and unique information defined in Ref. [11] have found wide acceptance (as a terminology note, it is common in the literature to refer to PID decompositions for systems containing a target and two sources as bivariate decompositions. That is, while the system is trivariate, the decomposition is sometimes referred to as bivariate based on the number of sources). In this paper, we will in fact make use of these measures when a concrete implementation of the PID will be required.

Even in the trivariate case with two sources and one target, however, there are open problems regarding the understanding of the PID atoms in relation to the interdependencies within the system. First, Harder et al. [22] pointed out that the redundant information shared between the sources about the target can intuitively arise from the following two qualitatively different modes of three-wise interdependence:

- the *source redundancy*, which is redundancy which 'must already manifest itself in the mutual information between the sources' (Note that Ref. [22] interchangeably refers to the sources as 'inputs': we will discuss this further in Section 4 when addressing the characterization of source redundancy);
- the *mechanistic redundancy*, which can be larger than zero even if there is no mutual information between the sources.

As pointed out by Harder and colleagues [22], a more precise conceptual and formal separation of these two kinds of redundancy still needs to be achieved, and presents fundamental challenges. The very notion that two statistically independent sources can nonetheless share information about a target was not captured by some earlier definitions of redundancy [31]. However, Ref. [3] provided a game-theoretic argument to show intuitively how independent variables may share information about another variable. Nonetheless, several studies [23,32] described the property that the PID measures of redundancy can be positive even when there are no correlations between the sources as undesired. On a different note, other authors [20] pointed out that the two different notions of redundancy can define qualitatively different modes of information processing in (neural) input-output networks.

Other issues were recently pointed out by James and Crutchfield [21], who indicated that the very definition of the PID lattice prevents its use as a general tool for assessing the full structure of trivariate (let alone multi-variate) statistical dependencies. In particular, Ref. [21] considered dyadic and triadic systems, which underlie quite interesting and common modes of multivariate interdependencies. They showed that, even though the PID atoms are among the very few measures that can distinguish between the two kinds of systems, a PID lattice with two source variables and one target variable cannot allot the full joint entropy $H(X, Y, Z)$ of either system. The decomposition of the joint entropy in terms of information components that reflect qualitatively different interactions within the system has also been subject of recent research, that however relies on constructions differing substantially from the PID lattice [29,33].

In summary, the PID framework, in its current form, does not yet provide a satisfactorily fine and complete description of the distribution of information in trivariate systems. The PID atoms do assess trivariate dependencies better than Shannon's measures, but they cannot quantify interesting finer interdependencies within the system, such as the source redundancy that the sources share about the target and the mechanistic redundancy that even two independent sources can share about the target. In addition, they are limited to describing the dependencies between the chosen sources and target, thus enforcing a certain perspective on the system that does not naturally suit all systems.

3. More PID Diagrams Unveil Finer Structure in the PID Framework

To address the open problems described above, we begin by pointing out the feature of the PID lattice that underlies all the issues in the characterization of trivariate systems outlined in Section 2. As we illustrate in Figure 1a, while a single PID diagram involves the mutual information quantities that one or both of the sources (in the figure, Y and Z) carry about the target X, it does not contain the mutual information between the sources $I(Y : Z)$ and their conditional mutual information given the target $I(Y : Z|X)$. This precludes the characterization of source redundancy with a single PID diagram, as it prevents any comparison between the elements of the PID and $I(Y : Z)$. Moreover, it also signals that a single PID lattice with two sources and one target cannot in general account for the total entropy $H(X, Y, Z)$ of the system.

These considerations suggest that the inability of the PID framework to provide a complete information-theoretic description of trivariate systems is not a consequence of the axiomatic construction underlying the PID lattice. Instead, it follows from restricting the analysis to the limited perspective on the system that is enforced by classifying the variables into sources and target when defining a PID lattice. We thus elaborate that significant progress can be achieved, *without the addition of further axioms or assumptions to the PID framework*, if one just considers, alongside the PID diagram in Figure 1a, the other two PID diagrams that are induced respectively by labeling Y or Z as the target in the system. When considering the resulting three PID diagrams (Figure 2), the previously missing mutual information $I(Y : Z)$ between the original sources of the left-most diagram is now decomposed into PID atoms of the middle and the right-most diagrams in Figure 2, and the same happens with $I(Y : Z|X)$.

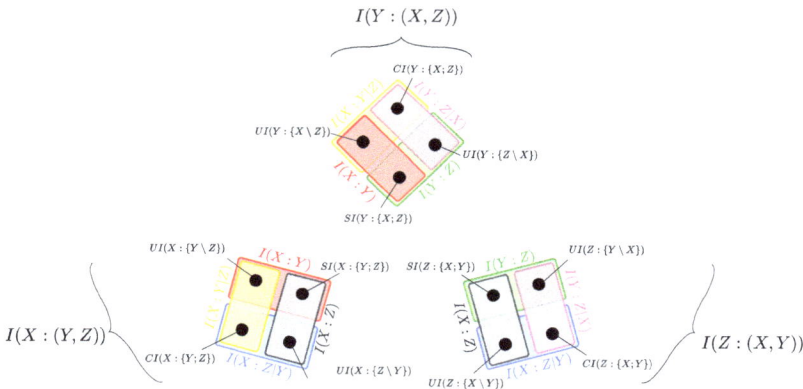

Figure 2. The three possible PIDs of a trivariate probability distribution $p(x, y, z)$ that follow from the three possible choices for the target variable: on the left the target is X, in the middle it is Y and on the right it is Z. The coloured rectangles highlight the linear relationships between the twelve PID atoms and the six Shannon information quantities. Note that the orientations of the PIDs for $I(X : (Y, Z))$ and $I(Z : (X, Y))$ are rotated with respect to the PID for $I(Y : (X, Z))$ to highlight their reciprocal relations, as will become more apparent in Figure 4.

In the following we take advantage of this shift in perspective to resolve the finer structure of the PID diagrams and, at the same time, to generalize its descriptive power going beyond the current limited framework, where only the information that two (source) variables carry about the third (target) variable is decomposed. More specifically, even though the PID relies on setting a partial point of view about the system, we will show that describing how the PID atoms change when we systematically rotate the choice of the PID target variable effectively overcomes the limitations intrinsic to one PID alone.

3.1. The Relationship between PID Diagrams with Different Target Selections

To identify the finer structure underlying all the PID diagrams in Figure 2, we first focus on the relationships between the PID atoms of two different diagrams, with the goal of understanding how to move from one perspective on the system to another. The key observation here is that, for each pair of variables in the system, their mutual information and their conditional mutual information given the third variable appear in two of the PID diagrams. This imposes some constraints relating the PID atoms in two different diagrams. For example, if we consider $I(X : Y)$ and $I(X : Y|Z)$, we find that:

$$I(X : Y) = SI(X : \{Y; Z\}) + UI(X : \{Y \backslash Z\}) = SI(Y : \{X; Z\}) + UI(Y : \{X \backslash Z\}), \tag{1}$$

$$I(X : Y|Z) = CI(X : \{Y; Z\}) + UI(X : \{Y \backslash Z\}) = CI(Y : \{X; Z\}) + UI(Y : \{X \backslash Z\}), \tag{2}$$

where the first and second equality in each equation result from the decomposition of $I(X : (Y, Z))$ (left-most diagram in Figure 2) and $I(Y : (X, Z))$ (middle diagram in Figure 2), respectively. From Equation (1) we see that, when the roles of a target and a source are reversed (here, the roles of X and Y), the difference in redundancy is the opposite of the difference in unique information with respect to the other source (here, Z). Similarly, Equation (2) shows that the difference in synergy is the opposite of the difference in unique information with respect to the other source. Combining these two equalities, we also see that the difference in redundancy is equal to the difference in synergy. Therefore, the equalities impose relationships across some PID atoms appearing in two different diagrams.

These relationships are depicted in Figure 3. The eight PID atoms appearing in the two diagrams can be expressed in terms of only six subatoms, due to the constraints of the form of Equations (1) and (2). In particular, to select the smallest nonnegative pieces of information resulting from the constraints, we define:

$$RSI(X \overset{Z}{\leftrightarrow} Y) := \min[SI(X : \{Y; Z\}), SI(Y : \{X; Z\})], \tag{3a}$$

$$RCI(X \overset{Z}{\leftrightarrow} Y) := \min[CI(X : \{Y; Z\}), CI(Y : \{X; Z\})], \tag{3b}$$

$$RUI(X \overset{Z}{\leftrightarrow} Y) := \min[UI(X : \{Y \backslash Z\}), UI(Y : \{X \backslash Z\})]. \tag{3c}$$

The above terms are called the Reversible Shared Information of X and Y considering Z ($RSI(X \overset{Z}{\leftrightarrow} Y)$; the orange block in Figure 3), the Reversible Complementary Information of X and Y considering Z ($RCI(X \overset{Z}{\leftrightarrow} Y)$; the gray block in Figure 3), and the Reversible Unique Information of X and Y considering Z ($RUI(X \overset{Z}{\leftrightarrow} Y)$; the magenta block in Figure 3). The attribute *reversible* highlights that, when we reverse the roles of target and source between the two variables at the endpoints of the arrow in RSI, RCI, or RUI (here, X and Y), the reversible pieces of information are still included in the same type of PID atom (redundancy, synergy, or unique information with respect to the third variable). For example, the orange block in Figure 3 indicates a common amount of redundancy in both PID diagrams: as such, $RSI(X \overset{Z}{\leftrightarrow} Y)$ contributes both to redundant information that Y and Z share about X, and to redundant information that X and Z share about Y. By construction, these reversible components are symmetric in the reversed variables. Note that, when we reverse the role of two variables, the third variable (here, Z) remains a source and is thus put in the middle of our notation in Equation (3a)–(3c). We also define the Irreversible Shared Information $IRSI(X \overset{Z}{\leftarrow} Y)$ between X and Y considering Z (the light blue block in Figure 3) as follows:

$$IRSI(X \overset{Z}{\leftarrow} Y) := SI(X : \{Y; Z\}) - RSI(X \overset{Z}{\leftrightarrow} Y). \tag{4}$$

The attribute *irreversible* in the above definition indicates that this piece of redundancy is specific to one of the two PIDs alone. More precisely, the uni-directional arrow in $IRSI(X \overset{Z}{\leftarrow} Y)$ indicates that this piece of information is a part of the redundancy with X as a target, but it is not a part of the redundancy with Y as a target (In this paper, directional arrows never represent any kind of *causal* directionality: the PID framework is only capable to quantify statistical (correlational) dependencies). Correspondingly, at least one between $IRSI(X \overset{Z}{\leftarrow} Y)$ and $IRSI(Y \overset{Z}{\leftarrow} X)$ is always zero. More generally, $IRSI$ quantifies asymmetries between two different PIDs: for example, when moving from the left to the right PID in Figure 3, the light blue block $IRSI(X \overset{Z}{\leftarrow} Y)$ indicates an equivalent amount of information that is lost for the redundancy $SI(X : \{Y; Z\})$ and the synergy $CI(X : \{Y; Z\})$ atoms, and is instead counted as a part of the unique information $UI(Y : \{X \backslash Z\})$ atom. In other words, assuming that the two redundancies are ranked as in Figure 3, we find that:

$$IRSI(X \overset{Z}{\leftarrow} Y) = SI(X : \{Y; Z\}) - SI(Y : \{X; Z\}) \tag{5a}$$

$$= CI(X : \{Y; Z\}) - CI(Y : \{X; Z\}) \tag{5b}$$

$$= UI(Y : \{X \backslash Z\}) - UI(X : \{Y \backslash Z\}). \tag{5c}$$

While the coarser Shannon information quantities that are decomposed in both diagrams in Figure 3, namely $I(X : Y)$ and $I(X : Y | Z)$, are symmetric under swap of $X \leftrightarrow Y$, their PID decompositions (see Equations (1) and (2)) are not: Equation (5) show that $IRSI$ quantifies the amount of this asymmetry. More precisely, the PID decompositions of $I(X : Y)$ and $I(X : Y | Z)$ will preserve the $X \leftrightarrow Y$ symmetry if and only if $IRSI(X \overset{Z}{\leftarrow} Y) = IRSI(Y \overset{Z}{\leftarrow} X) = 0$. Note that the differences of redundancies, of synergies, and of unique information terms are always constrained by equations such as Equation (5a)–(5c). Hence, unlike for the reversible measures, we do not need to consider independent notions of irreversible synergy or irreversible unique information.

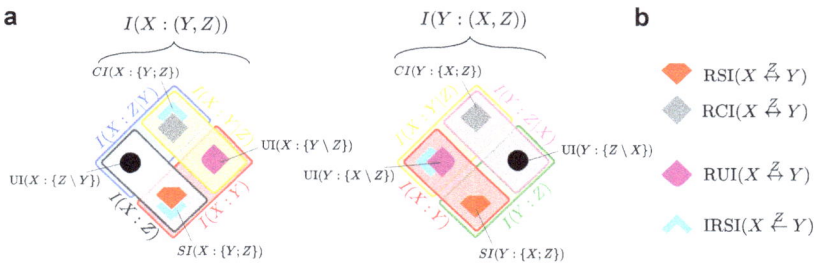

Figure 3. (**a**) The relationships between the PID atoms from two diagrams with different target selections. When we swap a target and a source, the differences in the amount of redundancy, synergy, and unique information with respect to the third variable are not independent due to equations of the type of Equations (1) and (2). Here, we consider the PID diagram of $I(X : (Y, Z))$ (left) and $I(Y : (X, Z))$ (right) from Figure 2, under the assumption that $SI(Y : \{X; Z\}) \leq SI(X : \{Y; Z\})$; (**b**) The reversible pieces of information $RSI(X \overset{Z}{\leftrightarrow} Y)$ (orange block), $RCI(X \overset{Z}{\leftrightarrow} Y)$ (gray block) and $RUI(X \overset{Z}{\leftrightarrow} Y)$ (magenta block) contribute to the same kind of atom across the two PID diagrams. The irreversible piece of information $IRSI(X \overset{Z}{\leftarrow} Y)$ (light blue block) contributes to different kinds of atom across the two PID diagrams. The remaining two unique information atoms (black dots) are not constrained by equations of the type of Equations (1) and (2) when only X or Y are considered as target.

In summary, the four subatoms in Equations (3a)–(3c) and (4), together with the two remaining unique information terms (the black dots in Figure 3), allow us to characterize both PIDs in Figure 3 and to understand how the PID atoms change when moving from one PID to another. We remark that in all cases the subatoms indicate *amounts* of information, while the same subatom can be interpreted in different ways depending on the specific PID atom to which it contributes. In other words, the fact that two atoms contain the same subatom does not indicate that there is some qualitatively equivalent information contained in both atoms. For example, both the redundancy $SI(X : \{Y; Z\})$ and the synergy $CI(X : \{Y; Z\})$ in Figure 3 contain the same subatom $IRSI(X \overset{Z}{\leftarrow} Y)$, even though the PID construction ensures that the redundancy and the synergy contain qualitatively different pieces of information.

3.2. Unveiling the Finer Structure of the PID Framework

So far we have examined the relationships among the PID atoms corresponding to two different perspectives we hold about the system, whereby we reverse the roles of target and source between two variables in the system. We have seen that the PID atoms of different diagrams are not independent, as they are constrained by equations of the type of Equations (1) and (2). More specifically, the eight PID atoms of two diagrams can be expressed in terms of only six independent quantities, including reversible and irreversible pieces of information. The next question is how many subatoms we need to describe all the three possible PIDs (see Figure 2). Since there are six constraints, three equations of the type of Equation (1) and three equations of the type of Equation (2), one may be tempted to think that the twelve PID atoms of all three PID diagrams can be expressed in terms of only six independent quantities. However, the six constraints are not independent: this is most easily seen from the symmetry of the co-information measure [34], which is defined as the mutual information of two variables minus their conditional information given the third (e.g., Equation (1) minus Equation (2)). The co-information is invariant to any permutation of the variables, and this property highlights that only five of the six constraints are linearly independent. Accordingly, we will now detail how seven subatoms are sufficient to describe the relationships among the atoms of all three PIDs: we call these subatoms the *minimal subatoms' set* of the PID diagrams.

In Figure 4 we see how the minimal subatoms' set builds the PID diagrams. We assume, without loss of generality, that $SI(Y : \{X; Z\}) \leq SI(X : \{Y; Z\}) \leq SI(Z : \{X; Y\})$. Then, we consider the three possible instances of Equation (5) for the three possible choices of the target variable, and we find that the same ordering also holds for the synergy atoms: $CI(Y : \{X; Z\}) \leq CI(X : \{Y; Z\}) \leq CI(Z : \{X; Y\})$. This property is related to the invariance of the co-information: indeed, Ref. [10] indicated that the co-information can be expressed as the difference between the redundancy and the synergy within each PID diagram, i.e.,

$$col(X; Y; Z) = SI(i : \{j; k\}) - CI(i : \{j; k\}), \tag{6}$$

for any assignment of X, Y, Z, to i, j, k.

These ordering relations are enough to understand the nature of the minimal subatoms' set: we start with the construction of the three redundancies, which can all be expressed in terms of the smallest RSI and two subsequent increments. In Figure 4, these correspond respectively to $RSI(X \overset{Y}{\leftrightarrow} Y) = SI(Y : \{X; Z\})$ (orange block), $IRSI(X \overset{Z}{\leftarrow} Y)$ (light blue block) and $IRSI(Z \overset{Y}{\leftarrow} X)$ (yellow block). In parallel, we can construct the three synergies with the smallest RCI and the same increments used for the redundancies. In Figure 4, these correspond respectively to $RCI(X \overset{Y}{\leftrightarrow} Y) = CI(Y : \{X; Z\})$ (gray block) and the same two $IRSI$ used before. To construct the unique information atoms, it is sufficient to further consider the three independent RUI defined by taking all possible permutations of X, Y and Z in Equation (3c). In Figure 4, these correspond to $RUI(X \overset{Z}{\leftrightarrow} Y) = UI(X : \{Y \backslash Z\})$ (magenta block), $RUI(X \overset{Y}{\leftrightarrow} Z) = UI(Z : \{X \backslash Y\})$ (brown block), and $RUI(Y \overset{X}{\leftrightarrow} Z) = UI(Z : \{Y \backslash X\})$ (blue block). We thus see that, in total, seven minimal subatoms are enough to describe the underlying structure of the three PID diagrams of any system. Among these seven building blocks, five are reversible pieces of information, i.e., they contribute to the same kind of PID atom across different PID diagrams; the other two are irreversible pieces of information, that contribute to different kinds of PID atom across different diagrams. The complete minimal set can only be determined when all three PIDs are jointly considered and compared. That is, after the PID atoms have been evaluated, the subatoms provide a complete description of the structure of the PID atoms. As shown in Figure 3, pairwise PIDs' comparisons can at most distinguish two subatoms in any redundancy (or synergy), while the three-wise PIDs' comparison discussed above allowed us to discern three subatoms in $SI(Z : \{X; Y\})$ (and $CI(Z : \{X; Y\})$; see also Figure 4). The full details of the mathematical construction of the decomposition presented in Figure 4 is described in Appendix A.

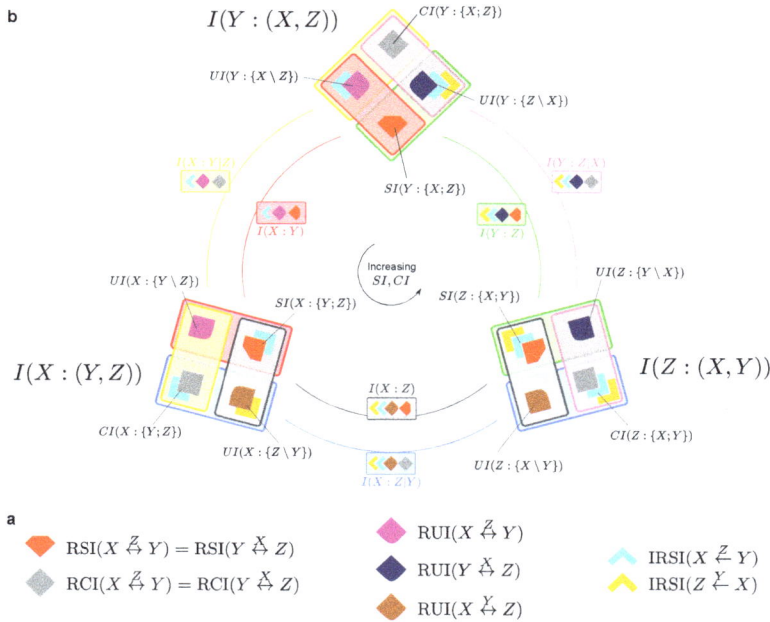

Figure 4. Constructing the full structure of the three PID diagrams in Figure 2 in terms of a minimal set of information subatoms. (**a**) All the 12 PID atoms in Figure 2 can be expressed as sums of seven independent PID subatoms that are displayed as coloured blocks in (**b**) (as in Figure 2, the orientations of the PIDs for $I(X : (Y, Z))$ and $I(Z : (X, Y))$ are rotated with respect to the PID for $I(Y : (X, Z))$ to highlight their reciprocal relations). Five of these subatoms are reversible pieces of information, that are included in the same kind of PID atom across different PID diagrams; the other two subatoms are the irreversible pieces of information, that can be included in different kinds of PID atom across different diagrams. Assuming, without loss of generality, that $SI(Y : \{X; Z\}) \le SI(X : \{Y; Z\}) \le SI(Z : \{X; Y\})$, the five reversible subatoms are: $RSI(X \overset{Z}{\leftrightarrow} Y)$ (orange), $RCI(X \overset{Z}{\leftrightarrow} Y)$ (gray), $RUI(X \overset{Z}{\leftrightarrow} Y)$ (magenta), $RUI(Y \overset{X}{\leftrightarrow} Z)$ (blue), and $RUI(X \overset{Y}{\leftrightarrow} Z)$ (brown). The two irreversible subatoms are $IRSI(X \overset{Z}{\leftarrow} Y)$ (light blue) and $IRSI(Z \overset{Y}{\leftarrow} X)$ (yellow).

Importantly, while the definition of a single PID lattice relies on the specific perspective adopted on the system, which labels two variables as the sources and one variable as the target, the decomposition in Figure 4 is invariant with respect to the classification of the variables. As described above, it only relies on computing all three PID diagrams and then using the ordering relations of the atoms, without any need to classify the variables *a priori*. As illustrated in Figure 4, the decomposition of the mutual information and conditional mutual information quantities in terms of the subatoms is also independent of the PID adopted. Our invariant minimal set in Figure 4 thus extends the descriptive power of the PID framework beyond the limitations that were intrinsic to considering an individual PID diagram. In the next sections, we will show how the invariant minimal set can be used to identify the part of the redundant information about a target that specifically arises from the mutual information between the sources (the source redundancy), and to decompose the total entropy of any trivariate system.

Remarkably, the decomposition in Figure 4 does not rely on any extension of Williams and Beer's axioms. We unveiled finer structure underlying the PID lattices just by considering more PID lattices at a time and comparing PID atoms across different lattices. The operation of taking the minimum between different PID atoms that defines the reversible pieces of information in Equation (3a)–(3c) might be reminiscent of the minimum operation that underlies the definition of the redundancy measures $I_{\min}$ in Ref. [10] and I_{red} in Ref. [22]. However, the minimum in Equation (3a)–(3c) operates on pieces of information about different variables and identifies a common amount of information between different PID atoms, while the minimum in $I_{\min}$ and I_{red} always operates on pieces of information about the same variable (the target) and aims at identifying information about the target that is qualitatively common between the two sources. We further remark that the decomposition in Figure 4 does not rely in any respect on the specific definition of the PID measures that is used to calculate the PID atoms: it only relies on the axiomatic PID construction presented in Ref. [10].

4. Quantifying Source Redundancy

The structure of the three PID diagrams that was unveiled with the construction in Figure 4 enables a finer characterization of the modes of information distribution among three variables than what has previously been possible. In particular, we will now address the open problem of quantifying the *source redundancy*, i.e., the part of the redundancy that 'must already manifest itself in the mutual information between the sources' [22]. Consider for example the redundancy $SI(X : \{Y; Z\})$ in Figure 4: it is composed by $RSI(X \overset{Z}{\leftrightarrow} Y)$ (orange block) and $IRSI(X \overset{Z}{\leftarrow} Y)$ (light blue block). We can check which of these subatoms are shared with the mutual information of the sources $I(Y : Z)$. To do this, we have to move from the middle PID diagram in Figure 4, that contains $SI(X : \{Y; Z\})$, to any of the other two diagrams, that both contain $I(Y : Z)$. Consistently, in these other two diagrams $I(Y : Z)$ is composed by the same four subatoms (the orange, the light blue, the yellow and the blue block), and the only difference across diagrams is that these subatoms are differently distributed between unique information and redundancy PID atoms. In particular, we can see that both the orange and the light blue block which make up $SI(X : \{Y; Z\})$ are contained in $I(Y : Z)$. Thus, whenever any of them is nonzero, we know *at the same time* that Y and Z share some information about X (i.e., $SI(X : \{Y; Z\}) > 0$) and that there are correlations between Y and Z (i.e., $I(Y : Z) > 0$). Accordingly, in the scenario of Figure 4 the entire redundancy $SI(X : \{Y; Z\})$ is explained by the mutual information of the sources: all the redundant information that Y and Z share about X arises from the correlations between Y and Z.

If we then consider the redundancy $SI(Y : \{X; Z\})$, that coincides with the orange block, we also find that it is totally explained in terms of the mutual information between the corresponding sources $I(X : Z)$, which indeed contains an orange block. However, if we consider the third redundancy $SI(Z : \{X; Y\})$, that is composed by an orange, a light blue and a yellow block, we find that only the orange and the light blue block contribute to $I(X : Y)$, while the yellow does not. The source redundancy that X and Y share about Z should thus equal the sum of the orange and the light blue block, that are common to both the redundancy $SI(Z : \{X; Y\})$ and the mutual information $I(X : Y)$. To quantify in full generality the amounts of information that contribute both to the redundancy SI and to the mutual information of its sources, we define the source redundancy that two sources S_1 and S_2 share about a target T as:

$$SR(T : \{S_1; S_2\}) := \max\{RSI(T \overset{S_2}{\leftrightarrow} S_1), \, RSI(T \overset{S_1}{\leftrightarrow} S_2)\}$$
$$= \max\{\min[SI(T : \{S_1; S_2\}), SI(S_1 : \{S_2; T\})], \quad\quad (7)$$
$$\min[SI(T : \{S_1; S_2\}), SI(S_2 : \{S_1; T\})]\}.$$

One can easily verify that Equation (7) identifies the blocks that contribute to both $SI(T : \{S_1; S_2\})$ and $I(S_1 : S_2)$ in Figure 4, for any choice of sources and target (for instance, $T = Z$, $S_1 = X$ and $S_2 = Y$). This definition can be justified as follows: each RSI measure in Equation (7) compares $SI(T : \{S_1; S_2\})$ with one of the other two redundancies that are contained in the mutual information between the sources $I(S_1 : S_2)$ (namely, $SI(S_1 : \{S_2; T\})$ and $SI(S_2 : \{S_1; T\})$). Some of the subatoms included in $I(S_1 : S_2)$ are contained in one of these two redundancies, but not in the other, as they move to the unique information mode when we change PID. Therefore, by taking the maximum in

Equation (7) we ensure that $SR(T : \{S_1; S_2\})$ precisely captures all the common subatoms of $I(S_1 : S_2)$ and $SI(T : \{S_1; S_2\})$, without double-counting. In a complementary way, we can define the non-source redundancy that two sources share about a target:

$$NSR(T : \{S_1; S_2\}) := SI(T : \{S_1; S_2\}) - SR(T : \{S_1; S_2\}). \tag{8}$$

In particular, if we consider the redundancy $SI(Z : \{X; Y\})$ in Figure 4, $NSR(Z : \{X; Y\})$ corresponds to the yellow block. As desired, whenever this block is larger than zero, X and Y can share information about Z (i.e., $SI(Z : \{X; Y\}) > 0$) even if there is no mutual information between the sources X and Y (i.e., $I(X : Y) = 0$).

4.1. The Difference between Source and Non-Source Redundancy

Equations (7) and (8) show how we can split the redundant information that two sources share about a target into two nonnegative information components: when the source-redundancy SR is larger than zero there are also correlations between the sources, while the non-source redundancy NSR can be larger than zero even when the sources are independent. SR is thus seen to quantify the *pairwise correlations between the sources that also produce redundant information about the target*: this discussion is pictorially summarized in Figure 5. In particular, the source redundancy SR is clearly upper-bound by the mutual information between the sources, i.e.,

$$SR(T : \{S_1; S_2\}) \leq I(S_1 : S_2). \tag{9}$$

On the other hand, NSR does not arise from the pairwise correlations between the sources: let us calculate NSR in a paradigmatic example that was proposed in Ref. [22] to remark the subtle possibility that two statistically independent variables can share information about a third variable. Suppose that Y and Z are uniform binary random variables, with $Y \perp Z$, and X is deterministically fixed by the relationship $X = Y \wedge Z$, where $\wedge$ represents the AND logical operation. Here, $SI(X : \{Y; Z\}) \approx 0.311$ bits (according to different measures of redundancy [11,22]) even if $I(Y : Z) = 0$. Indeed, from our definitions in Equations (7) and (8) we find that, since $SI(Y : \{X; Z\}), SI(Z : \{X; Y\}) \leq I(Y : Z) = 0$, here $NSR(X : \{Y; Z\}) = SI(X : \{Y; Z\}) > 0$ even though $I(Y : Z) = 0$. We will comment more extensively on this instructive example in Section 6.

Interestingly, the non-source redundancy NSR is a part of the redundancy that is related to the synergy of the same PID diagram. Indeed, two of the three possible NSR defined in Equation (8) are always zero, and the third can be larger than zero if and only if the yellow block in Figure 4 is larger than zero. From Figures 4 and 5 we can thus see that, whenever we find positive non-source redundancy in a PID diagram, the same amount of information (the yellow block) is also present in the synergy of that diagram. Thus, while there is source redundancy if and only if there is mutual information between the sources, the existence of non-source redundancy is a sufficient (though not necessary) condition for the existence of synergy. We can thus interpret NSR as *redundant information about the target that implies that the sources carry synergistic information about the target*: we give a graphical characterization of NSR in Figure 5. In particular, the non-source redundancy is clearly upper-bound by the corresponding synergy:

$$NSR(T : \{S_1; S_2\}) \leq CI(T : \{S_1; S_2\}). \tag{10}$$

In the specific examples considered in Harder et al. [22], where the underlying causal structure of the system is such that the sources always generate the target, the non-source redundancy can indeed be associated with the notion of 'mechanistic redundancy' that was introduced in that work: the causal mechanisms connecting the target with the sources induce a non-zero NSR that contributes to the redundancy independently of the correlations between the sources. In general, since the causal structure of the analyzed system is unknown, it is impossible to quantify 'mechanistic redundancy' with statistical measures, while it is always possible to quantify and interpret the non-source redundancy as described in Section 4.

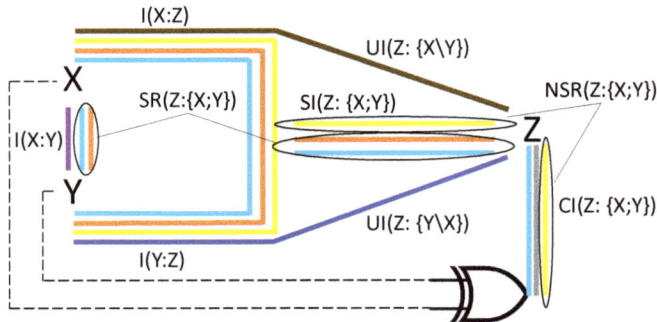

Figure 5. Exploded view of the information that two variables X, Y carry about a third variable Z. The mutual informations of each source about the target are decomposed into PID atoms as in Figure 1b, but the PID atoms are now further decomposed in terms of the minimal subatoms' set: the thick colored lines represent the subatoms with the same colour code as in Figure 4. Here, we assume that the variables are ordered as in Figure 4. The finer structure of the PID atoms allows us to identify the source redundancy $SR(Z : \{X;Y\})$ (orange line + light blue line) as the part of the full redundancy $SI(Z : \{X;Y\})$ that is apparent in the mutual information between the sources $I(X : Y)$. Instead, the amount of information in the non-source redundancy $NSR(Z : \{X;Y\})$ (yellow line) also appears in the synergy $CI(Z : \{X;Y\})$.

In Section 6 we will examine concrete examples to show how our definitions of source and non-source redundancy refine the information-theoretic description of trivariate systems, as they quantify qualitatively different ways that two variables can share information about a third.

We conclude this Section with more general comments about our quantification of source redundancy. We note that the arguments used to define the source redundancy in Section 4 can be equally used to study common or exclusive information components of other PID terms. For example, we can identify the magenta subatom as the component of the mutual information between the sources X and Y that cannot be related to their redundant information about Z. Similarly, we could consider which part of a synergy is related to the conditional mutual information between the sources.

5. Decomposing the Joint Entropy of a Trivariate System

Understanding how information is distributed in trivariate systems should also provide a descriptive allotment of all parts of the joint entropy $H(X, Y, Z)$ [21,33]. For comparison, Shannon's mutual information enables a semantic decomposition of the bivariate entropy $H(X, Y)$ in terms of univariate conditional entropies and $I(X : Y)$, that quantifies shared fluctuations (or covariations) between the two variables [33]:

$$H(X, Y) = I(X : Y) + H(X|Y) + H(Y|X). \tag{11}$$

However, in spite of recent proposals to decompose the joint entropy [29,33], a univocal descriptive decomposition of the trivariate entropy $H(X, Y, Z)$ in terms of nonnegative information-theoretic quantities is still missing to date. Since the PID axioms in Ref. [10] decompose mutual information quantities, one might hope that the PID atoms could also provide a descriptive entropy decomposition. Yet, at the beginning of Section 3, we pointed out that a single PID lattice does not include the mutual information between the sources and their conditional mutual information given the target: this suggests that a single PID lattice with two sources and one target cannot in general contain the full $H(X,Y,Z)$. More concretely, Ref. [21] has recently suggested precise examples of trivariate dependencies, i.e., dyadic and triadic dependencies, where a single PID lattice with two sources and one target cannot account for, and thus describe the parts of, the full $H(X,Y,Z)$.

We will now show how the novel finer PID structure unveiled in Section 3 can be used to decompose the full entropy $H(X,Y,Z)$ of any trivariate system in terms of nonnegative information-theoretic quantities. Then, to show the descriptive power of this decomposition, we will analyze the dyadic and triadic dependencies that were considered in Ref. [21], and that are described in Figure 6. Both kinds of dependencies underlie common modes of information sharing among three and more variables, but in both cases the atoms of a single PID diagram only sum up to two of the three bits of the full entropy $H(X,Y,Z)$ [21]. We will illustrate how the missing allotment of the third bit of entropy in those systems is not due to intrinsic limitations of the PID axioms, but just to the limitations of considering a single PID diagram at a time—the common practice in the literature so far.

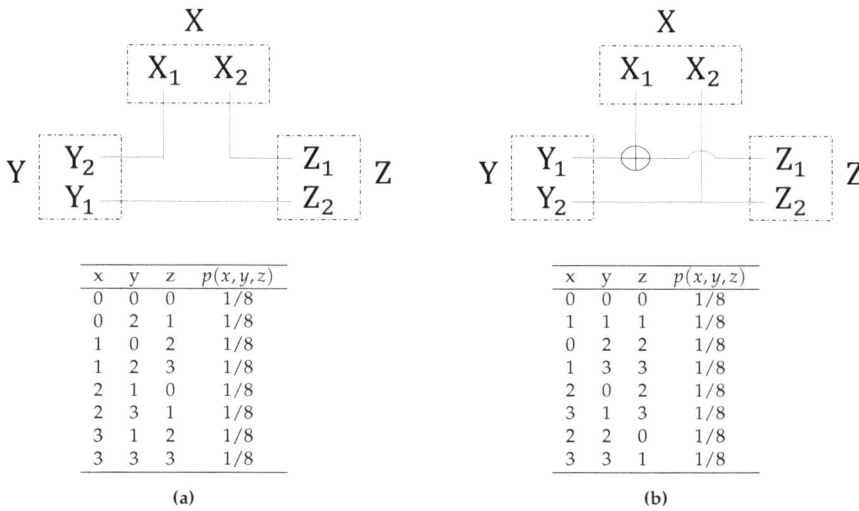

x	y	z	$p(x,y,z)$
0	0	0	1/8
0	2	1	1/8
1	0	2	1/8
1	2	3	1/8
2	1	0	1/8
2	3	1	1/8
3	1	2	1/8
3	3	3	1/8

(a)

x	y	z	$p(x,y,z)$
0	0	0	1/8
1	1	1	1/8
0	2	2	1/8
1	3	3	1/8
2	0	2	1/8
3	1	3	1/8
2	2	0	1/8
3	3	1	1/8

(b)

Figure 6. Dyadic and triadic statistical dependencies in a trivariate system, as defined in Ref. [21]. The tables display the non-zero probability values $p(x,y,z)$ as a function of the possible outcomes of the three stochastic variables X,Y,Z with domain $\{0,1,2,3\}$. $X \sim (X_1,X_2)$, $Y \sim (Y_1,Y_2)$, $Z \sim (Z_1,Z_2)$ (the symbol $\sim$ here means 'is distributed as'), where $X_1, X_2, Y_1, Y_2, Z_1, Z_2$ are binary uniform random variables. (**a**) The underlying rules that give rise to dyadic dependencies are $X_1 = Y_2, Y_1 = Z_2, Z_1 = X_2$; (**b**) The underlying rules that give rise to triadic dependencies are $X_1 = Y_1 \oplus Z_1, X_2 = Y_2 = Z_2$.

5.1. The Finer Structure of the Entropy $H(X,Y,Z)$

The minimal subatoms' set that we illustrated in Figure 4 allowed us to decompose all three PID lattices of a generic system. However, to fully describe the distribution of information in trivariate systems, we also wish to find a generalization of Equation (11) to the trivariate case, i.e., to decompose the full trivariate entropy $H(X,Y,Z)$ in terms of univariate conditional entropies and PID quantities. With this goal in mind, we first subtract from $H(X,Y,Z)$ the terms which describe statistical fluctuations of only one variable (conditioned on the other two). The sum of these terms was indicated as $H_{(1)}$ in Ref. [33], and there quantified as

$$H_{(1)} = H(X|Y,Z) + H(Y|X,Z) + H(Z|Y,X). \tag{12}$$

This subtraction is useful because $H_{(1)}$ is a part of the total entropy which does not overlap with any of the 12 PID atoms in Figure 2. The remaining entropy $H(X, Y, Z) - H_{(1)}$ was defined as the dual total correlation in Ref. [35] and recently considered in Ref. [33]:

$$DTC := H(X, Y, Z) - H_{(1)}. \tag{13}$$

DTC quantifies joint statistical fluctuations of more than one variable in the system. A simple calculation yields

$$DTC = I(X : Y|Z) + I(Y : Z|X) + I(X : Z|Y) + coI(X; Y; Z), \tag{14}$$

which is manifestly invariant under permutations of X, Y and Z, and shows that DTC can be written as a sum of some of the 12 PID atoms. For example, expressing the co-information as the difference $I(X : Z) - I(X : Z|Y)$, we can arbitrarily use the four atoms from the left-most diagram in Figure 2 to decompose the sum $I(X : Y|Z) + I(X : Z)$ and then add $UI(Y : \{Z \backslash X\}) + CI(Y : \{X; Z\})$ from the middle diagram to decompose $I(Y : Z|X)$. If we then plug this expression of DTC in Equation (13), we achieve a decomposition of the full entropy of the system in terms of $H_{(1)}$ and PID quantities:

$$H(X, Y, Z) = H_{(1)} + SI(X : \{Y; Z\}) + UI(X : \{Y \backslash Z\}) + UI(X : \{Z \backslash Y\})$$
$$+ CI(X : \{Y; Z\}) + UI(Y : \{Z \backslash X\}) + CI(Y : \{X; Z\}), \tag{15}$$

which provides a nonnegative decomposition of the total entropy of any trivariate system. However, this decomposition is not unique, since the co-information can be expressed in terms of different pairs of conditional and unconditional mutual informations, according to Equation (6). This arbitrariness strongly limits the descriptive power of this kind of entropy decompositions, because the PID atoms on the RHS can only be interpreted within individual PID perspectives.

To address this issue, we construct a less arbitrary entropy decomposition by using the invariant minimal subatoms' set that was presented in Figure 4: importantly, that set can be interpreted without specifying an individual, and thus partial, PID point of view that we hold about the system.

Thus, after we name the variables of the system such that $SI(Y : \{X; Z\}) \leq SI(X : \{Y; Z\}) \leq SI(Z : \{X; Y\})$, we express the coarser PID atoms in Equation (15) in terms of the minimal set to obtain:

$$H(X, Y, Z) - H_{(1)} = RSI(Y \overset{Z}{\leftrightarrow} X) + 2 \, RCI(Y \overset{Z}{\leftrightarrow} X) \tag{16}$$
$$+ RUI(X \overset{Z}{\leftrightarrow} Y) + RUI(Y \overset{X}{\leftrightarrow} Z) + RUI(X \overset{Y}{\leftrightarrow} Z)$$
$$+ 2 \, IRSI(Z \overset{Y}{\leftarrow} X) + 3 \, IRSI(X \overset{Z}{\leftarrow} Y).$$

Unlike Equation (15), the entropy decomposition expressed in Equation (16) and illustrated in Figure 7 fully describes the distribution of information in trivariate systems without the need of a specific perspective about the system. Importantly, this decomposition is unique: even though the co-information can be expressed in different ways in terms of conditional and unconditional mutual informations, in terms of the subatoms it is uniquely represented as the orange block minus the gray block (see Figure 7). Similarly, the conditional mutual information terms of Equation (14) are composed by the same blocks independently of the PID, as highlighted in Figure 4.

$$H(X,Y,Z) = H_{(1)} + \boxed{\;\blacklozenge\;} + \boxed{\;\blacklozenge\;} + \boxed{\;\blacklozenge\;} + \boxed{\;\blacklozenge\; - \;\blacklozenge\;}$$

$$\underset{I(X:Y|Z)}{} \quad \underset{I(Y:Z|X)}{} \quad \underset{I(X:Z|Y)}{} \qquad \underset{coI(X;Y;Z)}{}$$

$$= H_{(1)} + \blacklozenge + \blacklozenge + \blacklozenge + \blacklozenge + \blacklozenge + \blacklozenge + \blacklozenge$$

Figure 7. The joint entropy of the system $H(X,Y,Z)$ is decomposed in terms of the minimal set identified in Figure 4, once the univariate fluctuations quantified with $H_{(1)}$ have been subtracted out (see Equation (16)). As in Figure 4, we assume without loss of generality $SI(Y : \{X;Z\}) \leq SI(Z : \{X;Y\}) \leq SI(X : \{Y;Z\})$. The finer PID structure unveiled in this work enables a general entropy decomposition in terms of quantities that can be interpreted without relying on a specific PID point of view, even though they have been defined within the PID framework. The colored areas represent the Shannon information quantities that are included in the DTC of Equation (14), with the same color code of Figure 2.

We remark that the entropy decomposition in Equation (16) does not rely in any respect on the choice of a specific redundancy measure, but it follows directly after the calculation of the minimal subatoms' set according to any definition of the PID.

5.2. Describing $H(X,Y,Z)$ for Dyadic and Triadic Systems

To show an application of the finer entropy decomposition in Equation (16), we now compute its terms for the dyadic and the triadic dependencies considered in Ref. [21] and defined in Figure 6. For this application, we make use of the definitions of PID that were proposed in Ref. [11]. In both cases $H_{(1)} = 0$. For the dyadic system, there are only three positive quantities in the minimal set: the three reversible unique informations $RUI(X \overset{Y}{\leftrightarrow} Z) = RUI(Y \overset{X}{\leftrightarrow} Z) = RUI(X \overset{Z}{\leftrightarrow} Y) = 1$ bit. For the triadic system, there are only two positive quantities in the minimal set: $RSI(X \overset{Z}{\leftrightarrow} Y) = RCI(X \overset{Z}{\leftrightarrow} Y) = 1$ bit, but $RCI(X \overset{Z}{\leftrightarrow} Y)$ is counted twice in the DTC. We illustrate the resulting entropy decompositions, according to Equation (16), in Figure 8.

The decompositions in Figure 8 enable a clear interpretation of how information is finely distributed within dyadic and triadic dependencies. The three bits of the total $H(X,Y,Z)$ in the dyadic system are seen to be distributed equally among unique information modes: each variable contains 1 bit of unique information with respect to the second variable about the third variable. Further, these unique information terms are all reversible, which reflects the symmetry of the system under pairwise swapping of the variables. This description provides a simple and accurate summary of the total entropy of the dyadic system, which matches the dependency structure illustrated in Figure 6a.

The three bits of the total $H(X,Y,Z)$ in the triadic system consist of one bit of the smallest reversible redundancy and two bits of the smallest reversible synergy, since the latter appears twice in $H(X,Y,Z)$. Again, the reversible nature of these pieces of information reflects the symmetry of the system under pairwise swapping of the variables. Further, the bit of reversible redundancy represents the bit of information that is redundantly available to all three variables, while the two bits of reversible synergy are due to the three-wise XOR structure (see Figure 6b). Why does the XOR structure provide two bits of synergistic information? Because if $X = Y \oplus Z$ then the only positive quantity in the set of subatoms in Figure 4b is the smallest RCI, which however appears twice in the entropy $H(X,Y,Z)$. Importantly, these two bits of synergy do not come from the same PID diagram: our entropy decomposition in Equation (16) could account for both bits only because it fundamentally relies on cross-comparisons between different PID diagrams, as illustrated in Figure 4.

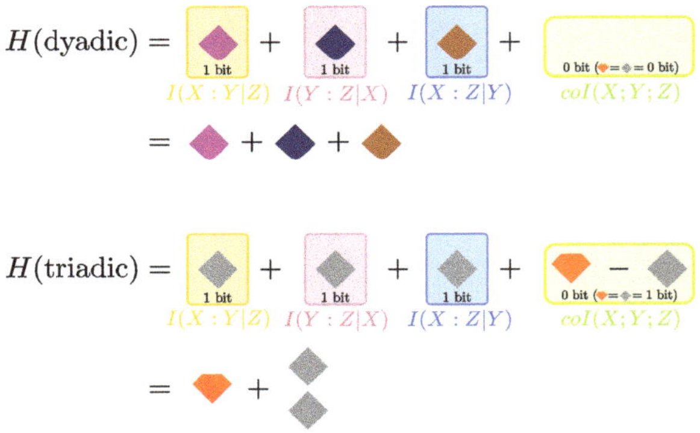

Figure 8. The joint entropy $H(X, Y, Z) = 3$ bits of a dyadic (upper panel) and a triadic (lower panel) system, as defined in Figure 6, is decomposed in terms of the minimal set as illustrated in Figure 7. In the dyadic system, $H(X, Y, Z)$ is decomposed into three pieces of (reversible) unique information: each variable contains 1 bit of unique information with respect to the second variable about the third variable. In the triadic system, $H(X, Y, Z)$ is decomposed into one bit of information shared among all three variables (the reversible redundancies) and two bits of (reversible) synergistic information due to the three-wise XOR structure.

We remark that the results and the interpretations presented in this subsection depend on our use of the PID definition presented in Ref. [11]. If a different redundancy measure is adopted, the subatoms that decompose the entropy $H(X, Y, Z)$ in Figure 8 would still be the same, but their value could change.

6. Applications of the Finer Structure of the PID Framework

The aim of this Section is to show the additional insights that the finer structure of the PID framework, unveiled in Section 3.2 and Figure 4, can bring to the analysis of trivariate systems. We examine paradigmatic examples of trivariate systems and calculate the novel PID quantities of source and non-source redundancy that we described in Section 4. Most of these examples have been considered in the literature [11,22,23,26,33] to validate the definitions, or to suggest interpretations, of the PID atoms. We also discuss how SR matches the notion of source redundancy introduced in Ref. [22] and discussed in Ref. [20]. Finally, we suggest and motivate a practical interpretation of the reversible redundancy subatom RSI.

The decomposition of the PID lattices in terms of the minimal subatoms' set that we illustrated in Figure 4 is to be calculated after the traditional PID atoms have been computed. Since it relies only on the lattice relations, the subatomic decomposition can be attained for any of the several measures that have been proposed to underpin the PID [10,11,22,26]. In some cases the value of the atoms can be derived from axiomatic arguments, hence it does not depend on the specific measure selected; in other cases it has to be actually calculated choosing a particular measure, and it can differ for different measures. In the latter cases, our computations rely on the definitions of PID that were proposed in Ref. [11], which are widely accepted in the literature for trivariate systems. Whenever we needed a numerical computation of the PID atoms, we performed it with a Matlab package we specifically developed for this task, which is freely available for download and reuse through Zenodo and Github (https://doi.org/10.5281/zenodo.850362).

6.1. Computing Source and Non-Source Redundancy

6.1.1. Copying—The Redundancy Arises Entirely from Source Correlations

Consider a system where Y and Z are random binary variables that are correlated according to a control parameter λ [22]. For example, consider a uniform binary random variable W that 'drives' both Y and Z with the same strength λ. More precisely, $p(y|w) = \lambda/2 + (1 - \lambda)\delta_{yw}$ and $p(z|w) = \lambda/2 + (1 - \lambda)\delta_{zw}$ [22]. This system is completed by taking $X = (Y, Z)$, i.e., a two-bit random variable that reproduces faithfully the joint outcomes of the generating variables Y, Z.

We consider the inputs Y and Z as the PID sources and the output X as the PID target, thus selecting the left-most PID diagram in Figure 2. Figure 9a shows our calculations of the full redundancy $SI(X : \{Y; Z\})$, the source redundancy $SR(X : \{Y; Z\})$ and the non-source redundancy $NSR(X : \{Y; Z\})$, based on the definitions in Ref. [11]. The parameter λ is varied between $\lambda = 0$, corresponding to $Y = Z$, and $\lambda = 1$, corresponding to $Y \perp Z$. Since $NSR(X : \{Y; Z\}) = 0$ for any $0 \leq \lambda \leq 1$, we interpret that all the redundancy $SI(X : \{Y; Z\})$ arises from the correlations between Y and Z (which are tuned with λ). This property was presented as the identity axiom in Ref. [22], where the authors argued that in the 'copying' example the entire redundancy should be already apparent in the sources. Indeed, the PID definitions of Ref. [11] that we use here abide by the identity axiom, and these results would not change if we used other PID definitions that still satisfy the identity axiom.

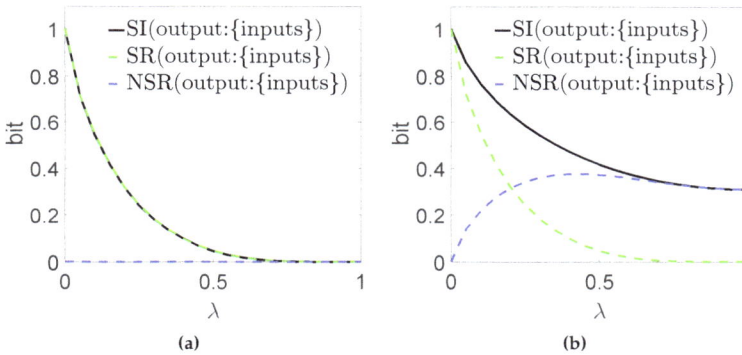

Figure 9. The binary random variables Y and Z are uniformly distributed inputs that determine the output X as $X = (Y, Z)$ in panel (**a**) and as $X = Y \wedge Z$ in panel (**b**). Correlations between Y and Z decrease with increasing λ, from perfect correlation ($\lambda = 0$) to perfect independence ($\lambda = 1$). The full redundancy $SI(X : \{Y; Z\})$, the source redundancy $SR(X : \{Y; Z\})$ and the non-source redundancy $NSR(X : \{Y; Z\})$ of the inputs about the output are plotted as a function of λ. (**a**) Since the output variable X copies the inputs (Y, Z), all of $SI(X : \{Y; Z\})$ can only come from the correlations between the inputs, which is reflected in $NSR(X : \{Y; Z\})$ being identically 0 for all values of λ. (**b**) For all values of $\lambda > 0$ we find $NSR(X : \{Y; Z\}) > 0$, i.e., there is a part of the redundancy $SI(X : \{Y; Z\})$ which does not arise from the correlations between the inputs Y and Z. Accordingly, $NSR(X : \{Y; Z\})$ also appears in the synergy $CI(X : \{Y; Z\})$.

6.1.2. AND Gate: The Redundancy is not Entirely Related to Source Correlations

Consider a system where the correlations between two binary random variables, the inputs Y and Z, are described by the control parameter λ as in Section 6.1.1, but the output X is determined by the AND function as $X = Y \wedge Z$ [22]. As the causal structure of the system would suggest, we consider the inputs Y and Z as the PID sources and the output X as the PID target, thus selecting the left-most PID diagram in Figure 2. Figure 9b shows our calculations of the full redundancy $SI(X : \{Y; Z\})$, the source redundancy $SR(X : \{Y; Z\})$ and the non-source redundancy $NSR(X : \{Y; Z\})$, based on the definitions in Ref. [11].

SR and NSR now show a non-trivial behavior as a function of the $Y - Z$ correlation parameter λ. If $\lambda = 0$ and thus $Y = Z$, the full redundancy is made up entirely of source redundancy—trivially, both $SI(X : \{Y; Z\})$ and $SR(X : \{Y; Z\})$ equal the mutual information $I(Y : Z) = H(Y) = 1$ bit. When λ increases, the full redundancy decreases monotonically to its minimum value of ≈ 0.311 bits for $\lambda = 1$ (when $Y \perp Z$). Importantly, $SR(X : \{Y; Z\})$ decreases monotonically as a function of λ to its minimum value of zero bits when $\lambda = 1$: this behavior is indeed expected from a measure that quantifies correlations between the sources that also produce redundant information about the target (see Section 4). On the other hand, $NSR(X : \{Y; Z\}) > 0$ for $\lambda > 0$, and it increases as a function of λ. When $\lambda = 1$, i.e., when $Y \perp Z$, NSR corresponds to the full redundancy. This is compatible with our description of non-source redundancy (see Section 4) as redundancy that is not related to the source correlations; indeed, $NSR > 0$ implies that the sources also carry synergistic information about the target (here, due to the relationship $X = Y \wedge Z$). NSR thus also quantifies the notion of mechanistic redundancy that was introduced in Ref. [22] with reference to this scenario.

6.1.3. Dice Sum: Tuning Irreversible Redundancy

Consider a system where Y and Z are two uniform random variables, each representing the outcome of a die throw [22]. A parameter λ controls the correlations between the two dice: $p(y, z) = \lambda/36 + (1 - \lambda)/6 \, \delta_{yz}$. Thus, for $\lambda = 0$ the dice throws always match, for $\lambda = 1$ the outcomes are completely independent. Further, the output X combines each pair (y, z) of input outcomes as $x = y + \alpha z$, where $\alpha \in \{1, 2, 3, 4, 5, 6\}$. This example was suggested in Ref. [22] specifically to point out the conceptual difficulties in quantifying the interplay between the redundancy and the source correlations, that we addressed with the identification of SR and NSR (see Section 4).

Harder et al. calculated redundancy by using their own proposed measure I_{red}, which also abides by the PID axioms in Ref. [10] but is different than the measure $SI(X : \{Y; Z\})$ introduced in Ref. [11]. However, Ref. [11] also calculated $SI(X : \{Y; Z\})$ for this example: they showed that the differences between $SI(X : \{Y; Z\})$ and I_{red} in this example are only quantitative (and there is no difference at all for some values of α), while the qualitative behaviour of both measures as a function of λ is very similar.

Figure 10 shows our calculations of the full redundancy $SI(X : \{Y; Z\})$, the source redundancy $SR(X : \{Y; Z\})$ and the non-source redundancy $NSR(X : \{Y; Z\})$, based on the definitions in Ref. [11]. We display the two 'extreme' cases $\alpha = 1$, $\alpha = 6$ for illustration.

With $\alpha = 6$, X is isomorphic to the joint variable (Y, Z): for any λ, this implies on one hand $SI(X : \{Y; Z\}) = SI((Y; Z) : \{Y; Z\}) = I(Y : Z)$, and on the other $UI(Y : \{Z \backslash X\}) = 0$ and thus $SI(Y : \{X; Z\}) = I(Y : Z)$ (see Equation (1)). According to Equation (7), we thus find that the source redundancy $SR(X : \{Y; Z\})$ saturates, for any λ, the general inequality in Equation (9): all correlations between the inputs Y, Z also produce redundant information about X (see Section 4). Further, according to Equation (8), we find $NSR(X : \{Y; Z\}) = 0$ for any λ (see Figure 10b): we thus interpret that all the redundancy $SI(X : \{Y; Z\})$ arises from correlations between the inputs. Instead, if we fix λ and decrease α, the two inputs Y, Z are more and more symmetrically combined in the output X: with $\alpha = 1$, the pieces of information respectively carried by each input about the output overlap maximally. Correspondingly, the full redundancy $SI(X : \{Y; Z\})$ increases [22]. However, keeping λ fixed does not change the inputs' correlations. Thus, we expect that the relative contribution of the inputs' correlations to the full redundancy $SI(X : \{Y; Z\})$ should decrease proportionally. Indeed, in Figure 10a we find $NSR(X : \{Y; Z\}) > 0$ for $\lambda > 0$, which signals that a part of $SI(X : \{Y; Z\})$ is not related to the inputs' correlations.

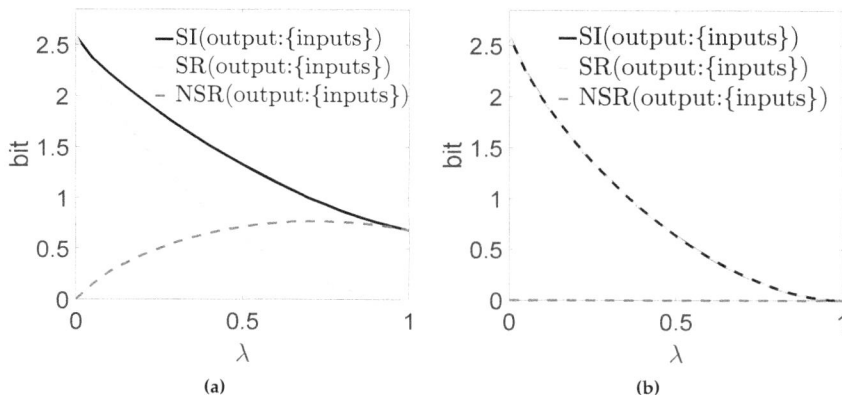

Figure 10. Two dice with uniformly distributed outcomes (y, z) ranging from 1 to 6 are the inputs fed to an output third variable X as follows: (a) $x = y + z$; (b) $x = y + 6 z$. The parameter λ controls the correlations between Y and Z (from complete correlation for $\lambda = 0$ to complete independence for $\lambda = 1$). In panel (a) the inputs Y, Z are symmetrically combined to determine the output X, while in panel (b) this symmetry is lost: accordingly, for any fixed λ the full redundancy $SI(X : \{Y; Z\})$ is larger in (a) than in (b). Since the value of α does not change the inputs' correlations, the relative contribution of the inputs' correlations to the full redundancy $SI(X : \{Y; Z\})$ increases from (a) to (b). Indeed, in panel (a) $NSR(X : \{Y; Z\}) > 0$ for $\lambda > 0$, while in (b) $SI(X : \{Y; Z\}) = SR(X : \{Y; Z\})$ for any λ, i.e., the redundancy arises entirely from inputs' correlations.

We finally note that also in this paradigmatic example the splitting of the redundancy into SR and NSR addresses the challenge of separating the two kinds of redundancy outlined in Ref. [22].

6.1.4. Trivariate Jointly Gaussian Systems

Barrett considered in detail the application of the PID to trivariate jointly Gaussian systems (X, Y, Z) in which the target is univariate [23]: he showed that several specific proposals for calculating the PID atoms all converge, in this case, to the same following measure of redundancy:

$$SI(X : \{Y; Z\}) = \min[I(X : Y), I(X : Z)]. \tag{17}$$

We note that Equation (17) highlights the interesting property that, in trivariate Gaussian systems with a univariate target, the redundancy is as large as it can be, since it saturates the general inequalities $SI(X : \{Y; Z\}) \leq I(X : Y), I(X : Z)$.

Direct application of our definitions in Equations (7) and (8) to such systems yields:

$$SR(X : \{Y; Z\}) = \min[I(X : Y), I(X : Z), I(Y : Z)],$$
$$NSR(X : \{Y; Z\}) = \min[I(X : Y), I(X : Z)] - \min[I(X : Y), I(X : Z), I(Y : Z)]. \tag{18}$$

Thus, we find that in these systems the source redundancy is also as large as it can be, since it also saturates the general inequalities $SR(X : \{Y; Z\}) \leq I(X : Y), I(X : Z), I(Y : Z)$ (which follow immediately from its definition in Equation (7)). Further, combining Equation (17) with Equation (18) gives:

$$SR(X : \{Y; Z\}) = \begin{cases} SI(X : \{Y; Z\}), & \text{if } I(Y : Z) \geq SI(X : \{Y; Z\}), \\ I(Y : Z), & \text{otherwise;} \end{cases} \quad (19)$$

$$NSR(X : \{Y; Z\}) = \begin{cases} 0, & \text{if } I(Y : Z) \geq SI(X : \{Y; Z\}), \\ SI(X : \{Y; Z\}) - I(Y : Z), & \text{otherwise.} \end{cases} \quad (20)$$

This identification of source redundancy, which quantifies *pairwise correlations between the sources that also produce redundant information about the target* (see Section 4), provides more insight about the distribution of information in Gaussian systems. Indeed, the property that source redundancy is maximal indicates that the correlations between any pair of source variables (for example, $\{Y; Z\}$ as considered above) produces as much redundant information as possible about the corresponding target (X as considered above). Accordingly, when $I(Y : Z) < SI(X : \{Y; Z\})$, the redundancy also includes some non-source redundancy $NSR(X : \{Y; Z\}) > 0$ that implies the existence of synergy, i.e., $CI(X : \{Y; Z\}) > 0$.

6.2. RSI Quantifies Information between two Variables That also Passes Monotonically Through the Third

In this section, we discuss a practical interpretation of the reversible redundancy subatom RSI defined in Equation (3a). We note that $RSI(X \overset{Z}{\leftrightarrow} Y)$ appears both in $SI(X : \{Y; Z\})$ and in $SI(Y : \{X; Z\})$, i.e., it quantifies a common amount of information that Z shares with each of the two other variables about the third variable. We shorten this description into the statement that $RSI(X \overset{Z}{\leftrightarrow} Y)$ quantifies 'information between X and Y that also passes through Z'. Reversible redundancy is further characterized by the following Proposition:

Proposition 1. $RSI(X \overset{(Z,Z')}{\leftrightarrow} Y) \geq RSI(X \overset{Z}{\leftrightarrow} Y).$

Proof. Both $SI(X : \{Y; (Z, Z')\}) \geq SI(X : \{Y; Z\})$ and $SI(Y : \{X; (Z, Z')\}) \geq SI(Y : \{X; Z\})$ [10]. Thus, $RSI(X \overset{Z}{\leftrightarrow} Y) = \min[SI(X : \{Y; Z\}), SI(Y : \{X; Z\})] \leq \min[SI(X : \{Y; (Z, Z')\}), SI(Y : \{X; (Z, Z')\})] = RSI(X \overset{(Z,Z')}{\leftrightarrow} Y)$. $\square$

Indeed, the fact that RSI always increases whenever we expand the middle variable corresponds to the increased capacity of the entropy of the middle variable to host information between the endpoint variables. We discuss this interpretation of RSI by examining several examples, where we can motivate *a priori* our expectations about this novel mode of information sharing.

6.2.1. Markov Chains

Consider the most generic Markov chain $X \to Z \to Y$, defined by the property that $p(x, y|z) = p(x|z)p(y|z)$, i.e., that X and Y are conditionally independent given Z. The Markov structure allows us to formulate a clear *a priori* expectation about the amount of information between the endpoints of the chain, X and Y, that also 'passes through the middle variable Z': this information should clearly equal $I(X : Y)$, because whatever information is established between X and Y *must* pass through Z.

Indeed, the Markov property $I(X : Y|Z) = 0$ implies, as we see immediately from Figure 2, that $SI(X : \{Y; Z\}) = SI(Y : \{X; Z\}) = I(X : Y)$. By virtue of Equations (3a) and (4), in accordance with our expectation, we find

$$RSI(X \overset{Z}{\leftrightarrow} Y) = I(X : Y), \quad (21)$$

which holds true independently of the marginal distributions of X, Y and Z. Notably, the symmetry of $RSI(X \overset{Z}{\leftrightarrow} Y)$ under swap of the endpoint variables $X \leftrightarrow Y$ is also compatible with the property that $X \to Z \to Y$ is a Markov chain if and only if $Y \to Z \to X$ is a Markov chain. In words, whatever information flows from X through Z to Y equals the information flowing from Y through Z to X.

More generally, the Markov property $I(X : Y|Z) = 0$ implies that $RCI(Y \overset{Z}{\leftrightarrow} X) = IRSI(X \overset{Z}{\leftarrow} Y)$ $= RUI(X \overset{Z}{\leftrightarrow} Y) = 0$: thus, only four of the seven subatoms of the minimal set in Figure 4 can be larger than zero. Following the procedure illustrated in Section 3.2, we decompose the three PIDs of the system in terms of the minimal subatoms' set as shown in Figure 11. In particular, we see that also $RSI(X \overset{Y}{\leftrightarrow} Z) = I(X : Y)$, thus matching our expectation that in the Markov chain $X \to Z \to Y$ the information between X and Z that also passes through Y still equals $I(X : Y)$. We finally note that none of the results regarding Markov chains depends on specific definitions of the PID atoms: they were derived only on the basis of the PID axioms in Ref. [10].

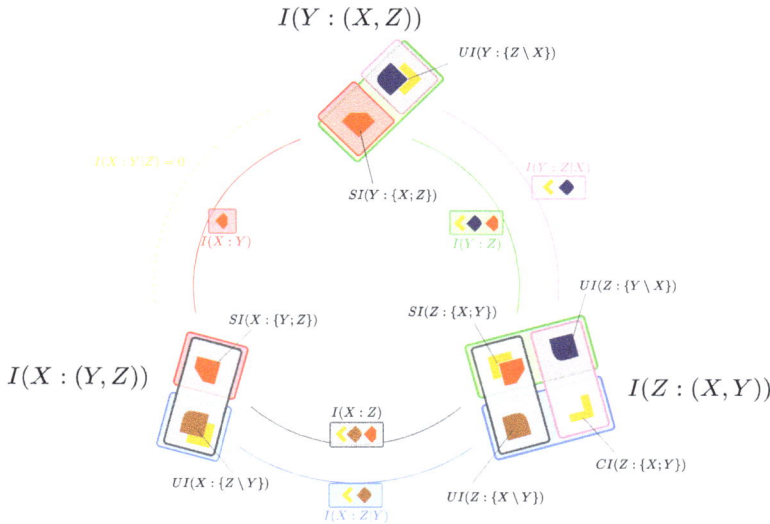

Figure 11. The finer structure of the three PID diagrams, as defined in Figure 4, in the case of a generic Markov chain $X \to Z \to Y$. Three of the seven subatoms of the minimal set described in Figure 4 are forced to zero by the Markov property $I(X : Y|Z) = 0$.

6.2.2. Two Parallel Communication Channels

Consider five binary uniform random variables X_1, X_2, Y_1, Y_2, Z with three parameters $0 \le \lambda_1, \lambda_2, \lambda_3 \le 1$ controlling the correlations $X_1 \overset{\lambda_1}{\leftrightarrow} Z \overset{\lambda_2}{\leftrightarrow} Y_1$, $X_2 \overset{\lambda_3}{\leftrightarrow} Y_2$ (in the same way λ controls the $Y \overset{\lambda}{\leftrightarrow} Z$ correlations in Section 6.1.1). We consider the trivariate system (X, Y, Z) with $X = (X_1, X_2)$ and $Y = (Y_1, Y_2)$ (see Figure 12). Given that $(X_1, Z, Y_1) \perp\!\!\!\perp (X_2, Y_2)$, we intuitively expect that the information between X and Y that also passes through Z, in this case, should equal $I(X_1 : Y_1)$, which in general will be smaller than $I(X : Y) = I(X_1 : Y_1) + I(X_2 : Y_2)$. Indeed, if we use the PID definitions of Ref. [11], $(X_1, Z, Y_1) \perp\!\!\!\perp (X_2, Y_2)$ implies that $SI((X_1 X_2) : \{(Y_1 Y_2); Z\}) = SI(X_1 : \{Y_1; Z\}) + SI(X_2 : \{Y_2; \varnothing\}) = SI(X_1 : \{Y_1; Z\})$ [11] and, similarly, $SI((Y_1 Y_2) : \{(X_1 X_2); Z\}) = SI(Y_1 : \{X_1; Z\})$. However, from the previous subsection we find that $SI(X_1 : \{Y_1; Z\}) = I(X_1 : Y_1)$ and $SI(Y_1 : \{X_1; Z\}) = I(X_1 : Y_1)$. Thus, we find that $RSI(X \overset{Y}{\leftrightarrow} Z) = I(X_1 : Y_1)$, in agreement with our expectations.

$$X_1 \xrightarrow{\lambda_1} Z \xrightarrow{\lambda_2} Y_1$$
$$X_2 \xrightarrow{\qquad \lambda_3 \qquad} Y_2$$

$$RSI((X_1, X_2) \overset{Z}{\leftrightarrow} (Y_1, Y_2)) = I(X_1 : Y_1)$$

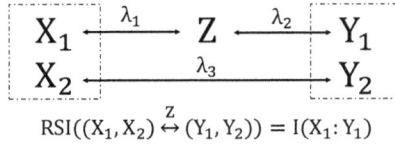

Figure 12. $X = (X_1, X_2)$ and $Y = (Y_1, Y_2)$ share information via two parallel channels: one passes through Z, the other does not. The parameters $\lambda_1, \lambda_2, \lambda_3$ control the correlations as depicted with the arrows. If we use the PID definitions in Ref. [11], we find that, in agreement with the interpretation in Section 6.2, $RSI(X \overset{Z}{\leftrightarrow} Y)$ here quantifies information between X and Y that also passes through Z, as it is equal to $I(X_1 : Y_1)$.

6.2.3. Other Examples

To further describe the interpretation of $RSI(X \overset{Z}{\leftrightarrow} Y)$ as information between X and Y that also passes through Z, we here reconsider the examples, among those discussed before Section 6.2, where we can formulate intuitive expectations about this information sharing mode. In the dyadic system described in Figure 6a, our expectation is that there should be no information between two variables that also passes through the other: indeed, $RSI(X \overset{Z}{\leftrightarrow} Y) = 0$ in this case. Instead, in the triadic system described in Figure 6b, we expect that the information between two variables that also passes through the other should equal the information that is shared among all three variables, which amounts to 1 bit. Indeed, we find $RSI(X \overset{Z}{\leftrightarrow} Y) = 1$ bit. In the 'copying' example in Section 6.1.1 $X = (Y, Z)$, thus we expect that $I(Y : Z)$ corresponds to information between Y, Z that also passes through X, but also to information between X and Z that also passes through Y. Indeed, $RSI(X \overset{Z}{\leftrightarrow} Y) = RSI(X \overset{Y}{\leftrightarrow} Z) = I(Y : Z)$. We remark that all the values of RSI in these examples depend on the specific definitions of the PID atoms in Ref. [11].

7. Discussion

The Partial Information Decomposition (PID) pioneered by Williams and Beer has provided an elegant construction to quantify redundant, synergistic, and unique information contributions to the mutual information that a set of sources carries about a target variable [10]. More generally, it has generated considerable interest as it addresses the difficult yet practically important problem of extending information theory beyond the classic bivariate measures of Shannon to fully characterize multivariate dependencies.

However, the axiomatic PID construction, as originally formulated by Williams and Beer, fundamentally relied on the separation of the variables into a target and a set of sources. While this classification was developed to study systems in which it is natural to identify a target, in general it introduces a partial perspective that prevents a complete and general information-theoretic description of the system. More specifically, the original PID framework could not quantify some important modes of information sharing, such as source redundancy [22], and could not allot the system's full joint entropy [21]. The work presented here addresses these issues focusing on trivariate systems, by extending the original PID framework in two respects. First, we decomposed the original PID atoms in terms of finer information subatoms with a well defined interpretation that is invariant to different variables' classifications. Then, we constructed an extended framework to completely decompose the distribution of information within any trivariate system.

Importantly, our formulation did not require the addition of further axioms to the original PID construction. We proposed that distinct PIDs for the same system, corresponding to different target selections, should be evaluated and then compared to identify how the decomposition of information changes across different perspectives. More specifically, we identified reversible pieces of information (RSI, RUI, RCI) that contribute to the same kind of PID atom if we reverse the roles of target and source between two variables. The complementary subatomic components of the PID lattices are the irreversible pieces of information ($IRSI$), that contribute to different kinds of PID atom for different target selections. These subatoms thus measure asymmetries between different decompositions of Shannon quantities pertaining to two different PIDs of the same system, and such asymmetries reveal the additional detail with which the PID atoms assess trivariate dependencies as compared to the coarser and more symmetric Shannon information quantities.

The crucial result of this approach was unveiling the finer structure underlying the PID lattices: we showed that an invariant minimal set of seven information subatoms is sufficient to decompose the three PIDs of any trivariate system. In the remainder of this section, possible uses of these subatoms and their implications for the understanding of systems of three variables are discussed.

7.1. Use of the Subatoms to Map the Distribution of Information in Trivariate Systems

Our minimal subatoms' set was first used to characterize more finely the distribution of information among three variables. We clarified the interplay between the redundant information shared between two variables A and B about a third variable C, on one side, and the correlations between A and B, on the other. We decomposed the redundancy into the sum of source-redundancy SR, which *quantifies the part of the redundancy which arises from the pairwise correlations between the sources A and B*, and non-source redundancy NSR, which can be larger than zero even if the sources A and B are statistically independent. Interestingly, we found that NSR *quantifies the part of the redundancy which implies that A and B also carry synergistic information about C*. The separation of these qualitatively different components of redundancy promises to be useful in the analysis of any complex system where several inputs are combined to produce an output [20].

Then, we used our minimal subatoms' set to extend the descriptive power of the PID framework in the analysis of any trivariate system. We constructed a general, unique, and nonnegative decomposition of the joint entropy $H(X, Y, Z)$ in terms of information-theoretic components that can be clearly interpreted without arbitrary variable classifications. This construction parallels the decomposition of the bivariate entropy $H(X, Y)$ in terms of Shannon's mutual information $I(X : Y)$. We demonstrated the descriptive power of this approach by decomposing the complex distribution of information in dyadic and triadic systems, which was shown not to be possible within the original PID framework [21].

We gave practical examples of how the finer structure underlying the PID atoms provides more insight into the distribution of information within important and well-studied trivariate systems. In this spirit, we put forward a practical interpretation of the reversible redundancy RSI, and future work will address additional interpretations of the components of the minimal subatoms' set.

7.2. Possible Extensions of the Formalism to Multivariate Systems with Many Sources

The insights that derive from our extension of the PID framework suggest that the PID lattices could also be useful to characterize the statistical dependencies in multivariate systems with more than two sources. Our approach does not rely on the adoption of specific PID measures, but only on the axiomatic construction of the PID lattice. Thus, it could be extended to the multivariate case by embedding trivariate lattices within larger systems' lattices [12]: a further breakdown of the minimal subatoms' set could be obtained if the current system were embedded as part of a bigger system. However, the implementation of these generalizations is left for future work. We anticipate that such generalizations would be computationally as expensive as the implementation of the multivariate PID method proposed by Williams and Beer. Indeed, after the computation of the traditional PID lattices, our approach only requires the comparison of different PID lattices, that relies on computing pairwise minima between PID atoms—a computationally trivial operation.

Further, when systems with more than two sources are considered, the definition of source redundancy might be extended as to determine which subatoms of a redundancy can be explained by dependencies among the sources (for example, by replacing the mutual information between two sources with a measure of the overall dependencies among all sources, such as the *total correlation* introduced in Ref. [36]). More generally, the idea of comparing different PID diagrams that partially decompose the same information can also be generalized to identify finer structure underlying higher-order PID lattices with different numbers of variables. These identifications might also help addressing specific questions about the distribution of information in complex multivariate systems.

7.3. Potential Implications for Systems Biology and Systems Neuroscience

A common problem in system biology is to characterize how the function of the whole biological system is shaped by the dependencies among its many constituent biological variables. In many cases, ranging from gene regulatory networks [37] to metabolic pathways [14] and to systems neuroscience [38–40], an information-theoretic decomposition of how information is distributed and processed within different parts of the system would allow a model-free characterization of these dependencies. The work discussed here can be used to shed more light on these issues by allowing to tease apart qualitatively different modes of interaction, as a first necessary step to understanding the causal structure of the observed phenomena [41].

In systems neuroscience, the decomposition introduced here may be important for studying specific and timely questions about neural information processing. This work can contribute to the study of neural population coding, that is the study of how the concerted activity of many neurons encodes information about ecologically relevant variables such as sensory stimuli [40,42]. In particular, a key characteristic of a neural population code is the degree to which pairwise or higher-order cross-neuron statistical dependencies are used by the brain to encode and process information, in a potentially redundant or synergistic way, across neurons [7–9] and across time [43]. Moreover, characterizing different types of redundancy may also be relevant to further study the relation between the information of neural responses and neural connectivity [44]. Our work is also of potential relevance to study another timely issue in neuroscience, that is the relevance for perception of the information about sensory variables carried by neural activity [19,45]. This is a crucial issue to resolve the diatribe about the nature of the neural code, that is the set of symbols used by neurons to encode information and produce brain function [46–48]. Addressing this problem requires mapping the information in the multivariate distribution of variables such as the stimuli presented to the subject, the neural activity elicited by the presentation of such stimuli, and the behavioral reports of the subject's perception. More specifically, it requires characterizing the information between the presented stimulus and the behavioral report of the perceived stimulus that can be extracted from neural activity [19]. It is apparent that developing general decompositions of the information exchanged in multivariate systems, as we did here, is key to succeeding in rigorously addressing these fundamental systems-level questions.

Acknowledgments: We are grateful to members of Panzeri's Laboratory for useful feedback, and to P. E. Latham, A. Brovelli and C. de Mulatier for useful discussions. This research was supported in part by the Fondation Bertarelli.

Author Contributions: All authors conceived the research; G.P., E.P. and D.C. performed the research; G.P., E.P. and D.C. wrote a first draft of the manuscript; all authors edited and approved the final manuscript; S.P. supervised the research.

Conflicts of Interest: The authors declare no conflict of interest.

Appendix A. Summary of the Relationships Among Distinct PID Lattices

In this Section we report the full set of equations and definitions that determine the finer structure of the three PID diagrams, as described in Section 3.2 and Figure 4 in the main text. Without loss of generality, we assume that

$$SI(Y : \{X; Z\}) \leq SI(X : \{Y; Z\}) \leq SI(Z : \{X; Y\}).$$

We start by writing down explicitly all the permutations of Equations (1), (2), and (6), relating traditional information quantities to the PID:

$$I(X:Y) = SI(X:\{Y;Z\}) + UI(X:\{Y\backslash Z\}) = SI(Y:\{X;Z\}) + UI(Y:\{X\backslash Z\}), \tag{A1a}$$

$$I(X:Z) = SI(X:\{Y;Z\}) + UI(X:\{Z\backslash Y\}) = SI(Z:\{X;Y\}) + UI(Z:\{X\backslash Y\}), \tag{A1b}$$

$$I(Y:Z) = SI(Y:\{X;Z\}) + UI(Y:\{Z\backslash X\}) = SI(Z:\{X;Y\}) + UI(Z:\{Y\backslash X\}); \tag{A1c}$$

$$I(X:Y|Z) = CI(X:\{Y;Z\}) + UI(X:\{Y\backslash Z\}) = CI(Y:\{X;Z\}) + UI(Y:\{X\backslash Z\}), \tag{A2a}$$

$$I(X:Z|Y) = CI(X:\{Y;Z\}) + UI(X:\{Z\backslash Y\}) = CI(Z:\{X;Y\}) + UI(Z:\{X\backslash Y\}), \tag{A2b}$$

$$I(Y:Z|X) = CI(Y:\{X;Z\}) + UI(Y:\{Z\backslash X\}) = CI(Z:\{X;Y\}) + UI(Z:\{Y\backslash X\}); \tag{A2c}$$

$$\begin{aligned}
CoI(X;Y;Z) &= SI(X:\{Y;Z\}) - CI(X:\{Y;Z\}) \\
&= SI(Y:\{X;Z\}) - CI(Y:\{X;Z\}) \\
&= SI(Z:\{X;Y\}) - CI(Z:\{X;Y\}).
\end{aligned} \tag{A3}$$

By using only the relations above, together with assumption (*) and the definitions of *RSI*, *RCI*, *RUI* and *IRSI* (Equations (3a)–(3c) and (4)), we will now show how all PID elements, and hence all information quantities above, can be decomposed in terms of the minimal set of seven subatoms depicted as elementary blocks in Figure 4.

By applying the definitions of *RSI* (Equation (3a)) and *IRSI* (Equation (4)), assumption (*) implies the following:

$$\begin{aligned}
IRSI(Y \overset{Z}{\nleftarrow} X) &= SI(Y:\{X;Z\}) - RSI(Y \overset{Z}{\nleftrightarrow} X) \\
&= SI(Y:\{X;Z\}) - \min[SI(X:\{Y;Z\}), SI(Y:\{X;Z\})] \\
&= SI(Y:\{X;Z\}) - SI(Y:\{X;Z\}) \\
&= 0,
\end{aligned} \tag{A4}$$

as well as

$$\begin{aligned}
IRSI(X \overset{Z}{\nleftarrow} Y) &= SI(X:\{Y;Z\}) - RSI(X \overset{Z}{\nleftrightarrow} Y) \\
&= SI(X:\{Y;Z\}) - \min[SI(X:\{Y;Z\}), SI(Y:\{X;Z\})] \\
&= SI(X:\{Y;Z\}) - SI(Y:\{X;Z\}) \\
&= CI(X:\{Y;Z\}) - CI(Y:\{X;Z\}) \\
&= UI(Y:\{X\backslash Z\}) - UI(X:\{Y\backslash Z\}),
\end{aligned} \tag{A5}$$

where the fourth equality follows from Equation (A3), and the fifth from Equation (A2a). Analogously, using Equations (A3) and (A2b), we get

$$\begin{aligned}
IRSI(Z \overset{Y}{\nleftarrow} X) &= SI(Z:\{X;Y\}) - RSI(Z \overset{Y}{\nleftrightarrow} X) \\
&= SI(Z:\{X;Y\}) - \min[SI(X:\{Y;Z\}), SI(Z:\{X;Y\})] \\
&= SI(Z:\{X;Y\}) - SI(X:\{Y;Z\}) \\
&= CI(Z:\{X;Y\}) - CI(X:\{Y;Z\}) \\
&= UI(X:\{Z\backslash Y\}) - UI(Z:\{X\backslash Y\}).
\end{aligned} \tag{A6}$$

Using the relations above, we can express the *SI* atoms of the three PIDs in terms of the *RSI* and *IRSI* subatoms:

$$SI(Y:\{X;Z\}) = RSI(X \overset{Z}{\nleftrightarrow} Y) + IRSI(Y \overset{Z}{\nleftarrow} X) = RSI(X \overset{Z}{\nleftrightarrow} Y), \tag{A7}$$

where we used Equation (A4),

$$SI(X : \{Y; Z\}) = RSI(X \overset{Z}{\leftrightharpoons} Y) + IRSI(X \overset{Z}{\leftharpoondown} Y), \tag{A8}$$

where we used Equation (A5), and

$$
\begin{aligned}
SI(Z : \{X; Y\}) &= SI(X : \{Y; Z\}) + [CI(Z : \{X; Y\}) - CI(X : \{Y; Z\})] \\
&= RSI(X \overset{Z}{\leftrightharpoons} Y) + IRSI(X \overset{Z}{\leftharpoondown} Y) + [CI(Z : \{X; Y\}) - CI(X : \{Y; Z\})] \\
&= RSI(X \overset{Z}{\leftrightharpoons} Y) + IRSI(X \overset{Z}{\leftharpoondown} Y) + IRSI(Z \overset{Y}{\leftharpoondown} X),
\end{aligned}
\tag{A9}
$$

where the first equality follows from Equation (A3), the second from Equation (A8) and the third from Equation (A6). Analogously, we can express the *CI* atoms in terms of *RCI* and *IRSI*:

$$CI(Y : \{X; Z\}) = RCI(X \overset{Z}{\leftrightharpoons} Y) + IRSI(Y \overset{Z}{\leftharpoondown} X) = RCI(X \overset{Z}{\leftrightharpoons} Y), \tag{A10}$$

$$CI(X : \{Y; Z\}) = RCI(X \overset{Z}{\leftrightharpoons} Y) + IRSI(X \overset{Z}{\leftharpoondown} Y), \tag{A11}$$

and

$$
\begin{aligned}
CI(Z : \{X; Y\}) &= CI(X : \{Y; Z\}) + [SI(Z : \{X; Y\}) - SI(X : \{Y; Z\})] \\
&= RCI(X \overset{Z}{\leftrightharpoons} Y) + IRSI(X \overset{Z}{\leftharpoondown} Y) + [SI(Z : \{X; Y\}) - SI(X : \{Y; Z\})] \\
&= RCI(X \overset{Z}{\leftrightharpoons} Y) + IRSI(X \overset{Z}{\leftharpoondown} Y) + IRSI(Z \overset{Y}{\leftharpoondown} X).
\end{aligned}
\tag{A12}
$$

Note how the same *IRSI* terms appear in Equations (A8) and (A11), and in Equations (A9) and (A12), in agreement with the invariance of the co-information (Equation (A3)).

Finally, to study the *UI* atoms, we observe that, given (*), Equation (A1a) implies $UI(X : \{Y \backslash Z\}) \leq UI(Y : \{X \backslash Z\})$, Equation (A1b) implies $UI(Z : \{X \backslash Y\}) \leq UI(X : \{Z \backslash Y\})$, and Equation (A1c) implies $UI(Z : \{Y \backslash X\}) \leq UI(Y : \{Z \backslash X\})$. These inequalities, in turn, imply that

$$RUI(X \overset{Z}{\leftrightharpoons} Y) = \min[UI(X : \{Y \backslash Z\}), UI(Y : \{X \backslash Z\})] = UI(X : \{Y \backslash Z\}),$$

$$RUI(X \overset{Y}{\leftrightharpoons} Z) = \min[UI(X : \{Z \backslash Y\}), UI(Z : \{X \backslash Y\})] = UI(Z : \{X \backslash Y\}),$$

$$RUI(Y \overset{X}{\leftrightharpoons} Z) = \min[UI(Y : \{Z \backslash X\}), UI(Z : \{Y \backslash X\})] = UI(Z : \{Y \backslash X\}),$$

which we can rewrite as

$$UI(X : \{Y \backslash Z\}) = RUI(X \overset{Z}{\leftrightharpoons} Y), \tag{A13}$$

$$UI(Z : \{X \backslash Y\}) = RUI(X \overset{Y}{\leftrightharpoons} Z), \tag{A14}$$

$$UI(Z : \{Y \backslash X\}) = RUI(Y \overset{X}{\leftrightharpoons} Z). \tag{A15}$$

Hence,

$$
\begin{aligned}
UI(Y : \{X \backslash Z\}) &= UI(X : \{Y \backslash Z\}) + [SI(X : \{Y; Z\}) - SI(Y : \{X; Z\})] \\
&= RUI(X \overset{Z}{\leftrightharpoons} Y) + IRSI(X \overset{Z}{\leftharpoondown} Y),
\end{aligned}
\tag{A16}
$$

where the first equality follows from Equation (A1a), and the second from Equations (A5) and (A13). Similarly,

$$UI(X : \{Z\backslash Y\}) = UI(Z : \{X\backslash Y\}) + [SI(Z : \{X; Y\}) - SI(X : \{Y; Z\})]$$
$$= RUI(X \overset{Y}{\leftrightarrow} Z) + IRSI(Z \overset{Y}{\leftarrow} X),$$

(A17)

where we used Equations (A1b), (A6) and (A14), and

$$UI(Y : \{Z\backslash X\}) = UI(Z : \{Y\backslash X\}) + [SI(Z : \{X; Y\}) - SI(Y : \{X; Z\})]$$
$$= RUI(Y \overset{X}{\leftrightarrow} Z) + IRSI(Z \overset{Y}{\leftarrow} X) + IRSI(Z \overset{Y}{\leftarrow} X),$$

(A18)

where we used Equations (A1c), (A7), (A9) and (A15).

Equations (A7)–(A18) thus describe all atoms of the three PIDs using the following seven subatoms:

$$IRSI(X \overset{Z}{\leftarrow} Y)$$
$$IRSI(Z \overset{Y}{\leftarrow} X)$$
$$RSI(X \overset{Z}{\leftrightarrow} Y)$$
$$RCI(X \overset{Z}{\leftrightarrow} Y)$$
$$RUI(X \overset{Z}{\leftrightarrow} Y)$$
$$RUI(X \overset{Y}{\leftrightarrow} Z)$$
$$RUI(Y \overset{X}{\leftrightarrow} Z).$$

This description is illustrated graphically in Figure 4, where each subatom is represented as a coloured block.

References

1. Shannon, C.E. A Mathematical Theory of Communication. *Bell Syst. Tech. J.* **1948**, *27*, 379–423.
2. Ay, N.; Olbrich, E.; Bertschinger, N.; Jost, J. A unifying framework for complexity measures of finite systems. In Proceedings of the European Conference Complex Systems, Oxford, UK, 25–29 September 2006; Volume 6.
3. Bertschinger, N.; Rauh, J.; Olbrich, E.; Jost, J. Shared Information—New Insights and Problems in Decomposing Information in Complex Systems. In Proceedings of the Proceedings of the ECCS 2012, Brussels, Belguim, 3–7 September 2012.
4. Tononi, G.; Sporns, O.; Edelman, G.M. Measures of degeneracy and redundancy in biological networks. *Proc. Natl. Acad. Sci. USA* **1999**, *96*, 3257–3262.
5. Tikhonov, M.; Little, S.C.; Gregor, T. Only accessible information is useful: Insights from gradient-mediated patterning. *R. Soc. Open Sci.* **2015**, *2*, 150486, doi: 10.1098/rsos.150486.
6. Timme, N.; Alford, W.; Flecker, B.; Beggs, J.M. Synergy, redundancy, and multivariate information measures: An experimentalist's perspective. *J. Comput. Neurosci.* **2014**, *36*, 119–140.
7. Pola, G.; Thiele, A.; Hoffmann, K.P.; Panzeri, S. An exact method to quantify the information transmitted by different mechanisms of correlational coding. *Network* **2003**, *14*, 35–60.
8. Schneidman, E.; Bialek, W.; Berry, M.J. Synergy, redundancy, and independence in population codes. *J. Neurosci.* **2003**, *23*, 11539–11553.
9. Latham, P.E.; Nirenberg, S. Synergy, redundancy, and independence in population codes, revisited. *J. Neurosci.* **2005**, *25*, 5195–5206.
10. Williams, P.L.; Beer, R.D. Nonnegative Decomposition of Multivariate Information. *arXiv* **2010**, arXiv:1004.2515.
11. Bertschinger, N.; Rauh, J.; Olbrich, E.; Jost, J.; Ay, N. Quantifying Unique Information. *Entropy* **2014**, *16*, 2161–2183.

12. Chicharro, D.; Panzeri, S. Synergy and Redundancy in Dual Decompositions of Mutual Information Gain and Information Loss. *Entropy* **2017**, *19*, 71, doi:10.3390/e19020071.

13. Anastassiou, D. Computational analysis of the synergy among multiple interacting genes. *Mol. Syst. Biol.* **2007**, *3*, 83, doi:10.1038/msb4100124.

14. Lüdtke, N.; Panzeri, S.; Brown, M.; Broomhead, D.S.; Knowles, J.; Montemurro, M.A.; Kell, D.B. Information-theoretic sensitivity analysis: A general method for credit assignment in complex networks. *J. R. Soc. Interface* **2008**, *5*, 223–235.

15. Watkinson, J.; Liang, K.C.; Wang, X.; Zheng, T.; Anastassiou, D. Inference of Regulatory Gene Interactions from Expression Data Using Three-Way Mutual Information. *Ann. N. Y. Acad. Sci.* **2009**, *1158*, 302–313.

16. Faes, L.; Marinazzo, D.; Nollo, G.; Porta, A. An Information-Theoretic Framework to Map the Spatiotemporal Dynamics of the Scalp Electroencephalogram. *IEEE Trans. Biomed. Eng.* **2016**, *63*, 2488–2496.

17. Pitkow, X.; Liu, S.; Angelaki, D.E.; DeAngelis, G.C.; Pouget, A. How Can Single Sensory Neurons Predict Behavior? *Neuron* **2015**, *87*, 411–423.

18. Haefner, R.M.; Gerwinn, S.; Macke, J.H.; Bethge, M. Inferring decoding strategies from choice probabilities in the presence of correlated variability. *Nat. Neurosci.* **2013**, *16*, 235–242.

19. Panzeri, S.; Harvey, C.D.; Piasini, E.; Latham, P.E.; Fellin, T. Cracking the Neural Code for Sensory Perception by Combining Statistics, Intervention, and Behavior. *Neuron* **2017**, *93*, 491–507.

20. Wibral, M.; Priesemann, V.; Kay, J.W.; Lizier, J.T.; Phillips, W.A. Partial information decomposition as a unified approach to the specification of neural goal functions. *Brain Cogn.* **2017**, *112*, 25–38.

21. James, R.G.; Crutchfield, J.P. Multivariate Dependence Beyond Shannon Information. *arXiv* **2016**, arXiv:1609.01233v2.

22. Harder, M.; Salge, C.; Polani, D. Bivariate measure of redundant information. *Phys. Rev. E* **2013**, *87*, 012130.

23. Barrett, A.B. Exploration of synergistic and redundant information sharing in static and dynamical Gaussian systems. *Phys. Rev. E* **2015**, *91*, 052802.

24. Williams, P.L. Information Dynamics: Its Theory and Application to Embodied Cognitive Systems. Ph.D. Thesis, Indiana University, Bloomington, IN, USA, 2011.

25. Ince, R.A.A. Measuring Multivariate Redundant Information with Pointwise Common Change in Surprisal. *Entropy* **2017**, *19*, 318, doi:10.3390/e19070318.

26. Griffith, V.; Koch, C. Quantifying Synergistic Mutual Information. In *Guided Self-Organization: Inception*; Springer: Berlin/Heidelberg, Germany, 2014; pp. 159–190.

27. Griffith, V.; Chong, E.K.P.; James, R.G.; Ellison, C.J.; Crutchfield, J.P. Intersection Information Based on Common Randomness. *Entropy* **2014**, *16*, 1985–2000.

28. Banerjee, P.K.; Griffith, V. Synergy, Redundancy and Common Information. *arXiv* **2015**, arXiv:1509.03706.

29. Ince, R.A.A. The Partial Entropy Decomposition: Decomposing multivariate entropy and mutual information via pointwise common surprisal. *arXiv* **2017**, arXiv:1702.01591v2.

30. Rauh, J.; Banerjee, P.K.; Olbrich, E.; Jost, J.; Bertschinger, N. On extractable shared information. *arXiv* **2017**, arXiv:1701.07805.

31. Griffith, V.; Koch, C. Quantifying synergistic mutual information. *arXiv* **2013**, arXiv:1205.4265v3.

32. Stramaglia, S.; Angelini, L.; Wu, G.; Cortes, J.M.; Faes, L.; Marinazzo, D. Synergetic and Redundant Information Flow Detected by Unnormalized Granger Causality: Application to Resting State fMRI. *IEEE Trans. Biomed. Eng.* **2016**, *63*, 2518–2524.

33. Rosas, F.; Ntranos, V.; Ellison, C.J.; Pollin, S.; Verhelst, M. Understanding Interdependency Through Complex Information Sharing. *Entropy* **2016**, *18*, 38, doi:10.3390/e18020038.

34. McGill, W.J. Multivariate information transmission. *Psychometrika* **1954**, *19*, 97–116.

35. Han, T.S. Nonnegative entropy measures of multivariate symmetric correlations. *Inf. Control* **1978**, *36*, 133–156.

36. Watanabe, S. Information Theoretical Analysis of Multivariate Correlation. *IBM J. Res. Dev.* **1960**, *4*, 66–82.

37. Margolin, A.A.; Nemenman, I.; Basso, K.; Wiggins, C.; Stolovitzky, G.; Favera, R.D.; Califano, A. ARACNE: An Algorithm for the Reconstruction of Gene Regulatory Networks in a Mammalian Cellular Context. *BMC Bioinform.* **2006**, *7*, doi:10.1186/1471-2105-7-S1-S7.

38. Averbeck, B.B.; Latham, P.E.; Pouget, A. Neural correlations, population coding and computation. *Nat. Rev. Neurosci.* **2006**, *7*, 358–366.

39. Quian, R.Q.; Panzeri, S. Extracting information from neuronal populations: information theory and decoding approaches. *Nat. Rev. Neurosci.* **2009**, *10*, 173–185.

40. Panzeri, S.; Macke, J.H.; Gross, J.; Kayser, C. Neural population coding: combining insights from microscopic and mass signals. *Trends Cogn. Sci.* **2015**, *19*, 162–172.
41. Pearl, J. *Causality: Models, Reasoning and Inference*, 2nd ed.; Cambridge University Press: Cambridge, UK, 2009.
42. Shamir, M. Emerging principles of population coding: In search for the neural code. *Curr. Opin. Neurobiol.* **2014**, *25*, 140–148.
43. Runyan, C.A.; Piasini, E.; Panzeri, S.; Harvey, C.D. Distinct timescales of population coding across cortex. *Nature* **2017**, *548*, 92–96.
44. Timme, N.M.; Ito, S.; Myroshnychenko, M.; Nigam, S.; Shimono, M.; Yeh, F.C.; Hottowy, P.; Litke, A.M.; Beggs, J.M. High-Degree Neurons Feed Cortical Computations. *PLoS Comput. Biol.* **2016**, *12*, e1004858.
45. Jazayeri, M.; Afraz, A. Navigating the Neural Space in Search of the Neural Code. *Neuron* **2017**, *93*, 1003–1014.
46. Gallego, J.A.; Perich, M.G.; Miller, L.E.; Solla, S.A. Neural Manifolds for the Control of Movement. *Neuron* **2017**, *94*, 978–984.
47. Sharpee, T.O. Optimizing Neural Information Capacity through Discretization. *Neuron* **2017**, *94*, 954–960.
48. Pitkow, X.; Angelaki, D.E. Inference in the Brain: Statistics Flowing in Redundant Population Codes. *Neuron* **2017**, *94*, 943–953.

entropy

MDPI

Article

Secret Sharing and Shared Information

Johannes Rauh

Max Planck Institute for Mathematics in the Sciences, 04103 Leipzig, Germany; jrauh@mis.mpg.de;
Tel.: +49-341-9959-602

Received: 21 June 2017; Accepted: 5 November 2017; Published: 9 November 2017

Abstract: Secret sharing is a cryptographic discipline in which the goal is to distribute information about a secret over a set of participants in such a way that only specific authorized combinations of participants together can reconstruct the secret. Thus, secret sharing schemes are systems of variables in which it is very clearly specified which subsets have information about the secret. As such, they provide perfect model systems for information decompositions. However, following this intuition too far leads to an information decomposition with negative partial information terms, which are difficult to interpret. One possible explanation is that the partial information lattice proposed by Williams and Beer is incomplete and has to be extended to incorporate terms corresponding to higher-order redundancy. These results put bounds on information decompositions that follow the partial information framework, and they hint at where the partial information lattice needs to be improved.

Keywords: information decomposition; partial information lattice; shared information; secret sharing

MSC: 94A17; 94A62

1. Introduction

Williams and Beer [1] have proposed a general framework to decompose the multivariate mutual information $I(S; X_1, \ldots, X_n)$ between a target random variable S and predictor random variables $X_1, \ldots, X_n$ into different terms (called *partial information terms*) according to different ways in which combinations of the variables $X_1, \ldots, X_n$ provide unique, shared, or synergistic information about S. Williams and Beer argue that such a decomposition can be based on a measure of shared information. The underlying idea is that any information can be classified according to "who knows what". However, is this true?

A situation where the question "who knows what?" is easy to answer very precisely is secret sharing—a part of cryptography in which the goal is to distribute information (the *secret*) over a set of participants such that the secret can only be reconstructed if certain *authorized* combinations of participants join their information (see [2] for a survey). The set of authorized combinations is called the *access structure*. Formally, the secret is modelled as a random variable S, and a *secret sharing scheme* assigns a random variable X_i to each participant i in such a way that if $\{i_1, \ldots, i_k\}$ is an authorized set of participants, then S is a function of $X_{i_1}, \ldots, X_{i_k}$; that is, $H(S|X_{i_1}, \ldots, X_{i_k}) = 0$; and, conversely, if $\{i_1, \ldots, i_k\}$ is not authorized, then $H(S|X_{i_1}, \ldots, X_{i_k}) > 0$. It is assumed that the participants know the scheme, and so any authorized combination of participants can reconstruct the secret if they join their information. A secret sharing scheme is *perfect* if non-authorized sets of participants know nothing about the secret; i.e., $H(S|X_{i_1}, \ldots, X_{i_k}) = H(S)$. Thus, in a perfect secret sharing scheme, it is very clearly specified "who knows what". In this sense, perfect secret sharing schemes provide model systems for which it should be easy to write down an information decomposition.

One connection between secret sharing and information decompositions is that the set of access structures of secret sharing schemes with n participants is in one-to-one correspondence with the partial information terms of Williams and Beer. This correspondence makes it possible to give another interpretation to all partial information terms: namely, the partial information term is a measure of how similar a given system of random variables is to a secret sharing scheme with a given access structure.

This correspondence also allows the introduction of the *secret sharing property* that makes the above intuition precise: An information decomposition satisfies this property if and only if any perfect secret sharing scheme has just a single partial information term (which corresponds to its access structure). Lemma 2 states that the secret sharing property is implied by the Williams and Beer axioms, which shows that the secret sharing property plays well together with the ideas of Williams and Beer. Proposition 1 shows that in an information decomposition that satisfies a natural generalization of this property, it is possible to prescribe arbitrary nonnegative values to all partial information terms.

These results suggest that perfect secret sharing schemes fit well together with the ideas of Williams and Beer. However, following this intuition too far leads to inconsistencies. As Theorem 4 shows, extending the secret sharing property to pairs of perfect secret sharing schemes leads to negative partial information terms. While other authors have started to build an intuition for negative partial terms and argue that they may be unavoidable in information decompositions, the concluding section collects arguments against such claims and proposes as another possible solutions that the Williams and Beer framework is incomplete and is missing nodes that represent higher-order redundancy.

Cryptography, where the goal is not only to transport information (as in coding theory) but also to keep it concealed from unauthorized parties, has initiated many interesting developments in information theory; for example, by introducing new information measures and re-interpreting older ones (see, for example, [3,4]). This manuscript focuses on another contribution of cryptography: probabilistic systems with well-defined distribution of information.

The remainder of this article is organized as follows: Section 2 summarizes definitions and results of secret sharing schemes. Section 3 introduces different secret sharing properties that fix the values that a measure of shared information assigns to perfect secret sharing schemes and combinations thereof. The main result of Section 4 is that the pairwise secret sharing property leads to negative partial information terms. Section 5 discusses the implications of this incompatibility result.

2. Perfect Secret Sharing Schemes

We consider n participants among whom we want to distribute information about a secret in such a way that we can control which subsets of participants together can decrypt the secret.

Definition 1. *An* access structure $\mathcal{A}$ *is a family of subsets of* $\{1, \ldots, n\}$, *closed to taking supersets. Elements of* $\mathcal{A}$ *are called* authorized sets.

A secret sharing scheme *with access structure* $\mathcal{A}$ *is a family of random variables* $S, X_1, \ldots, X_n$ *such that:*

- $H(X_A, S) = H(X_A)$, *whenever* $A \in \mathcal{A}$.

  Here, $X_A = (X_i)_{i \in A}$ *for all subsets* $A \subseteq \{1, \ldots, n\}$. *A secret sharing scheme is* perfect *if*

- $H(X_A, S) = H(X_A) + H(S)$, *whenever* $A \notin \mathcal{A}$.

The condition for perfection is equivalent to $H(S|X_A) = H(S)$. See [2] for a survey on secret sharing.

Theorem 1. *For any access structure* $\mathcal{A}$ *and any* $h > 0$, *there exists a perfect secret sharing scheme with access structure* $\mathcal{A}$ *for which the entropy of the secret* S *equals* $H(S) = h$.

Proof. Perfect secret sharing schemes for arbitrary access structures were first constructed by Ito et al. [5]. In this construction, the entropy of the secret equals 1 bit. Combining n copies of such a secret sharing scheme gives a secret sharing scheme with a secret of n bit. As explained in [2] (Claim 1), the distribution of the secret may be perturbed arbitrarily (as long as the support of the distribution remains the same). In this way it is possible to prescribe the entropy of the secret in a perfect secret sharing scheme. □

Example 1. *Let* Y_1, Y_2, Y_3, S *be independent uniform binary random variables, and let* $A = (Y_1, Y_2 \oplus S)$, $B = (Y_2, Y_3 \oplus S)$, $C = (Y_3, Y_1 \oplus S)$, *where* $\oplus$ *denotes addition modulo 2 (or the XOR operation). Then* (S, A, B, C) *is a perfect secret sharing scheme with access structure*

$$\{A, B\}, \{A, C\}, \{B, C\}, \quad \{A, B, C\}.$$

It may be of little surprise that integer addition modulo k is an important building block in many secret sharing schemes.

While the existence of perfect secret sharing schemes is solved, there remains the problem of finding efficient secret sharing schemes in the sense that the variables $X_1, \ldots, X_n$ should be as small as possible (in the sense of a small entropy), given a fixed entropy of the secret. For instance, in Example 1, $H(X_i)/H(S) = 2$ for all i (see [2] for a survey).

Since an access structure $\mathcal{A}$ is closed to taking supersets, it is uniquely determined by its inclusion-minimal elements

$$\underline{\mathcal{A}} := \{A \in \mathcal{A} : \text{if } B \subseteq A \text{ and } B \neq A, \text{ then } B \notin \mathcal{A}\}.$$

For instance, in Example 1, the first three elements belong to $\underline{\mathcal{A}}$. The set $\underline{\mathcal{A}}$ has the property that no element of $\underline{\mathcal{A}}$ is a subset of another element of $\underline{\mathcal{A}}$. Such a collection of sets is called an *antichain*. Conversely, any such antichain equals the set of inclusion-minimal elements of a unique access structure.

The antichains have a natural lattice structure, which was used by Williams and Beer to order the different values of shared information and organize them into what they call the *partial information lattice*. The same lattice also has a description in terms of secret sharing.

Definition 2. *Let* $(A_1, \ldots, A_k)$ *and* $(B_1, \ldots, B_l)$ *be antichains. Then*

$$(A_1, \ldots, A_k) \preceq (B_1, \ldots, B_l) \quad :\Longleftrightarrow \quad \text{for any } B_i \text{ there exists } A_j \text{ with } A_j \subseteq B_i.$$

The partial information lattice for the case $n = 3$ is depicted in Figure 1.

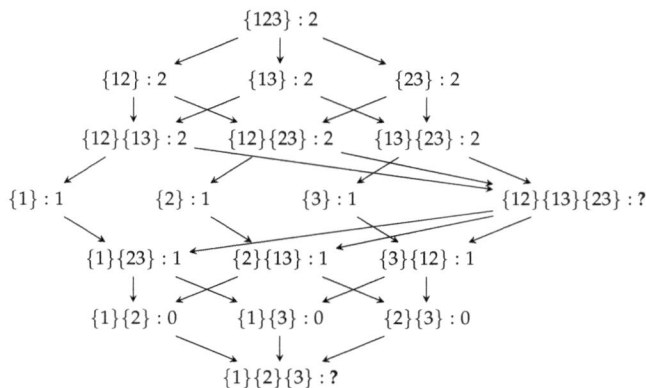

Figure 1. The partial information lattice for $n = 3$. Each node is indexed by an antichain. The values (in bit) of the shared information in the XOR example from the proof of Theorem 2 according to the pairwise secret sharing property are given after the colon.

Lemma 1. *Let $\mathcal{A}$ be an access structure on $\{1, \ldots, n\}$, and let $(B_1, \ldots, B_l)$ be an antichain. Then $B_1, \ldots, B_l$ are all authorized for $\mathcal{A}$ if and only if $\underline{\mathcal{A}} \preceq (B_1, \ldots, B_l)$.*

Proof. The statement directly follows from the definitions. □

3. Information Decompositions of Secret Sharing Schemes

Williams and Beer [1] proposed to decompose the total mutual information $I(S; X_1, \ldots, X_n)$ between a target random variable S and predictor random variables $X_1, \ldots, X_n$ according to different ways in which combinations of the variables $X_1, \ldots, X_n$ provide unique, shared, or synergistic information about S. One of their main ideas is to base such a decomposition on a single *measure of shared information* $I_\cap$, which is a function $I(S; Y_1, \ldots, Y_k)$ that takes as arguments a list of random variables, of which the first, S, takes a special role. To arrive at a decomposition of $I(S; X_1, \ldots, X_n)$, the variables $Y_1, \ldots, Y_k$ are taken to be combinations $X_A = (X_i)_{i \in A}$ of $X_1, \ldots, X_n$, corresponding to subsets A of $\{1, \ldots, n\}$. For simplicity, $I_\cap(S; X_{A_1}, \ldots, X_{A_k})$ is denoted by $I_\cap(S; A_1, \ldots, A_k)$ for all $A_1, \ldots, A_k \subseteq \{1, \ldots, n\}$.

Williams and Beer proposed a list of axioms that such a measure $I_\cap$ should satisfy. It follows from these axioms that it suffices to consider the function $I_\cap(S; A_1, \ldots, A_k)$ in the case that $(A_1, \ldots, A_k)$ is an antichain. Moreover, $I_\cap(S; \cdot)$ is a monotone function on the partial information lattice (Definition 2).

Thus, it is natural to write each value $I_\cap(S; A_1, \ldots, A_k)$ on the lattice as a sum of local terms I_∂ corresponding to the antichains that lie below $(A_1, \ldots, A_k)$ in the lattice:

$$I_\cap(S; A_1, \ldots, A_k) = \sum_{(B_1, \ldots, B_l) \preceq (A_1, \ldots, A_k)} I_\partial(S; B_1, \ldots, B_l).$$

The terms I_∂ are called *partial information terms*. This representation always exists, and the partial information terms are uniquely defined (using a Möbius inversion). However, it is not guaranteed that I_∂ is always nonnegative. If I_∂ is nonnegative, then $I_\cap$ is called *locally positive*.

Williams and Beer also defined a function denoted by $I_{\min}$ that satisfies their axioms and that is locally positive. While the framework is intriguing and has attracted a lot of further research (as this special issue illustrates), the function $I_{\min}$ has been criticized as not measuring the right thing. The difficulty of finding a reasonable measure of shared information that is locally positive [6,7] has led some to argue that maybe local positivity is not a necessary requirement for an information decomposition. This issue is discussed further in Section 5.

The goal of this section is to present additional natural properties for a measure of shared information that relate secret sharing with the intuition behind information decompositions. In a perfect secret sharing scheme, any combination of participants knows either nothing or everything about S. This motivates the following definition:

Definition 3. *A measure of shared information $I_\cap$ has the* secret sharing property *if and only if for any access structure $\mathcal{A}$ and any perfect secret sharing scheme $(X_1, \ldots, X_n, S)$ with access structure $\mathcal{A}$, the following holds:*

$$I_\cap(S; A_1, \ldots, A_k) = \begin{cases} H(S), & \text{if } A_1, \ldots, A_k \text{ are all authorized,} \\ 0, & \text{otherwise,} \end{cases} \qquad \text{for all } A_1, \ldots, A_k \subseteq \{1, \ldots, n\}.$$

Lemma 2. *The secret sharing property is implied by the Williams and Beer axioms.*

Proof. The Williams and Beer axioms imply that

$$I_\cap(S; A_1, \ldots, A_k) \leq I(S; A_i) = 0$$

whenever A_i is not authorized. On the other hand, when $A_1, \ldots, A_k$ are all authorized, then the monotonicity axiom implies

$$I_\cap(S; A_1, \ldots, A_k) \geq I_\cap(S; A_1, \ldots, A_k, S) = I_\cap(S; S) = H(S). \quad \square$$

Perfect secret sharing schemes lead to information decompositions with a single nonzero partial information term:

Lemma 3. *If $I_\cap$ has the secret sharing property and if $(X_1, \ldots, X_n, S)$ is a perfect secret sharing scheme with access structure $\mathcal{A}$, then*

$$I_\partial(S; A_1, \ldots, A_k) = \begin{cases} H(S), & \text{if } \underline{A} = \{A_1, \ldots, A_k\}, \\ 0, & \text{otherwise,} \end{cases} \qquad \text{for all } A_1, \ldots, A_k \subseteq \{1, \ldots, n\}. \quad (1)$$

Proof. Suppose that $\underline{A} = \{A'_1, \ldots, A'_{k'}\}$, and let $J_\partial(S; A_1, \ldots, A_k)$ be the right hand side of Equation (1). We need to show that $I_\partial = J_\partial$. Since the Möbius inversion is unique, it suffices to show that $J_\cap = I_\cap$, where

$$J_\cap(S; A_1, \ldots, A_k) = \sum_{(B_1, \ldots, B_l) \preceq (A_1, \ldots, A_k)} J_\partial(S; B_1, \ldots, B_l).$$

By Lemma 1,

$$J_\cap(S; A_1, \ldots, A_k) = \begin{cases} H(S), & \text{if } A_1, \ldots, A_k \text{ are all authorized,} \\ 0, & \text{otherwise,} \end{cases}$$

for any $A_1, \ldots, A_k \subseteq \{X_1, \ldots, X_n\}$, from which the claim follows. $\quad \square$

What happens when we have several secret sharing schemes involving the same participants? In order to have a clear intuition, assume that the secret sharing schemes satisfy the following definition:

Definition 4. *Let $\mathcal{A}_1, \ldots, \mathcal{A}_l$ be access structures on $\{1, \ldots, n\}$. A combination of (perfect) secret sharing schemes with access structures $\mathcal{A}_1, \ldots, \mathcal{A}_l$ consists of random variables $S_1, \ldots, S_l, X_1, \ldots, X_n$ such that $(S_i, X_1, \ldots, X_n)$ is a (perfect) secret sharing scheme with access structure $\mathcal{A}_i$ for $i = 1, \ldots, l$ and such that*

$$H(S_i | S_1, \ldots, S_{i-1}, S_{i+1}, \ldots, S_l, X_A) = H(S_i) \text{ if } A \notin \mathcal{A}_i.$$

This definition ensures that the secrets are independent in the sense that knowing some of the secrets provides no information about the other secrets. Formally, one can see that the secrets are probabilistically independent as follows: For any $A \notin \mathcal{A}_i$ (for example, $A = \emptyset$),

$$H(S_i | S_1, \ldots, S_{i-1}, S_{i+1}, \ldots, S_l) \geq H(S_i | S_1, \ldots, S_{i-1}, S_{i+1}, \ldots, S_l, X_A) = H(S_i).$$

In Definition 4, if two access structures $\mathcal{A}_i, \mathcal{A}_j$ are identical, then we can replace S_i and S_j by a single random variable (S_i, S_j) and obtain a smaller combination of (perfect) secret sharing schemes.

In a combination of perfect secret sharing schemes, it is very clear who knows what: Namely, a group of participants knows all secrets for which it is authorized, while it knows nothing about the remaining secrets. This motivates the following definition:

Definition 5. *A measure of shared information $I_\cap$ has the combined secret sharing property if and only if for any combination of perfect secret sharing schemes with access structures $\mathcal{A}_1, \ldots, \mathcal{A}_l$,*

$$I_\cap((S_1, \ldots, S_l); A_1, \ldots, A_k) = H(\{S_i : A_1, \ldots, A_k \in \mathcal{A}_i\}) \qquad (2)$$

(the entropy of those secrets for which $A_1, \ldots, A_k$ are all authorized). $I_\cap$ *has the* pairwise secret sharing property *if and only if the same holds true in the special case $l = 2$.*

The combined secret sharing property implies the pairwise secret sharing property. The pairwise secret sharing property does not follow from the Williams and Beer axioms. For example, $I_{\min}$ satisfies the Williams and Beer axioms, but not the pairwise secret sharing property (as will become apparent in Theorem 2). So one can ask whether the pairwise and combined secret sharing properties are compatible with the Williams and Beer axioms. This question is difficult to answer, since currently there are only two proposed measures of shared information that satisfy the Williams and Beer axioms, namely $I_{\min}$ and the *minimum of mutual informations* [8]:

$$I_{\mathrm{MMI}}(S; A_1, \ldots, A_k) := \min_{i=1,\ldots,k} I(S; A_i).$$

Both measures do not satisfy the pairwise secret sharing property.

While there has been no further proposal for a function that satisfies the Williams and Beer axioms for arbitrarily many arguments, several measures have been proposed for the "bivariate case" $k = 2$, notably I_{red} of Harder et al. [9] and $\widehat{SI}$ of [10]. The appendix shows that $\widehat{SI}$ at least satisfies the combined secret sharing property "as far as possible".

Combinations of l perfect secret sharing schemes lead to information decompositions with at most l nonzero partial information terms.

Lemma 4. *Assume that $I_\cap$ has the combined secret sharing property. If $(S_1, \ldots, S_l, X_1, \ldots, X_n)$ is a combination of perfect secret sharing schemes with pairwise different access structures $\mathcal{A}_1, \ldots, \mathcal{A}_l$, then*

$$I_\partial\big((S_1, \ldots, S_l); A_1, \ldots, A_k\big) = \begin{cases} H(S_i), & \text{if } \mathcal{A}_i = \{A_1, \ldots, A_k\} \text{ for some } i \in \{1, \ldots, l\}, \\ 0, & \text{otherwise,} \end{cases}$$

for any $A_1, \ldots, A_k \subseteq \{X_1, \ldots, X_n\}$.

The proof is similar to the proof of Lemma 3 and is omitted.

The combined secret sharing property implies that any combination of nonnegative values can be prescribed as partial information values.

Proposition 1. *Suppose that a nonnegative number $h_{\mathcal{A}}$ is given for any antichain $\mathcal{A}$. For any measure of shared information that satisfies the combined secret sharing property, there exist random variables $S, X_1, \ldots, X_n$ such that the corresponding partial measure I_∂ satisfies $I_\partial(S; A_1, \ldots, A_k) = h_{A_1, \ldots, A_k}$ for all antichains $\mathcal{A} = (A_1, \ldots, A_k)$.*

Proof. By Theorem 1, for each antichain $\mathcal{A}$ there exists a perfect secret sharing scheme $S_{\mathcal{A}}$, $X_{1,\mathcal{A}}, \ldots, X_{n,\mathcal{A}}$ with $H(S_{\mathcal{A}}) = h_{\mathcal{A}}$. Combine independent copies of these perfect secret sharing schemes and let

$$S = (S_{\mathcal{A}})_{\mathcal{A}}, \quad X_1 = (X_{1,\mathcal{A}})_{\mathcal{A}}, \quad \ldots, \quad X_n = (X_{n,\mathcal{A}})_{\mathcal{A}},$$

where $\mathcal{A}$ runs over all antichains. Then $S, X_1, \ldots, X_n$ is an independent combination of perfect secret sharing schemes, and the statement follows from Lemma 4. □

Unfortunately, not every random variable S can be decomposed in such a way as a combination of secret sharing schemes. However, Proposition 1 suggests that, given a measure $I_\cap$ of shared information that satisfies the combined secret sharing property, $I_\partial(S; \underline{A})$ can informally be interpreted as a measure that quantifies how much $(X_1, \ldots, X_n, S)$ looks like a perfect secret sharing scheme with access structure $\mathcal{A}$.

Lemma 5. *Suppose that $I_\cap$ is a measure of shared information that satisfies the pairwise secret sharing property. If X_1 and X_2 are independent, then $I_\cap((X_1, X_2); X_1, X_2) = 0$.*

In the language of [11], the lemma says that the pairwise secret sharing property implies the *independent identity property*.

Proof. Let $S_1 = X_1$, $S_2 = X_2$. Then S_1, S_2, X_1, X_2 is a pair of perfect secret sharing schemes with access structures $\mathcal{A}_1 = \{\{1\}\}$ and $\mathcal{A}_2 = \{\{2\}\}$. The statement follows from Definition 5, since X_1 is not authorized for $\mathcal{A}_2$ and X_2 is not authorized for $\mathcal{A}_1$. $\square$

4. Incompatibility with Local Positivity

Unfortunately, although the combined secret sharing property very much fits the intuition behind the axioms of Williams and Beer, it is incompatible with a nonnegative decomposition according to the partial information lattice:

Theorem 2. *Let $I_\cap$ be a measure of shared information that satisfies the Williams–Beer axioms and has the pairwise secret sharing property. Then, I_∂ is not nonnegative.*

Proof. The XOR example, which was already used by Bertschinger et al. [6] and Rauh et al. [7] to prove incompatibility results for properties of information decompositions, can also be used here.

Let X_1, X_2 be independent binary uniform random variables, let $X_3 = X_1 \oplus X_2$, and let $S = (X_1, X_2, X_3)$. Observe that the situation is symmetric in X_1, X_2, X_3. In particular, X_2, X_3 are also independent, and $X_1 = X_2 \oplus X_3$. The following values of $I_\cap$ can be computed from the assumptions:

- $I_\cap(S; X_1, (X_2 X_3)) = I_\cap(S; X_1, (X_1 X_2 X_3)) = I_\cap(S; X_1) = 1\,\text{bit}$, since X_1 is a function of (X_2, X_3) and by the monotonicity axiom.
- $I_\cap(S; X_1, X_2) = I_\cap((X_1 X_2 X_3); X_1, X_2) = I_\cap((X_1 X_2); X_1, X_2) = 0$ by Lemma 5.

By monotonicity, $I_\cap(S; X_1, X_2, X_3) = 0$. Moreover, $I_\cap(S; (X_1 X_2), (X_1 X_3), (X_2 X_3)) \le 2\,\text{bit}$, since $2\,\text{bit}$ is the total entropy in the system. Then, however,

$$
\begin{aligned}
I_\partial(S; (X_1 X_2), (X_1 X_3), (X_2 X_3)) &= I_\cap(S; (X_1 X_2), (X_1 X_3), (X_2 X_3)) \\
&\quad - I_\cap(S; X_1, (X_2 X_3)) - I_\cap(S; X_2, (X_1 X_3)) - I_\cap(S; X_3, (X_1 X_2)) \pm 0 \\
&\le 2\,\text{bit} - 3\,\text{bit} = -1\,\text{bit},
\end{aligned}
$$

where ± 0 denotes values of $I_\cap$ that vanish. Thus, I_∂ is not nonnegative. $\square$

Note that the random variables $(S = (X_1, X_2, X_3), X_1, X_2, X_3)$ from the proof of Theorem 2 form three perfect secret sharing schemes that do not satisfy the definition of a combination of perfect secret sharing schemes. The three secrets X_1, X_2, X_3 are not independent, but they are pair-wise independent (and so Lemma 4 does not apply).

Remark 1. *The XOR example from the proof of Theorem 2 (which was already used by Bertschinger et al. [6] and Rauh et al. [7]) was criticized by Chicharro and Panzeri [12] on the grounds that it involves random variables that stand in a deterministic functional relation (in the sense that $X_3 = X_1 \oplus X_2$). Chicharro and Panzeri argue that in such a case it is not appropriate to use the full partial information lattice. Instead, the functional relationship should be used to eliminate (or identify) nodes from the lattice. Thus, while the monotonicity axiom of Williams and Beer implies $I_\cap(S; X_3, (X_2, X_3)) = I_\cap(S; X_3)$ (and so $\{3; 23\}$ is not part of the partial information lattice), the same axiom also implies that $I_\cap(S; X_3, (X_1, X_2)) = I_\cap(S; X_3)$ in the XOR example, and so $\{3; 12\}$ should similarly be excluded from the lattice when analyzing this particular example. Note, however, that the first argument is a formal argument that is valid for all joint distributions of S, X_1, X_2, X_3, while the second argument takes into account the particular underlying distribution.*

It is easy to work around this objection. The deterministic relationship disappears when an arbitrarily small stochastic noise is added to the joint distribution. To be precise, let X_1, X_2 be independent binary random variables, and let X_3 be binary with

$$P(X_3 = x_3 | X_1 = x_1, X_2 = x_2) = \begin{cases} 1 - \epsilon, & \text{if } x_3 = x_1 \oplus x_2, \\ \epsilon, & \text{otherwise,} \end{cases}$$

for $0 \le \epsilon \le 1$. For $\epsilon = 0$, the example from the proof is recovered. Assuming that the partial information terms depend continuously on this joint distribution, the partial information term $I_{\partial}(S; (X_1 X_2), (X_1 X_3), (X_2 X_3))$ will still be negative for small $\epsilon > 0$. Thus, assuming continuity, the conclusion of Theorem 2 still holds true when the information decomposition according to the full partial information lattice is only considered for random variables that do not satisfy any functional deterministic constraint.

Remark 2. *Analyzing the proof of Theorem 2, one sees that the independent identity axiom (Lemma 5) is the main ingredient to arrive at the contradiction. The same property also arises in the other uses of the XOR example [6,7].*

5. Discussion

Perfect secret sharing schemes correspond to systems of random variables in which it is very clearly specified "who knows what". In such a system, it is easy to assign intuitive values to the shared information nodes in the partial information lattice, and one may conjecture that the intuition behind this assignment is the same intuition that underlies the Williams and Beer axioms, which define the partial information lattice. Moreover, following the same intuition, independent combinations of perfect secret sharing schemes can be used as a tool to construct systems of random variables with prescribable (nonnegative) values of partial information.

Unfortunately, this extension to independent combinations of perfect secret sharing schemes is not without problems: By Theorem 2, it leads to decompositions with negative partial information terms. What does it mean, however, that the examples derived from the same intuition as the Williams and Beer axioms contradict the same axioms in this way? Is this an indication that the whole idea of information decomposition does not work (and that the question posed in the first paragraph of the introduction cannot be answered affirmatively)?

There are several ways out of this dilemma. The first solution is to assign different values to combinations of perfect secret sharing schemes. This solution will not be pursued further in this text, as it would change the interpretation of the information decomposition as measuring "who knows what".

The second solution is to accept negative partial values in the information decomposition. It has been argued that negative values of information can be given an intuitive interpretation in terms of confusing or misleading information. For event-wise (also called "local") information quantities, such as the event-wise mutual information $i(s; x) = \log(p(s)/p(s|x))$, this interpretation goes back to the early days of information theory [13]. Sometimes, this phenomenon is called "misinformation" [11,14]. However, in the usual language, misinformation refers to "false or incorrect information, especially when it is intended to trick someone" [15], which is not the effect that is modelled here. Thus, the word misinformation should be avoided, in order not to mislead the reader into the wrong intuition.

While negative event-wise information quantities are well-understood, the situation is more problematic for average quantities. When an agent receives side-information in the form of the value x of a relevant random variable X, she changes her strategy. While the prior strategy should be based on the prior distribution $p(S)$, the new strategy should be based on the posterior $p(S|X = x)$. Clearly, in a probabilistic setting, any change of strategy can lead to a better or worse result in a single instance. On average, though, side-information never hurts (and it is never advantageous on average to ignore side-information), which is why the mutual information is never negative. Similarly, it is natural to expect non-negativity of other information quantities. It is difficult to imagine how correct side-information (or an aspect thereof) can be misleading on average. The situation is different for incorrect information, where the interpretation of a negative value is much easier.

More conceptually, I would suspect that an (averaged) information quantity that may change its sign actually conflates different aspects of information, just as the interaction information (or co-information) conflates synergy and redundancy [1] (and one can argue whether the same should be true for event-wise quantities; cf. [16]).

In any case, allowing negative partial values alters the interpretation of an information decomposition to a point where it is questionable whether the word "decomposition" is still appropriate. When decomposing an object into parts, the parts should in some reasonable way be sub-objects. For example, in a Fourier decomposition of a function, the Fourier components are never larger than the function (in the sense of the L^2-norm), and the sum of the squared L^2-norms of the Fourier coefficients equals the squared L^2-norm of the original function. As another example, given a (positive) amount of money and two investment options, it may indeed be possible to invest a negative share of the total amount into one of the two options in order to increase the funds that can be invested in the second option. However, such short selling is regulated in many countries with much stronger rules than ordinary trading.

I do not claim that an information decomposition with negative partial information terms cannot possibly make sense. However, it has to be made clear precisely how to interpret negative terms, and it is important to distinguish between correct information that leads to a suboptimal decision due to unlikely events happening ("bad luck") and incorrect information that leads to decisions being based on the wrong posterior probabilities (as opposed to the "correct" conditional probabilities).

A third solution is to change the underlying lattice structure of the decomposition. A first step in this direction was done by Chicharro and Panzeri [12], who propose to decompose mutual information according to subsets of the partial information lattice. However, it is also conceivable that the lattice has to be enlarged.

Williams and Beer derived the partial information lattice from their axioms together with the assumption that everything can be expressed in terms of shared information (that is, according to "who knows what"). Shared information is sometimes equivalently called *redundant information*, but it may be necessary to distinguish the two. Information that is shared by several random variables is information that is accessible to each single random variable, but redundancy can also arise at higher orders. An example is the infamous XOR example from the proof of Theorem 2: In this example, each pair X_i, X_j is independent and contains of two bits, but the total system X_1, X_2, X_3 has only two bits. Therefore, there is one bit of redundancy. However, this redundancy bit is not located anywhere specifically: It is not contained in either of X_1, X_2, X_3, and thus it is not shared information. Since the redundant bit is not part of X_1, it is not "shared" by X_1 in this sense. This phenomenon corresponds to the fact that random variables can be pairwise independent without being independent.

This kind of higher-order redundancy does not have a place in the partial information lattice, so it may be that nodes corresponding to higher-order redundancy have to be added. When the lattice is enlarged in this way, the structure of the Möbius inversion is changed, and it is possible that the resulting lattice leads to nonnegative partial information terms, without changing those cumulative information values that are already present in the original lattice. If this approach succeeds, the answer to the question from the introduction will be negative: Simply classifying information according to "who knows what" (i.e., shared information) does not work, since it does not capture higher-order redundancy. The analysis of extensions of the partial information lattice is the scope of future work.

Acknowledgments: I thank Fero Matúš for teaching me about secret sharing schemes. I thank Guido Montúfar and Pradeep Kr. Banerjee for their remarks about the manuscript. I am grateful to Nils Bertschinger, Jürgen Jost and Eckehard Olbrich for many inspiring discussions on the topic. I thank the reviewers for many comments, in particular concerning the discussion.

Conflicts of Interest: The author declares no conflict of interest.

Appendix A. Combined Secret Sharing Properties for Small k

This section discusses the defining Equation (2) of the combined secret sharing property for $k = 1$ and $k = 2$. The case $k = 1$ is incorporated in the definition of a combination of perfect secret sharing schemes: The following lemma implies that any measure of shared information that satisfies self-redundancy satisfies Equation (2) for $k = 1$. Recall that Williams and Beer's self-redundancy axiom implies that $I_\cap(S; X_A) = I(S; X_A)$.

Lemma A1. *Let* $(S_1, \ldots, S_l, X_1, \ldots, X_n)$ *be a combination of perfect secret sharing schemes with access structures* $\mathcal{A}_1, \ldots, \mathcal{A}_l$. *Then*

$$I\big((S_1, \ldots, S_l); X_A\big) = H(\{S_i : A \in \mathcal{A}_i\}).$$

Proof. Suppose that the secret for which A is authorized are $S_1, \ldots, S_m$. Then

$$H(S_1, \ldots, S_l | X_A) = H(S_1, \ldots, S_m | X_A) + H(S_{m+1}, \ldots, S_l | S_1, \ldots, S_m, X_A)$$

$$= H(S_{m+1}, \ldots, S_l | S_1, \ldots, S_m, X_A) \leq H(S_{m+1}, \ldots, S_l) \leq \sum_{i=m+1}^{l} H(S_i).$$

On the other hand,

$$H(S_{m+1}, \ldots, S_l | S_1, \ldots, S_m, X_A) = \sum_{i=m+1}^{l} H(S_i | S_1, \ldots, S_{i-1}, X_A)$$

$$\geq \sum_{i=m+1}^{l} H(S_i | S_1, \ldots, S_{i-1}, S_{i+1}, \ldots, S_l, X_A) = \sum_{i=m+1}^{l} H(S_i).$$

By independence (remark after Definition 4), $\sum_{i=m+1}^{l} H(S_i) = H(S_{m+1}, \ldots, S_l)$ and $\sum_{i=1}^{m} H(S_i) = H(S_1, \ldots, S_m)$. Thus,

$$I\big((S_1, \ldots, S_l); X_A\big) = H(S_1, \ldots, S_l) - H(S_1, \ldots, S_l | X_A) = H(S_1, \ldots, S_m). \quad \square$$

The next result shows that the bivariate measure of shared information $\widetilde{SI}(S; X, Y)$ proposed by Bertschinger et al. [10] satisfies Equation (2) for $k \leq 2$. The reader is referred to *loc. cit.* for definitions and elementary properties of $\widetilde{SI}$.

Proposition A1. *Let* $(S_1, \ldots, S_l, X_1, \ldots, X_n)$ *be a combination of perfect secret sharing schemes with access structures* $\mathcal{A}_1, \ldots, \mathcal{A}_l$, *Then*

$$\widetilde{SI}\big((S_1, \ldots, S_l); X_{A_1}, X_{A_2}\big) = H(\{S_i : A \in \mathcal{A}_1 \cap \mathcal{A}_2\}).$$

Proof. For given A_1, A_2, suppose that $S_1, \ldots, S_m$ are the secrets for which at least one of A_1 or A_2 is authorized and that $S_{m+1}, \ldots, S_l$ are the secrets for which neither A_1 nor A_2 is authorized alone.

Let P be the joint distribution of $S_1, \ldots, S_l, X_{A_1}, X_{A_2}$. Let Δ_P be the set of alternative joint distributions for $S_1, \ldots, S_l, X_{A_1}, X_{A_2}$ that have the same marginal distributions as P on the subsets $(S_1, \ldots, S_l, X_{A_1})$ and $(S_1, \ldots, S_l, X_{A_2})$. According to the definition of $\widetilde{SI}$, we need to compare P with the elements of Δ_P and find the maximum of $H_Q\big((S_1, \ldots, S_l) | X_{A_1}, X_{A_2}\big)$ over $Q \in \Delta_P$, where the subscript to H indicates with respect to which of these joint distributions the conditional entropy is evaluated.

Define a distribution Q^* for $S_1, \ldots, S_l, X_{A_1}, X_{A_2}$ by

$$Q^*(s_1, \ldots, s_l, x_1, x_2) = P(s_1, \ldots, s_l) P(x_{A_1} = x_1 | s_1, \ldots, s_l) P(x_{A_2} = x_2 | s_1, \ldots, s_l).$$

Then $Q^* \in \Delta_P$. Under P, the secrets $S_{m+1}, \ldots, S_l$ are independent of X_{A_1} (marginally) and independent of X_{A_2}, and so $S_{m+1}, \ldots, S_l$ are independent of the pair (X_{A_1}, X_{A_2}) under Q^*. On the other hand, $S_1, \ldots, S_m$ are a function of either X_{A_1} or X_{A_2} under P, and so $S_1, \ldots, S_m$ is a function of (X_{A_1}, X_{A_2}) under Q^*. Thus,

$$H_{Q^*}(S_1, \ldots, S_l | X_{A_1}, X_{A_2}) = H_{Q^*}(S_{m+1}, \ldots, S_l) = H_P(S_{m+1}, \ldots, S_l).$$

On the other hand, under any joint distribution $Q \in \Delta_P$, the secrets $S_1, \ldots, S_m$ are functions of X_{A_1}, X_{A_2}, whence

$$H_Q(S_1, \ldots, S_l | X_{A_1}, X_{A_2}) \leq H_Q(S_{m+1}, \ldots, S_l) = H_P(S_{m+1}, \ldots, S_l).$$

It follows that Q^* solves the optimization problem in the definition of $\widetilde{SI}$.

Suppose that the secrets for which X_{A_1} is authorized are $S_1, \ldots, S_r$ and that the secrets for which X_{A_2} is authorized are $S_s, \ldots, S_m$ (with $1 \leq r, s \leq m$). One computes

$$I_{Q^*}((S_1, \ldots, S_l); X_{A_1} | X_{A_2}) = H(S_1, \ldots, S_{s-1}) = \sum_{i=1}^{s-1} H(S_i) \quad \text{and}$$

$$I_{Q^*}((S_1, \ldots, S_l); X_{A_1}) = H(S_1, \ldots, S_r) = \sum_{i=1}^{r} H(S_i),$$

whence

$$\widetilde{SI}((S_1, \ldots, S_l); X_{A_1}, X_{A_2}) = I_{Q^*}((S_1, \ldots, S_l); X_{A_1}) - I_{Q^*}((S_1, \ldots, S_l); X_{A_1} | X_{A_2})$$

$$= \sum_{i=s}^{r} H(S_i) = H(S_s, \ldots, S_r). \quad \square$$

References

1. Williams, P.; Beer, R. Nonnegative Decomposition of Multivariate Information. *arXiv* **2010**, arXiv: 1004.2515v1.
2. Beimel, A. Secret-Sharing Schemes: A Survey. In Proceedings of the Third International Conference on Coding and Cryptology, Qingdao, China, 30 May– 3 June 2011; Springer: Berlin/Heidelberg, Germany, 2011; pp. 11–46.
3. Maurer, U.; Wolf, S. The Intrinsic Conditional Mutual Information and Perfect Secrecy. In Proceedings of the 1997 IEEE International Symposium on Information Theory, Ulm, Germany, 29 June–4 July 1997.
4. Csiszar, I.; Narayan, P. Secrecy capacities for multiple terminals. *IEEE Trans. Inf. Theory* **2004**, *50*, 3047–3061.
5. Ito, M.; Saito, A.; Nishizeki, T. Secret Sharing Scheme Realizing General Access Structure. In Proceedings of the IEEE Global Telecommunication Conference, Tokyo, Japan, 15–18 November 1987; pp. 99–102.
6. Bertschinger, N.; Rauh, J.; Olbrich, E.; Jost, J. Shared Information—New Insights and Problems in Decomposing Information in Complex Systems. In *Proceedings of the European Conference on Complex Systems*; Springer: Berlin/Heidelberg, Germany, 2013; pp. 251–269.
7. Rauh, J.; Bertschinger, N.; Olbrich, E.; Jost, J. Reconsidering Unique Information: Towards a Multivariate Information Decomposition. In Proceedings of the 2014 IEEE International Symposium on 2014 Information Theory (ISIT), Honolulu, HI, USA, 29 June–4 July 2014; pp. 2232–2236.
8. Barrett, A.B. An exploration of synergistic and redundant information sharing in static and dynamical Gaussian systems. *Phys. Rev. E* **2014**, *91*, 52802.
9. Harder, M.; Salge, C.; Polani, D. A Bivariate measure of redundant information. *Phys. Rev. E* **2013**, *87*, 12130.

10. Bertschinger, N.; Rauh, J.; Olbrich, E.; Jost, J.; Ay, N. Quantifying unique information. *Entropy* **2014**, *16*, 2161–2183.

11. Ince, R. Measuring multivariate redundant information with pointwise common change in surprisal. *Entropy* **2017**, *19*, 38.

12. Chicharro, D.; Panzeri, S. Synergy and Redundancy in Dual Decompositions of Mutual Information Gain and Information Loss. *Entropy* **2017**, *19*, 71.

13. Fano, R.M. *Transmission of Information*; MIT Press: Cambridge, MA, USA, 1961.

14. Wibral, M.; Lizier, J.T.; Priesemann, V. Bits from Brains for Biologically Inspired Computing. *Front. Robot. AI* **2015**, *2*, 5.

15. Macmillan Publishers Limited. Macmillan Dictionary. Available online: http://www.macmillandictionary.com/ (accessed on 15 March 2012).

16. Ince, R. The Partial Entropy Decomposition: Decomposing multivariate entropy and mutual information via pointwise common surprisal. *arXiv* **2017**, arXiv:1702.01591.

Article

Coarse-Graining and the Blackwell Order

Johannes Rauh [1,*]**, Pradeep Kr. Banerjee** [1]**, Eckehard Olbrich** [1]**, Jürgen Jost** [1]**, Nils Bertschinger** [2]
and David Wolpert [3,4]

[1]   Max Planck Institute for Mathematics in the Sciences, 04103 Leipzig, Germany;
      pradeep@mis.mpg.de (P.K.B.); olbrich@mis.mpg.de (E.O.); jjost@mis.mpg.de (J.J.)
[2]   Frankfurt Institute for Advanced Studies, 60438 Frankfurt, Germany; bertschinger@fias.uni-frankfurt.de
[3]   Santa Fe Institute, Santa Fe, NM 87501, USA; dhw@santafe.edu
[4]   Massachusetts Institute of Technology, Cambridge, MA 02139, USA
*   Correspondence: jrauh@mis.mpg.de

Received: 15 June 2017; Accepted: 20 September 2017; Published: 6 October 2017

Abstract: Suppose we have a pair of information channels, κ_1, κ_2, with a common input. The Blackwell order is a partial order over channels that compares κ_1 and κ_2 by the maximal expected utility an agent can obtain when decisions are based on the channel outputs. Equivalently, κ_1 is said to be Blackwell-inferior to κ_2 if and only if κ_1 can be constructed by garbling the output of κ_2. A related partial order stipulates that κ_2 is more capable than κ_1 if the mutual information between the input and output is larger for κ_2 than for κ_1 for any distribution over inputs. A Blackwell-inferior channel is necessarily less capable. However, examples are known where κ_1 is less capable than κ_2 but not Blackwell-inferior. We show that this may even happen when κ_1 is constructed by coarse-graining the inputs of κ_2. Such a coarse-graining is a special kind of "pre-garbling" of the channel inputs. This example directly establishes that the expected value of the shared utility function for the coarse-grained channel is larger than it is for the non-coarse-grained channel. This contradicts the intuition that coarse-graining can only destroy information and lead to inferior channels. We also discuss our results in the context of information decompositions.

Keywords: Channel preorders; Blackwell order; degradation order; garbling; more capable; coarse-graining

MSC: 62C05; 62B15; 94A15

1. Introduction

Suppose we are given the choice of two channels that both provide information about the same random variable, and that we want to make a decision based on the channel outputs. Suppose that our utility function depends on the joint value of the input to the channel and our resultant decision based on the channel outputs. Suppose as well that we know the precise conditional distributions defining the channels, and the distribution over channel inputs. Which channel should we choose? The answer to this question depends on the choice of our utility function as well as on the details of the channels and the input distribution. So, for example, without specifying how we will use the channels, in general we cannot just compare their information capacities to choose between them.

Nonetheless, for certain pairs of channels we can make our choice, even without knowing the utility functions or the distribution over inputs. Let us represent the two channels by two (column) stochastic matrices κ_1 and κ_2, respectively. Then, if there exists another stochastic matrix λ such that $\kappa_1 = \lambda \cdot \kappa_2$, there is never any reason to strictly prefer κ_1; if we choose κ_2, we can always make our decision by chaining the output of κ_2 through the channel λ and then using the same decision function we would have used had we chosen κ_1. This simple argument shows that whatever the three stochastic matrices are and whatever the decision rule we would use if we chose channel κ_1, we can always get the same expected utility by instead choosing channel κ_2 with an appropriate decision rule. In this kind of situation, where $\kappa_1 = \lambda \cdot \kappa_2$, we say that κ_1 is a *garbling* (or *degradation*) of κ_2. It is much more difficult to prove that the converse also holds true:

Theorem 1. *(Blackwell's theorem [1]) Let κ_1, κ_2 be two stochastic matrices representing two channels with the same input alphabet. Then the following two conditions are equivalent:*

1. *When the agent chooses κ_2 (and uses the decision rule that is optimal for κ_2), her expected utility is always at least as big as the expected utility when she chooses κ_1 (and uses the optimal decision rule for κ_1), independent of the utility function and the distribution of the input S.*
2. *κ_1 is a garbling of κ_2.*

Blackwell formulated his result in terms of a statistical decision maker who reacts to the outcome of a *statistical experiment*. We prefer to speak of a decision problem instead of a statistical experiment. See [2,3] for an overview.

Blackwell's theorem motivates looking at the following partial order over channels κ_1, κ_2 with a common input alphabet:

$$\kappa_1 \leq \kappa_2 \quad :\Longleftrightarrow \quad \begin{cases} \text{one of the two statements} \\ \text{in Blackwell's theorem holds true.} \end{cases}$$

We call this partial order the *Blackwell order* (this partial order is called *degradation order* by other authors [4,5]). If $\kappa_1 \leq \kappa_2$, then κ_1 is said to be Blackwell-inferior to κ_2. Strictly speaking, the Blackwell order is only a preorder, since there are channels $\kappa_1 \neq \kappa_2$ that satisfy $\kappa_1 \leq \kappa_2 \leq \kappa_1$ (when κ_1 arises from κ_2 by permuting the output alphabet). However, for our purposes, such channels can be considered as equivalent. We write $\kappa_1 < \kappa_2$ if $\kappa_1 \leq \kappa_2$ and $\kappa_1 \not\geq \kappa_2$. By Blackwell's theorem, this implies that κ_2 performs at least as well as κ_1 in any decision problem and that there exist decision problems in which κ_2 outperforms κ_1.

For a given distribution of S, we can also compare κ_1 and κ_2 by comparing the two mutual informations $I(S; X_1)$, $I(S; X_2)$ between the common input S and the channel outputs X_1 and X_2. The data processing inequality shows that $\kappa_2 \geq \kappa_1$ implies $I(S; X_2) \geq I(S; X_1)$. However, the converse implication does not hold. The intuitive reason is that for the Blackwell order, not only the amount of information is important. Rather, the question is how much of the information that κ_1 or κ_2 preserve is relevant for a given fixed decision problem (that is, a given fixed utility function).

Given two channels κ_1, κ_2, suppose that $I(S; X_2) \geq I(S; X_1)$ for all distributions of S. In this case, we say that κ_2 is *more capable* than κ_1. Does this imply that $\kappa_1 \leq \kappa_2$? The answer is known to be negative in general [6]. In Proposition 2 we introduce a new surprising example of this phenomenon with a particular structure. In fact, in this example, κ_1 is a Markov approximation of κ_2 by a deterministic function, in the following sense: Consider another random variable $f(S)$ that arises from S by applying a (deterministic) function f. Given two random variables S, X, denote by $X \leftarrow S$ the channel defined by the conditional probabilities $P_{X|S}(x|s)$, and let $\kappa_2 := (X \leftarrow S)$ and $\kappa_1 := (X \leftarrow f(S)) \cdot (f(S) \leftarrow S)$. Thus, κ_1 can be interpreted as first replacing S by $f(S)$ and then sampling X according to the conditional distribution $P_{X|S}(x|f(s))$. Which channel is superior? Using the data processing inequality, it is easy to see that κ_1 is less capable than κ_2. However, as Proposition 2 shows, in general $\kappa_1 \not\leq \kappa_2$.

We call κ_1 a Markov approximation, because the output of κ_1 is independent of the input S given $f(S)$. The channel κ_1 can also be obtained from κ_2 by "pre-garbling" (Lemma 3); that is, there is another stochastic matrix λ^f that satisfies $\kappa_1 = \kappa_2 \cdot \lambda^f$. It is known that pre-garbling may improve the performance of a channel (but not its capacity) as we recall in Section 2. What may be surprising is that this can happen for pre-garblings of the form λ^f, which have the effect of coarse-graining according to f.

The fact that the more capable preorder does not imply the Blackwell order shows that "Shannon information," as captured by the mutual information, is not the same as "Blackwell information," as needed for the Blackwell decision problems. Indeed, our example explicitly shows that even though coarse-graining always reduces Shannon information, it need not reduce Blackwell information. Finally, let us mention that there are further ways of comparing channels (or stochastic matrices); see [5] for an overview.

Proposition 2 builds upon another effect that we find paradoxical: Namely, there exist random variables S, X_1, X_2 and there exists a function $f : S \to S'$ from the support of S to a finite set S' such that the following holds:

1. S and X_1 are independent given $f(S)$.
2. $(X_1 \leftarrow f(S)) \leq (X_2 \leftarrow f(S))$.
3. $(X_1 \leftarrow S) \not\leq (X_2 \leftarrow S)$.

Statement (1) says that everything X_1 knows about S, it knows through $f(S)$. Statement (2) says that X_2 knows more about $f(S)$ than X_1. Still, (3) says that we cannot conclude that X_2 knows more about S than X_1. The paradox illustrates that it is difficult to formalize what it means to "know more."

Understanding the Blackwell order is an important aspect of understanding information decompositions; that is, the quest to find new information measures that separate different aspects of the mutual information $I(S; X_1, \ldots, X_k)$ of k random variables $X_1, \ldots, X_k$ and a target variable S (see the other contributions of this special issue and references therein). In particular, [7] argues that the Blackwell order provides a natural criterion when a variable X_1 has unique information about S with respect to X_2. We hope that the examples we present here are useful in developing intuition on how information can be shared among random variables and how it behaves when applying a deterministic function, such as a coarse-graining. Further implications of our examples on information decompositions are discussed in [8]. In the converse direction, information decomposition measures (such as measures of unique information) can be used to study the Blackwell order and deviations from the Blackwell order. We illustrate this idea in Example 4.

The remainder of this work is organized as follows: In Section 2, we recall how pre-garbling can be used to improve the performance of a channel. We also show that the pre-garbled channel will always be less capable and that simultaneous pre-garbling of both channels preserves the Blackwell order. In Section 3, we state a few properties of the Blackwell order, and we explain why we find these properties counter-intuitive and paradoxical. In particular, we show that coarse-graining the input can improve the performance of a channel. Section 4 contains a detailed discussion of an example that illustrates these properties. In Section 5 we use the unqiue information measure from [7], which has properties similar to the Le Cam's deficiency, to illustrate deviations from the Blackwell relation.

2. Pre-Garbling

As discussed above (and as made formal in Blackwell's theorem (Theorem 1)), garbling the output of a channel ("post-garbling") never increases the quality of a channel. On the other hand, garbling the input of a channel ("pre-garbling") may increase the performance of a channel, as the following example shows.

Example 1. *Suppose that an agent can choose an action from a finite set $\mathcal{A}$. She then receives a utility $u(a, s)$ that depends both on the chosen action $a \in \mathcal{A}$ and on the value s of a random variable S. Consider the channels*

$$\kappa_1 = \begin{pmatrix} 0.9 & 0 \\ 0.1 & 1 \end{pmatrix} \text{ and } \kappa_2 = \kappa_1 \cdot \begin{pmatrix} 0 & 1 \\ 1 & 0 \end{pmatrix} = \begin{pmatrix} 0 & 0.9 \\ 1 & 0.1 \end{pmatrix},$$

and the utility function

s	0	0	1	1
a	0	1	0	1
$u(s,a)$	2	0	0	1

For uniform input, the optimal decision rule for κ_1 is

$$a(0) = 0, \ a(1) = 1$$

and the opposite

$$a(0) = 1, \ a(1) = 0$$

for κ_2. The expected utility with κ_1 is 1.4, while using κ_2, it is slightly higher (1.45).

It is also not difficult to check that neither of the two channels is a garbling of the other (cf. Propsition 3.22 in [5]).

The intuitive reason for the difference in the expected utilities is that the channel κ_2 transmits one of the states without noise and the other state with noise. With a convenient pre-processing, it is possible to make sure that the relevant information for choosing an action and for optimizing expected utility is transmitted with less noise.

Note the symmetry of the example: each of the two channels arises from the other by a convenient pre-processing, since the pre-processing is invertible. Hence, the two channels are not comparable by the Blackwell order. In contrast, two channels that only differ by an invertible garbling of the output are equivalent with respect to the Blackwell order.

The pre-garbling in Example 1 is invertible, and so it is more aptly described as a pre-processing. In general, though, pure pre-garbling and pure pre-processing are not easily distinguishable, and it is easy to perturb Example 1 by adding noise without changing the conclusion. In Section 3, we will present an example in which the pre-garbling consists of coarse-graining. It is much more difficult to understand how coarse-graining can be used as sensible pre-processing.

Even though pre-garbling can make a channel better (or, more precisely, more suited for a particular decision problem at hand), pre-garbling cannot invert the Blackwell order:

Lemma 1. *If $\kappa_1 < \kappa_2 \cdot \lambda$, then $\kappa_1 \not\succeq \kappa_2$.*

Proof. Suppose that $\kappa_1 < \kappa_2 \cdot \lambda$. Then the capacity of κ_1 is less than the capacity of $\kappa_2 \cdot \lambda$, which is bounded by the capacity of κ_2. Therefore, the capacity of κ_1 is less than the capacity of κ_2. $\square$

Additionally, it follows directly from Blackwell's theorem that

$$\kappa_1 \leq \kappa_2 \text{ implies } \kappa_1 \cdot \lambda \leq \kappa_2 \cdot \lambda$$

for any channel λ, where the input and output alphabets of λ equal the input alphabet of κ_1, κ_2. Thus, pre-garbling preserves the Blackwell order when applied to both channels simultaneously.

Finally, let us remark that certain kinds of simultaneous pre-garbling can also be "hidden" in the utility function; namely, in Blackwell's theorem, it is not necessary to vary the distribution of S as long as the support of the (fixed) input distribution has full support S (that is, every state of the input alphabet of κ_1 and κ_2 appears with positive probability). In this setting, it suffices to look only at different utility functions. When the input distribution is fixed, it is more convenient to think in terms of random variables instead of channels, which slightly changes the interpretation of the decision problem. Suppose we are given random variables S, X_1, X_2 and a utility function $u(a,s)$ depending on the value of S and an action $a \in \mathcal{A}$ as above. If we cannot look at both X_1 and X_2, should we choose to look at X_1 or at X_2 to make our decision?

Theorem 2. *(Blackwell's theorem for random variables [7]) The following two conditions are equivalent:*

1. *Under the optimal decision rule, when the agent chooses X_2, her expected utility is always at least as large as the expected utility when she chooses X_1, independent of the utility function.*
2. $(X_1 \leftarrow S) \leq (X_2 \leftarrow S)$.

3. Pre-Garbling by Coarse-Graining

In this section we present a few counter-intuitive properties of the Blackwell order.

Proposition 1. *There exist random variables S, X_1, X_2 and a function $f : S \to S$ from the support of S to a finite set S' such that the following holds:*

1. *S and X_1 are independent given $f(S)$.*
2. $(X_1 \leftarrow f(S)) < (X_2 \leftarrow f(S))$.
3. $(X_1 \leftarrow S) \not\leq (X_2 \leftarrow S)$.

This result may at first seem paradoxical. After all, property (3) implies that there exists a decision problem involving S for which it is better to use X_1 than X_2. Property (1) implies that any information that X_1 has about S is contained in X_1's information about $f(S)$. One would therefore expect that, from the viewpoint of X_1, any decision problem in which the task is to predict S and to react on S looks like a decision problem in which the task is to react to $f(S)$. But property (2) implies that for such a decision problem, it may in fact be better to look at X_2.

Proof of Proposition 1. The proof is by Example 2, which will be given in Section 4. This example satisfies

1. *S and X_1 are independent given $f(S)$.*
2. $(X_1 \leftarrow f(S)) \leq (X_2 \leftarrow f(S))$.
3. $(X_1 \leftarrow S) \not\leq (X_2 \leftarrow S)$.

It remains to show that it is also possible to achieve the strict relation $(X_1 \leftarrow f(S)) < (X_2 \leftarrow f(S))$ in the second statement. This can easily be done by adding a small garbling to the channel $X_1 \leftarrow f(S)$ (e.g., by adding a binary symmetric channel with sufficiently small noise parameter ϵ). This ensures $(X_1 \leftarrow f(S)) < (X_2 \leftarrow f(S))$, and if the garbling is small enough, this does not destroy the property $(X_1 \leftarrow S) \not\leq (X_2 \leftarrow S)$. □

The example from Proposition 1 also leads to the following paradoxical property:

Proposition 2. *There exist random variables S, X and there exists a function $f : S \to S'$ from the support of S to a finite set S' such that the following holds:*

$$(X \leftarrow f(S)) \cdot (f(S) \leftarrow S) \not\leq X \leftarrow S.$$

Let us again give a heuristic argument for why we find this property paradoxical. Namely, the combined channel $(X \leftarrow f(S)) \cdot (f(S) \leftarrow S)$ can be seen as a Markov chain approximation of the direct channel $X \leftarrow S$ that corresponds to replacing the conditional distribution

$$P_{X|S}(x|s) = \sum_{f(s)} P_{X|Sf(S)}(x|s, f(s)) P_{f(S)|S}(f(s)|s).$$

by

$$\sum_{f(s)} P_{X|f(S)}(x|f(s)) P_{f(S)|S}(f(s)|s).$$

Proposition 2 together with Blackwell's theorem states that there exist situations where this approximation is better than the correct channel.

Proof of Proposition 2. Let S, X_1, X_2 be as in Example 2 in Section 4 that also proves Proposition 1, and let $X = X_2$. In that example, the two channels $X_1 \leftarrow f(S)$ and $X_2 \leftarrow f(S)$ are equal. Moreover, X_1 and S are independent given $f(S)$. Thus, $(X \leftarrow f(S)) \cdot (f(S) \leftarrow S) = (X_1 \leftarrow S)$. Therefore, the statement follows from $(X_1 \leftarrow S) \not\preceq (X_2 \leftarrow S)$. □

On the other hand, the channel $(X \leftarrow f(S)) \cdot (f(S) \leftarrow S)$ is always less capable than $X \leftarrow S$:

Lemma 2. *For any random variables S, X, and function $f : S \to S$, the channel $(X \leftarrow f(S)) \cdot (f(S) \leftarrow S)$ is less capable than $X \leftarrow S$.*

Proof. For any distribution of S, let X' be the output of the channel $(X \leftarrow f(S)) \cdot (f(S) \leftarrow S)$. Then, X' is independent of S given $f(S)$. On the other hand, since f is a deterministic function, X' is independent of $f(S)$ given S. Together, this implies $I(S; X') = I(f(S); X')$. Using the fact that the joint distributions of $(X, f(S))$ and $(X', f(S))$ are identical and applying the data processing inequality gives

$$I(S; X') = I(f(S); X') = I(f(S); X) \leq I(S; X). \quad \square$$

The setting of Proposition 2 can also be understood as a specific kind of pre-garbling. Namely, consider the channel λ^f defined by

$$\lambda^f_{s',s} := P_{S|f(S)}(s'|f(s)).$$

The effect of this channel can be characterized as a randomization of the input: the precise value of S is forgotten, and only the value of $f(S)$ is preserved. Then, a new value s' is sampled for S according to the conditional distribution of S given $f(S)$.

Lemma 3. $(X \leftarrow f(S)) \cdot (f(S) \leftarrow S) = (X \leftarrow S) \cdot \lambda^f$.

Proof. $\displaystyle\sum_{s_1} P_{X|S}(x|s_1)P_{S|f(S)}(s_1|f(s)) = \sum_{s_1,t} P_{X|S}(x|s_1)P_{S|f(S)}(s_1|t)P_{f(S)|S}(t|s)$

$$= \sum_t P_{X|f(S)}(x|t)P_{f(S)|S}(t|s),$$

where we have used that $X - S - f(S)$ forms a Markov chain. □

While it is easy to understand that pre-garbling can be advantageous in general (since it can work as preprocessing), we find it surprising that this can also happen in the case where the pre-garbling is done in terms of a function f; that is, in terms of a channel λ^f that does coarse-graining.

4. Examples

Example 2. *Consider the joint distribution*

$f(s)$	s	x_1	x_2	$P_{f(S)SX_1X_2}$
0	0	0	0	1/4
0	1	0	1	1/4
0	0	1	0	1/8
0	1	1	0	1/8
1	2	1	1	1/4

and the function f: {0, 1, 2} → {0, 1} with f(0) = f(1) = 0 and f(2) = 1. Then, X_1 and X_2 are independent uniform binary random variables, and $f(S) = \mathrm{AND}(X_1, X_2)$. By symmetry, the joint distributions of the pairs

$(f(S), X_1)$ and $(f(S), X_2)$ are identical, and so the two channels $X_1 \leftarrow f(S)$ and $X_2 \leftarrow f(S)$ are identical. In particular, $(X_1 \leftarrow f(S)) \leq (X_2 \leftarrow f(S))$.

On the other hand, consider the utility function

s	a	$u(s,a)$
0	0	0
0	1	0
1	0	1
1	1	0
2	0	0
2	1	1

To compute the optimal decision rule, let us look at the conditional distributions:

s	x_1	$P_{S|X_1}(s|x_1)$	s	x_2	$P_{S|X_2}(s|x_2)$
0	0	1/2	0	0	3/4
1	0	1/2	1	0	1/4
0	1	1/4	0	1	0
1	1	1/4	1	1	1/2
2	1	1/2	2	1	1/2

The optimal decision rule for X_1 is $a(0) = 0$, $a(1) = 1$, with expected utility

$$u_{X_1} := 1/2 \cdot 1/2 + 1/2 \cdot 1/2 = 1/2.$$

The optimal decision rule for X_2 is $a(0) = 0$, $a(1) \in \{0, 1\}$ (this is not unique in this case), with expected utility

$$u_{X_2} := 1/2 \cdot 1/4 + 1/2 \cdot 1/2 = 3/8 < 1/2.$$

How can we understand this example? Some observations:

- It is easy to see that X_2 has more irrelevant information than X_1: namely, X_2 can determine relatively precisely when $S = 0$. However, since $S = 0$ gives no utility independent of the action, this information is not relevant. It is more difficult to understand why X_2 has less relevant information than X_1. Surprisingly, X_1 can determine more precisely when $S = 1$: if $S = 1$, then X_1 "detects this" (in the sense that X_1 chooses action 0) with probability 2/3. For X_2, the same probability is only 1/3.
- The conditional entropies of S given X_2 are smaller than the conditional entropies of S given X_1:

$$H(S|X_1 = 0) = \log(2), \qquad\qquad H(S|X_1 = 1) = \tfrac{3}{2}\log(2),$$
$$H(S|X_2 = 0) = 2\log(2) - \tfrac{3}{2}\log(3) \approx 0.4150375\log(2), \qquad H(S|X_2 = 1) = \log(2).$$

- One can see in which sense $f(S)$ captures the relevant information for X_1, and indeed for the whole decision problem: knowing $f(S)$ is completely sufficient in order to receive the maximal utility for each state of S. However, when information is incomplete, it matters how the information about the different states of S is mixed, and two variables X_1, X_2 that have the same joint distribution with $f(S)$ may perform differently. It is somewhat surprising that it is the random variable that has less information about S and that is conditionally independent of S given $f(S)$ which actually performs better.

Example 2 is different from the pre-garbling Example 1 discussed in Section 2. In the latter, both channels had the same amount of information (mutual information) about S, but for the given decision problem the information provided by κ_2 was more relevant than the information provided by κ_1. The first difference in Example 2 is that X_1 has less mutual information about S than X_2 (Lemma 2). Moreover, both channels are identical with respect to $f(S)$; i.e., they provide the same information about $f(S)$, and for X_1 it is the only information it has about S. So, one could argue that X_2 has additional information that does not help, but decreases the expected utility instead.

We give another example which shows that X_2 can also be chosen as a deterministic function of S.

Example 3. *Consider the joint distribution*

$f(s)$	s	x_1	x_2	$P_{f(S)SX_1X_2}$
0	0	0	0	1/6
0	0	1	0	1/6
0	1	0	1	1/6
0	1	1	1	1/6
1	2	1	1	1/3

The function f is as above, but now also X_2 is a function of S. Again, the two channels $X_1 \leftarrow f(S)$ and $X_2 \leftarrow f(S)$ are identical, and X_1 is independent of S given $f(S)$. Consider the utility function

s	a	$u(s,a)$
0	0	0
0	1	0
1	0	0
1	1	1
2	0	0
2	1	-1

One can show that it is optimal for an agent who relies on X_2 to always choose action 0, which brings no reward (and no loss). However, when the agent knows that X_1 is zero, he may safely choose action 1 and has a positive probability of receiving a positive reward.

To add another interpretation to the last example, we visualize the situation in the following Bayesian network:

$$X \leftarrow S \to f(S) \to X',$$

where, as in Proposition 2 and its proof, we let $X = X_2$, and we consider $X' = X_1$ as an approximation of X. Then, S denotes the state of the system that we are interested in, and X denotes a given set of observables of interest. $f(S)$ can be considered as a "proxy" in situations where it is difficult to observe X directly. For example, in neuroimaging, instead of directly measuring the neural activity X, one might look at an MRI signal $f(S)$. In economic and social sciences, monetary measures like the GDP are used as a proxy for prosperity.

A decision problem can always be considered as a classification problem defined by the utility $u(s,a)$ by considering the optimal action as the class label of state S. Proposition 2 now says that there exist S, X, $f(S)$, and a classification problem $u(s,a)$, such that the approximated features X' (simulated from $f(S)$) allow for a better classification (higher utility) than the original features X.

In such a situation, looking at $f(S)$ will always be better than looking at either X or X'. Thus, the paradox will only play a role in situations where it is not possible to base the decision on $f(S)$ directly. For example, $f(S)$ might still be too large, or X might have a more natural interpretation, making it easier to interpret for the decision taker. However, when it is better to base a decision on a proxy rather than directly on the observable of interest, this interpretation may be erroneous.

5. Information Decomposition and Le Cam Deficiency

Given two channels κ_1, κ_2, how can one decide whether or not $\kappa_1 \leq \kappa_2$? The easiest way is to check whether the equation $\kappa_1 = \lambda \cdot \kappa_2$ has a solution λ that is a stochastic matrix. In the finite alphabet case, this amounts to checking the feasibility of a linear program, which is considered computationally easy. However, when the feasibility check returns a negative result, this approach does not give any more information (e.g., how far κ_1 is away from being a garbling of κ_2). A function that quantifies how far κ_1 is from being a garbling of κ_2 is given by the *(Le Cam) deficiency* and its various generalizations [9]. Another such function is given by UI defined in [7] that accounts for the fact that the channels we consider are of the form $\kappa_1 = (X_1 \leftarrow S)$ and $\kappa_2 = (X_2 \leftarrow S)$; that is, they are derived from conditional distributions of random variables. In contrast to the deficiencies, UI depends on the input distribution to these channels.

Let $P_{SX_1X_2}$ be a joint distribution of S and the outputs X_1 and X_2. Let Δ_P be the set of all joint distributions of the random variables S, X_1, X_2 (with the same alphabets) that are compatible with the marginal distributions of $P_{SX_1X_2}$ for the pairs (S, X_1) and (S, X_2); i.e.,

$$\Delta_P := \{Q_{SX_1X_2} \in \Delta : Q_{SX_1} = P_{SX_1}, Q_{SX_2} = P_{SX_2}\}.$$

In other words, Δ_P consists of all joint distributions that are compatible with κ_1 and κ_2 and that have the same distribution for S as $P_{SX_1X_2}$. Consider the function

$$UI(S; X_1 \backslash X_2) := \min_{Q \in \Delta_P} I_Q(S; X_1 | X_2),$$

where I_Q denotes the conditional mutual information evaluated with respect to the the the joint distribution Q. This function has the following property: $UI(S; X_1 \backslash X_2) = 0$ if and only if $\kappa_1 \leq \kappa_2$ [7]. Computing UI is a convex optimization problem. However, the condition number can be very bad, which makes the problem difficult in practice.

UI is interpreted in [7] as a measure of the *unique* information that X_1 conveys about S (with respect to X_2). So, for instance, with this interpretation Example 2 can be summarized as follows: neither X_1 nor X_2 has unique information about $f(S)$. However, both variables have unique information about S, although X_1 is conditionally independent of S given $f(S)$ and thus—in contrast to X_2—contains no "additional" information about S. We now apply UI to a parameterized version of the AND gate in Example 2.

Example 4. *Figure 1a shows a heat map of UI computed on the set of all distributions of the form*

$f(s)$	s	x_1	x_2	$P_{f(S)SX_1X_2}$
0	0	0	0	$1/8 + 2b$
0	1	0	0	$1/8 - 2b$
0	0	0	1	$1/8 + a$
0	1	0	1	$1/8 - a$
0	0	1	0	$1/8 + a/2 + b$
0	1	1	0	$1/8 - a/2 - b$
1	2	1	1	$1/4$

where $-1/8 \leq a \leq 1/8$ and $-1/16 \leq b \leq 1/16$. This is the set of distributions of S, X_1, X_2 that satisfy the following constraints:

1. *X_1, X_2 are independent;*
2. *$f(S) = \text{AND}(X_1, X_2)$, where f is as in Example 2; and*
3. *X_1 is independent of S given $f(S)$.*

Along the secondary diagonal $b = a/2$, the marginal distributions of the pairs (S, X_1) and (S, X_2) are identical. In such a situation, the channels $(X_1 \leftarrow S)$ and $(X_2 \leftarrow S)$ are Blackwell-equivalent, and so UI vanishes. Further away from the diagonal, the marginal distributions differ, and UI grows. The maximum value is achieved at the corners for $(a, b) = (-1/8, 1/16), (1/8, -1/16)$. At the upper left corner $(a, b) = \pm(-1/8, 1/16)$, we recover Example 2.

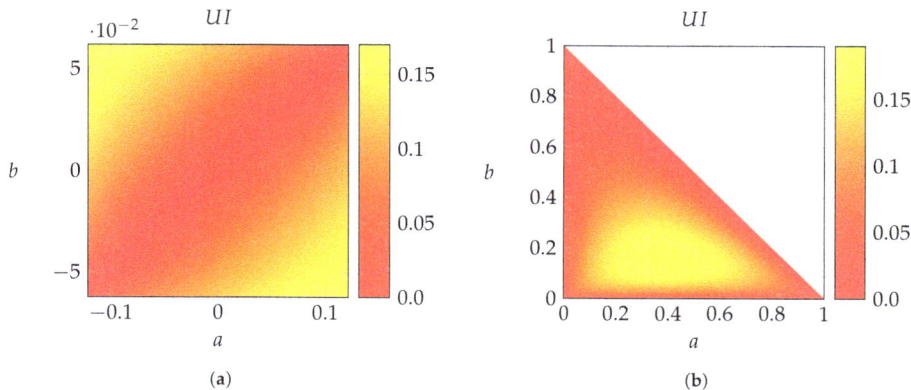

Figure 1. Heatmaps for the function UI in (**a**) Example 4, and (**b**) Example 5.

Example 5. *Figure 1b shows a heat map of UI computed on the set of all distributions of the form*

$f(s)$	s	x_1	x_2	$P_{f(S)SX_1X_2}$
0	0	0	0	$a^2/(a+b)$
0	0	1	0	$ab/(a+b)$
0	1	0	1	$ab/(a+b)$
0	1	1	1	$b^2/(a+b)$
1	2	1	1	$1-a-b$

where $a, b \geq 0$ and $a + b \leq 1$. This extends Example 3, which is recovered for $a = b = 1/3$. This is the set of distributions of S, X_1, X_2 that satisfy the following constraints:

1. *X_2 is a function of S, where the function is as in Example 3.*
2. *X_1 is independent of S given $f(S)$.*
3. *The channels $X_1 \leftarrow f(S)$ and $X_2 \leftarrow f(S)$ are identical.*

Acknowledgments: We thank the participants of the PID workshop at FIAS in Frankfurt in December 2016 for many stimulating discussions on this subject. Nils Bertschinger thanks h.c. Maucher for funding his position.

Author Contributions: The research was initiated by J.R. and carried out by all authors. Computer experiments to find and analyze the examples were done by P.K.B. D.W. simplified Example 1. J.J. and N.B. added interpretation. The manuscript was written by J.R., P.K.B., E.O. and D.W. Nobody played the synthesizer. All authors have read and approved the final manuscript.

Conflicts of Interest: The authors declare no conflict of interest.

References

1. Blackwell, D. Equivalent Comparisons of Experiments. *Ann. Math. Stat.* **1953**, *24*, 265–272.
2. Torgersen, E. *Comparison of Statistical Experiments*; Cambridge University Press: New York, NY, USA, 1991.
3. Le Cam, L. Comparison of Experiments—A Short Review. *Stat. Probab. Game Theory* **1996**, *30*, 127–138.
4. Bergmans, P. Random coding theorem for broadcast channels with degraded components. *IEEE Trans. Inf. Theory* **1973**, *19*, 197–207.
5. Cohen, J.; Kemperman, J.; Zbăganu, G. *Comparisons of Stochastic Matrices with Applications in Information Theory, Statistics, Economics, and Population Sciences*; Birkhäuser: Boston, MA, USA, 1998.
6. Körner, J.; Marton, K. Comparison of two noisy channels. In *Topics in Information Theory*; Colloquia Mathematica Societatis János Bolyai: Keszthely, Hungary, 1975; Volume 16, pp. 411–423.
7. Bertschinger, N.; Rauh, J.; Olbrich, E.; Jost, J.; Ay, N. Quantifying unique information. *Entropy* **2014**, *16*, 2161–2183.

8.   Rauh, J.; Banerjee, P.K.; Olbrich, E.; Jost, J.; Bertschinger, N. On extractable shared information. *arXiv* **2017**, arXiv:1701.07805.

9.   Raginsky, M. Shannon meets Blackwell and Le Cam: Channels, codes, and statistical experiments. In Proceedings of the 2011 IEEE International Symposium on Information Theory Proceedings, St. Petersburg, Russia, 31 July–5 August 2011; pp. 1220–1224.

MDPI

Article

Multiscale Information Decomposition: Exact Computation for Multivariate Gaussian Processes

Luca Faes [1,2,*], **Daniele Marinazzo** [3] **and Sebastiano Stramaglia** [4,5]

1   Bruno Kessler Foundation, 38123 Trento, Italy
2   BIOtech, Department of Industrial Engineering, University of Trento, 38123 Trento, Italy
3   Data Analysis Department, Ghent University, 9000 Ghent, Belgium; daniele.marinazzo@ugent.be
4   Dipartimento di Fisica, Universitá degli Studi Aldo Moro, 70126 Bari, Italy; sebastiano.stramaglia@ba.infn.it
5   Istituto Nazionale di Fisica Nucleare, 70126 Sezione di Bari, Italy
*   Correspondence: faes.luca@gmail.com; Tel.: +39-0461-282773

Received: 21 June 2017; Accepted: 7 August 2017; Published: 8 August 2017

Abstract: Exploiting the theory of state space models, we derive the exact expressions of the information transfer, as well as redundant and synergistic transfer, for coupled Gaussian processes observed at multiple temporal scales. All of the terms, constituting the frameworks known as interaction information decomposition and partial information decomposition, can thus be analytically obtained for different time scales from the parameters of the VAR model that fits the processes. We report the application of the proposed methodology firstly to benchmark Gaussian systems, showing that this class of systems may generate patterns of information decomposition characterized by prevalently redundant or synergistic information transfer persisting across multiple time scales or even by the alternating prevalence of redundant and synergistic source interaction depending on the time scale. Then, we apply our method to an important topic in neuroscience, i.e., the detection of causal interactions in human epilepsy networks, for which we show the relevance of partial information decomposition to the detection of multiscale information transfer spreading from the seizure onset zone.

Keywords: information dynamics; information transfer; multiscale entropy; multivariate time series analysis; redundancy and synergy; state space models; vector autoregressive models

1. Introduction

The information-theoretic treatment of groups of correlated degrees of freedom can reveal their functional roles as memory structures or information processing units. A large body of recent work has shown how the general concept of "information processing" in a network of multiple interacting dynamical systems described by multivariate stochastic processes can be dissected into basic elements of computation defined within the so-called framework of information dynamics [1]. These elements essentially reflect the new information produced at each moment in time about a target system in the network [2], the information stored in the target system [3,4], the information transferred to it from the other connected systems [5,6] and the modification of the information flowing from multiple source systems to the target [7,8]. The measures of information dynamics have gained more and more importance in both theoretical and applicative studies in several fields of science [9–18]. While the information-theoretic approaches to the definition and quantification of new information, information storage and information transfer are well understood and widely accepted, the problem of defining, interpreting and using measures of information modification has not been fully addressed in the literature.

Information modification in a network is tightly related to the concepts of redundancy and synergy between source systems sharing information about a target system, which refer to the existence of common information about the target that can be retrieved when the sources are used separately (redundancy) or when they are used jointly (synergy) [19]. Classical multivariate entropy-based approaches refer to the interaction information decomposition (IID), which reflects information modification through the balance between redundant and synergetic interaction among different source systems influencing the target [20–22]. The IID framework has the drawback that it implicitly considers redundancy and synergy as mutually exclusive concepts, because it quantifies information modification with a single measure of interaction information [23] (also called co-information [24]) that takes positive or negative values depending on whether the net interaction between the sources is synergistic or redundant. This limitation has been overcome by the elegant mathematical framework introduced by Williams and Beer [25], who proposed the so-called partial information decomposition (PID) as a nonnegative decomposition of the information shared between a target and a set of sources into terms quantifying separately unique, redundant and synergistic contributions. However, the PID framework has the drawback that the terms composing the PID cannot be obtained unequivocally from classic measures of information theory (i.e., entropy and mutual information), but a new definition of either redundant, synergistic or unique information needs to be provided to implement the decomposition. Accordingly, much effort has focused on finding the most proper measures to define the components of the PID, with alternative proposals defining new measures of redundancy [25,26], synergy [27,28] or unique information [29]. The proliferation of different definitions is mainly due to the fact that there is no full consensus on which axioms should be stated to impose desirable properties for the PID measures. An additional problem which so far has seriously limited the practical implementation of these concepts is the difficulty in providing reliable estimates of the information measures appearing in the IID and PID decompositions. The naive estimation of probabilities by histogram-based methods followed by the use of plug-in estimators leads to serious bias problems [30,31]. While the use of binless density estimators [32] and the adoption of schemes for dimensionality reduction [33,34] have been shown to improve the reliability of estimates of information storage and transfer [35], the effectiveness of these approaches for the computation of measures of information modification has not been demonstrated yet. Interestingly, both the problems of defining appropriate PID measures and of reliably estimating these measures from data are much alleviated if one assumes that the observed variables have a joint Gaussian distribution. Indeed, in such a case, recent studies have proven the equivalence between most of the proposed redundancy measures to be used in the PID [36] and have provided closed form solutions to the issue of computing any measure of information dynamics from the parameters of the vector autoregressive (VAR) model that characterizes an observed multivariate Gaussian process [17,37,38].

The second fundamental question that is addressed in this study is relevant to the computation of information dynamics for stochastic processes displaying multiscale dynamical structures. It is indeed well known that many complex physical and biological systems exhibit peculiar oscillatory activities, which are deployed across multiple temporal scales [39–41]. The most common way to investigate such activities is to resample at different scales, typically through low pass filtering and downsampling [42,43], the originally measured realization of an observed process, so as to yield a set of rescaled time series, which are then analyzed employing different dynamical measures. This approach is well established and widely used for the multiscale entropy analysis of individual time series measured from scalar stochastic processes. However, its extension to the investigation of the multiscale structure of the information transfer among coupled processes is complicated by theoretical and practical issues [44,45]. Theoretically, the procedure of rescaling alters the causal interactions between lagged components of the processes in a way that is not fully understood and, if not properly performed, may alter the temporal relations between processes and thus induce spurious detection of information transfer. In practical analysis, filtering and downsampling are known to degrade severely the estimation of information dynamics and to impact consistently the detectability, accuracy and data demand [46,47].

In recent works, we have started tackling the above problems within the framework of linear VAR modeling of multivariate Gaussian processes, with the focus on the multiscale computation of information storage and information transfer [48,49]. In this study, we aim at extending these recent theoretical advances to the multiscale analysis of information modification in multivariate Gaussian systems performed through the IID and PID decomposition frameworks. To this end, we exploit the theory of state space (SS) models [50] and build on recent theoretical results [44,45] to show that exact values of interaction transfer, as well as redundant and synergistic transfer can be obtained for coupled Gaussian processes observed at different time scales starting from the parameters of the VAR model that fits the processes and from the scale factor. The theoretical derivations are first used in examples of benchmark Gaussian systems, reporting that these systems may generate patterns of information decomposition characterized by prevalently redundant or synergistic information transfer persisting across multiple time scales or even by alternating the prevalence of redundant and synergistic source interaction depending on the time scale. The high computational reliability of the SS approach is then exploited in the analysis of real data by the application to a topic of great interest in neuroscience, i.e., the detection of information transfer in epilepsy networks.

The proposed framework is implemented in the msID MATLAB® toolbox, which is uploaded as Supplementary Material to this article and is freely available for download from www.lucafaes.net/msID.html and https://github.com/danielemarinazzo/multiscale_PID.

2. Information Transfer Decomposition in Multivariate Processes

Let us consider a discrete-time, stationary vector stochastic process composed of M real-valued zero-mean scalar processes, $Y_n = [Y_{1,n} \cdots Y_{M,n}]^T$, $-\infty < n < \infty$. In an information-theoretic framework, the information transfer between scalar sub-processes is quantified by the well-known transfer entropy (TE), which is a popular measure of the "information transfer" directed towards an assigned target process from one or more source processes. Specifically, the TE quantifies the amount of information that the past of the source provides about the present of the target over and above the information already provided by the past of the target itself [5]. Taking Y_j as target and Y_i as source, the TE is defined as:

$$T_{i \to j} = I(Y_{j,n}; Y_{i,n}^- | Y_{j,n}^-) \tag{1}$$

where $Y_{i,n}^- = [Y_{i,n-1}, Y_{i,n-2} \cdots]$ and $Y_{j,n}^- = [Y_{j,n-1}, Y_{j,n-2} \cdots]$ represent the past of the source and target processes and $I(\cdot; \cdot | \cdot)$ denotes conditional mutual information (MI). In the presence of two sources Y_i and Y_k and a target Y_j, the information transferred toward Y_j from the sources Y_i and Y_k taken together is quantified by the joint TE:

$$T_{ik \to j} = I(Y_{j,n}; Y_{i,n}^-, Y_{k,n}^- | Y_{j,n}^-). \tag{2}$$

Under the premise that the information jointly transferred to the target by the two sources is different than the sum of the amounts of information transferred individually, in the following, we present two possible strategies to decompose the joint TE (2) into amounts eliciting the individual TEs, as well as redundant and/or synergistic TE terms.

2.1. Interaction Information Decomposition

The first strategy, which we denote as interaction information decomposition (IID), decomposes the joint TE (2) as:

$$T_{ik \to j} = T_{i \to j} + T_{k \to j} + \mathcal{I}_{ik \to j}, \tag{3}$$

where $\mathcal{I}_{ik \to j}$ is denoted as interaction transfer entropy (ITE) because it is equivalent to the interaction information [23] computed between the present of the target and the past of the two sources, conditioned to the past of the target:

$$\mathcal{I}_{ik \to j} = I(Y_{j,n}; Y_{i,n}^-; Y_{k,n}^- | Y_{j,n}^-). \tag{4}$$

The interaction TE quantifies the modification of the information transferred from the source processes Y_i and Y_k to the target Y_j, being positive when Y_i and Y_k cooperate in a synergistic way and negative when they act redundantly. This interpretation is evident from the diagrams of Figure 1: in the case of synergy (Figure 1a), the two sources Y_i and Y_k taken together contribute to the target Y_j with more information than the sum of their individual contributions ($\mathcal{T}_{ik \to j} > \mathcal{T}_{i \to j} + \mathcal{T}_{k \to j}$), and the ITE is positive; in the case of redundancy (Figure 1b), the sum of the information amounts transferred individually from each source to the target is higher than the joint information transfer ($\mathcal{T}_{i \to j} + \mathcal{T}_{k \to j} > \mathcal{T}_{ik \to j}$), so that the ITE is negative.

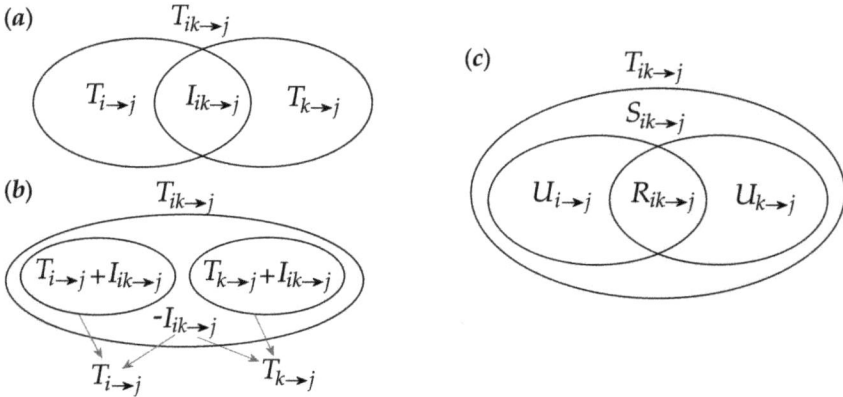

Figure 1. Venn diagram representations of the interaction information decomposition (IID) (**a,b**) and the partial information decomposition (PID) (**c**). The IID is depicted in a way such that all areas in the diagrams are positive: the interaction information transfer $\mathcal{I}_{ik \to j}$ is positive in (**a**), denoting net synergy, and is negative in (**b**), denoting net redundancy.

2.2. Partial Information Decomposition

An alternative expansion of the joint TE is that provided by the so-called partial information decomposition (PID) [25]. The PID evidences four distinct quantities measuring the unique information transferred from each individual source to the target, measured by the unique TEs $\mathcal{U}_{i \to j}$ and $\mathcal{U}_{k \to j}$, and the redundant and synergistic information transferred from the two sources to the target, measured by the redundant TE $\mathcal{R}_{ik \to j}$ and the synergistic TE $\mathcal{S}_{ik \to j}$. These four measures are related to each other and to the joint and individual TEs by the following equations (see also Figure 1c):

$$\mathcal{T}_{ik \to j} = \mathcal{U}_{i \to j} + \mathcal{U}_{k \to j} + \mathcal{R}_{ik \to j} + \mathcal{S}_{ik \to j}, \tag{5a}$$

$$\mathcal{T}_{i \to j} = \mathcal{U}_{i \to j} + \mathcal{R}_{ik \to j}, \tag{5b}$$

$$\mathcal{T}_{k \to j} = \mathcal{U}_{k \to j} + \mathcal{R}_{ik \to j}. \tag{5c}$$

In the PID defined above, the terms $\mathcal{U}_{i \to j}$ and $\mathcal{U}_{k \to j}$ quantify the parts of the information transferred to the target process Y_j, which are unique to the source processes Y_i and Y_k, respectively, thus reflecting contributions to the predictability of the target that can be obtained from one of the sources alone, but not from the other source alone. Each of these unique contributions sums up with the redundant transfer $\mathcal{R}_{ik \to j}$ to yield the information transfer from one source to the target as is known from the classic Shannon information theory. Then, the term $\mathcal{S}_{ik \to j}$ refers to the synergy between the two sources while they transfer information to the target, intended as the information that is uniquely obtained taking the two sources Y_i and Y_k together, but not considering them alone. Compared to the IID defined in (3), the PID (5) has the advantage that it provides distinct non-negative measures of redundancy and synergy, thereby accounting for the possibility that redundancy and synergy may coexist as separate elements of information modification. Interestingly, the IID and PID defined in Equations (3) and (5) are related to each other in a way such that:

$$\mathcal{I}_{ik \to j} = \mathcal{S}_{ik \to j} - \mathcal{R}_{ik \to j},$$ (6)

thus showing that the interaction TE is actually a measure of the 'net' synergy manifested in the transfer of information from the two sources to the target.

An issue with the PID (5) is that its constituent measures cannot be obtained through classic information theory simply subtracting conditional MI terms as done for the IID; an additional ingredient to the theory is needed to get a fourth defining equation to be added to (5) for providing an unambiguous definition of $\mathcal{U}_{i \to j}$, $\mathcal{U}_{k \to j}$, $\mathcal{R}_{ik \to j}$ and $\mathcal{S}_{ik \to j}$. While several PID definitions have been proposed arising from different conceptual definitions of redundancy and synergy [26,27,29], here, we make reference to the so-called minimum MI (MMI) PID [36]. According to the MMI PID, redundancy is defined as the minimum of the information provided by each individual source to the target. In terms of information transfer measured by the TE, this leads to the following definition of the redundant TE:

$$\mathcal{R}_{ik \to j} = \min\{\mathcal{T}_{i \to j}, \mathcal{T}_{k \to j}\}.$$ (7)

This choice satisfies the desirable property that the redundant TE is independent of the correlation between the source processes. Moreover, it has been shown that, if the observed processes have a joint Gaussian distribution, all previously-proposed PID formulations reduce to the MMI PID [36].

3. Multiscale Information Transfer Decomposition

3.1. Multiscale Representation of Multivariate Gaussian Processes

In the linear signal processing framework, the M-dimensional vector stochastic process $\mathbf{Y}_n = [Y_{1,n} \cdots Y_{M,n}]^T$ is classically described using a vector autoregressive (VAR) model of order p:

$$\mathbf{Y}_n = \sum_{k=1}^{p} A_k \mathbf{Y}_{n-k} + \mathbf{U}_n$$ (8)

where A_k are $M \times M$ matrices of coefficients and $\mathbf{U}_n = [U_{1,n} \cdots U_{M,n}]^T$ is a vector of M zero mean Gaussian processes with covariance matrix $\mathbf{\Sigma} \equiv \mathbb{E}[\mathbf{U}_n \mathbf{U}_n^T]$ ($\mathbb{E}$ is the expectation operator). To study the observed process Y at the temporal scale identified by the scale factor τ, we apply the following transformation to each constituent process $Y_m, m = 1, \ldots, M$:

$$\bar{Y}_{m,n} = \sum_{l=0}^{q} b_l Y_{m,n\tau - l}.$$ (9)

This rescaling operation corresponds to transforming the original process Y through a two-step procedure that consists of the following filtering and downsampling steps, yielding respectively the processes $\tilde{Y}$ and $\bar{Y}$:

$$\tilde{Y}_n = \sum_{l=0}^{q} b_l Y_{n-l},$$ (10a)

$$\bar{Y}_n = \tilde{Y}_{n\tau}, n = 1, \ldots, N/\tau$$ (10b)

The change of scale in (9) generalizes the averaging procedure originally proposed in [42], which sets $q = \tau - 1$ and $b_l = 1/\tau$ and, thus, realizes the step of filtering through the simple procedure of averaging τ subsequent samples. To improve the elimination of the fast temporal scales, in this study, we follow the idea of [43], in which a more appropriate low pass filter than averaging is employed. Here, we identify the b_l as the coefficients of a linear finite impulse response (FIR) low pass filter of order q; the FIR filter is designed using the classic window method with the Hamming window [51], setting the cutoff frequency at $f_\tau = 1/2\tau$ in order to avoid aliasing in the subsequent downsampling step. Substituting (8) in (10a), the filtering step leads to the process representation:

$$\tilde{Y}_n = \sum_{k=1}^{p} \mathbf{A}_k \tilde{Y}_{n-k} + \sum_{l=0}^{q} \mathbf{B}_l U_{n-l} \tag{11}$$

where $\mathbf{B}_l = b_l \mathbf{I}_M$ ($\mathbf{I}_M$ is the $M \times M$ identity matrix). Hence, the change of scale introduces a moving average (MA) component of order q in the original VAR(p) process, transforming it into a VARMA(p, q) process. As we will show in the next section, the downsampling step (10b) keeps the VARMA representation, altering the model parameters.

3.2. State Space Processes

3.2.1. Formulation of State Space Models

State space models are models that make use of state variables to describe a system by a set of first-order difference equations, rather than by one or more high-order difference equations [52,53]. The general linear state space (SS) model describing an observed vector process Y has the form:

$$X_{n+1} = \mathbf{A}X_n + W_n \tag{12a}$$
$$Y_n = \mathbf{C}X_n + V_n \tag{12b}$$

where the state Equation (12a) describes the update of the L-dimensional state (unobserved) process through the $L \times L$ matrix $\mathbf{A}$, and the observation Equation (12b) describes the instantaneous mapping from the state to the observed process through the $M \times L$ matrix $\mathbf{C}$. W_n and V_n are zero-mean white noise processes with covariances $\mathbf{Q} \equiv \mathbb{E}[W_n W_n^T]$ and $\mathbf{R} \equiv \mathbb{E}[V_n V_n^T]$ and cross-covariance $\mathbf{S} \equiv \mathbb{E}[W_n V_n^T]$. Thus, the parameters of the SS model (12) are $(\mathbf{A}, \mathbf{C}, \mathbf{Q}, \mathbf{R}, \mathbf{S})$.

Another possible SS representation is that evidencing the innovations $E_n = Y_n - \mathbb{E}[Y_n | Y_n^-]$, i.e., the residuals of the linear regression of Y_n on its infinite past $Y_n^- = [Y_{n-1}^T Y_{n-2}^T \cdots]^T$ [53]. This new SS representation, usually referred to as the "innovations form" SS model (ISS), is characterized by the state process $Z_n = \mathbb{E}[X_n | Y_n^-]$ and by the $L \times M$ Kalman gain matrix $\mathbf{K}$:

$$Z_{n+1} = \mathbf{A}Z_n + \mathbf{K}E_n \tag{13a}$$
$$Y_n = \mathbf{C}Z_n + E_n \tag{13b}$$

The parameters of the ISS model (13) are $(\mathbf{A}, \mathbf{C}, \mathbf{K}, \mathbf{V})$, where $\mathbf{V}$ is the covariance of the innovations, $\mathbf{V} \equiv \mathbb{E}[E_n E_n^T]$. Note that the ISS (13) is a special case of (12) in which $W_n = \mathbf{K}E_n$ and $V_n = E_n$, so that $\mathbf{Q} = \mathbf{K}\mathbf{V}\mathbf{K}^T$, $\mathbf{R} = \mathbf{V}$ and $\mathbf{S} = \mathbf{K}\mathbf{V}$.

Given an SS model in the form (12), the corresponding ISS model (13) can be identified by solving a so-called discrete algebraic Riccati equation (DARE) formulated in terms of the state error variance matrix $\mathbf{P}$ [45]:

$$\mathbf{P} = \mathbf{A}\mathbf{P}\mathbf{A}^T + \mathbf{Q} - (\mathbf{A}\mathbf{P}\mathbf{C}^T + \mathbf{S})(\mathbf{C}\mathbf{P}\mathbf{C}^T + \mathbf{R})^{-1}(\mathbf{C}\mathbf{P}\mathbf{A}^T + \mathbf{S}^T) \tag{14}$$

Under some assumptions [45], the DARE (14) has a unique stabilizing solution, from which the Kalman gain and innovation covariance can be computed as:

$$\mathbf{V} = \mathbf{C}\mathbf{P}\mathbf{C}^T + \mathbf{R}$$
$$\mathbf{K} = (\mathbf{A}\mathbf{P}\mathbf{C}^T + \mathbf{S})\mathbf{V}^{-1}, \tag{15}$$

thus completing the transformation from the SS form to the ISS form.

3.2.2. State Space Models of Filtered and Downsampled Linear Processes

Exploiting the close relation between VARMA models and SS models, first we show how to convert the VARMA model (11) into an ISS model in the form of (13) that describes the filtered process $\tilde{Y}_n$. To do this, we exploit Aoki's method [50] defining the state process $\tilde{Z}_n = [Y_{n-1}^T \cdots Y_{n-p}^T U_{n-1}^T \cdots U_{n-q}^T]^T$ that, together with $\tilde{Y}_n$, obeys the state Equation (13) with parameters $(\tilde{A}, \tilde{C}, \tilde{K}, \tilde{V})$, where:

$$\tilde{A} = \begin{bmatrix} A_1 & \cdots & A_{p-1} & A_p & B_1 & \cdots & B_{q-1} & B_q \\ I_M & \cdots & 0_M & 0_M & 0_M & \cdots & 0_M & 0_M \\ \vdots & & \vdots & \vdots & \vdots & & \vdots & \vdots \\ 0_M & \cdots & I_M & 0_M & 0_M & \cdots & 0_M & 0_M \\ 0_M & \cdots & 0_M & 0_M & 0_M & \cdots & 0_M & 0_M \\ 0_M & \cdots & 0_M & 0_M & I_M & \cdots & 0_M & 0_M \\ \vdots & & \vdots & \vdots & \vdots & & \vdots & \vdots \\ 0_M & \cdots & 0_M & 0_M & 0_M & \cdots & I_M & 0_M \end{bmatrix}$$

$$\tilde{C} = \begin{bmatrix} A_1 & \cdots & A_p & B_1 & \cdots & B_q \end{bmatrix}$$

$$\tilde{K} = \begin{bmatrix} I_M & 0_{M \times M(p-1)} & B_0^{-T} & 0_{M \times M(q-1)} \end{bmatrix}^T$$

and $\tilde{V} = B_0 \Sigma B_0^T$, where $\tilde{V}$ is the covariance of the innovations $\tilde{E}_n = B_0 U_n$.

Now, we turn to show how the downsampled process $\bar{Y}_n$ can be represented through an ISS model directly from the ISS formulation of the filtered process $\tilde{Y}_n$. To this end, we exploit recent theoretical findings providing the state space form of downsampled signals (Theorem III in [45]). Accordingly, the SS representation of the process downsampled at scale τ, $\bar{Y}_n = \tilde{Y}_{n\tau}$ has parameters $(\bar{A}, \bar{C}, \bar{Q}, \bar{R}, \bar{S})$, where $\bar{A} = \tilde{A}^\tau$, $\bar{C} = \tilde{C}$, $\bar{Q} = Q_\tau$, $\bar{R} = \tilde{V}$ and $\bar{S} = S_\tau$, with Q_τ and S_τ given by:

$$\begin{aligned} S_\tau &= \tilde{A}^{\tau-1} \tilde{K} \tilde{V} \\ Q_\tau &= \tilde{A} Q_{\tau-1} \tilde{A}^T + \tilde{K} \tilde{V} \tilde{K}^T, \tau \geq 2 \\ Q_1 &= \tilde{K} \tilde{V} \tilde{K}^T, \tau = 1. \end{aligned} \tag{16}$$

Therefore, the downsampled process has an ISS representation with state process $\bar{Z}_n = \tilde{Z}_{n\tau}$, innovation process $\bar{E}_n = \tilde{E}_{n\tau}$ and parameters $(\bar{A}, \bar{C}, \bar{K}, \bar{V})$, where $\bar{K}$ and $\bar{V}$ are obtained solving the DARE (14) and (15) for the SS model with parameters $(\bar{A}, \bar{C}, \bar{Q}, \bar{R}, \bar{S})$.

To sum up, the relations and parametric representations of the original process Y, the filtered process $\tilde{Y}$ and the downsampled process $\bar{Y}$ are depicted in Figure 2a. The step of low pass filtering (FLT) applied to a VAR(p) process yields a VARMA(p, q) process (where q is the filter order, and the cutoff frequency is $f_\tau = 1/2\tau$); this process is equivalent to an ISS process [50]. The subsequent downsampling (DWS) yields a different SS process, which in turn can be converted to the ISS form solving the DARE. Thus, both the filtered process $\tilde{Y}_n$ and the downsampled process $\bar{Y}_n$ can be represented as ISS processes with parameters $(\tilde{A}, \tilde{C}, \tilde{K}, \tilde{V})$ and $(\bar{A}, \bar{C}, \bar{K}, \bar{V})$ which can be derived analytically from the knowledge of the parameters of the original process $(A_1, \ldots, A_p, \Sigma)$ and of the filter (q, f_τ). In the next section, we show how to compute analytically any measure appearing in the information decomposition of a jointly Gaussian multivariate stochastic process starting from its associated ISS model parameters, thus opening the way to the analytical computation of these measures for multiscale (filtered and downsampled) processes.

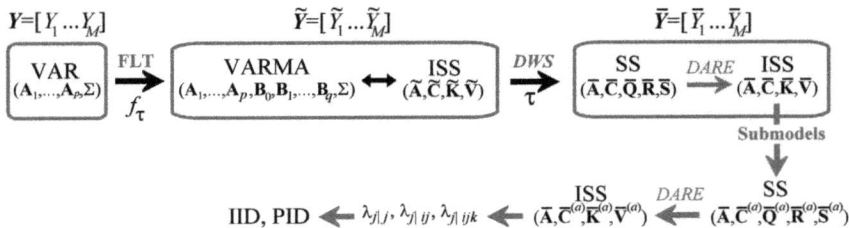

Figure 2. Schematic representation of a linear VAR process and of its multiscale representation obtained through filtering (FLT) and downsampling (DWS) steps. The downsampled process has an innovations form state space model (ISS) representation from which submodels can be formed to compute the partial variances needed for the computation of information measures appearing in the IID and PID decompositions. This makes it possible to perform multiscale information decomposition analytically from the original VAR parameters and from the scale factor.

3.3. Multiscale IID and PID

After introducing the general theory of information decomposition and deriving the multiscale representation of the parameters of a linear VAR model, in this section, we provide expressions for the terms of the IID and PID decompositions of the information transfer valid for multivariate jointly Gaussian processes. The derivations are based on the knowledge that the linear parametric representation of Gaussian processes given in (8) captures all of the entropy differences that define the various information measures [37] and that these entropy differences are related to the partial variances of the present of the target given its past and the past of one or more sources, intended as variances of the prediction errors resulting from linear regression [15,17]. Specifically, let us denote as $E_{j|j,n} = Y_{j,n} - \mathbb{E}[Y_{j,n}|Y_{j,n}^-]$, $E_{j|ij,n} = Y_{j,n} - \mathbb{E}[Y_{j,n}|Y_{i,n}^-, Y_{j,n}^-]$ the prediction error of a linear regression of $Y_{j,n}$ performed respectively on $Y_{j,n}^-$ and $(Y_{j,n}^-, Y_{i,n}^-)$ and as $\lambda_{j|j} = \mathbb{E}[E_{j|j,n}^2]$, $\lambda_{j|ij} = \mathbb{E}[E_{j|ij,n}^2]$, the corresponding prediction error variances. Then, the TE from Y_i to Y_j can be expressed as:

$$T_{i \to j} = \frac{1}{2} \ln \frac{\lambda_{j|j}}{\lambda_{j|ij}}. \qquad (17)$$

In a similar way, the joint TE from (Y_i, Y_k) to Y_j can be defined as:

$$T_{ik \to j} = \frac{1}{2} \ln \frac{\lambda_{j|j}}{\lambda_{j|ijk}}, \qquad (18)$$

where $\lambda_{j|ijk} = \mathbb{E}[E_{j|ijk,n}^2]$ is the variance of the prediction error of a linear regression of $Y_{j,n}$ on $(Y_{j,n}^-, Y_{i,n}^-, Y_{k,n}^-)$, $E_{j|ijk,n} = Y_{j,n} - \mathbb{E}[Y_{j,n}|Y_{i,n}^-, Y_{j,n}^-, Y_{k,n}^-]$. Based on these derivations, one can easily complete the IID decomposition of TE by computing $T_{k \to j}$ as in (17) and deriving the interaction TE from (3) and the PID decomposition, as well by deriving the redundant TE from (7), the synergistic TE from (6) and the unique TEs from (5).

Next, we show how to compute any partial variance from the parameters of an ISS model in the form of (13) [44,45]. The partial variance $\lambda_{j|a}$, where the subscript a denotes any combination of indexes $\in \{1, \ldots, M\}$, can be derived from the ISS representation of the innovations of a submodel obtained removing the variables not indexed by a from the observation equation. Specifically, we need to consider the submodel with state Equation (13b) and observation equation:

$$Y_n^{(a)} = C^{(a)} Z_n + E_n^{(a)}, \qquad (19)$$

where the superscript $^{(a)}$ denotes the selection of the rows with indices a of a vector or a matrix. It is important to note that the submodels (13a) and (19) are not in innovations form, but are rather an SS model with parameters $(\mathbf{A}, \mathbf{C}^{(a)}, \mathbf{KVK}^T, \mathbf{V}(a, a), \mathbf{KV}(:, a))$. This SS model can be converted to an ISS model with innovation covariance $\mathbf{V}^{(a)}$ solving the DARE (14) and (15), so that the partial variance $\lambda_{j|a}$ is derived as the diagonal element of $\mathbf{V}^{(a)}$ corresponding to the position of the target Y_j. Thus, with this procedure, it is possible to compute the partial variances needed for the computation of the information measures starting from a set of ISS model parameters; since any VAR process can be represented at scale τ as an ISS process, the procedure allows computing the IID and PID information decompositions for the rescaled multivariate process (see Figure 2).

It is worth remarking that, while the general formulation of IID and PID decompositions introduced in Section 2 holds for arbitrary processes, the multiscale extension detailed in Section 3 is exact only if the processes have a joint Gaussian distribution. In such a case, the linear VAR representation captures exhaustively the joint variability of the processes, and any nonlinear extension has no additional utility (a formal proof of the fact that a stationary Gaussian VAR process must be linear can be found in [37]). If, on the contrary, non-Gaussian processes are under scrutiny, the linear representation provided in Section 3.1 can still be adopted, but may miss important properties in the dynamics and thus provide only a partial description. Moreover, since the close correspondence between conditional entropies and partial variances reported in this subsection does not hold anymore for non-Gaussian processes, all of the obtained measures should be regarded as indexes of (linear) predictability rather than as information measures.

4. Simulation Experiment

To study the multiscale patterns of information transfer in a controlled setting with known dynamical interactions between time series, we consider a simulation scheme similar to some already used for the assessment of theoretical values of information dynamics [15,17]. Specifically, we analyze the following VAR process of order $M = 4$:

$$Y_{1,n} = 2\rho_1 cos2\pi f_1 Y_{1,n-1} - \rho_1^2 Y_{1,n-2} + U_{1,n}, \tag{20a}$$

$$Y_{2,n} = 2\rho_2 cos2\pi f_2 Y_{2,n-1} - \rho_2^2 Y_{2,n-2} + cY_{1,n-1} + U_{2,n}, \tag{20b}$$

$$Y_{3,n} = 2\rho_3 cos2\pi f_3 Y_{3,n-1} - \rho_3^2 Y_{3,n-2} + cY_{1,n-1} + U_{3,n}, \tag{20c}$$

$$Y_{4,n} = bY_{2,n-1} + (1-b)Y_{3,n-1} + U_{4,n}, \tag{20d}$$

where $\mathbf{U}_n = [U_{1,n} \cdots U_{4,n}]^T$ is a vector of zero mean white Gaussian noises with unit variance and uncorrelated with each other ($\Sigma = \mathbf{I}$). The parameter design in Equation (20) is chosen to allow autonomous oscillations in the processes Y_i, $i = 1, \ldots, 3$, obtained placing complex-conjugate poles with modulus ρ_i and frequency f_i in the complex plane representation of the transfer function of the vector process, as well as causal interactions between the processes at a fixed time lag of one sample and with strength modulated by the parameters b and c (see Figure 3). In this study, we set the coefficients related to self-dependencies to values generating well-defined oscillations in all processes ($\rho_1 = \rho_2 = \rho_3 = 0.95$) and letting Y_1 fluctuate at slower time scales than Y_2 and Y_3 ($f_1 = 0.1, f_2 = f_3 = 0.025$). We consider four configurations of the parameters, chosen to reproduce paradigmatic conditions of interaction between the processes:

(a)   isolation of Y_1 and Y_2 and unidirectional coupling $Y_3 \rightarrow Y_4$, obtained setting $b = c = 0$;

(b)   common driver effects $Y_2 \leftarrow Y_1 \rightarrow Y_3$ and unidirectional coupling $Y_3 \rightarrow Y_4$, obtained setting $b = 0$ and $c = 1$;

(c)   isolation of Y_1 and unidirectional couplings $Y_2 \rightarrow Y_4$ and $Y_3 \rightarrow Y_4$, obtained setting $b = 0.5$ and $c = 0$;

(d)   common driver effects $Y_2 \leftarrow Y_1 \rightarrow Y_3$ and unidirectional couplings $Y_2 \rightarrow Y_4$ and $Y_3 \rightarrow Y_4$, obtained setting $b = 0.5$ and $c = 1$.

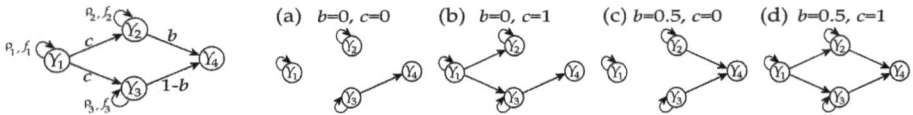

Figure 3. Graphical representation of the four-variate VAR process of Equation (20) that we use to explore the multiscale decomposition of the information transferred to Y_4, selected as the target process, from Y_2 and Y_3, selected as the source processes, in the presence of Y_1, acting as the exogenous process. To favor such exploration, we set oscillations at different time scales for Y_1 ($f_1 = 0.1$) and for Y_2 and Y_3 ($f_2 = f_3 = 0.025$), induce common driver effects from the exogenous process to the sources modulated by the parameter c and allow for varying strengths of the causal interactions from the sources to the target as modulated by the parameter b. The four configurations explored in this study are depicted in (**a–d**).

With this simulation setting, we compute all measures appearing in the IID and PID decompositions of the information transfer, considering Y_4 as the target process and Y_2 and Y_3 as the source processes. The theoretical values of these measures, computed as a function of the time scale using the IID and the PID, are reported in Figure 4. In the simple case of unidirectional coupling $Y_3 \rightarrow Y_4$ ($b = c = 0$, Figure 4a), the joint information transferred from (Y_2, Y_3) to Y_4 is exclusively due to the source Y_3 without contributions from Y_2 and without interaction effects between the sources ($\mathcal{T}_{23 \rightarrow 4} = \mathcal{T}_{3 \rightarrow 4} = \mathcal{U}_{3 \rightarrow 4}, \mathcal{T}_{2 \rightarrow 4} = \mathcal{U}_{2 \rightarrow 4} = 0, \mathcal{I}_{23 \rightarrow 4} = \mathcal{S}_{23 \rightarrow 4} = \mathcal{R}_{23 \rightarrow 4} = 0$).

When the causal interactions towards Y_4 are still due exclusively to Y_3, but the two sources Y_2 and Y_3 share information arriving from Y_1 ($b = 0, c = 1$; Figure 4b), the IID evidences that the joint information transfer coincides again with the transfer from Y_3 ($\mathcal{T}_{23 \rightarrow 4} = \mathcal{T}_{3 \rightarrow 4}$), but a non-trivial amount of information transferred from Y_2 to Y_4 emerges, which is fully redundant ($\mathcal{T}_{2 \rightarrow 4} = -\mathcal{I}_{23 \rightarrow 4}$). The PID highlights that the information from Y_3 to Y_4 is not all unique, but is in part transferred redundantly with Y_2, while the unique transfer from Y_2 and the synergistic transfer are negligible.

In the case of two isolated sources equally contributing to the target ($b = 0.5, c = 0$, Figure 4c), the IID evidences the presence of net synergy and of identical amounts of information transferred to Y_4 from Y_2 or Y_3 ($\mathcal{I}_{23 \rightarrow 4} > 0, \mathcal{T}_{2 \rightarrow 4} = \mathcal{T}_{3 \rightarrow 4}$). The PID documents that there are no unique contributions, so that the two amounts of information transfer from each source to the target coincide with the redundant transfer, and the remaining part of the joint transfer is synergistic ($\mathcal{U}_{2 \rightarrow 4} = \mathcal{U}_{3 \rightarrow 4} = 0$, $\mathcal{T}_{2 \rightarrow 4} = \mathcal{T}_{3 \rightarrow 4} = \mathcal{R}_{23 \rightarrow 4}, \mathcal{S}_{23 \rightarrow 4} = \mathcal{T}_{23 \rightarrow 4} - \mathcal{R}_{23 \rightarrow 4}$).

Finally, when the two sources share common information and contribute equally to the target ($b = 0.5, c = 1$; Figure 4d), we find that they send the same amount of information as before, but in this case, no unique information is sent by any of the sources ($\mathcal{T}_{2 \rightarrow 4} = \mathcal{T}_{3 \rightarrow 4}, \mathcal{U}_{2 \rightarrow 4} = \mathcal{U}_{3 \rightarrow 4} = 0$). Moreover, the nature of the interaction between the sources is not trivial and is scale dependent: at low time scales, where the dynamics are likely dominated by the fast oscillations of Y_1, the IID reveals net redundancy, and the PID shows that the redundant transfer prevails over the synergistic ($\mathcal{I}_{23 \rightarrow 4} < 0, \mathcal{R}_{23 \rightarrow 4} > \mathcal{S}_{23 \rightarrow 4}$); at higher time scales, where fast dynamics are filtered out and the slow dynamics of Y_2 and Y_3 prevail, the IID reveals net synergy, and the PID shows that the synergistic transfer prevails over the redundant ($\mathcal{I}_{23 \rightarrow 4} > 0, \mathcal{S}_{23 \rightarrow 4} > \mathcal{R}_{23 \rightarrow 4}$).

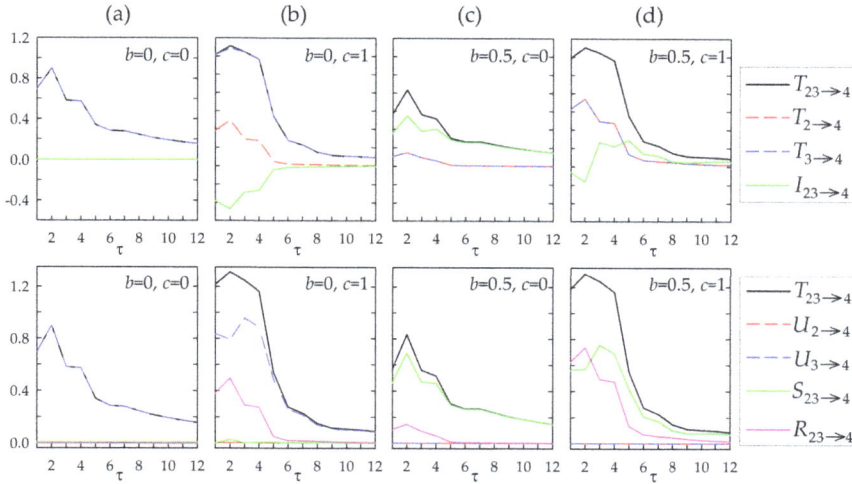

Figure 4. Multiscale information decomposition for the simulated VAR process of Equation (20). Plots depict the exact values of the entropy measures forming the interaction information decomposition (IID, upper row) and the partial information decomposition (PID, lower row) of the information transferred from the source processes Y_2 and Y_3 to the target process Y_4 generated according to the scheme of Figure 3 with four different configurations of the parameters. We find that linear processes may generate trivial information patterns with the absence of synergistic or redundant behaviors (**a**); patterns with the prevalence of redundant information transfer (**b**) or synergistic information transfer (**c**) that persist across multiple time scales; or even complex patterns with the alternating prevalence of redundant transfer and synergistic transfer at different time scales (**d**).

5. Application

As a real data application, we analyze intracranial EEG recordings from a patient with drug-resistant epilepsy measured by an implanted array of 8×8 cortical electrodes and two left hippocampal depth electrodes with six contacts each. The data are available in [54], and further details on the dataset are given in [55]. Data were sampled at 400 Hz and correspond to 10-s segments recorded in the pre-ictal period, just before the seizure onset, and 10 s during the ictal stage of the seizure, for a total of eight seizures. Defining and locating the seizure onset zone, i.e., the specific location in the brain where the synchronous activity of neighboring groups of cells becomes so strong so as to be able to spread its own activity to other distant regions, is an important issue in the study of epilepsy in humans. Here, we focus on the information flow from the sub-cortical regions, probed by depth electrodes, to the brain cortex. In [21], it has been suggested that Contacts 11 and 12, in the second depth electrode, are mostly influencing the cortical activity; accordingly, in this work, we consider Channels 11 and 12 as a pair of source variables for all of the cortical electrodes and decompose the information flowing from them using the multiscale IID and PID here proposed, both in the pre-ictal stage and in the ictal stage. An FIR filter with $q = 12$ coefficients is used, and the order p of the VAR model is fixed according to the Bayesian information criterion. In the analyzed dataset, the model order assessed in the pre-ictal phase was $p = 14.61 \pm 1.07$ (mean $\pm$ std. dev.across 64 electrodes and eight seizures) and during the ictal phase decreased significantly to $p = 11.09 \pm 3.95$.

In Figure 5, we depict the terms of the IID applied from the two sources (Channels $\{11, 12\}$) to any of the electrodes as a function of the scale τ, averaged over the eight seizures. We observe a relevant enhancement of the joint TE during the seizure, w.r.t. the pre-ictal period. This enhancement is determined by a marked increase of both the individual TEs from Channels 11 and 12 to all of the cortical electrodes; the patterns of the two TEs are similar to each other in both stages. The pattern of interaction information transfer displays prevalent redundant transfer for low values of τ and prevalent synergistic transfer for high τ, but the values of the interaction TE have relatively low magnitude and are only slightly different in pre-ictal and ictal conditions. It is worth stressing that at scale τ, the algorithm analyzes oscillations, in the time series, slower than $\frac{1}{2\tau f_s}$ s, where $f_s = 400$ Hz.

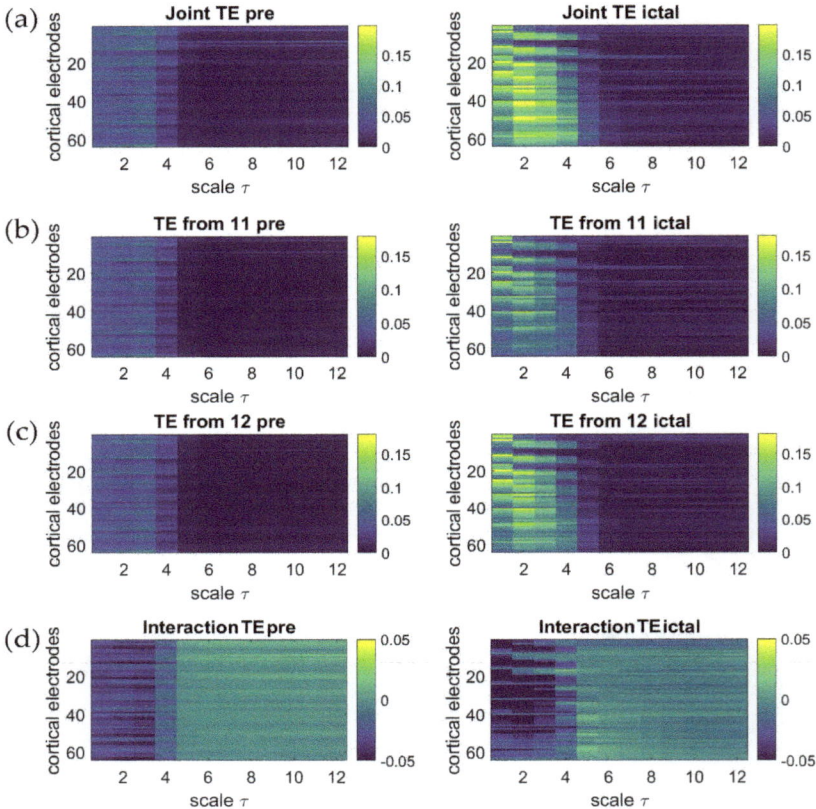

Figure 5. Interaction information decomposition (IID) of the intracranial EEG information flow from subcortical to cortical regions in an epileptic patient. The joint transfer entropy from depth Channels 11 and 12 to cortical electrodes (**a**); the transfer entropy from depth Channel 11 to cortical electrodes (**b**); the transfer entropy from depth Channel 12 to cortical electrodes (**c**) and the interaction transfer entropy from depth Channels 11 and 12 to cortical electrodes (**d**) are depicted as a function of the scale τ, after averaging over the eight pre-ictal segments (left column) and over the eight ictal segments (right column). Compared with pre-ictal periods, during the seizure, the IID evidences marked increases of the joint and individual information transfer from depth to cortical electrodes and low and almost unvaried levels of interaction transfer.

In Figure 6, we depict, on the other hand, the terms of the PID computed for the same data. This decomposition shows that the increased joint TE across the seizure transition seen in Figure 5a is in large part the result of an increase of both the synergistic and the redundant TE, which are markedly higher during the ictal stage compared with the pre-ictal. This explains why the interaction TE of Figure 5d, which is the difference between two quantities that both increase, is nearly constant moving from the pre-ictal to the ictal stage. The quantity that, instead, clearly differentiates between Channels 11 and 12 is the unique information transfer: indeed, only the unique TE from Channel 12 increases in the ictal stage, while the unique TE from Channel 13 remains at low levels.

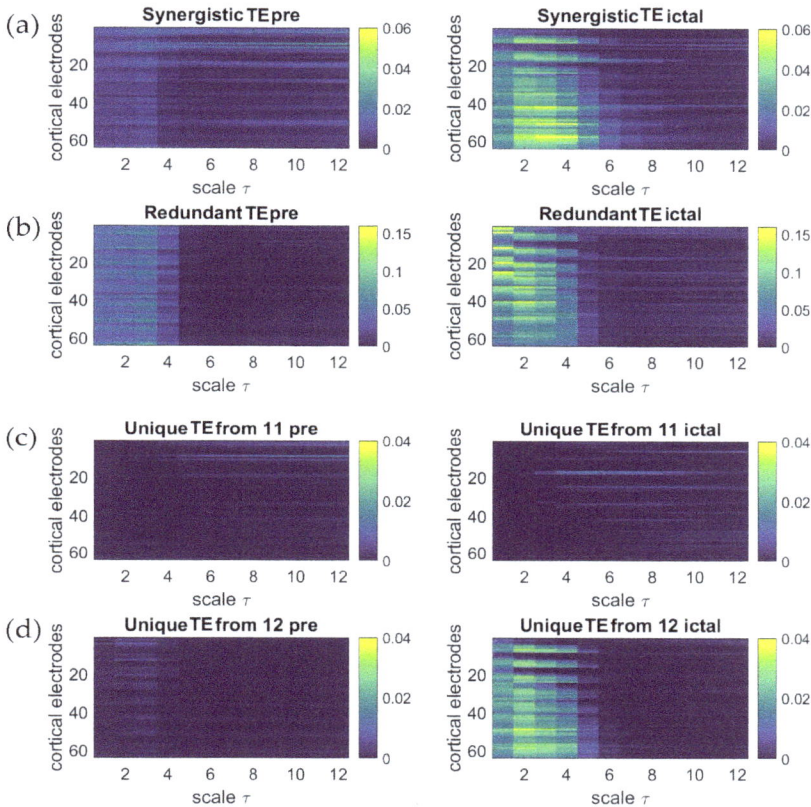

Figure 6. Partial information decomposition (PID) of the intracranial EEG information flow from subcortical to cortical regions in an epileptic patient. The synergistic transfer entropy from depth Channels 11 and 12 to cortical electrodes (**a**); the redundant transfer entropy from depth Channels 11 and 12 to cortical electrodes (**b**); the unique transfer entropy from depth Channel 11 to cortical electrodes (**c**) and the unique transfer entropy from depth Channel 12 to cortical electrodes (**d**) are depicted as a function of the scale τ, after averaging over the eight pre-ictal segments (left column) and over the eight ictal segments (right column). Compared with pre-ictal periods, during the seizure, the PID evidences marked increases of the information transferred synergistically and redundantly from depth to cortical electrodes and of the information transferred uniquely from one of the two depth electrodes, but not from the other.

In order to investigate the variability across trials of the estimates of the various information measures, in Figure 7, we depict the terms of both IID and PID expressed for each ictal episode as average values over all 64 cortical electrodes. The analysis shows that the higher average values observed in Figures 5 and 6 at Scales 1–4 during the ictal state for the joint TE, the two individual TEs, the redundant and synergistic TEs and the unique TE from depth Channel 12 are the result of an increase of the measures for almost all of the observed seizure episodes.

These findings are largely in agreement with the increasing awareness that epilepsy is a network phenomenon that involves aberrant functional connections across vast parts of the brain on virtually all spatial scales [56,57]. Indeed, our results document that the occurrence of seizures is associated with a relevant increase of the information flowing from the subcortical regions (associated with the depth electrode) to the cortex and that the character of this information flow is mostly redundant both in the pre-ictal and in the ictal state. Here, the need for a multiscale approach is testified by the fact that several quantities in the ictal state (e.g., the joint TE, the synergistic ITand the unique ITfrom Channel 12) attain their maximum at scale $\tau > 1$.

Moreover, the approaches that we propose for information decomposition appear useful to improve the localization of epileptogenic areas in patients with drug-resistant epilepsy. Indeed, our analysis suggests that Contact 12 is the closest to the seizure onset zone, and it is driving the cortical oscillations during the ictal stage, as it sends unique information to the cortex. On the other hand, to disentangle this effect, it has been necessary to include also Channel 11 in the analysis and to make the PID of the total information from the pair of depth channels to the cortex; indeed, the redundancy between Channels 11 and 12 confounds the informational pattern unless the PID is performed.

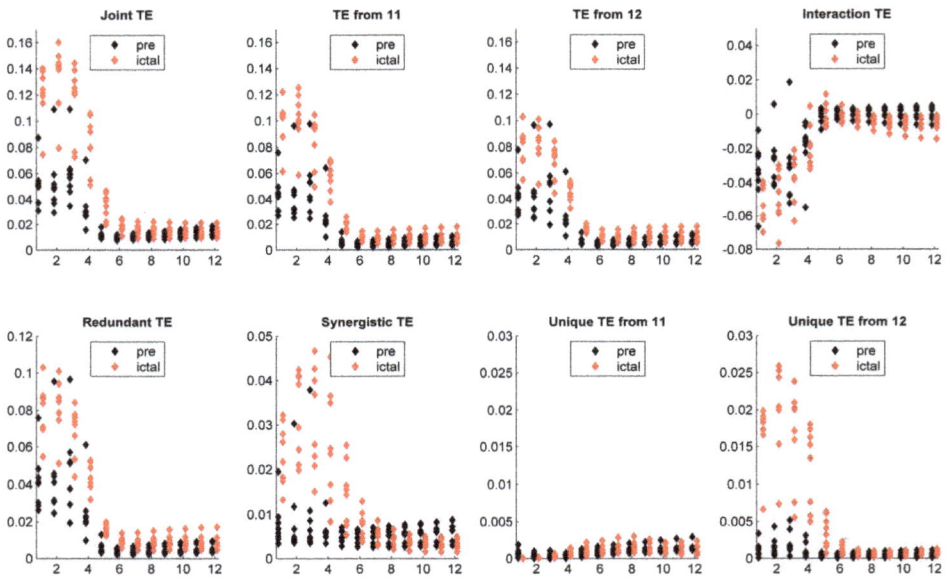

Figure 7. Multiscale representation of the measures of interaction information decomposition (IID, top) and partial information decomposition (PID, bottom) computed as a function of the time scale for each of the eight seizures during the pre-ictal period (black) and the ictal period (red). Values of joint transfer entropy (TE), individual TE, interaction TE, redundant TE, synergistic TE and unique TE are obtained taking the depth Channels 11 and 12 as sources and averaging over all 64 target cortical electrodes. Increases during seizure of the joint TE, individual TEs from both depth electrodes, redundant and synergistic TE and unique TE from the depth electrode 12 are evident at low time scales for almost all considered episodes.

6. Conclusions

Understanding how multiple inputs may combine to create the output of a given target is a fundamental challenge in many fields, in particular in neuroscience. Shannon's information theory is the most suitable frame to cope with this problem and thus to assess the informational character

Entropy **2017**, *19*, 408

of multiplets of variables describing complex systems; IID indeed measures the balance between redundant and synergetic interaction within the classical multivariate entropy-based approach. Recently Shannon's information theory has been extended, in the PID, so as to provide specific measures for the information that several variables convey individually (unique information), redundantly (shared information) or only jointly (synergistic information) about the output.

The contribution of the present work is the proposal of an analytical frame where both IID and PID can be exactly evaluated in a multiscale fashion, for multivariate Gaussian processes, on the basis of simple vector autoregressive identification. In doing this, our work opens the way for both the theoretical analysis and the practical implementation of information modification in processes that exhibit multiscale dynamical structures. The effectiveness of the proposed approach has been demonstrated both on simulated examples and on real publicly-available intracranial EEG data. Our results provide a firm ground to the multiscale evaluation of PID, to be applied in all applications where causal influences coexist at multiple temporal scales.

Future developments of this work include the refinement of the SS model structure to accommodate the description of long-range linear correlations [58] or its expansion to the description of nonstationary processes [59] and the formalization of exact cross-scale computation of information decomposition within and between multivariate processes [60]. A major challenge in the field remains the generalization of this type of analysis to non-Gaussian processes, for which exact analytical solutions or computationally-reliable estimation approaches are still lacking. This constitutes a main direction for further research, because real-world processes display very often non-Gaussian distributions, which would make an extension to nonlinear models or model-free approaches beneficial. The questions that are still open in this respect include the evaluation of proper theoretical definitions of synergy or redundancy for nonlinear processes [25–29], the development of reliable entropy estimators for multivariate variables with different dimensions [6,35,61] and the assessment of the extent to which non-linear model-free methods really outperform the linear model-based approach adopted here and in previous investigations [62].

Supplementary Materials: Supplementary Material to this article and is freely available for download from www.lucafaes.net/msID.html and https://github.com/danielemarinazzo/multiscale_PID.

Acknowledgments: The study was supported in part by the IRCS-Healthcare Research Implementation Program, Autonomous Province of Trento.

Author Contributions: Luca Faes, Daniele Marinazzo and Sebastiano Stramaglia conceived of the study, participated in he critical discussion on all aspects and contributed to writing the paper. Luca Faes designed the theoretical framework, realized the MATLAB codes and performed the simulation study. Sebastiano Stramaglia analyzed the experimental data. All authors have approved the final manuscript.

Conflicts of Interest: The authors declare no conflict of interest.

References

1. Lizier, J.T.; Prokopenko, M.; Zomaya, A.Y. A framework for the local information dynamics of distributed computation in complex systems. In *Guided Self-Organization: Inception*; Springer: Berlin/Heidelberg, Germany, 2014; pp. 115–158.
2. Pincus, S. Approximate entropy (ApEn) as a complexity measure. *Chaos Interdiscip. J. Nonlinear Sci.* **1995**, *5*, 110–117.
3. Lizier, J.T.; Prokopenko, M.; Zomaya, A.Y. Local measures of information storage in complex distributed computation. *Inf. Sci.* **2012**, *208*, 39–54.
4. Wibral, M.; Lizier, J.T.; Vögler, S.; Priesemann, V.; Galuske, R. Local active information storage as a tool to understand distributed neural information processing. *Front. Neuroinform.* **2014**, *8*, doi:10.3389/fninf.2014.00001.
5. Schreiber, T. Measuring information transfer. *Phys. Rev. Lett.* **2000**, *85*, 461.
6. Wibral, M.; Vicente, R.; Lizier, J.T. *Directed Information Measures in Neuroscience*; Springer: Berlin/Heidelberg, Germany, 2014.

7. Lizier, J.T.; Prokopenko, M.; Zomaya, A.Y. Information modification and particle collisions in distributed computation. *Chaos Interdiscip. J. Nonlinear Sci.* **2010**, *20*, 037109.

8. Wibral, M.; Lizier, J.T.; Priesemann, V. Bits from brains for biologically inspired computing. *Front. Robot. Artif. Intell.* **2015**, *2*, doi:10.3389/frobt.2015.00005.

9. Lizier, J.T.; Pritam, S.; Prokopenko, M. Information dynamics in small-world Boolean networks. *Artif. Life* **2011**, *17*, 293–314.

10. Wibral, M.; Rahm, B.; Rieder, M.; Lindner, M.; Vicente, R.; Kaiser, J. Transfer entropy in magnetoencephalographic data: Quantifying information flow in cortical and cerebellar networks. *Prog. Biophys. Mol. Biol.* **2011**, *105*, 80–97.

11. Hlinka, J.; Hartman, D.; Vejmelka, M.; Runge, J.; Marwan, N.; Kurths, J.; Paluš, M. Reliability of inference of directed climate networks using conditional mutual information. *Entropy* **2013**, *15*, 2023–2045.

12. Barnett, L.; Lizier, J.T.; Harré, M.; Seth, A.K.; Bossomaier, T. Information flow in a kinetic Ising model peaks in the disordered phase. *Phys. Rev. Lett.* **2013**, *111*, 177203.

13. Marinazzo, D.; Pellicoro, M.; Wu, G.; Angelini, L.; Cortés, J.M.; Stramaglia, S. Information transfer and criticality in the ising model on the human connectome. *PLoS ONE* **2014**, *9*, e93616.

14. Faes, L.; Nollo, G.; Jurysta, F.; Marinazzo, D. Information dynamics of brain–heart physiological networks during sleep. *New J. Phys.* **2014**, *16*, 105005.

15. Faes, L.; Porta, A.; Nollo, G. Information decomposition in bivariate systems: Theory and application to cardiorespiratory dynamics. *Entropy* **2015**, *17*, 277–303.

16. Porta, A.; Faes, L.; Nollo, G.; Bari, V.; Marchi, A.; de Maria, B.; Takahashi, A.C.; Catai, A.M. Conditional self-entropy and conditional joint transfer entropy in heart period variability during graded postural challenge. *PLoS ONE* **2015**, *10*, e0132851.

17. Faes, L.; Porta, A.; Nollo, G.; Javorka, M. Information Decomposition in Multivariate Systems: Definitions, Implementation and Application to Cardiovascular Networks. *Entropy* **2017**, *19*, 5.

18. Wollstadt, P.; Sellers, K.K.; Rudelt, L.; Priesemann, V.; Hutt, A.; Fröhlich, F.; Wibral, M. Breakdown of local information processing may underlie isoflurane anesthesia effects. *PLoS Comput. Biol.* **2017**, *13*, e1005511.

19. Schneidman, E.; Bialek, W.; Berry, M.J. Synergy, redundancy, and independence in population codes. *J. Neurosci.* **2003**, *23*, 11539–11553.

20. Stramaglia, S.; Wu, G.R.; Pellicoro, M.; Marinazzo, D. Expanding the transfer entropy to identify information circuits in complex systems. *Phys. Rev. E* **2012**, *86*, 066211.

21. Stramaglia, S.; Cortes, J.M.; Marinazzo, D. Synergy and redundancy in the Granger causal analysis of dynamical networks. *New J. Phys.* **2014**, *16*, 105003.

22. Stramaglia, S.; Angelini, L.; Wu, G.; Cortes, J.; Faes, L.; Marinazzo, D. Synergetic and Redundant Information Flow Detected by Unnormalized Granger Causality: Application to Resting State fMRI. *IEEE Trans. Biomed. Eng.* **2016**, *63*, 2518–2524.

23. McGill, W. Multivariate information transmission. *Trans. IRE Prof. Group Inf. Theory* **1954**, *4*, 93–111.

24. Bell, A.J. The co-information lattice. In Proceedings of the Fourth International Symposium on Independent Component Analysis and Blind Signal Separation (ICA), Nara, Japan, 1–4 April 2003.

25. Williams, P.L.; Beer, R.D. Nonnegative decomposition of multivariate information. *arXiv* **2010**, arXiv:1004.2515.

26. Harder, M.; Salge, C.; Polani, D. Bivariate measure of redundant information. *Phys. Rev. E* **2013**, *87*, 012130.

27. Griffith, V.; Chong, E.K.; James, R.G.; Ellison, C.J.; Crutchfield, J.P. Intersection information based on common randomness. *Entropy* **2014**, *16*, 1985–2000.

28. Quax, R.; Har-Shemesh, O.; Sloot, P.M.A. Quantifying Synergistic Information Using Intermediate Stochastic Variables. *Entropy* **2017**, *19*, 85.

29. Bertschinger, N.; Rauh, J.; Olbrich, E.; Jost, J.; Ay, N. Quantifying unique information. *Entropy* **2014**, *16*, 2161–2183.

30. Panzeri, S.; Senatore, R.; Montemurro, M.A.; Petersen, R.S. Correcting for the sampling bias problem in spike train information measures. *J. Neurophysiol.* **2007**, *98*, 1064–1072.

31. Faes, L.; Porta, A. Conditional entropy-based evaluation of information dynamics in physiological systems. In *Directed Information Measures in Neuroscience*; Springer: Berlin/Heidelberg, Germany, 2014; pp. 61–86.

32. Kozachenko, L.; Leonenko, N.N. Sample estimate of the entropy of a random vector. *Probl. Peredachi Inf.* **1987**, *23*, 9–16.

33. Vlachos, I.; Kugiumtzis, D. Nonuniform state-space reconstruction and coupling detection. *Phys. Rev. E* **2010**, *82*, 016207.
34. Marinazzo, D.; Pellicoro, M.; Stramaglia, S. Causal information approach to partial conditioning in multivariate data sets. *Comput. Math. Methods Med.* **2012**, *2012*, 303601.
35. Faes, L.; Kugiumtzis, D.; Nollo, G.; Jurysta, F.; Marinazzo, D. Estimating the decomposition of predictive information in multivariate systems. *Phys. Rev. E* **2015**, *91*, 032904
36. Barrett, A.B. Exploration of synergistic and redundant information sharing in static and dynamical Gaussian systems. *Phys. Rev. E* **2015**, *91*, 052802.
37. Barrett, A.B.; Barnett, L.; Seth, A.K. Multivariate Granger causality and generalized variance. *Phys. Rev. E* **2010**, *81*, 041907.
38. Porta, A.; Bari, V.; de Maria, B.; Takahashi, A.C.; Guzzetti, S.; Colombo, R.; Catai, A.M.; Raimondi, F.; Faes, L. Quantifying Net Synergy/Redundancy of Spontaneous Variability Regulation via Predictability and Transfer Entropy Decomposition Frameworks. *IEEE Trans. Biomed. Eng.* **2017**, doi:10.1109/TBME.2017.2654509.
39. Ivanov, P.; Nunes Amaral, L.; Goldberger, A.; Havlin, S.; Rosenblum, M.; Struzik, Z.; Stanley, H. Multifractality in human heartbeat dynamics. *Nature* **1999**, *399*, 461–465.
40. Chou, C.M. Wavelet-based multi-scale entropy analysis of complex rainfall time series. *Entropy* **2011**, *13*, 241–253.
41. Wang, J.; Shang, P.; Zhao, X.; Xia, J. Multiscale entropy analysis of traffic time series. *Int. J. Mod. Phys. C* **2013**, *24*, 1350006.
42. Costa, M.; Goldberger, A.L.; Peng, C.K. Multiscale entropy analysis of complex physiologic time series. *Phys. Rev. Lett.* **2002**, *89*, 068102.
43. Valencia, J.; Porta, A.; Vallverdú, M.; Clariá, F.; Baranowski, R.; Orłowska-Baranowska, E.; Caminal, P. Refined multiscale entropy: Application to 24-h holter recordings of heart period variability in healthy and aortic stenosis subjects. *IEEE Trans. Biomed. Eng.* **2009**, *56*, 2202–2213.
44. Barnett, L.; Seth, A.K. Granger causality for state-space models. *Phys. Rev. E* **2015**, *91*, 040101.
45. Solo, V. State-space analysis of Granger-Geweke causality measures with application to fMRI. *Neural Comput.* **2016**, *28*, 914–949.
46. Florin, E.; Gross, J.; Pfeifer, J.; Fink, G.R.; Timmermann, L. The effect of filtering on Granger causality based multivariate causality measures. *Neuroimage* **2010**, *50*, 577–588.
47. Barnett, L.; Seth, A.K. Detectability of Granger causality for subsampled continuous-time neurophysiological processes. *J. Neurosci. Methods* **2017**, *275*, 93–121.
48. Faes, L.; Montalto, A.; Stramaglia, S.; Nollo, G.; Marinazzo, D. Multiscale Analysis of Information Dynamics for Linear Multivariate Processes. *arXiv* **2016**, arXiv:1602.06155.
49. Faes, L.; Nollo, G.; Stramaglia, S.; Marinazzo, D. Multiscale Granger causality. *arXiv* **2017**, arXiv:1703.08487.
50. Aoki, M.; Havenner, A. State space modeling of multiple time series. *Econom. Rev.* **1991**, *10*, 1–59.
51. Oppenheim, A.V.; Schafer, R.W. *Digital Signal Processing*; Prentice-Hall: Englewood Cliffs, NJ, USA, 1975.
52. Hannan, E.J.; Deistler, M. *The Statistical Theory of Linear Systems*; Society for Industrial and Applied Mathematics: Philadelphia, PA, USA, 2012; Volume 70.
53. Aoki, M. *State Space Modeling of Time Series*; Springer Science & Business Media: Berlin/Heidelberg, Germany, 2013.
54. Earth System Research Laboratory. Available online: http://math.bu.edu/people/kolaczyk/datasets.html (accessed on 5 May 2017).
55. Kramer, M.A.; Kolaczyk, E.D.; Kirsch, H.E. Emergent network topology at seizure onset in humans. *Epilepsy Res.* **2008**, *79*, 173–186.
56. Richardson, M.P. Large scale brain models of epilepsy: Dynamics meets connectomics. *J. Neurol Neurosurg. Psychiatry* **2012**, *83*, 1238–1248.
57. Dickten, H.; Porz, S.; Elger, C.E.; Lehnertz, K. Weighted and directed interactions in evolving large-scale epileptic brain networks. *Sci. Rep.* **2016**, *6*, 34824.
58. Sela, R.J.; Hurvich, C.M. Computationally efficient methods for two multivariate fractionally integrated models. *J. Time Ser. Anal.* **2009**, *30*, 631–651.
59. Kitagawa, G. Non-gaussian state—Space modeling of nonstationary time series. *J. Am. Stat. Assoc.* **1987**, *82*, 1032–1041.
60. Paluš, M. Cross-scale interactions and information transfer. *Entropy* **2014**, *16*, 5263–5289.

61. Papana, A.; Kugiumtzis, D.; Larsson, P. Reducing the bias of causality measures. *Phys. Rev. E* **2011**, *83*, 036207.

62. Porta, A.; de Maria, B.; Bari, V.; Marchi, A.; Faes, L. Are Nonlinear Model-Free Conditional Entropy Approaches for the Assessment of Cardiac Control Complexity Superior to the Linear Model-Based One? *IEEE Trans. Biomed. Eng.* **2017**, *64*, 1287–1296.

entropy

MDPI

Article

Bivariate Partial Information Decomposition: The Optimization Perspective

Abdullah Makkeh *, Dirk Oliver Theis and Raul Vicente

Institute of Computer Science, University of Tartu, 51014 Tartu, Estonia; dotheis@ut.ee (D.O.T.); raul.vicente.zafra@ut.ee (R.V.)
* Correspondence: makkeh@ut.ee

Received: 7 July 2017; Accepted: 28 September 2017; Published: 7 October 2017

Abstract: Bertschinger, Rauh, Olbrich, Jost, and Ay (Entropy, 2014) have proposed a definition of a decomposition of the mutual information $MI(\mathbf{X} : \mathbf{Y}, \mathbf{Z})$ into shared, synergistic, and unique information by way of solving a convex optimization problem. In this paper, we discuss the solution of their Convex Program from theoretical and practical points of view.

Keywords: partial information decomposition; bivariate information decomposition; applications of convex optimization

1. Introduction

Bertschinger, Rauh, Olbrich, Jost, and Ay [1] have proposed to compute a decomposition of the mutual information $MI(\mathbf{X} : \mathbf{Y}, \mathbf{Z})$ into shared, synergistic, and unique contributions by way of solving a convex optimization problem. It is important to mention that William and Beer in [2] were the first to propose a measure for information decomposition. That measure suffered from serious flaws, which prompted a series of other papers [3–5] trying to improve these results. Denote by X the range of the random variable $\mathbf{X}$, by Y the range of $\mathbf{Y}$, and by Z the range of $\mathbf{Z}$. Further, let

$$b_{x,y}^{\mathsf{y}} = \mathbb{P}(\mathbf{X} = x \wedge \mathbf{Y} = y) \qquad \text{for all } x \in X, y \in Y$$
$$b_{x,z}^{\mathsf{z}} = \mathbb{P}(\mathbf{X} = x \wedge \mathbf{Z} = z) \qquad \text{for all } x \in X, y \in Y$$

the marginals. Then the Bertschinger et al. definition of synergistic information is

$$\widetilde{\mathrm{CI}}(\mathbf{X} : \mathbf{Y}; \mathbf{Z}) := \mathrm{MI}(\mathbf{X} : \mathbf{Y}, \mathbf{Z}) - \min_{q} \mathrm{MI}_{(x,y,z) \sim q}(x : y, z),$$

where MI denotes mutual information, and $(x, y, z) \sim q$ stands for (x, y, z) is distributed according to the probability distribution q. The minimum ranges over all probability distributions $q \in [0, 1]^{X \times Y \times Z}$ which satisfy the marginal equations

$$q_{x,y,*} = b_{x,y}^{\mathsf{y}} \qquad \text{for all } (x, y) \in X \times Y$$
$$q_{x,*,z} = b_{x,z}^{\mathsf{z}} \qquad \text{for all } (x, z) \in X \times Z.$$

We use the notational convention that an asterisk "$*$" is to be read as "sum over everything that can be plugged in here"; e.g.,

$$q_{x,y,*} := \sum_{w} q_{x,y,w}.$$

(We don't use the symbol $*$ in any other context).

As pointed out in [1], the function $q \mapsto \mathrm{MI}_{(x,y,z) \sim q}(x : y, z)$ is convex and smooth in the interior of the convex region, and hence, by textbook results on convex optimization, for every $\varepsilon > 0$, an ε-approximation of the optimum—i.e., a probability distribution $q^\star$ for which $\mathrm{MI}_{(x,y,z)\sim q^\star}(x : y, z)$ is at most ε larger than the minimum—can be found in at most $O(\alpha^2/\varepsilon^2)$ rounds of an iterative algorithm (e.g., Proposition 7.3.1 in [6]), with a parameter α (the asphericity of the feasible region) depending on the marginals b^y, b^z. Each iteration involves evaluating the gradient of $\mathrm{MI}_{(x,y,z)\sim q^\star}(x : y, z)$ with respect to q, and $O(|X \times Y \times Z|)$ additional arithmetic operations.

In Appendix A of their paper, Bertschinger et al. discuss practical issues related to the solution of the convex optimization problem: They analyze the feasible region, solve by hand some of the problems, and complain that Mathematica could not solve the optimization problem out of the box.

Our paper is an expansion of that appendix in [1]. Driven by the need in application areas (e.g., [7]) to have a robust, fast, out-of-the-box method for computing $\widetilde{\mathrm{CI}}$, we review the required convex optimization background; discuss different approaches to computing $\widetilde{\mathrm{CI}}$ and the theoretical reasons contributing to the poor performance of some of them; and finally compare several of these approaches on practical problem instance. In conclusion, we offer two different practical ways of computing $\widetilde{\mathrm{CI}}$.

The convex optimization problem in [1] is the following:

$$
\begin{aligned}
&\text{minimize } \mathrm{MI}_{(x,y,z)\sim q}(x : y, z) \quad \text{over } q \in \mathbb{R}^{X\times Y\times Z} \\
&\text{subject to} \quad q_{x,y,*} = b^y_{x,y} \qquad\qquad \text{for all } (x, y) \in X \times Y \\
&\qquad\qquad\quad\ q_{x,*,z} = b^z_{x,z} \qquad\qquad \text{for all } (x, z) \in X \times Z \\
&\qquad\qquad\quad\ q_{x,y,z} \geq 0 \qquad\qquad\quad\ \text{for all } (x, y, z) \in X \times Y \times Z
\end{aligned}
\tag{CP}
$$

This is a Convex Program with a non-linear, convex continuous objective function $q \mapsto \mathrm{MI}_{(x,y,z)\sim q}(x : y, z)$, which is smooth in all points q with no zero entry. The feasible region (i.e., the set of all q satisfying the constraints) is a compact convex polyhedron, which is nonempty, as the distribution of the original random variables $(\mathbf{X}, \mathbf{Y}, \mathbf{Z})$ is a feasible solution. Clearly, the condition that q should be a probability distribution is implied by it satisfying the marginal equations.

In the next section, we review some background about convex functions and convex optimization, in particular with respect to non-smooth functions. Section 3 aims to shed light on the theoretical properties of the convex program (CP). In Section 4, we present the computer experiments which we conducted and their results. Some result tables are in the Appendix A.

2. Convex Optimization Basics

Since the target audience of this paper is not the optimization community, for the sake of easy reference, we briefly review the relevant definitions and facts, tailored for our problem: the feasible region is polyhedral, the objective function is (convex and continuous but) not everywhere smooth.

Let f be a convex function defined on a convex set $C \subseteq \mathbb{R}^n$; let $x \in C$ and $d \in \mathbb{R}^n$. The *directional derivative* of f at x in the direction of d is defined as the limit

$$
f'(x; d) := \lim_{\varepsilon \searrow 0} \frac{f(x + \varepsilon d) - f(x)}{\varepsilon} \in [-\infty, \infty].
\tag{1}
$$

Remark 1 ([8], Lemma 2.71). *The limit in (1) exists in $[-\infty, \infty]$, by the convexity of f.*

A *subgradient* of f at a point $x \in C$ is a vector $g \in \mathbb{R}^n$ such that, for all $y \in C$, we have $f(y) \geq f(x) + g^\top(y - x)$.

There can be many subgradients of f at a point. However:

Remark 2. *If f is differentiable at x, then $\nabla f(x)$ is the only subgradient of f at x.*

Lemma 1 ([8], Lemma 2.73). *If g is a subgradient of f at $x \in C$, and $y \in C$, then $g^\top(y - x) \leq f'(x; y - x)$.*

We state the following lemma for convenience (it will be used in the proof of Proposition 2). It is a simplification of the Moreau-Rockafellar Theorem ([8], Theorem 2.85).

Lemma 2. *Let $C_i \subset \mathbb{R}^\ell$, $i = 1, \ldots, k$ be a closed convex sets, $f_i \colon C_i \to \mathbb{R}$ continuous convex functions, and*

$$f(x) = \sum_{i=1}^{k} f_i(x_i)$$

for $x = (x_1, \ldots, x_k) \in \prod_{i=1}^{k} C_i \subset \mathbb{R}(^\ell)^k$. If, for $i = 1, \ldots, k$, g_i is a subgradient of f_i at x_i, then $g := (g_1, \ldots, g_k)$ is a subgradient of f at x. Moreover, all subgradients of f at x are of this form.

For the remainder of this section, we reduce the consideration to convex sets C of the form given in the optimization problem (CP). Let A be an $m \times n$ matrix, and b an m-vector. Assume that $C = \{x \in \mathbb{R}^n \mid Ax = b \wedge x \geq 0\}$. Suppose f is a convex function defined on C. Consider the convex optimization problem $\min_{x \in C} f(x)$.

Let $x \in C$. A vector $d \in \mathbb{R}^n$ is called a *feasible descent direction* of f at x, if

(a)   for some $\varepsilon > 0$, $x + \varepsilon d \in C$ ("feasible direction"); and
(b) $f'(x; d) < 0$ ("descent direction").

Clearly, if a feasible descent direction of f at x exists, then x is not a minimum of f over C. The following theorem is a direct consequence of ([8], Theorem 3.34) adapted to our situation.

Theorem 1 (Karush-Kuhn-Tucker). *Suppose that for every $j = 1, \ldots, n$, there is an $x \in C$ with $x_j > 0$. The function f attains a minimum over C at a point $x \in C$ if, and only if, there exist*

- *a subgradient g of f at x*
- *and an m-vector $\lambda \in \mathbb{R}^m$*

such that $A^\top \lambda \leq g$, and for all $j \in \{1, \ldots, n\}$ with $x_j > 0$, equality holds: $(A^\top \lambda)_j = g_j$.

The condition "$x_j = 0$ or $(A^\top \lambda)_j = g_j$" is called *complementarity*.

2.1. Convex Optimization Through Interior Point Methods

For reasons which will become clear in Section 3, it is reasonable to try to solve (CP) using a so-called *Interior Point* or *Barrier* Method (we gloss over the (subtle) distinction between IPMs and BMs). The basic idea of these iterative methods is that all iterates are in the "interior" of something. We sketch what goes on.

From a *theoretical* point of view, a closed convex set C is given in terms of a *barrier function*, i.e., a convex function $F \colon C \to \mathbb{R}$ with $F(x) \to \infty$ whenever x approaches the boundary of C. The prototypical example is $C := [0, \infty]^n$, and $F(x) := -\sum_j \ln x_j$. The goal is to minimize a *linear* objective function $x \mapsto \sum_j c_j x_j$ over C.

To fit problems with nonlinear (convex) objective f into this paradigm, the "epigraph form" is used, which means that a new variable s is added along with a constraint $f(q) \leq s$, and the objective is "minimize s".

The algorithm maintains a *barrier parameter* $t > 0$, which is increased gradually. In iteration k, it solves the unconstrained optimization problem

$$x^{(k)} := \operatorname*{argmin}_{x \in \mathbb{R}^n} \left(t_k \cdot c^\top x + F(x) \right) \qquad (2)$$

The barrier parameter is then increased, and the algorithm proceeds with the next iteration. The fact that the barrier function tends to infinity towards the boundary of C makes sure that the iterates $x^{(k)}$ stay in the interior of C for all k.

The unconstrained optimization problem (2) is solved using Newton's method, which is itself iterative. If x is the current iterate, Newton's method finds the minimum to the second order Taylor expansion about x of the objective function $g\colon x \mapsto t_k \cdot c^\top x + F(x)$:

$$\underset{y \in \mathbb{R}^n}{\operatorname{argmin}}\left(g(x) + \nabla g(x)^\top(y - x) + \tfrac{1}{2}(y - x)Hg(x)(y - x)\right)$$
$$= x - Hg(x)^{-1}\nabla g(x).$$

where $Hg(x)$ denotes the Hessian of g at x. Then Newton's method updates x (e.g., by simply replacing it with the argmin).

Note that $Hg(x) = HF(x)$, so the convergence properties (as well as, whether it works at all or not), depend on the properties of the barrier function. Suitable barrier functions are known to exist for all compact convex sets [9,10]. However, for a given set, finding one which can be quickly evaluated (along with the gradient and the Hessian) is sometimes a challenge.

A more "concrete" point of view is the following. Consider f and C as above: $C := \{x \mid Ax = b, x \geq 0\}$ and f is convex and continuous on C, and smooth in the interior of C (here: the points $x \in C$ with $x_j > 0$ for all j) which we assume to be non-empty. A simple barrier-type algorithm is the following. In iteration k, solve the equality-constrained optimization problem

$$x^{(k)} := \underset{\substack{x \in \mathbb{R}^n \\ Ax = b}}{\operatorname{argmin}}\left(t_k \cdot f(x) - \sum_{j=1}^{n} \ln(x_j)\right). \tag{3}$$

The equality-constrained problem is solved by a variant of Newton's method: If x is the current iterate, Newton's method finds the minimum to the second order Taylor expansion about x of the function $g\colon x \mapsto t_k f(x) - \sum \ln(x_j)$ *subject to* the equations $Ax = b$, using Lagrange multipliers, i.e., it solves the linear system

$$Hg(x)y + A^\top \lambda = -\nabla g(x)$$
$$Ax = b,$$

with $Hg(x) = t_k Hf(x) + \operatorname{Diag}_j(1/x_j^2)$ and $\nabla g(x) = t_k \nabla f(x) - \operatorname{Diag}_j(1/x_j)$, where $\operatorname{Diag}(\cdot)$ denotes a diagonal matrix of appropriate size with the given diagonal.

By convexity of f, $Hf(x)$ is positive semidefinite, so that adding the diagonal matrix results in a (strictly) positive definite matrix. Hence, the system of linear equations always has a solution.

The convergence properties of this simple barrier method now depend on properties of f. We refer the interested reader to ([11], Chapter 11) for the details.

Generally speaking, today's practical Interior Point Methods are "Primal-Dual Interior Point Methods": They solve the "primal" optimization problem and the dual—the system in Theorem 1—simultaneously. After (successful) termination, they return not only an ε-approximation x of the optimum, but also the Lagrange multipliers λ which *certify* optimality.

3. Theoretical View on Bertschinger et al.'s Convex Program

We will use the following running example in this section.

Example 1 (Part 1/4). *Let* $X := \{0, 1, 2, 3\}$, $Y, Z := \{0, 1\}$.

$$
\begin{array}{llll}
b^y_{0,0} := 1/4 & b^y_{0,1} := 0 & b^z_{0,0} := 1/4 & b^z_{0,1} := 0 \\
b^y_{1,0} := 0 & b^y_{1,1} := 1/4 & b^z_{1,0} := 0 & b^z_{1,1} := 1/4 \\
b^y_{2,0} := 1/8 & b^y_{2,1} := 1/8 & b^z_{2,0} := 1/8 & b^z_{2,1} := 1/8 \\
b^y_{3,0} := 1/8 & b^y_{3,1} := 1/8 & b^z_{3,0} := 1/8 & b^z_{3,1} := 1/8
\end{array}
$$

Among all $q \in \mathbb{R}^{X \times Y \times Z}$ (16 dimensions) satisfying the 16 equations

$$q_{x,y,*} = b^y_{x,y} \; \forall x, y, \qquad\qquad\qquad q_{x,*,z} = b^z_{x,z} \; \forall x, z,$$

and the 16 inequalities $q_{x,y,z} \geq 0$ for all x, y, z, we want to find one which minimizes $\mathrm{MI}_{(x,y,z) \sim q}(x : y, z)$.

3.1. The Feasible Region

In this section, we discuss the feasible region:

$$\P(b) := \left\{ q \in \mathbb{R}^{X \times Y \times Z}_+ \;\Big|\; \forall x, y, z \colon \; q_{x,y,*} = b^y_{x,y} \;\&\; q_{x,*,z} = b^z_{x,z} \right\}.$$

We will omit the "(b)" when no confusion can arise. We will always make the following assumptions:

1. b^y and b^z are probability distributions.
2. $\P(b)$ is non-empty. This is equivalent to $b^y_{x,*} = b^z_{x,*}$ for all $x \in X$.
3. No element of X is "redundant", i.e., for every $x \in X$ we have both $b^y_{x,*} > 0$ and $b^z_{x,*} > 0$.

First of all, recall the vectors $\bar{d} \in \mathbb{R}^{X \times Y \times Z}$ from Appendix A.1 of [1], defined by

$$\bar{d}^{x,y,z,y',z'} := 1^{x,y,z} + 1^{x,y',z'} - 1^{x,y,z'} - 1^{x,y',z}, \tag{4}$$

(where $1^{\cdots}$ is the vector with exactly one non-zero entry in the given position, and 0 otherwise) satisfy $\bar{d}_{x,y,*} = \bar{d}_{x,*,z} = 0$ for all x, y, z (we omit the superscripts for convenience, when possible).

Our first proposition identifies the triples (x, y, z) for which the equation $q_{x,y,z} = 0$ holds for all $q \in \P(b)$.

Proposition 1. *If $q_{x,y,z} = 0$ holds for all $q \in \P(b)$, then $b^y_{x,y} = 0$ or $b^z_{x,z} = 0$.*

Proof. Let $x, y, z \in X \times Y \times Z$, and assume that $b^y_{x,y} \neq 0$ and $b^z_{x,z} \neq 0$. Take any $p \in \P(b)$ with $p_{x,y,z} = 0$—if no such p exists, we are done. Since the marginals are not zero, there exist $y' \in Y \setminus \{y\}$ and $z' \in Z \setminus \{z\}$ with $p_{x,y',z} > 0$ and $p_{x,y,z'} > 0$. Let $\bar{d} := \bar{d}^{x,y,z,y',z'}$. By the remarks preceding the proposition, $p + \delta \bar{d}$ satisfies the marginal equations for all $\delta \in \mathbb{R}$. Since $p_{x,y',z} > 0$ and $p_{x,y,z'} > 0$, there exists a $\delta > 0$ such that $q := p + \delta \bar{d} \geq 0$. Hence, we have $q \in \P(b)$ and $q_{x,y,z} > 0$, proving that the equation $q_{x,y,z} = 0$ is not satisfied by all elements of $\P(b)$. $\square$

The proposition allows us to restrict the set of variables of the Convex Program to those triples (x, y, z) for which a feasible q exists with $q_{x,y,z} \neq 0$ (thereby avoiding unneccessary complications in the computation of the objective function and derivatives; see the next section); the equations with RHS 0 become superfluous. We let

$$\mathcal{J}(b) := \left\{ (x, y, z) \in X \times Y \times Z \;\Big|\; b^y_{x,y} > 0 \;\wedge\; b^z_{x,z} > 0 \right\} \tag{5}$$

denote the set of remaining triplets. We will omit the "(b)" when no confusion can arise.

Example 2 (Part 2/4). *Continuing Example 1, we see that the equations with RHS 0 allow to omit the variables $q_{x,y,z}$ for the following triples (x, y, z):*

$$(0, 1, 0) \quad (0, 0, 1)$$
$$(0, 1, 1) \quad (0, 1, 1)$$
$$(1, 0, 0) \quad (1, 0, 0)$$
$$(1, 0, 1) \quad (1, 1, 0)$$

Hence, $\mathcal{J} := \left\{ (0, 0, 0), (1, 1, 1), (2, y, z), (3, y', z') \mid y, z, y', z' \in \{0, 1\} \right\}$ (10 variables), and we are left with the following 12 equations:

$$q_{0,0,*} = b_{0,0}^y = 1/4 \qquad\qquad q_{0,*,0} = b_{0,0}^z = 1/4$$

$$q_{1,1,*} = b_{1,1}^y = 1/4 \qquad\qquad q_{1,*,1} = b_{1,1}^z = 1/4$$

$$q_{2,0,*} = b_{2,0}^y = 1/8 \quad q_{2,1,*} = b_{2,1}^y = 1/8 \qquad q_{2,*,0} = b_{2,0}^z = 1/8 \quad q_{2,*,1} = b_{2,1}^z = 1/8$$

$$q_{3,0,*} = b_{3,0}^y = 1/8 \quad q_{3,1,*} = b_{3,1}^y = 1/8 \qquad q_{3,*,0} = b_{3,0}^z = 1/8 \quad q_{3,*,1} = b_{3,1}^z = 1/8.$$

We can now give an easy expression for the dimension of the feasible region (with respect to the number of variables, $|\mathcal{J}(b)|$).

Corollary 1. *The dimension of* $\mathbb{¶}(b)$ *is*

$$|\mathcal{J}(b)| + |X| - |\{(x,y) \mid b_{x,y}^y > 0\}| - |\{(x,z) \mid b_{x,z}^z > 0\}|.$$

Proof. For $x \in X$, the set $Y_x := \{y \in Y \mid b_{x,y}^y > 0\}$ is non-empty, by assumption 3; the same is true for $Z_x := \{z \in Z \mid b_{x,z}^z > 0\}$. Note that $\mathcal{J}(b)$ is the disjoint union of the sets $\{x\} \times Y_x \times Z_x, x \in X$.

Lemmas 26 and 27 of [1] now basically complete the proof of our corollary: It implies that the dimension is equal to

$$\sum_{x \in X} (|Y_x| - 1)(|Z_x| - 1).$$

As $\mathcal{J}(b)$ is the disjoint union of the sets $\{x\} \times Y_x \times Z_x, x \in X$, we find this quantity to be equal to

$$|\mathcal{J}(b)| + |X| - \sum_{x \in X} (|Y_x| + |Z_x|).$$

Finally,

$$\sum_{x \in X} (|Y_x| + |Z_x|) = |\{(x,y) \mid b_{x,y}^y > 0\}| + |\{(x,z) \mid b_{x,z}^z > 0\}|,$$

which concludes the proof of the corollary. $\square$

Example 3 (Part 3/4). *Continuing Example 1, we find that the values for the variables $q_{0,0,0}$ and $q_{1,1,1}$ are fixed to $1/4$, whereas each of two sets of four variables $q_{2,y,z}$, $y, z \in \{0, 1\}$, and $q_{3,y,z}$, $y, z \in \{0, 1\}$, offers one degree of freedom, so the dimension should be 2. And indeed:* $|\mathcal{J}| + |X| - |\{(x,y) \mid b_{x,y}^y > 0\}| - |\{(x,z) \mid b_{x,z}^z > 0\}| = 10 + 4 - 12 = 2.$

3.2. The Objective Function and Optimality

We now discuss the objective function. The goal is to minimize

$$\mathrm{MI}_{(x,y,z)\sim q}(x : y, z) = H_{(x,y,z)\sim q}(x) - H_{(x,y,z)\sim q}(x \mid y, z).$$

Since the distribution of x is fixed by the marginal equations, the first term in the sum is a constant, and we are left with minimizing negative conditional entropy. We start by discussing negative conditional entropy as a function on its full domain (We don't assume q to be a probability distrbution, let alone to satisfy the marginal equations): $f \colon \mathbb{R}_+^{\mathcal{J}} \to \mathbb{R}$ (with $\mathcal{J} = \mathcal{J}(b)$ for arbitrary but fixed b)

$$f \colon q \mapsto \sum_{(x,y,z)\in\mathcal{J}} q_{x,y,z} \ln\left(\frac{q_{x,y,z}}{q_{*,y,z}}\right), \tag{6}$$

where we set $0 \ln(\dots) := 0$, as usual. The function f is continuous on its domain, and it is smooth on $[0,\infty]^{\mathcal{J}}$. Indeed, we have the gradient

$$\left(\nabla f(q)\right)_{x,y,z} = \partial_{x,y,z} f(q) = \ln\left(\frac{q_{x,y,z}}{q_{*,y,z}}\right), \tag{7}$$

and the Hessian

$$\left(Hf(q)\right)_{(x,y,z),(x',y',z')} = \partial_{x,y,z}\partial_{x',y',z'}f(q) = \begin{cases} 0, & \text{if } (y',z') \neq (y,z) \\[2mm] \dfrac{-1}{q_{*,y,z}}, & \text{if } (y',z') = (y,z),\, x \neq x' \\[3mm] \dfrac{q_{*,y,z} - q_{x,y,z}}{q_{x,y,z}\, q_{*,y,z}}, & \text{if } (x',y',z') = (x,y,z). \end{cases} \tag{8}$$

It is worth pointing out that the Hessian, while positive semidefinite, is not positive definite, and, more pertinently, it is not in general positive definite on the tangent space of the feasible region, i.e., $r^\top Hf(q)r = 0$ is possible for $r \in \mathbb{R}^\mathcal{J}$ with $r(x,y,*) = r(x,*,z) = 0$ for all (x,y,z). Indeed, if, e.g., $\mathcal{J} = X \times Y \times Z$, it is easy to see the kernel of $Hf(q)$ has dimension $|Y \times Z|$, whereas the feasible region has dimension $|X|(|Y|-1)(|Z|-1) = |X \times Y \times Z| - |X||Y| - |X||Z| + 1$. Hence, if $|Y \times Z| > |X||Y| + |X||Z| - 1$, then for every $q(!)$ the kernel of $Hf(q)$ the must have a non-empty intersection with the tangent space of the feasible region.

Optimality condition and boundary issues. In the case of points q which lie on the boundary of the domain, i.e., $q_{x,y,z} = 0$ for at least one triplet (x,y,z), some partial derivatives don't exist. For y, z, denote $X_{yz} := \{x \mid (x,y,z) \in \mathcal{J}\}$. The situation is as follows.

Proposition 2. *Let $q \in \P$.*

(a) *If there is a $(x,y,z) \in \mathcal{J}$ with $q_{x,y,z} = 0$ but $q_{*,y,z} > 0$, then f does not have a subgradient at q. Indeed, there is a feasible descent direction of f at q with directional derivative $-\infty$.*

(b) *Otherwise—i.e., for all $(x,y,z) \in \mathcal{J}$, $q_{x,y,z} = 0$ only if $q_{*,y,z} = 0$—subgradients exist. For all y, z, let $\varrho^{y,z} \in [0,1]^{X_{yz}}$ be a probability distribution on X_{yz}. Suppose that, for all y, z with $q_{*,y,z} > 0$,*

$$\varrho_x^{y,z} = \frac{q_{x,y,z}}{q_{*,y,z}} \qquad \text{for all } x \in X_{yz}. \tag{9}$$

Then g defined by $g_{x,y,z} := \ln(\varrho_x^{y,z})$, for all $(x,y,z) \in \mathcal{J}$, is a subgradient of f at q.

Moreover, g' is a subgradient iff there exists such a g with

- $g'_{x,y,z} \leq g$ *for all* $(x,y,z) \in \mathcal{J}$ *with* $q_{x,y,z} = 0$;
- $g_{x,y,z} = g$ *for all* $(x,y,z) \in \mathcal{J}$ *with* $q_{x,y,z} > 0$.

Proof. For Proposition 2, let $(y,z) \in Y \times Z$ with $q_{*,y,z} > 0$, and $x \in X_{yz}$ with $q_{x,y,z} = 0$. There exist y', z' such that $q_{x,y',z'}, q_{x,y,z'} > 0$. This means that $\bar{d} := \bar{d}^{x,y,z,y',z'}$ as defined in (4) is a feasible direction. Direct calculations (written down in [12]) show that $f'(q;\bar{d}) = -\infty$. Invoking Lemma 1 yields non-existence of the subgradient.

As to Proposition 2, we prove the statement for every pair $(y,z) \in Y \times Z$ for the function

$$f_{yz} \colon \mathbb{R}_+^{X_{yz}} \to \mathbb{R} \colon q \mapsto \sum_{x \in X_{yz}} q_x \ln(q_x/q_*),$$

and then use Lemma 2.

Let us fix one pair (y,z). If $q_{x,y,z} > 0$ for all $x \in X_{yz}$ holds, then we $f_{y,z}$ is differentiable at q, so we simply apply Remark 2.

Now assume $q_{*,y,z} = 0$. A vector $g \in \mathbb{R}^{X_{yz}}$ is a subgradient of f_{yz}, iff

$$\sum_{x \in X_{yz}} r_{x,y,z} \ln(r_{x,y,z}/r_{*,y,z}) = f_{yz}(q+r) \overset{!}{\geq} f_{yz}(q) + g^\top r = f_{yz}(q) + \sum_{x \in X_{yz}} r_x g_x \tag{10}$$

holds for all $r \in \mathbb{R}^\mathcal{J}$ with $r_{x,y,z} \geq$ for all $(x,y,z) \in \mathcal{J}$, and $r_* > 0$.

We immediately deduce $g_x \leq 0$ for all x. Let $\varrho'_x := e^{g_x}$ for all x, and $\varrho_x := \varrho'_x / C$, with $C := \varrho'_*$. Clearly, ϱ is a probability distribution on X_{yz}. Moreover, the difference LHS-RHS of (10) is equal to

$$D_{\text{KL}}(r/r_* \| \rho) + \ln C,$$

with D_{KL} denoting Kullback-Leibler divergence. From the usual properties of the Kullback-Leibler divergence, we see that this expression is greater-than-or-equal to 0 for all r, if and only if $C \geq 1$, which translates to

$$\sum_x e^{g_x} \leq 1.$$

From this, the statements in Proposition 2 follow. $\square$

From the proposition, we derive the following corollary.

Corollary 2. *Suppose a minimum of f over $\P(b)$ is attained in a point q with $q_{x,y,z} = 0$ for a triple (x,y,z) with $b^y_{x,y} > 0$ and $b^z_{x,z} > 0$. Then $q_{u,y,z} = 0$ for all $u \in X$.*

Proof. This follows immediately from the fact, expressed in item Proposition 2 of the lemma, that a negative feasible descent direction exists at a point q which with $q_{x,y,z} = 0$ for a triple $(x,y,z) \in \mathcal{J}(b)$ with $q_{*,y,z} > 0$. $\square$

Based on Proposition 2 and the Karush-Kuhn-Tucker conditions, Theorem 1, we can now write down the optimality condition.

Corollary 3. *Let $q \in \P(b)$. The minimum of f over $\P(b)$ is attained in q if, and only if, (a) $q_{*,y,z} = 0$ holds whenever there is an $(x,y,z) \in \mathcal{J}(b)$ with $q_{x,y,z} = 0$; and (b) there exist*

- $\lambda_{x,y} \in \mathbb{R}$, *for each (x,y) with $b^y_{x,y} > 0$;*
- $\mu_{x,z} \in \mathbb{R}$, *for each (x,z) with $b^z_{x,z} > 0$;*

satisfying the following: For all y, z with $q_{*,y,z} > 0$,

$$\lambda_{x,y} + \mu_{x,z} = \ln\left(\frac{q_{x,y,z}}{q_{*,y,z}}\right) \qquad \text{holds for all } x \in X_{yz};$$

for all y, z with $q_{*,y,z} = 0$ (but $X_{yz} \neq \varnothing$), there is a probability distribution $\varrho \in]0,1]^{X_{yz}}$ on X_{yz} such that

$$\lambda_{x,y} + \mu_{x,z} \leq \ln(\varrho^{y,z}_x) \qquad \text{holds for all } x \in X_{yz}.$$

Example 4 (Part 4/4). *Continuing Example 1, let us now find the optimal solution "by hand". First of all, note that $q_{0,0,0} = 1/4 = q_{1,1,1}$ holds for all feasible solutions. By Corollary 2, if q is an optimal solution, since $q_{0,0,0} > 0$, we must have $q_{x,0,0} > 0$ for $x = 2, 3$; and similarly, $q_{x,1,1} > 0$ for $x = 2, 3$.*

To verify whether or not a given q is optimal, we have to find solutions $\lambda_{x,y}$, $\mu_{x,z}$ to the following system of equations and/or inequalities:

$$\lambda_{0,0} + \mu_{0,0} = \ln(1/4/(1/4 + q_{2,0,0} + q_{3,0,0}))$$
$$\lambda_{2,0} + \mu_{2,0} = \ln(q_{2,0,0}/(1/4 + q_{2,0,0} + q_{3,0,0}))$$
$$\lambda_{3,0} + \mu_{3,0} = \ln(q_{3,0,0}/(1/4 + q_{2,0,0} + q_{3,0,0}))$$

$$\lambda_{2,0} + \mu_{2,1} \begin{cases} = \ln(q_{2,0,1}/(q_{2,0,1} + q_{3,0,1})), & \text{if } q_{*,0,1} > 0 \\ \leq \ln(\varrho_2^{01}), & \text{if } q_{*,0,1} = 0 \end{cases}$$

$$\lambda_{3,0} + \mu_{3,1} \begin{cases} = \ln(q_{3,0,1}/(q_{2,0,1} + q_{3,0,1})), & \text{if } q_{*,0,1} > 0 \\ \leq \ln(\varrho_3^{01}), & \text{if } q_{*,0,1} = 0 \end{cases}$$

$$\lambda_{2,1} + \mu_{2,0} \begin{cases} = \ln(q_{3,1,0}/(q_{2,1,0} + q_{3,1,0})), & \text{if } q_{*,1,0} > 0 \\ \leq \ln(\varrho_2^{10}), & \text{if } q_{*,1,0} = 0 \end{cases} \qquad (11)$$

$$\lambda_{3,1} + \mu_{3,0} \begin{cases} = \ln(q_{3,1,0}/(q_{2,1,0} + q_{3,1,0})), & \text{if } q_{*,1,0} > 0 \\ \leq \ln(\varrho_3^{10}), & \text{if } q_{*,1,0} = 0 \end{cases}$$

$$\lambda_{1,1} + \mu_{1,1} = \ln(1/4/(1/4 + q_{2,1,1} + q_{3,1,1}))$$
$$\lambda_{2,1} + \mu_{2,1} = \ln(q_{2,1,1}/(1/4 + q_{2,1,1} + q_{3,1,1}))$$
$$\lambda_{3,1} + \mu_{3,1} = \ln(q_{3,1,1}/(1/4 + q_{2,1,1} + q_{3,1,1})).$$

The ϱ's, if needed, have to be found: both ϱ^{01} and ϱ^{10} must be probability distributions on $\{2, 3\}$.

From the conditions on the zero-nonzero pattern of an optimal solution, we can readily guess an optimal q. Let's guess wrongly, first, though: $q_{2,y,z} = q_{3,y,z} = 1/16$ for all y, z. In this case, all of the constraints in (11) become equations, and the ϱ's don't occur. We have a system of equations in the variables $\lambda_{.,.}$ and $\mu_{.,.}$. It is easy to check that the system does not have a solution.

Our next guess for an optimal solution q is: $q_{2,0,1} = q_{3,0,1} = 0$; $q_{2,1,0} = q_{3,1,0} = 0$; $q_{2,0,0} = q_{3,0,0} = 1/8$; $q_{2,1,1} = q_{3,1,1} = 1/8$. In (11), the constraints involving $(y, z) = (0, 1), (1, 0)$ are relaxed to inequalities, with the freedom to pick arbitrary probability distributions ϱ^{01} and ϱ^{10} in the RHSs. We choose the following: $\lambda_{x,y} = \mu_{x,z} = \frac{1}{2} \ln(1/2)$ for all x, y, z. The equations are clearly satisfied. The inequalities are satisfied if we take $\varrho_0^{01} = \varrho_1^{01} = 1/2$; $\varrho_0^{01} = \varrho_1^{01} = 1/2$.

3.3. Algorithmic Approaches

We now discuss several possibilities of solving the convex program (CP): Gradient descent and interior point methods; geometric programming; cone programming over the so-called exponential cone. We'll explain the terms in the subsections.

3.3.1. Gradient Descent

Proposition 2 and Corollary 2 together with the running example make clear that the boundary is "problematic". On the one hand, the optimal point can sometimes be on the boundary. (In fact, this is already the case for the AND-gate, as computed in Appendix A.2 of [1].) On the other hand, by Corollary 2, optimal boundary points lie in a lower dimensional subset inside the boundary (codimension $\approx |X|$), and the optimal points on the boundary are "squeezed in" between boundary regions which are "infinitely strongly repellent" (which means to express that the feasible descent direction has directional derivative $-\infty$).

From the perspective of choosing an algorithm, it is pertinent that subgradients do not exist everywhere on the boundary. This rules out the use of algorithms which rely on evaluating (sub-)gradients on the boundary, such as projected (sub-)gradient descent. (And also generic active set and sequential quadratic programming methods (We refer the interested reader to [13] for background on these optimization methods); the computational result in Section 4 illustrate that.

Due to the huge popularity of gradient descent, the authors felt under social pressure to present at least one version of it. Thus, we designed an ad-hoc quick-and-dirty gradient descent algorithm which does its best to avoid the pitfalls of the feasible region: it's boundary. We now describe this algorithm.

Denote by A the matrix representing the LHS of the equations in (CP); also reduce A by removing rows which are linear combinations of other rows. Now, multiplication by the matrix $P := A^\top(A^\top A)^{-1}A$ amounts to projection onto the tangent space $\{d \mid Ad = 0\}$ of the feasible region, and $P\nabla f(q)$ is the gradient of f in the tanget space, taken at the point q.

The strategy by which we try to avoid the dangers of approaching the boundary of the feasible region in the "wrong" way is by never reducing the smallest entry of the current iterate q by more then 10%. Here is the algorithm.

Algorithm 1: Gradient Descent

---

1 Construct the matrix A
2 Compute $P := A^\top(A^\top A)^{-1}A$
3 Initialize q to a point in the interior of the feasible region
4 **repeat**
5   Compute $f(q)$ **if** $f(q)$ *better than all previous solutions* **then**
6   store q
7   Compute the gradient $\nabla f(q)$
8   Compute the projection of the gradient $g := P\nabla f(q)$
9   Determine a step size η, ensuring $q_{xyz} > \eta g_{xyz}$ for all xyz
10   Update $q = q - \eta g$
11 **until** *stopping criterion is reached*

There are lots of challenges with this approach, not the least of which is deciding on the step size η. Generally, a good step size for gradient descent is 1 over the largest eigenvalue of the Hessian—but the eigenvalues of the Hessian tend to infinity.

The stopping criterion is also not obvious: we use a combination of the norm of the projected gradient, the distance to the boundary, and a maximum of 1000 iterations.

None of these decisions are motivated by careful thought.

3.3.2. Interior Point Methods

Using Interior Point Methods (IPMs) appears to be the natural approach: While the iterates can converge to a point on the boundary, none of the iterates actually lie on the boundary, and that is an inherent property of the method (not a condition which you try to enforce artificially as in the gradient descent approach of the previous section). Consequently, problems with gradients, or even non-existing subgradients, never occur.

Even here, however, there are caveats involving the boundary. Let us consider, as an example, the simple barrier approach sketched in Section 2.1 (page 4). The analysis of this method (see [11]) requires that the function $F\colon q \mapsto tf(q) - \sum_{xyz} \ln(q_{xyz})$ be *self-concordant*, which means that, for some constant C, for all q, h

$$D^3 F(q)[h, h, h] \leq C \cdot \left(h^\top HF(q)h\right)^{3/2}, \tag{12}$$

where D^3 denotes the tensor of third derivatives. The following is proven in [12]:

Proposition 3. *Let $n \geq 2$, and consider the function*

$$F: q \mapsto t \sum_{x=1}^{n} q_x \ln(q_x/q_*) - \sum_x \ln(q_x)$$

There is no C and no t such that (12) holds for all $q \in [0,\infty]^n$ and all h.

The proposition explains why, even for some IPMs, approaching the boundary can be problematic. We refer to [12] for more discussions about self-concordancy issues of the (CP).

Corollary 4. *The complexity of the interior point method via self-concordant barrier for solving (CP) is $O(M \cdot \log 1/\varepsilon)$, where M is the complexity of computing the Newton step (which can be done by computing and inverting the Hessian of the barrier)*

Proof. This follows immediately from the fact that a 3-self concordant barrier exists for (CP) (see [12]) and the complexity analysis of barrier method in [11], Chapter 11. □

Still, we find the IPM approach (with the commercial Interior Point software "Mosek") to be the most usable of all the approaches which we have tried.

3.3.3. Geometric Programming

Geometric Programs form a sub-class of Convex Programs; they are considered to be more easily solvable than general Convex Programs: Specialized algorithms for solving Geometric Programs have been around for a half-century (or more). We refer to [14] for the definition and background on Geometric Programming. The Langrange dual of (CP) can be written as the following Geometric Program in the variables $\lambda \in \mathbb{R}^{X \times Y}$, $\mu \in \mathbb{R}^{X \times Y}$ (for simplicity, assume that all the right-hand-sides $b^\cdot$ are strictly positive):

$$\text{minimize} \quad \sum_{(x,y)\in X\times Y} b^y_{x,y}\lambda_{x,y} + \sum_{(x,z)\in X\times Z} b^z x, z \cdot \mu_{x,z}$$

$$\text{subject to} \quad \ln\left(\sum_{x\in X} \exp\left(\lambda_{xy} + \mu_{xz}\right) \right) \leq 0 \qquad \text{for all } (y,z) \in Y \times Z. \tag{GP}$$

3.3.4. Exponential Cone Programming

Cone Programming is a far-reaching generalization of Linear Programming, which may contain so-called *generalized inequalities:* For a fixed closed convex cone $\mathcal{K}$, the generalized inequality "$a \leq_K b$" simply translates into $b - a \in K$. There is a duality theory similar to that for Linear Programming.

Efficient algorithms for Cone Programming exist for some closed convex cones; for example for the *exponential cone* [15], $\mathcal{K}_{\exp}$, which is the closure of all triples $(r, p, q) \in \mathbb{R}^3$ satisfying

$$q > 0 \text{ and } qe^{r/q} \leq p.$$

For $q > 0$, the condition on the right-hand side is equivalent to $r \leq q \ln(p/q)$, from which it can be easily verified that the following "Exponential Cone Program" computes (CP). The variables are $r, q, p \in \mathbb{R}^{X \times Y \times Z}$.

$$\text{maximize} \sum_{x,y,z} r_{x,y,z}$$

$$
\begin{array}{llr}
\text{subject to} & q_{x,y,*} = b^{y}_{x,y} & \text{for all } (x,y) \in X \times Y \\
& q_{x,*,z} = b^{z}_{x,z} & \text{for all } (x,z) \in X \times Z \\
& q_{*,y,z} - p_{x,y,z} = 0 & \text{for all } (x,y,z) \in X \times Y \times Z \\
& (r_{x,y,z}, q_{x,y,z}, p_{x,y,z}) \in \mathcal{K}_{\exp} & \text{for all } (x,y,z) \in X \times Y \times Z.
\end{array}
\tag{EXP}
$$

The first two constraints are just the marginal equations; the third type of equations connects the p-variables with the q-variables, and the generalized inequality connects these to the variables forming the objective function.

There are in fact several ways of modeling (CP) as an Exponential Cone Program; here we present the one which is most pleasant both theoretically (the duality theory is applicable) and in practice (it produces the best computational results). For the details, as well as for another model, we refer to [12].

4. Computational Results

In this section, we present the computational results obtained by solving Bertschinger et al.'s Convex Program (CP) on a large number of instances, using readily available software out-of-the-box. First we discuss the problem instances, then the convex optimization solvers, then discuss the results. Detailed tables are in the Appendix A.

4.1. Data

In all our instances, the vectors b^{y} and b^{z} are marginals computed form an input probability distribution p on $X \times Y \times Z$. Occasionally, "noise" (explained below) in the creation p leads to the phenomenon that the sum of all probabilities is not 1. This was not corrected before computing b^{y} and b^{z}, as it is irrelevant for the Convex Program (CP).

We have three types of instances, based on (1) "gates"—the "paradigmatic examples" listed in Table 1 of [1]; (2) Example 31 and Figure A.1 of [1]; (3) discretized 3-dimensional Gaussians.

(1) Gates. The instances of the type (1) are based on the "gates" (RDN, UNQ, XOR, AND, RDNXOR, RDNUNQXOR, XORAND) described Table 1 of [1]:

RDN $X = Y = Z$ uniformly distributed.
UNQ $X = (Y, Z)$, Y, Z independent, uniformly distributed in $\{0,1\}$.
XOR $X = Y \text{ XOR } Z$, Y, Z independent, uniformly distributed in $\{0,1\}$.
AND $X = Y \text{ AND } Z$, Y, Z independent, uniformly distributed in $\{0,1\}$.
RDNXOR $X = (Y_1 \text{ XOR } Z_1, W)$, $Y = (Y_1, W)$, $Z = (Z_1, W)$, Y_1, Z_1, W independent, uniformly distributed in $\{0,1\}$.
RDNUNQXOR $X = (Y_1 \text{ XOR } Z_1, (Y_2, Z_2), W)$, $Y = (Y_1, Y_2, W)$; $Z = (Z_1, Z_2, W)$, Y_1, Y_2, Z_1, Z_2, W independent, uniformly distributed in $\{0,1\}$.
XORAND $X = (Y \text{ XOR } Z, Y \text{ AND } Z)$, Y, Z independent, uniformly distributed in $\{0,1\}$.

Each gate gives rise to two sets of instances: (1a) the "unadulterated" probability distribution computed from the definition of the gate (see Table A1); (1b) empirical distributions generated by randomly sampling from W and computing (x, y, z). In creating the latter instances, we incorporate "noise" by perturbing the output randomly with a certain probability. We used 5 levels of noise, corresponding to increased probabilities of perturbing the output. For example, AND 0 refers to the empirical distribution of the AND gate without perturbation (probability $= 0$), AND 4 refers to the empirical distribution of the AND gate with output perturbed with probability 0.1. The perturbation probabilities are: 0, 0.001, 0.01, 0.05, 0.1. (See Tables A2 and A3).

(2) Example-31 instances. In Appendix A.2 of their paper, Bertschinger et al. discuss the following input probability distribution: $\mathbf{X}, \mathbf{Y}, \mathbf{Z}$ are independent uniformly random in $\{0,1\}$. They present

this as an example that the optimization problem can be "ill-conditioned". We have derived a large number of instances based on that idea, by taking $(\mathbf{X}, \mathbf{Y}, \mathbf{Z})$ uniformly distributed on $X \times Y \times Z$, with $|X|$ ranging in $\{2, \ldots, 5\}$ and $|Y| = |Z|$ in $\{|X|, \ldots, 5 \cdot |X|\}$. These instances are referred to as "independent" in Table A4. We also created "noisy" versions of the probability distributions by perturbing the probabilities. In the results, we split the instances into two groups according to whether the fraction $|Y|/|X|$ is at most 2 or greater than 2. (The rationale behind the choice of the sizes of $|X|, |Y|, |Z|$ is the fact mentioned in Section 3.2, that the Hessian has a kernel in the tangent space if the ratio is high.)

(3) Discretized gaussians. We wanted to have a type of instances which was radically different from the somewhat "combinatorial" instances (1) and (2). We generated (randomly) twenty 3×3 covariance matrices between standard Gaussian random variables $\mathbf{X}$, $\mathbf{Y}$, and $\mathbf{Z}$. We then discretized the resulting continuous 3-dimensional Gaussian probability distribution by integrating numerically (We used the software Cuba [16] in version 4.2 for that) over boxes $[0,1]^3$, $[0,0.75]^3$, $[0,0.5]^3$, $[0,0.25]^3$; and all of their translates which held probability mass at least 10^{-20}.

We grouped these instances according to the number of variables, see Table A5: The instances in the "Gauss-1" group are the ones with at most 1000 variables; "Gauss-2" have between 1000 and 5000; "Gauss-3" have between 5000 and 250,000, "Gauss-4" between 25,000 and 75,000.

4.2. Convex Optimization Software We Used

For our computational results, we made an effort to use as many software toolboxes as we could get our hands on. (The differences in the results are indeed striking.) Most of our software was coded in the Julia programming language (version 0.6.0) using the "MathProgBase" package see [17] which is part of JuliaOpt. (The only the CVXOPT interior point algorithm and for the Geometric Program did we use Python.) On top of that, the following software was used.

- CVXOPT [18] is written and maintained by Andersen, Dahl, and Vandenberghe. It transforms the general Convex Problems with nonlinear objective function into an epigraph form (see Section 2.1), before it deploys an Interior Point method. We used version 1.1.9.
- Artelys Knitro [19] is an optimization suite which offers four algorithms for general Convex Programming, and we tested all of them on (CP). The software offers several different convex programming solvers: We refer by "Knitro_Ip" to their standard Interior Point Method [20]; by "Knitro_IpCG" to their IPM with conjugate gradient (uses projected cg to solve the linear system [21]); "Knitro_As" is their Active Set Method [22]; "Knitro_SQP" designates their Sequential Quadratic Programming Method [19]. We used version 10.2.
- Mosek [23] is an optimization suite which offers algorithms for a vast range of convex optimization problems. Their Interior Point Method is described in [24,25]. We used version 8.0.
- Ipopt [20] is a software for nonlinear optimization which can also deal with non-convex objectives. At its heart is an Interior Point Method (as the name indicates), which is enhanced by approaches to ensure convergence even in the non-convex case. We used version 3.0.
- ECOS. We are aware of only two Conic Optimization software toolboxes which allow to solve Exponential Cone Programs: ECOS and SCS.
  ECOS is a lightweigt numerical software for solving convex cone programs [26], using an Interior Point approach. We used the version from Nov 8, 2016.
- SCS [27,28] stands for Splitting Conic Solver. It is a numerical optimization package for solving large-scale convex cone problems. It is a first-order method, and generally very fast. We used version 1.2.7.

4.3. Results

We now discuss the computational results of every solver. The tables in Appendix A give three data points for each pair of instance (group) and solver: Whether the instance was solved, or, respectively, which fraction of the instances in the group were solved ("solved"); the time ("time") or, respectively, average time "avg. tm" needed (All solvers had a time limit 10,000.0 s. We used a computer server with Intel(R) Core(TM) i7-4790K CPU (4 cores) and 16GB of RAM) to solve it/them. The third we call the "optimality measure": When a solver reports having found (optimal) primal and dual solutions, we computed:

- the maximum amount by which any of the marginal equations is violated;
- the maximum amount by which any of the nonnegativity inequalities is violated;
- the maximum amount by which the inequality "$A^\top \lambda \leq g$" in Theorem 1 is violated;
- the maximum amount by which the complementarity $|x_j| |(A^\top \lambda)_j - g_j|$ is violated.

Of all these maxima we took that maximum to have an indication of how reliable the returned solution is, with respect to feasibility and optimality. In the tables, the maximum is then taken over all the instances in the group represented in each row of the table. We chose this "worst-case" approach to emphasize the need for a *reliable* software.

4.3.1. Gradient Descent

The Gradient Descent algorithm sketched in the previous section finds its limits when the matrix P becomes too large to fit into memory, which happens for the larger ones of the Gaussian instances.

Even before that, the weakness of the algorithm are clear in the fact that the best solution it finds is often not optimal (the other methods produce better solutions). The running times are mid-field compared to the other methods.

It is probable that a more carefully crafted gradient descent would preform better, but the conceptual weaknesses of the approach remain.

4.3.2. Interior Point algorithms

Cvxopt It terminated with optimal solution on all the noiseless discrete gates except "RDN"see Tables A1–A3. But whenever the noise was added it started failing except on "XOR". So as the number of samples increased even when decreasing the noise it still failed. The optimality measure was always bad except on some of the noiseless gates. Mainly the dual solution was weakly feasible on the solved ones. It bailed out 56% due to KKT system problems and the rest was iteration and time limit.

For the uniformly binary gates again Cvxopt failed on the noisy distributions with 34% success on Noisy 1 and 0% on Noisy 2. It solved all the independent correctly with acceptable optimality measure.

For the Gaussian, it totally failed with 40% computational problems in solving the KKT system and 38% time limit and the rest were iteration limit. The solver was slowest compared to others and required high memory to store the model.

Knitro_Ip It terminated with optimal solution on all the discrete gates with good optimality measure see Tables A1–A4. But for the Gaussian, it failed most of the time. It solved only 5% of all Gaussian instances where 75% of the solved have less than 1000 variable Table A5. 25% of the unsolved instances couldn't get a dual feasible solution. The rest stopped due to iteration or time limit but still they had either infeasible dual or the KKT conditions were violated.

Kintro_Ip was a little bit slower than Mosek on most of the discrete gates. On the Gaussian, it very slow. For most of the instances, Mosek had its optimality measure 1000 times better than Knitro_Ip. Even though it worked well for the discrete gates we can't rely on it for this problem since it is not stable with the variables number.

Knitro IpCG. It terminated with optimal solution on all the discrete gates except "XOR" and "RDNXOR" see Tables A2 and A3. For those two particular gates, it reached the iteration limit. For the Gaussian, it did a little better than Knitro_Ip since the projected conjugate gradient helps with large scale problems. It was able to solve only 6.2% none of which had less than 1000 variables and with 50% having more than 8000 variables see Table A5. 25% of the unsolved instances couldn't get a dual feasible solution. The rest stopped due to iteration or time limit with similar optimality measure violation as Knitro_Ip.

Knitro_IpCG was considerably slower than Mosek and Knitro_Ip. And it had worse optimality measure than Mosek on most of its solved instances. Since Knitro_IpCG couldn't solve a big set of instances this solver is unreliable for this problem.

Mosek terminated with reporting an optimal solution almost on all instances. For the gates, with sufficient optimality measure see Tables A1–A4. For the Gaussian gates, it terminated optimally for 70% of the instances see Table A5. 25% bailed out mainly since the gap between the primal and dual is negligible with the dual solution and primal are weakly feasible. Most of the latter instances had more than 10,000 variables. For the last 5% Mosek reached its iteration limit.

For all discrete gates Mosek terminated within milliseconds except "RDNUNQXOR" it terminated in at most 1 s. For the Gaussians only those with more than 25,000 variables took up to 5 s to be solved, the rest was done in less than a second.

Ipopt failed on the unadulterated "AND", "UNQ", and "XORAND" distributions since the optimal solution lays on the boundary and Ipopt hits the boundary quit often. Also, it couldn't solve any of the noiseless with sampling "AND" and "UNQ" gates, but managed to solve 30% of those of "XORAND". When the noise was applied, it worked 90% with some noisy "RDNXOR" gates, but bailed out on "RDNUNQXOR" gate with 10%, 5%, and 1%. Note that the optimality measure was highly violated even when Ipopt terminates optimally see Tables A1–A4 The violation mainly was in the dual feasibility and for the instances which had the solution on the boundary.

With the Gaussians, Ipopt couldn't solve any instance see Table A5. In 55% of the instances it terminated with computational errors or search direction being to small. 25% terminated with iteration limit and the rest with time limit. Same as for discrete case the optimality measure was not good. For discrete gates Ipopt was as fast a Mosek. Overall Ipopt is an unreliable solver for this problem.

4.3.3. Geometric Programming: CVXOPT

CVXOPT has a dedicated Geometric Programming interface, which we used to solve the Geometric Program (GP). The approach cannot solve a single instance to optimality: CVXOPT always exceeds the limit on the number of iterations. Hence, we have not listed it in the tables.

4.3.4. Exponential Cone Programming: ECOS & SCS

SCS failed on each and every instance. For ECOS, however, on average the running times were fast, but the results were mildly good. For types 1 and 2 instances, ECOS was most of the time the fastest solver. It terminated optimally for all the instances of types 1 and 2.

ECOS terminated suboptimal on 90% of the Gaussian instances. On the rest of the Gaussian gates it ran into numerical problems, step size became negligible, and terminated with unfeasible solutions.

ECOS handled types 1 and 2 instances pretty well. It was mostly slow on the Gaussian instances. For example, some Gaussian instances took ECOS more than seven hours to terminate sub-optimally. Modeling the problem as an exponential cone programming seems to be a good choice. ECOS vigorous performance makes it a reliable candidate for this optimization problem.

4.3.5. Miscellaneous Approaches

We include the following in the computational results for completeness. From what we discussed in the previous section, it is no surprise at all that these methods perform badly.

Knitro_SQP failed on the unadulterated "AND" and "RDNUNQAND" distributions see Table A1. When the noise was added to the discrete gates it couldn't solve more than 35% per gate see Tables A2 and A3. It didn't have better optimality measure than Mosek except on "RDN 1" which is can't be built on since Knitro_SQP solved only 8% of those instances. Note that on the unsolved instances, Knitro_SQP was the only solver which gave wrong answers (30% of the unsolved) i.e., claimed that the problem is infeasible.

Similarly, on Gaussian it failed in 78% of the instances due to computational problems and the rest were iteration or time limit see Table A5. This confirms that such type of algorithms is unsuitable for our problem; see Section 3.

Knitro_As. It failed on all the instances, trying to evaluate the Hessian at boundary points.

5. Conclusions

A closer look at the Convex Program proposed by Bertschinger et al. [1] reveals some subtleties which makes the computation of the information decomposition challenging. We have tried to solve the convex program using a number of software packages out-of-the-box (including, BTW, [29]).

Two of the solvers, namely, Mosek and ECOS work very satisfactorily. Even though, ECOS sometimes was 1000 times faster than Mosek, on some of types 1 and 2 instances, the time difference was no more than 5 s making this advantage rather dispensable. Nevertheless, on these instances, Mosek had better optimality measures. Note that on the hard problems of type 1, namely, "RDNUNQXOR", Mosek was faster than ECOS see Figure 1.

On the other hand, ECOS was slower than Mosek on the Gaussian instances especially when the (CP) had a huge number of variables. For these instances, the time difference is significant (hours) see Figure 2, which is rather problematic for ECOS.

Hence each of the two solvers has its pros and cones. This means that using both solvers is an optimal strategy when approaching the problem. One suggestion would be to use ECOS as a surrogate when Mosek fails to give a solution. Earlier ad-hoc approaches, which were based on CVXOPT and on attempts to "repair" a situation when the solver failed to converge, appear to be redundant.

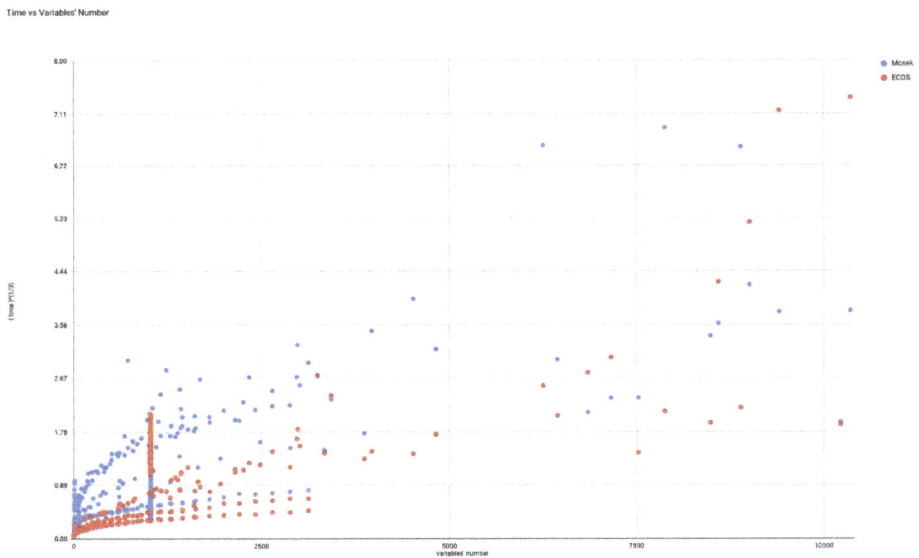

Figure 1. Comparison Between the running times of Mosek and ECOS for problems whose number of variables is at below 10,500.

Time vs Variables' Number

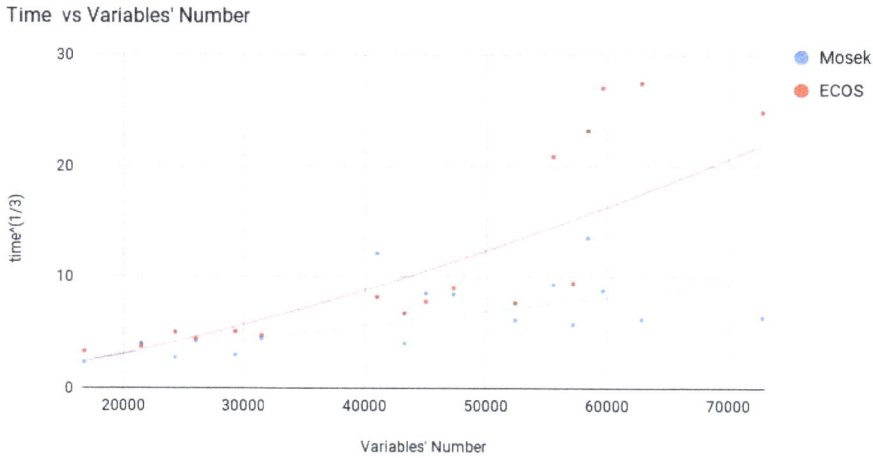

Figure 2. Comparison Between the running times of Mosek and ECOS for problems whose number of variables is at least 16,000.

Supplementary Materials: The source code in Julia used for the computations can be found here: github.com/dot-at/BROJA-Bivariate-Partial_Information_Decomposition.

Acknowledgments: This research was supported by the Estonian Research Council, ETAG (*Eesti Teadusagentuur*), through PUT Exploratory Grant #620. We also gratefully acknowledge funding by the European Regional Development Fund through the Estonian Center of Excellence in Computer Science, EXCS.

Author Contributions: RV+DOT developed the first algorithms and wrote the code. DOT+AM contributed the propositions and proofs, AM performed the computations and created the tables.

Conflicts of Interest: The authors declare no conflict of interest.

Appendix A. Tables of Instances and Computational Results

Table A1. Results for type (1a) instances: Gates—unadulterated probability distributions.

Instance	CVXOPT			Knitro_Ip			Knitro_IpCG			Knitro_As			Knitro_SQP			Mosek			Ipopt			ECOS			GD
	% solved	time (10^{-2})	opt. meas. (10^{-6})	% solved	time (10^{-2})	opt. meas. (10^{-6})	% solved	time (10^{-2})	opt. meas. (10^{-6})	% solved	time (10^{-2})	opt. meas. (10^{-6})	% solved	time (10^{-2})	opt. meas. (10^{-6})	% solved	time (10^{-2})	opt. meas. (10^{-6})	% solved	time (10^{-2})	opt. meas. (10^{-6})	% solved	time (10^{-2})	opt. meas. (10^{-6})	time (10^{-2})
XOR	y	1	0.06	y	0.7	0.31	y	98	0.8	n			y	2	0.63	y	0.18	0.28	y	0.15	1.5	y	0.03	0.001	25
AND	y	0.5	$21e^7$	y	30	0.5	y	52	13	n			n			y	0.06	0.15	y	0.09	10	y	0.03	0.01	25
UNQ	y	0.5	0.02	y	0.1	$81e^{-4}$	y	0.17	0.04	n			y	0.8	$21e^{-12}$	n	0.07	$1e^{-4}$				y	0.02	0.01	14
RDN	y	0.4	0.007	y	0.07	$1e^{-7}$	y	0.09	0.38	n			y	0.1	$41e^{-12}$	y	0.03	0.025	n			y	0.02	0.002	15
XORAND	y	0.6	$21e^6$	y	0.2	0.53	y	0.9	0.14	n			y	0.6	$1e^{-7}$	y	0.09	0.2	n			y	0.03	0.005	14
RDNXOR	y	3	0.12	y	0.2	0.06	y	271	0.3	n			y			y	0.2	$2e^{-4}$	y	0.2	$2e^4$	y	0.05	0.002	24
RDNUNQXOR	y	14	0.4	y	1.3	0.001	y	537	0.09	n			n			y	0.01	$6e^{-7}$	y	0.5	150	y	0.16	0.008	77

Table A2. Results for type (1b) instances: Gates (XOR, AND, UNQ, RDN) with noise.

Instance	CVXOPT			Knitro_Ip			Knitro_IpCG			Knitro_As			Knitro_SQP			Mosek			Ipopt			ECOS			GD
	% solved	avg. tm 10^{-2}	Opt meas. 10^{-6}	% solved	avg. tm 10^{-2}	Opt meas. 10^{-6}	% solved	avg. tm 10^{-2}	Opt meas. 10^{-6}	% solved	avg. tm 10^{-2}	Opt meas. 10^{-6}	% solved	avg. tm 10^{-2}	Opt meas. 10^{-6}	% solved	avg. tm 10^{-2}	Opt meas. 10^{-6}	% solved	avg. tm 10^{-2}	Opt meas. 10^{-6}	% solved	avg. tm 10^{-2}	Opt meas. 10^{-6}	avg. tm 10^{-2}
XOR 0	100	0.01	3e^6	100	0.9	0.46	100	150	1.06	0			30	4	0.87	100	4	0.004	100	0.1	2e^6	100	0.04	0.01	28
XOR 1	100	2.2	3e^6	100	1	0.98	89	254	2.02	0			17	4	0.67	100	1.9	0.0066	100	0.2	3e^6	100	0.03	0.01	35
XOR 2	100	2	3e^6	100	1	0.92	63	259	1	0			26	9	0.91	100	12	0.0053	100	0.3	2e^6	100	0.04	0.01	36
XOR 3	99	2	4e^6	100	1	0.96	70	363	1	0			27	7	0.99	100	7	0.0056	100	0.2	2e^6	100	0.03	0.01	36
XOR 4	100	1	4e^6	100	1.3	1	98	40	1	0			20	13	0.98	100	8	0.0052	100	0.1	1e^6	100	0.03	0.01	32
AND 0	100	300	0.6	100	40	0.7	100	76	22.5	0			0			100	0.07	0.41	0			100	0.03	0.01	27
AND 1	65	2	2e^7	100	1	2.9	100	17	3.2	0			0			100	0.9	24	100	0.3	2e^6	100	0.04	0.01	42
AND 2	67	3	1e^6	100	2	1.7	100	4.9	4	0			0			100	1.2	0.11	100	0.3	1e^7	100	0.04	0.01	40
AND 3	51	3	2e^7	100	1.9	1.9	100	2	18	0			0			100	1.4	0.04	100	4	2e^6	100	0.04	0.01	39
AND 4	54	2	2e^7	100	1	5.4	100	2	9	0			0			100	0.5	41	100	0.5	4e^4	100	0.04	0.01	40
UNQ 0	100	0.4	2e^4	100	0.4	3e^{-3}	100	0.5	0.09	0			100	1.4	6e^{-12}	100	0.04	1e^{-4}	0			100	0.02	0.007	16
UNQ 1	100	4	4e^6	100	1	6	100	1.9	9	0			33	6	5	100	0.6	0.036	100	0.7	3e^5	100	0.07	0.01	42
UNQ 2	100	5	5e^6	100	2	0.04	100	2	0.84	0			26	13	0.46	100	0.69	0.12	100	0.6	4e^5	100	0.07	0.008	44
UNQ 3	100	5	7e^7	100	2	6	100	1	40	0			38	9	10	100	0.7	0.022	100	0.6	4e^5	100	0.07	0.004	44
UNQ 4	20	4	0.09	100	1.5	0.04	100	1	21	0			18	8	2.6	100	0.6	0.0019	100	0.7	3e^5	100	0.08	0.005	42
RDN 0	–	2	2e^6	100	0.4	6e^{-4}	100	0.4	0.51	0			100	0.6	1e^{-12}	100	1	0.0029	30	0.5	1e^5	100	0.02	0.01	16
RDN 1	0			100	1	2.8	100	1.2	2.8	0			8	2	2e^{-12}	100	0.3	0.005	100	0.3	2e^7	100	0.04	0.01	41
RDN 2	0			100	1	6.7	100	1.9	6.2	0			14	4.5	2e^{-10}	100	1.1	0.01	100	0.5	2e^6	100	0.04	0.01	40
RDN 3	0			100	1.7	4.1	100	1.8	9.8	0			35	4	2.7	100	0.4	2e^{-4}	100	0.6	4e^6	100	0.05	0.01	41
RDN 4	1	12	2e^6	100	0.9	0.07	100	0.8	1	0			3	3	2e^{-6}	100	0.2	1e^{-5}	100	0.4	4e^6	100	0.04	0.008	41

Table A3. Results for type (1b) instances: Gates (XORAND, RDNXOR, RDNUNQXOR) with noise.

Instance	CVXOPT			Knitro_Ip			Knitro_IpCG			Knitro_As			Knitro_SQP			Mosek			Ipopt			ECOS			GD
	% solved	avg. tm 10^{-2}	Opt meas. 10^{-6}	% solved	avg. tm 10^{-2}	Opt meas. 10^{-6}	% solved	avg. tm 10^{-2}	Opt meas. 10^{-6}	% solved	avg. tm 10^{-2}	Opt meas. 10^{-6}	% solved	avg. tm 10^{-2}	Opt meas. 10^{-6}	% solved	avg. tm 10^{-2}	Opt meas. 10^{-6}	% solved	avg. tm 10^{-2}	Opt meas. 10^{-6}	% solved	time (10^{-2})	opt. meas. (10^{-6})	time (10^{-2})
XORAND 0	100	0.6	20	100	0.6	0.74	100	0.9	0.2	0			100	2	2e^{-6}	100	0.1	0.4	0			100	0.03	0.01	27
XORAND 1	94	6	1e^7	100	1.1	5	100	5	6.2	0			35	9	4.9	100	1	0.5	100	0.6	8e^5	100	0.08	18	43
XORAND 2	67	8	12.6	100	2	6.2	100	10	8.2	0			15	22	5.4	100	1.3	0.4	100	0.8	1e^3	100	0.09	0.01	42
XORAND 3	12			100	2	5.7	100	2	8.7	0			15	16	6.9	100	1.3	2.2	100	1	6e^4	100	0.1	0.03	42
XORAND 4	0			100	2	7.2	100	3	8	0			93	2		100	0.7	40	100	0.8	4e^4	100	0.1	0.04	43
RDNXOR 0	100	3	3.11	100	1	0.46	70	407	1	0			30	6	0.23	100	1	2e^{-9}	100	0.2	2e^4	100	0.06	0.01	27
RDNXOR 1	0			100	3.5	18.2	100	2	30	0			0			100	25	5	1	261	0.016	100	0.3	0.01	40
RDNXOR 2	1	190	1e^6	100	9	60	100	3	250	0			0			100	250	50	90	10	1.9	100	0.4	0.01	41
RDNXOR 3	0			100	8	93	100	3	30	0			1			100	15	0.01	99	4.7	0.02	100	0.3	0.01	40
RDNXOR 4	1	145	2e^7	100	6	90	100	5	31	0			0			100	15	0.01	100	5	0.02	100	0.4	0.01	43
RDNUNQXOR 0	90	25	3e^6	100	2	0.34	100	636	1	0			0			100	6	0.2	100	0.7	234	100	0.3	0.008	75
RDNUNQXOR 1	0			100	57	297	100	34	107	0			0			100	269	4	0			100	200	0.01	1156
RDNUNQXOR 2	0			100	110	364	100	60	128	0			0			100	483	4	0			100	195	0.01	1170
RDNUNQXOR 3	0			100	130	561	100	91	597	0			0			100	307	20	0			100	224	0.01	1140
RDNUNQXOR 4	0			100	129	422	100	132	1e^4	0			0			100	388	0.003	92	437	207	100	555	0.024	1160

Table A4. Results for type (2) instances: Example-31.

Instance	CVXOPT			Mosek			Ipopt			ECOS			GD
	% solved	time (10^{-2})	opt. meas. (10^{-6})	% solved	time (10^{-2})	opt. meas. (10^{-6})	% solved	time (10^{-2})	opt. meas. (10^{-6})	% solved	time (10^{-2})	opt. meas. (10^{-6})	time (10^{-2})
Independent 1	100	28	1.15	100	4	91e^{-4}	100	18	31e^{-3}	100	0.5	0.01	100
Independent 2	100	255	0.2	100	22	0.003	92	437	207	100	2	0.02	802
Noisy 1	34	113	21e^7	100	87	0.008	38	6.2	21e^3	100	2.8	0.014	100
Noisy 2	0			100	648	0.011	7	602	21e^{-3}	100	8	0.016	813

Table A5. Results for type (3) instances: Discretized 3-dimensional Gaussians.

Instance	CVXOPT			Knitro_Ip			Knitro_IpCG			Knitro_As			Knitro_SQP			Mosek			Ipopt			ECOS			GD
	% solved	time (10^{-2})	opt. meas. (10^{-6})	% solved	time (10^{-2})	opt. meas. (10^{-6})	% solved	time (10^{-2})	opt. meas. (10^{-6})	% solved	time (10^{-2})	opt. meas. (10^{-6})	% solved	time (10^{-2})	opt. meas. (10^{-6})	% solved	time (10^{-2})	opt. meas. (10^{-6})	% solved	time (10^{-2})	opt. meas. (10^{-6})	% solved	time (10^{-2})	opt. meas. (10^{-6})	time (10^{-2})
Gauss-1	0			21	820	4	0			0			0			71	0.27	0.4	0			100	4.2	0.19	1940
Gauss-2	0			2.7	950	10	5	1020	300	0			0			75	7.6	4	0			96	52	1.06	1890
Gauss-3	0			0			11	2170	900	0			0			65	551	2	0			96	632	6	–
Gauss-4	0			0			7	2560	2000	0			0			57	$4e^4$	60	0			75	$3e^5$	400	–

References

1. Bertschinger, N.; Rauh, J.; Olbrich, E.; Jost, J.; Ay, N. Quantifying unique information. *Entropy* **2014**, *16*, 2161–2183.
2. Williams, P.L.; Beer, R.D. Nonnegative decomposition of multivariate information. *arXiv* **2010**, arXiv:1004.2515.
3. Griffith, V.; Koch, C. Quantifying synergistic mutual information. In *Guided Self-Organization: Inception*; Springer: New York, NY, USA, 2014; pp. 159–190.
4. Harder, M.; Salge, C.; Polani, D. Bivariate measure of redundant information. *Phys. Rev. E* **2013**, *87*, 012130.
5. Bertschinger, N.; Rauh, J.; Olbrich, E.; Jost, J. Shared information—New insights and problems in decomposing information in complex systems. In Proceedings of the European Conference on Complex Systems, Brussels, Belgium, 2–7 September 2012; Springer: Cham, Switzerland, 2013; pp. 251–269.
6. Nemirovski, A. Efficient Methods in Convex Programming. Available online: http://www2.isye.gatech.edu/~nemirovs/Lect_EMCO.pdf (accessed on 6 October 2017).
7. Wibral, M.; Priesemann, V.; Kay, J.W.; Lizier, J.T.; Phillips, W.A. Partial information decomposition as a unified approach to the specification of neural goal functions. *Brain Cognit.* **2017**, *112*, 25–38.
8. Ruszczyński, A.P. *Nonlinear Optimization*; Princeton University Press: Princeton, NJ, USA, 2006; Volume 13.
9. Bubeck, S.; Eldan, R. The entropic barrier: A simple and optimal universal self-concordant barrier. *arXiv* **2014**, arXiv:1412.1587.
10. Bubeck, S.; Eldan, R. The entropic barrier: A simple and optimal universal self-concordant barrier. In Proceedings of the 28th Conference on Learning Theory, Paris, France, 3–6 July 2015.
11. Boyd, S.; Vandenberghe, L. *Convex Optimization*; Cambridge University Press: Cambridge, UK, 2004.
12. Makkeh, A. Applications of Optimization in Some Complex Systems. Ph.D. Thesis, University of Tartu, Tartu, Estonia, 2017.
13. Nocedal, J.; Wright, S. *Numerical Optimization*; Springer: New York, NY, USA, 2006.
14. Boyd, S.; Kim, S.J.; Vandenberghe, L.; Hassibi, A. A tutorial on geometric programming. *Optim. Eng.* **2007**, *8*, 67–127.
15. Chares, R. Cones and interior-point algorithms for structured convex optimization involving powers and exponentials, Doctoral dissertation, UCL-Université Catholique de Louvain, Louvain-la-Neuve, Belgium, 2009.
16. Hahn, T. CUBA—A library for multidimensional numerical integration. *Comput. Phys. Commun.* **2005**, *168*, 78–95. [hep-ph/0404043].
17. Lubin, M.; Dunning, I. Computing in operations research using Julia. *INFORMS J. Comput.* **2015**, *27*, 238–248.
18. Sra, S.; Nowozin, S.; Wright, S.J. *Optimization for Machine Learning*; MIT Press: Cambridge, MA, USA, 2012.
19. Byrd, R.; Nocedal, J.; Waltz, R. KNITRO: An Integrated Package for Nonlinear Optimization. In *Large-Scale Nonlinear Optimization*; Springer: New York, NY, USA, 2006; pp. 35–59.
20. Waltz, R.A.; Morales, J.L.; Nocedal, J.; Orban, D. An interior algorithm for nonlinear optimization that combines line search and trust region steps. *Math. Program.* **2006**, *107*, 391–408.
21. Byrd, R.H.; Hribar, M.E.; Nocedal, J. An interior point algorithm for large-scale nonlinear programming. *SIAM J. Optim.* **1999**, *9*, 877–900.
22. Byrd, R.H.; Gould, N.I.; Nocedal, J.; Waltz, R.A. An algorithm for nonlinear optimization using linear programming and equality constrained subproblems. *Math. Program.* **2004**, *100*, 27–48.

23. ApS, M. *Introducing the MOSEK Optimization Suite 8.0.0.94*. Available online: http://docs.mosek.com/8.0/pythonfusion/intro_info.html (accessed on 6 October 2017).
24. Andersen, E.D.; Ye, Y. A computational study of the homogeneous algorithm for large-scale convex optimization. *Comput. Optim. Appl.* **1998**, *10*, 243–269.
25. Andersen, E.D.; Ye, Y. On a homogeneous algorithm for the monotone complementarity problem. *Math. Program.* **1999**, *84*, 375–399.
26. Domahidi, A.; Chu, E.; Boyd, S. ECOS: An SOCP solver for embedded systems. In Proceedings of the European Control Conference (ECC), Zurich, Switzerland, 17–19 July 2013; pp. 3071–3076.
27. O'Donoghue, B.; Chu, E.; Parikh, N.; Boyd, S. Conic Optimization via Operator Splitting and Homogeneous Self-Dual Embedding. *J. Optim. Theory Appl.* **2016**, *169*, 1042–1068.
28. O'Donoghue, B.; Chu, E.; Parikh, N.; Boyd, S. SCS: Splitting Conic Solver, Version 1.2.7. Available online: https://github.com/cvxgrp/scs (accessed on 6 October 2017).
29. Grant, M.; Boyd, S. CVX: Matlab Software for Disciplined Convex Programming, Version 2.1. Available online: http://cvxr.com/cvx (accessed on 6 October 2017).

![entropy logo] *entropy*

MDPI

Article

Partial and Entropic Information Decompositions of a Neuronal Modulatory Interaction

Jim W. Kay [1,*]**, Robin A. A. Ince** [2]**, Benjamin Dering** [3] **and William A. Phillips** [3]

[1] Department of Statistics, University of Glasgow, Glasgow G12 8QQ, UK
[2] Institute of Neuroscience and Psychology, University of Glasgow, Glasgow G12 8QQ, UK;
     robin.ince@glasgow.ac.uk
[3] Faculty of Natural Sciences, University of Stirling, Stirling FK9 4LA, UK; b.r.dering@stir.ac.uk (B.D.);
     w.a.phillips@stir.ac.uk (W.A.P.)
*   Correspondence: jimkay049@gmail.com; Tel.: +44-141-391-3288

Received: 30 June 2017; Accepted: 23 October 2017; Published: 26 October 2017

Abstract: Information processing within neural systems often depends upon selective amplification of relevant signals and suppression of irrelevant signals. This has been shown many times by studies of contextual effects but there is as yet no consensus on how to interpret such studies. Some researchers interpret the effects of context as contributing to the selective receptive field (RF) input about which neurons transmit information. Others interpret context effects as affecting transmission of information about RF input without becoming part of the RF information transmitted. Here we use partial information decomposition (PID) and entropic information decomposition (EID) to study the properties of a form of modulation previously used in neurobiologically plausible neural nets. PID shows that this form of modulation can affect transmission of information in the RF input without the binary output transmitting any information unique to the modulator. EID produces similar decompositions, except that information unique to the modulator and the mechanistic shared component can be negative when modulating and modulated signals are correlated. Synergistic and source shared components were never negative in the conditions studied. Thus, both PID and EID show that modulatory inputs to a local processor can affect the transmission of information from other inputs. Contrary to what was previously assumed, this transmission can occur without the modulatory inputs becoming part of the information transmitted, as shown by the use of PID with the model we consider. Decompositions of psychophysical data from a visual contrast detection task with surrounding context suggest that a similar form of modulation may also occur in real neural systems.

Keywords: information theory; partial information decomposition; entropic information decomposition; synergy; redundancy; contextual modulation; neural information processing

1. Introduction

Amplifiers, such as hearing aids, for example, are designed to increase signal strength without distorting the informative content that it transmits, i.e., its "semantics". Though independence of semantics has been a truism of information theory since its inception, information decomposition may help distinguish the effects of amplifying inputs from driving inputs which determine what the output transmits information about, which is what we will refer to here as its "semantics". It may seem intuitively obvious that any output must necessarily transmit information about all inputs that affect it, but that intuition is misleading. Here, we use information decomposition to show that a modulatory input can influence the transmission of information about other inputs while remaining distinct from that information.

This may help resolve a long-standing controversy within the cognitive neurosciences concerning the nature of "contextual modulation". Many see the wide variety of psychophysical and physiological phenomena that are grouped under this heading as demonstrating that the concept of a neuron's receptive field, i.e., what the cell transmits information about, needs to be extended to include an extra-classical receptive field; see e.g., [1]. In contrast to that many others see these phenomena as evidence that contextual modulation does not change the cell's receptive field semantics; see e.g., [2–4].

Resolution of this issue requires an adequate definition of "modulation", which is used in several different, and often undefined, ways. It is frequently used to mean simply that one thing affects another. That unnecessary use of the term introduces substantial confusion, however, because the term is also often used to refer to a three-term interaction. It could be used to refer to any three-way interaction in which A effects the transmission of information about B by C. Our use is more specific than that, however. The essence of the modulatory interaction that we study here is that the modulator affects transmission of information about something else without becoming part of the information transmitted. The effect of the volume control on a radio provides a simple example. It changes signal strength without becoming part of the message conveyed. The use of the term "modulation" in telecommunications potentially adds further confusion, however, because in either amplitude modulation (AM) or frequency modulation (FM) it is the "modulatory" signal that is used to convey the message to be transmitted. That is the opposite of what we and many others in the cognitive and neurosciences refer to as "modulation". While awaiting a consensus that resolves this terminological confusion we define our usage of the term "modulation" as explicitly and as clearly as we can. Modulation that increases output signal strength is referred to as "amplification" or "facilitation". Modulation that decreases output signal strength is referred to as "disamplification", "suppression", or "attenuation".

Information decomposition could help clarify the notion of "modulation" as used within the cognitive and neurosciences in at least three ways. First, by requiring formal specifications to which decompositions can be applied it enforces adequate definition. Second, by being applied to a transfer function explicitly designed to be modulatory, it deepens our understanding of the information processing operations performed by such interactions. Third, decomposition of a modulatory interaction that is formally specified shows the conditions under which it can be distinguished from additive interactions and provides patterns of decomposition to which empirically observed patterns can be compared.

In this paper we apply information decomposition to a transfer function specifically designed to operate as a modulator within a formal neural network that uses contextually guided learning to discover latent statistical structure within its inputs [5]. We show that this transfer function has the properties required of a modulator, and that they can be clearly distinguished from additive interactions that do contribute to output semantics. A thorough understanding of this modulatory transfer function is of growing importance to neuroscience because recent advances suggest that something similar occurs at an intracellular level in neocortical pyramidal cells, and may be closely related to consciousness [6,7]. It is also important to machine learning because the information processing capabilities of networks such as those used for deep learning might be greatly enhanced if given the context-sensitivity that such modulatory interactions can provide.

Modulatory interactions distinguish the contributions of two distinct inputs to an output, so they imply some form of multivariate mutual information decomposition. Various forms of decomposition have been proposed, however, and they may offer different resolutions to this issue. We therefore compare resolutions that arise from two proposals discussed elsewhere in this Special Issue. One is Partial Information Decomposition [8–11]. The other is Entropic Information Decomposition [12,13]. We find that though there are important differences between these two proposed forms of decomposition, they are in agreement with respect to their implications for the issue of distinguishing between additive and modulatory interactions.

The notion of modulation is essentially a three-term interaction in which one input variable modulates transmission of information about a second input variable by an output. The two inputs therefore make fundamentally different kinds of contribution to the output. In contrast to that, additive interactions do not require the two inputs to remain distinct because their contributions can be summarized via a single integrated value. Many information decomposition spectra and surfaces are displayed in the following, demonstrating their expressive power and the variety of information processing operations that a single transfer function can perform.

2. Notation and Definitions

In this section we describe our notation and define the information-theoretic concepts which are used in the sequel. A generic "p" is used to denote a probability mass function, with the argument of the function signifying which distribution is being described. Capital letters are used to denote random variables, with their realised values appearing in lower-case. We denote the conditional probability that $Y = y$, given that $X_1 = x_1$ and $X_2 = x_2$ by the conditional mass function $p(y|x_1, x_2)$ for $y \in B$, and $(x_1, x_2) \in B^2$, where $B = \{-1, +1\}$.

In [14], the RF and contextual field (CF) inputs were multivariate, but here we consider the special case of the local processor in [14] having two binary inputs, X_1 and X_2, and one binary output, Y, with all three random variables having range space B. The joint distribution of (Y, X_1, X_2) is given by the probability mass function (p.m.f.) $p(y, x_1, x_2)$, where

$$p(y, x_1, x_2) = \Pr(Y = y, X_1 = x_1, X_2 = x_2), \quad (y, x_1, x_2) \in B^3.$$

This distribution will be considered in the form

$$p(y, x_1, x_2) = p(y|x_1, x_2)p(x_1, x_2), \tag{1}$$

and we will separately specify a joint p.m.f $p(x_1, x_2)$ and a conditional p.m.f. $p(y|x_1, x_2)$.

In the local processor in Figure 1, the value of X_1 provides the receptive field (RF) input to the local processor, while the value of X_2 is the input from the contextual field (CF). The value of the RF input, X_1, is multiplied by the signal strength s_1 to form the integrated RF input and similarly for the CF input, X_2. Therefore, the values taken by the integrated RF and CF inputs are $r = s_1 x_1$ and $c = s_2 x_2$. These integrated values have both strength and a sign. The strength is a constant property of the defined system, while the sign can change from sample to sample. The signal strengths, s_i, are positive real numbers. The manner in which these signals are combined in the output unit will be described in Section 3.

In this study, it is assumed that $\Pr(X_1 = 1) = \Pr(X_2 = 1) = \frac{1}{2}$ and that the correlation between X_1 and X_2 is d, where $-1 < d < 1$. This means that

$$\lambda \equiv \Pr(X_1 = 1, X_2 = 1) \quad = \Pr(X_1 = -1, X_2 = -1) = \frac{1+d}{4}, \tag{2}$$

$$\mu \equiv \Pr(X_1 = 1, X_2 = -1) = \Pr(X_1 = -1, X_2 = 1) \quad = \frac{1-d}{4}. \tag{3}$$

It is also assumed that the conditional output probability has a logistic form, with

$$\Pr(Y = 1|X_1 = x_1, X_2 = x_2) = 1/(1 + \exp(-T(x_1, x_2))), \tag{4}$$

where T is a transfer function which depends also on the signal strengths, s_1, s_2. In Section 3, the two transfer functions that are used in this study are specified. It should be noted that we are actually considering a class of trivariate probability distributions that are indexed by (s_1, s_2, d), where $s_1 > 0, s_2 > 0, -1 < d < 1$, although this indexation is suppressed in the sequel for ease of notation. The various classical measures of information and measures of partial information used are calculated using a member of the class of trivariate probability distributions, defined in (1)–(4), that is given by a particular choice of (s_1, s_2, d).

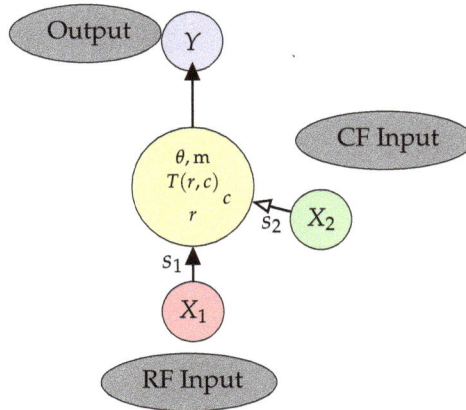

Figure 1. A local processor with binary receptive field (RF) input X_1, contextual field (CF) input X_2 and output Y. The weights on the connections from the RF and CF inputs into the output unit are s_1, s_2, which represent the strengths given to the input signals. The integrated RF input, r, and the integrated CF input, c, are passed through a transfer function T and a logistic nonlinearity within the output unit to produce the conditional output probability, θ, as well as the output conditional mean, m.

We now define the standard information theoretic terms that are required in this work and based on results in [15]. We denote by the function H the usual Shannon entropy, and note that any term with zero probabilities makes no contribution to the sums involved. The total mutual information that is shared by Y and the pair (X_1, X_2) is given by,

$$I[Y;(X_1, X_2)] = H(Y) + H(X_1, X_2) - H(Y, X_1, X_2). \tag{5}$$

The information that is shared between Y and X_1 but not with X_2 is

$$I[Y; X_1 | X_2] = H(Y, X_2) + H(X_1, X_2) - H(X_2) - H(Y, X_1, X_2), \tag{6}$$

and the information that is shared between Y and X_2 but not with X_1 is

$$I[Y; X_2 | X_1] = H(Y, X_1) + H(X_1, X_2) - H(X_1) - H(Y, X_1, X_2). \tag{7}$$

Finally, the co-information of (Y, X_1, X_2) has several equivalent forms

$$I[Y; X_1; X_2] = I[Y; X_1] - I[Y; X_1 | X_2] = I[Y; X_2] - I[Y; X_2 | X_1] = I[X_1; X_2] - I[X_1; X_2 | Y], \tag{8}$$

where, for $i = 1, 2$,

$$I[Y; X_i] = H(Y) + H(X_i) - H(Y, X_i), \text{ and } I[X_1; X_2] = H(X_1) + H(X_2) - H(X_1, X_2). \tag{9}$$

We note that classical Shannon information measures have been used in neural coding studies to investigate measures of synergy and redundancy; see for example [16].

When we come to define measures of partial information it will be necessary to calculate these information quantities with respect to another p.m.f., say $q(y, x_1, x_2)$, and to denote this we add the subscript "q" to such terms, e.g., $I_q(Y; X_1; X_2)$. This means that the p.m.f. $q(y, x_1, x_2)$ is used in the computation rather than the original p.m.f. $p(y, x_1, x_2)$.

3. An Interaction Designed to Be Modulatory

Our concern here is with variables that can take either positive or negative values, which can be seen as being analogous to excitation and inhibition in neural systems. We model that decision as a probabilistic binary variable that chooses between the values 1 and -1. The criteria to be met by a modulatory transfer function in this case have been stated and discussed in many previous papers; see e.g., [17–19]. The criteria for a modulatory interaction were stated for a local processor receiving two inputs: the integrated RF input, r, and the integrated CF input, c. The requirements were stated in terms of the level of activation within the local processor, although in this paper we use this term to denote the value of the transfer function, and they are amended slightly here. Please note that the term 'integrated' was used in previous work to refer to the weighting and summing of the components of a multivariate input; we continue to use this term here even though the input to each field is univariate. The value of the transfer function is fed into a logistic function to compute the conditional probability that a 1 will be transmitted. Stated in those terms the CF input modulates transmission of information about the RF input if four criteria are met:

1. If the integrated RF input is extremely weak, then the value of the transfer function is close to zero.
2. If the integrated CF input is extremely weak, then the value of the transfer function should be close to the integrated RF input.
3. If the integrated RF and CF inputs have the same sign, then the absolute value of the transfer function should be greater than when based on the RF input alone. On the other hand, if the RF and CF inputs are of opposite sign then the absolute value of the transfer function should be less than when based on the RF input alone.
4. The sign of the value of the transfer function is that of the integrated RF, so that the context cannot change the sign of the conditional mean of the output.

In general terms, the CF input would have no modulatory effect on the output when the output and the CF input are conditionally independent given the value of the RF input, which is equivalent to the conditional mutual information $I[Y; X_2|X_1]$ being equal to zero. One case where this happens for any member of the class of trivariate binary distributions defined in (1)–(4) is when the correlation between the inputs X_1, X_2 is ± 1, for then $I[Y; X_2|X_1] = 0$; see Theorem 5. On the other hand, in situations where this conditional mutual information is non-zero then X_2 influences the prediction of the output Y by the input X_1 in the sense that

$$\Pr(Y = y | X_1 = x_1, X_2 = x_2) \neq \Pr(Y = 1 | X_1 = x_1),$$

for at least one $(y, x_1, x_2) \in B^3$. This is a very general form of modulation, but the type of modulation defined in requirements 1–4 is very specific and we call it "contextual modulation". This contextual modulation is relevant within the local processor at the level of individual system inputs and outputs. On the other hand, the following conditions express the notion of contextual modulation for the whole ensemble of inputs and outputs:

M1: If the *RF* signal is strong enough and the CF input is extremely weak then $I[Y; X_1|X_2]$ can have its maximum value, $I[Y; X_1]$ can be maximised and $I[Y; X_2|X_1]$ is close to zero. This shows that the RF input is sufficient, thus allowing the information in the *RF* to be transmitted, and that the CF input is not necessary.

M2: $I[Y; X_2|X_1]$ and $I[Y; X_1]$ are close to zero when the RF input is extremely weak no matter how strong the CF input. This shows that the RF input is necessary for information to be transmitted, and that the CF input is not sufficient to transmit the information in the RF input.

M3: When $s_1 < s_2$ and when the RF input is weak, $I[Y; X_1]$ and $I[Y; X_1|X_2]$ are both larger when the CF input is moderate than when the CF input is weak. Thus the CF input modulates the transmission of information about the RF input.

One might expect that these two definitions of contextual modulation are linked. In the limiting situation of $s_1 \to 0$ it is possible to show that requirement 1 implies M1, and as $s_2 \to 0$ one finds that requirement 2 implies M2. It seems difficult to prove more general connections and so this matter is considered computationally in Section 3.1.

Multivariate binary processors were also considered in [5], thus allowing for choice between many more than two alternatives. It was also shown that the coherent infomax learning rule also applies to this multivariate case such that the contextually guided learning discovers variables defined on the RF input space that are statistically related to variables specified in, or discovered by, other streams of processing within the network. Thus it implements a multi-stream, non-linear, form of latent structure analysis. There are two distinct aspects of semantics in this system, i.e., the receptive field selectivity of each unit within a local processor and the positivity or negativity of its output. Here we are primarily concerned with that latter aspect. We show below that:

(i)   the modulatory input affects output only when the primary driving integrated RF input is non-zero but weak;

(ii)   that even when it does have an effect it has no effect on the sign of the conditional mean output, and

(iii)   that it can have those modulatory effects without the binary output transmitting any unique information about the modulator.

In the case where the processor has a binary output, the transfer function has the form

$$T(x_1, x_2) = r\left[k_1 + (1 - k_1) \exp\left(k_2 rc\right)\right], \quad (k_2 > 0, 0 < k_1 < 1),$$

where $r = s_1 x_1, c = s_2 x_2$, k_1 and k_2 are constants, and here we take $k_1 = \frac{1}{2}$ and $k_2 = 1$.

This transfer function was designed to effect a modulatory interaction between two input sources, with one source being the primary driver while the role of the the second "contextual" source is to modulate transmission of information about the primary source. The effect of the contextual source is to amplify or disamplify the strength of the signal from the primary source in such a way that the semantic content (the sign) of the primary source is not changed. Neither of the PID and EID considered in this paper has previously been applied to this kind of signal and we now show this to be possible.

In this paper, the version of the modulatory transfer function we use takes the form

$$T_M(x_1, x_2) = \tfrac{1}{2}r\left[1 + \exp\left(rc\right)\right] = \tfrac{1}{2}s_1 x_1 \left[1 + \exp\left(s_1 x_1 \times s_2 x_2\right)\right], \tag{10}$$

for given values x_1, x_2 of the random variables X_1, X_2, and given signal strengths s_1, s_2. Here the integrated RF input is $r = s_1 x_1$ and the integrated CF input is $c = s_2 x_2$, and they both have a sign and a strength. The output conditional probability is given by

$$\theta = \Pr(Y = 1 | X_1 = x_1, X_2 = x_2) = 1/\left[1 + \exp\left(-T_M(x_1, x_2)\right)\right]. \tag{11}$$

Whether this probability is greater than or less than $\frac{1}{2}$ is determined solely by the value of $x_1(\pm 1)$, and the form of T_M ensures that the contextual signal cannot change the sign of the output conditional mean. Thus the output produced has semantic content, and also the value of the output conditional probability, θ, gives the semantic content a measure of strength in the sense that values of θ closer to 0 or 1 indicate a more definite decision. The conditional variance of Y is $4\theta(1 - \theta)$, and so uncertainty in the output decision is largest when $\theta = 1/2$ and zero when $\theta = 0$ or 1. An alternative description is to say that the precision (reciprocal variance) is least when $\theta = 1/2$ and it tends to infinity as θ approaches 0 or 1. Within the local processor the conditional mean of the output, $m = 2\theta - 1$, is also computed. It has both a sign and a strength.

Given the form of T_M, the integrated RF will be amplified in magnitude whenever the signs of x_1 and x_2 agree, and it will be disamplified when these signs do not agree. The role of the integrated CF is to modify the strength of the conditional mean output without conveying its own semantic content (i.e., its sign). This form of transfer function ensures that the maximum extent of any disamplification of the primary signal is by a factor of 2.

By way of contrast, we also consider an additive transfer function by simply adding together the integrated RF and CF inputs, r, c, to give

$$T_A(x_1, x_2) = r + c = s_1 x_1 + s_2 x_2, \tag{12}$$

with the output conditional probability given by

$$\Pr(Y = 1 | X_1 = x_1, X_2 = x_2) = 1/[1 + \exp(-T_A(x_1, x_2))]. \tag{13}$$

The use of this transfer function also affects the values of θ and m but, unlike the modulatory transfer function, this additive transfer function can change the sign of the output conditional mean m, which is not consistent with the fourth condition for a modulatory transfer function described above. The additive transfer function does satisfy condition M1 but does not satisfy condition M2 or M3. This additive transfer function can be seen as a simple version of the common assumption within neurobiology that neurons function as integrate-and-fire point processors. While this assumption does not imply that all integration is linear it does mean that such integration computes a single value per local processor. The results produced using the these two different transfer functions will be discussed in Sections 5–8.

Please note that in the sequel we normally abbreviate the terms "integrated RF input" and "integrated CF input" by using just "RF input" and "CF input", respectively. In particular, whenever a strength is implied for the RF or CF input, then we mean that the 'integrated' values of these inputs are being considered.

3.1. Analysis Using Classical Shannon Measures

We start in this section by presenting results involving the classical Shannon measures for the system defined in Sections 2 and 3. First we recall that λ and μ are defined in (2) and (3) and set up some further simplifying notation which is used in the results. We set

$$u = \Pr(Y = 1 | X_1 = 1, X_2 = 1), \quad \text{and} \quad v = \Pr(Y = 1 | X_1 = 1, X_2 = -1). \tag{14}$$

The parameters u and v are function of s_1 and s_2, and u takes the value u_M or u_A depending on which transfer function is being used; similarly for v. From (10) for transfer function T_M

$$u_M = 1/(1 + \exp(-\tfrac{1}{2}s_1(1 + \exp(s_1 s_2)))), \quad \text{and} \quad v_M = 1/(1 + \exp(-\tfrac{1}{2}s_1(1 + \exp(-s_1 s_2)))), \tag{15}$$

whereas, from (12), for transfer function T_A

$$u_A = 1/(1 + \exp(-(s_1 + s_2))), \quad \text{and} \quad v_A = 1/(1 + \exp(-(s_1 - s_2))). \tag{16}$$

Finally, we define

$$z = 2\lambda u + 2\mu v, \quad w = 2\lambda u + 2\mu(1 - v) \quad \text{and} \quad h(v) = -v\log(v) - (1 - v)\log(1 - v), \tag{17}$$

where $0 < v < 1$. We note also that the value of z has two forms: z_M when transfer function T_M is used and z_A when transfer function T_A is employed; similarly for w. We now collect together our results in the following theorem, proof of which is relegated to the appendix.

Theorem 1. *It is assumed that $s_1 > 0, s_2 > 0$. For the probability distribution defined in (1)–(4), the following results hold.*

(a) $I[Y; X_1|X_2] = h(w) - 2\lambda h(u) - 2\mu h(v)$;

(b) $I[Y; X_2|X_1] = h(z) - 2\lambda h(u) - 2\mu h(v)$;

(c) $I[Y; X_1] = 1 - h(z)$;

(d) $I[Y; X_2] = 1 - h(w)$;

(e) $I[Y; X_1; X_2] = 1 - h(z) - h(w) + 2\lambda h(u) + 2\mu h(v)$;

(f) $I[Y; (X_1, X_2)] = 1 - 2\lambda h(u) - 2\mu h(v)$,

where from (15) and (16), $u = u_M, v = v_M$ when the transfer function T_M is employed and $u = u_A, v = v_A$ when the transfer function T_A is used.

Since we are particularly interested in interactions among the three variables, Y, X_1, X_2, we now show the classic Shannon information measures defined in (6)–(9), with surface plots given in Figures 2 and 3. A correlation between the inputs of 0.78 was considered to ensure that these measures have the same maximum possible value of 0.5 bits, and a zero correlation was considered to represent the case of independent inputs. One purpose is to discuss the general links between requirements 1–4 and conditions M1–M3 from Section 3 and also the use of the transfer functions defined in (10) and (12).

First, we notice in Figure 2 that the modulatory and additive transfer functions produce very different surfaces. In Figure 2a,b, the surface for T_M rises more quickly to its maximum than the surface for T_A, and in Figure 2a sections parallel to the s_1 axis are similar for $s_2 \geq 2$, whereas the surface for T_A is symmetric about the line $s_1 = s_2$. Figures 2d,f,h,j and 3d,f,h,j for T_A show clear asymmetry about the line $s_1 = s_2$.

When the strength of the CF input, s_2, is very small we notice in Figures 2e and 3e that $I[Y; X_2|X_1]$ is close to zero. Figure 2c shows that $I[Y; X_1|X_2]$ rises quickly, then gradually, towards its maximum at 0.5 as the strength of the RF input, s_1, increases, as does the surface in Figure 3c although there the maximum value is higher at 1. Figures 2g and 3g show that $I[Y; X_1]$ rises towards a maximum value of 1; this rise is much steeper when the correlation is 0.78 than when it is zero. These observations provide support for condition M1 when the modulatory transfer function is used. Similar observations on the corresponding figures based on the use of the additive transfer function show that condition M1 is satisfied in this case also.

Figures 2e,g and 3e,g show, when s_1 is close to zero, that $I[Y; X_2|X_1]$ and $I[Y; X_1]$ are both close to zero, thus supporting condition M2 when the modulatory transfer function is used. This is not the case when the additive transfer function is employed, as can be seen from Figures 2f,h and 3f,h. It is important to note that these figures do not all use the same scales for the heights of the surface. For example, the scales of Figures 2e and 3e are expanded because $I[Y; X2|X1]$ is always small when the transfer function is modulatory.

Also, when the strength of the RF input is weak (say $s_1 = 1$), we notice in Figures 2c,g and 3c,g that both $I[Y; X_1]$ and $I[Y; X_1|X_2]$ are larger for moderate CF strengths (say $s_2 = 5$) than when the the strength of the CF input is extremely weak ($s_2 = 0.05$, say), with this effect being stronger when the correlation between inputs is 0.78. This provides support for condition M3 when the modulatory transfer function is used. Inspection of the corresponding plots based on the additive transfer function show this effect only for $I[Y; X_1]$ in Figure 2h, and so condition M3 does not hold for the additive function.

(**a**) Modulatory, $I[Y; X_1; X_2]$

(**b**) Additive, $I[Y; X_1; X_2]$

(**c**) Modulatory, $I[Y; X_1 | X_2]$

(**d**) Additive, $I[Y; X_1 | X_2]$

(**e**) Modulatory, $I[Y; X_2 | X_1]$

(**f**) Additive, $I[Y; X_2 | X_1]$

(**g**) Modulatory, $I[Y; X_1]$

(**h**) Additive, $I[Y; X_1]$

(**i**) Modulatory, $I[Y; X_2]$

(**j**) Additive, $I[Y; X_2]$

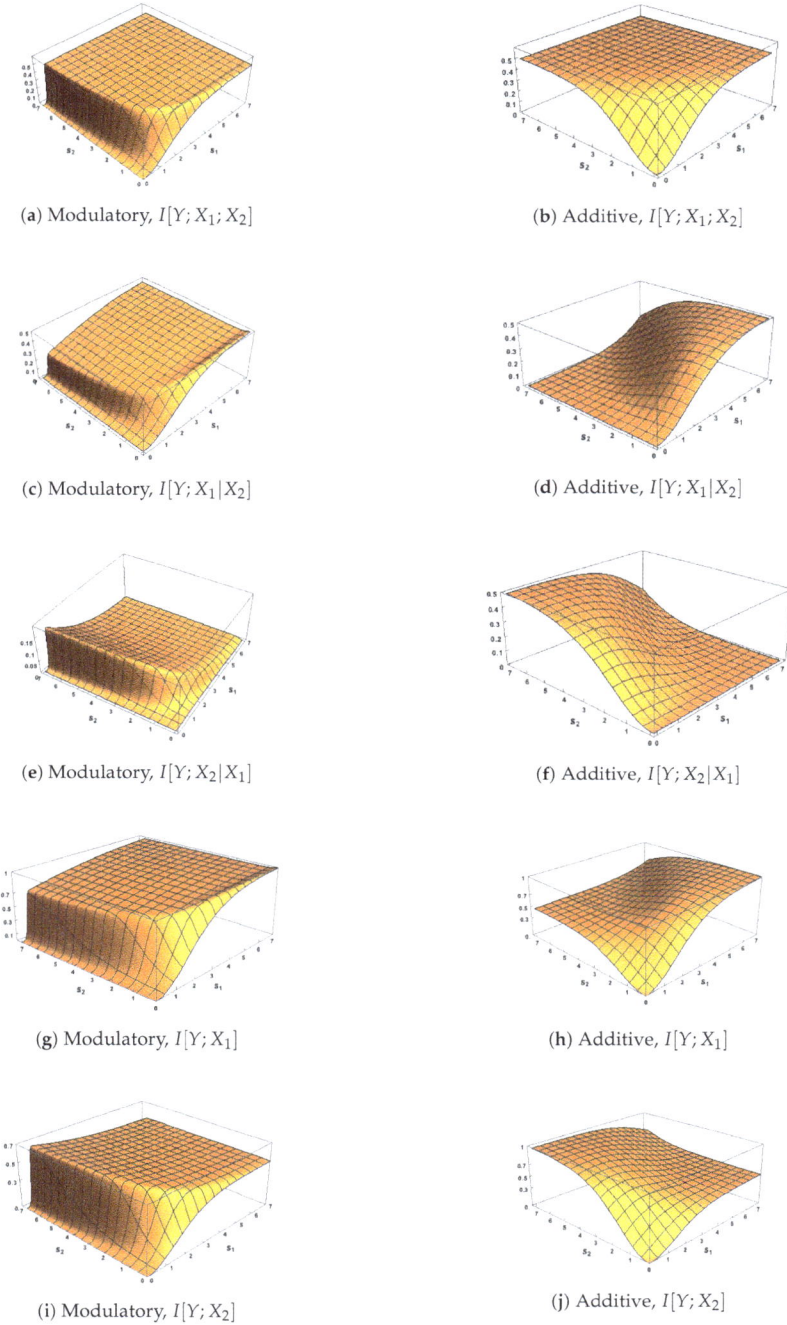

Figure 2. Classical Shannon measures (in bits), based on additive and modulatory transfer functions, and a correlation between inputs of 0.78.

In Figure 3a,b, the surfaces of the co-information $I[Y; X_1; X_2]$ are negative, as expected from (8), since the correlation between X_1 and X_2 is zero and so their mutual information is zero.

Finally we focus discussion on the phenomenon of particular relevance to the subject of this paper by considering the surface plots of $I(Y; X_2|X_1)$. In Figure 2e, an interesting pattern emerges. There is a steep rise for small values of s_1 and for all values of $s_2 \geq 2$, and then the surface quickly dies away. This pattern is repeated in Figure 3e. This suggests that X_2 is affecting the information shared between Y and X_1, indicating that modulation of some form might be taking place.

It could be argued, however, that X_2 is part of the output semantics in the sense that the output contains information specifically about X_2 itself. Since $I[Y; X_2|X_1]$ is clearly positive for these values of s_1, s_2, it is impossible to know whether or not this is the case based on this classical Shannon measure. It was shown in [8], that $I[Y; X_2|X_1]$ could be decomposed into two terms: the unique information that X_2 conveys about Y as well as synergistic information that is not available from X_2 alone, but rather gives the information that X_1 and X_2, acting jointly, have about the output Y. We now apply information decompositions in order to resolve these different interpretations. For discussion of some limitations of classical Shannon measures and the need for new measures of information, see [20].

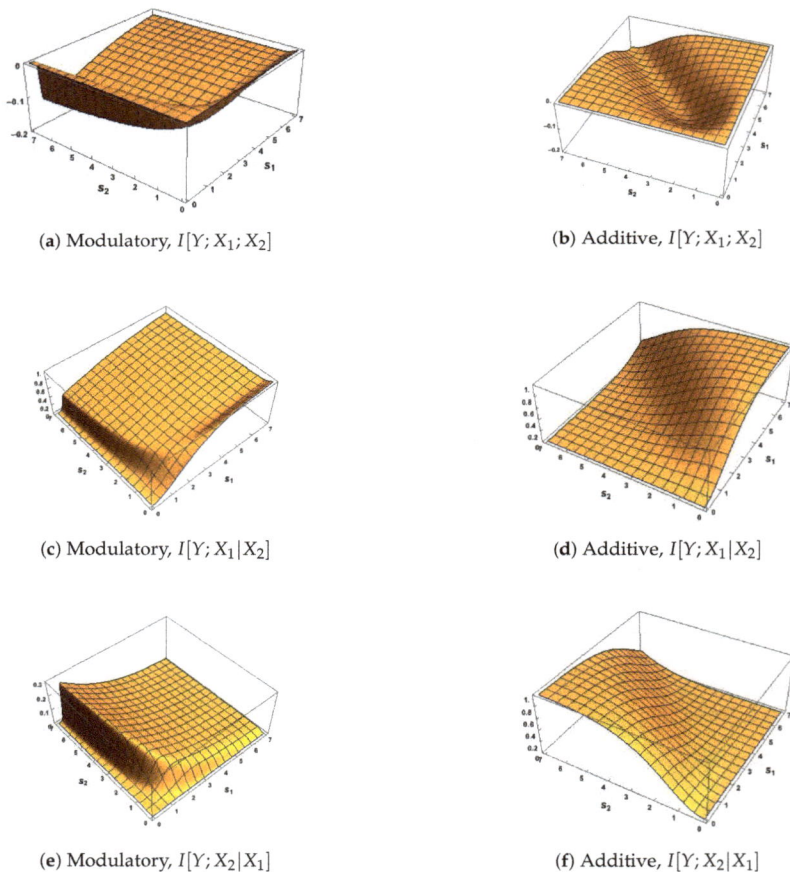

(a) Modulatory, $I[Y; X_1; X_2]$

(b) Additive, $I[Y; X_1; X_2]$

(c) Modulatory, $I[Y; X_1|X_2]$

(d) Additive, $I[Y; X_1|X_2]$

(e) Modulatory, $I[Y; X_2|X_1]$

(f) Additive, $I[Y; X_2|X_1]$

Figure 3. *Cont.*

(**g**) Modulatory, $I[Y; X_1]$

(**h**) Additive, $I[Y; X_1]$

(**i**) Modulatory, $I[Y; X_2]$

(**j**) Additive, $I[Y; X_2]$

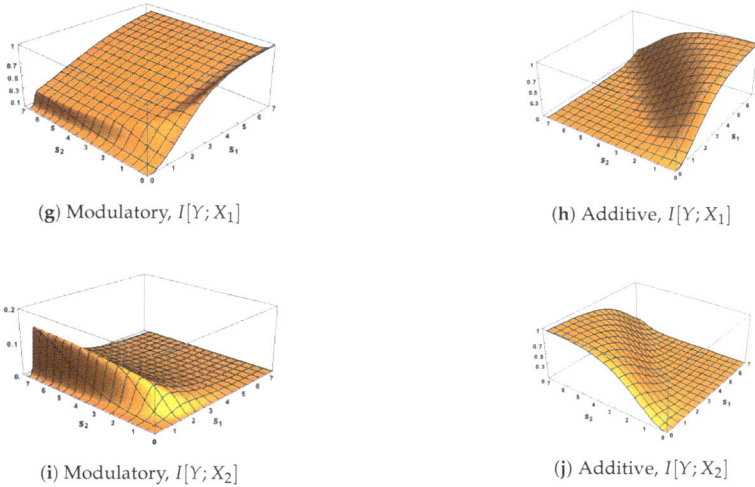

Figure 3. Classical Shannon measures (in bits), based on additive and modulatory transfer functions, and a zero correlation between inputs.

4. Information Decompositions

Williams and Beer [8] introduce a framework called the Partial Information Decomposition (PID) which decomposes mutual information between a target and a set of multiple predictor variables into a series of terms reflecting information which is shared, unique or synergistically available within and between subsets of predictors. Here we focus on the case of two input predictor variables, denoted X_1, X_2, and an output target Y. The information decomposition can be expressed as

$$I[Y; (X_1, X_2)] = I_{unq}[Y; X_1|X_2] + I_{unq}[Y; X_2|X_1] + I_{shdS+M}[Y; (X_1, X_2)] + I_{syn}[Y; (X_1, X_2)]$$

and it is the basis of both the information decompositions described in Sections 4.1 and 4.2. Adapting the notation of [21] we express our joint input mutual information in four terms as follows:

$\mathrm{Unq}X_1 \equiv I_{unq}[Y; X_1|X_2]$ denotes the unique information that X_1 conveys about Y;

$\mathrm{Unq}X_2 \equiv I_{unq}[Y; X_2|X_1]$ is the unique information that X_2 conveys about Y;

$\mathrm{Shar}_{\mathrm{S+M}} \equiv I_{shdS+M}[Y; (X_1, X_2)]$ gives the common (or redundant or shared) information that both X_1 and X_2 have about Y;

$\mathrm{Syn} \equiv I_{syn}[Y; (X_1, X_2)]$ is the synergy or information that the joint variable (X_1, X_2) has about Y that cannot be obtained by observing X_1 and X_2 separately.

It is possible to make deductions about a PID by using the following four equations which give a link between the components of a PID and certain classical Shannon measures of mutual information. The following are from Equations (4) and (5) in [21], with amended notation; see also [8].

$$I[Y; X_1] = \mathrm{Unq}X_1 + \mathrm{Shar}_{\mathrm{S+M}}, \tag{18}$$

$$I[Y; X_2] = \mathrm{Unq}X_2 + \mathrm{Shar}_{\mathrm{S+M}}, \tag{19}$$

$$I[Y; X_1|X_2] = \mathrm{Unq}X_1 + \mathrm{Syn}, \tag{20}$$

$$I[Y; X_2|X_1] = \mathrm{Unq}X_2 + \mathrm{Syn}. \tag{21}$$

We will refer to these results in Section 5 and use them in Section 6.

We consider here two different information decompositions. Although there are clear conceptual differences between the two, where they agree we can have some confidence we are accurately decomposing information as we would like. Where they disagree, we hope this may shed light on particular properties of the modulatory systems we study here, and also provide interesting comparisons of the two approaches.

It has been noted [22] that there are two different ways shared information can emerge. *Source* shared information refers to shared information that arises simply because the two inputs are correlated. For example, if $Y = X_1$ but X_1 and X_2 are correlated then there will be some $I(Y; X_2)$ and some redundancy $I_{shdS+M}[Y; (X_1 X_2)]$, even though X_2 plays no role in the computation implemented by the local processor. However, redundancy can also occur in systems where the inputs are statistically independent—in this case, it is referred to as *mechanistic* shared information, since it arises as a property of the function of the local processor. We denote I_{shdS+M} as the standard PID measure of shared information which quantifies both of these types together. However, both decompositions we consider provide a way to separately quantify these two types of shared information, which we denote by I_{shdS} and I_{shdM} for source and mechanistic respectively.

4.1. The Ibroja PID

In the Ibroja PID [9,10], the shared information component is based on an assumption that the information shared between two predictors about a target should not be affected by the marginal distribution of the two inputs (X_1, X_2) when the output is ignored. Instead, the shared information is a function only of the individual input-output marginal distributions of (Y, X_1) and (Y, X_2). In other words, the information about the output which is shared between the two inputs is independent of the correlation between the two inputs. In [9], this is motivated with an operational definition of unique information based on decision theory. It is claimed that unique information in input X_1 should correspond to the existence of a decision problem where two agents must try to guess the value of the output Y in which an agent acting optimally on evidence from X_1 can do systematically better (higher expected utility) than an agent acting optimally based on evidence from X_2; see also Appendix B2 in [21].

Following notation in [9], we consider a given joint distribution p for (Y, X_1, X_2), we let Δ be the set of all joint distributions of Y, X_1 and X_2, and define

$$\Delta_p = \{q \in \Delta : q(y, x_1) = p(y, x_1) \text{ and } q(y, x_2) = p(y, x_2), \text{ for all } (y, x_1, x_2) \in B^3\} \qquad (22)$$

as the set of all joint distributions which have the same (Y, X_1) and (Y, X_2) marginal distributions as p.

In Lemma 4 in [9] five equivalent optimisation problems are defined involving various information components. In this work we chose to minimise the total mutual information $I[Y; (X_1, X_2)]$ in order to find the optimal distribution q, denoted by $\hat{q}$. For the description of EID in Section 4.2, we note that this is equivalent to finding the distribution in Δ_p which maximizes the co-information $I[Y; X_1; X_2]$. This optimal distribution $\hat{q}$ is then used to calculate the four partial information measures:

$$\text{UnqX}_1 = I_{\hat{q}}[Y; X_1 | X_2], \qquad (23)$$
$$\text{UnqX}_2 = I_{\hat{q}}[Y; X_2 | X_1], \qquad (24)$$
$$\text{Shar}_{S+M} = I_{\hat{q}}[Y; X_1; X_2], \qquad (25)$$
$$\text{Syn} = I_p[Y; (X_1, X_2)] - I_{\hat{q}}[Y; (X_1, X_2)], \qquad (26)$$

and the information quantities, except $I_p[Y; (X_1, X_2)]$, are calculated with respect to the optimal distribution $\hat{q}$.

Using equations (7) & (8) from [23], the shared information can be split into non-negative *source* and *mechanistic* components that are defined as follows (in amended notation).

$$I_{shdS}[Y;(X_1,X_2)] = \max\{\min(I_{shdS+M}[Y;(X_1,X_2)], I_{shdS+M}[X_1;(X_2,Y)]),$$
$$\min(I_{shdS+M}[Y;(X_1,X_2)], I_{shdS+M}[X_2;(X_1,Y)])\}$$
$$I_{shdM}[Y;(X_1,X_2)] = I_{shdS+M}[Y;(X_1,X_2)] - I_{shdS}[Y;(X_1,X_2)]$$

A particular advantage of the Ibroja approach is that it results in a decomposition consisting of non-negative terms. A possibly counter-intuitive feature is that in our two input, one output local processor context, one might expect that $I_{shdS+M}[Y;(X_1,X_2)]$ should change depending on the marginal distribution of the inputs, (X_1,X_2), in that source shared information should increase as the correlation between the inputs increases (assuming the individual input-output marginals are fixed). In the systems defined in Section 2, however, the marginal distributions of (Y,X_1) and (Y,X_2) do depend on the correlation between the inputs, and so the Ibroja PID does change as this correlation changes.

4.2. The EID Using I_{ccs}

An alternative measure of shared information was recently proposed in [12]. Since at a local or pointwise level [24–28] (i.e., the terms inside the expectation), information is equal to change in surprisal, I_{ccs} seeks to measure shared information as the change in surprisal that is common to the input variables (hence CCS, Common Change in Surprisal). For two inputs, I_{ccs} is defined as:

$$I_{ccs}[Y;(X_1,X_2)] = \sum_{y,x_1,x_2} p(y,x_1,x_2)h_y^{com}(x_1,x_2)$$

$$h_y^{com}(x_1,x_2) = \begin{cases} i_{\tilde{q}}(y;x_1;x_2) & \text{if } \operatorname{sgn} i_{\tilde{q}}(y;x_1;x_2) = \operatorname{sgn} i_{\tilde{q}}(y;x_1) = \operatorname{sgn} i_{\tilde{q}}(y;x_2) = \operatorname{sgn} i_{\tilde{q}}(y;x_1,x_2) \\ 0 & \text{otherwise} \end{cases}$$

$$i_{\tilde{q}}(y;x_1;x_2) = i_{\tilde{q}}(y;x_1) + i_{\tilde{q}}(y;x_2) - i_{\tilde{q}}(y;x_1,x_2)$$

$$\tilde{q} = \arg\max_{q \in \Delta_p^2} \sum_{y,x_1,x_2} -q(y,x_1,x_2)\log q(y,x_1,x_2)$$

$$\Delta_p^2 = \left\{ q \in \Delta : \begin{array}{l} q(y,x_1) = p(y,x_1), q(y,x_2) = p(y,x_2) \\ q(x_1,x_2) = p(x_1,x_2), \text{for all } (y,x_1,x_2) \in B^3 \end{array} \right\}$$

where lower case symbols indicate the local or pointwise values of the corresponding information measures, i.e., $I_{\tilde{q}}(Y;X_1) = \sum_{y,x_1} p(y,x_1)i_{\tilde{q}}(y,x_1)$. The sign conditions ensure that only terms corresponding to genuine shared information are included; terms not meeting the sign equivalence represent either synergistic or ambiguous effects [12].

This approach has two fundamental conceptual differences from the Ibroja PID. The first is that in [12] a game theoretic operational definition of unique information is introduced. This is very similar to the decision theoretic argument in [9] but extends the considered situations to include games where the utility function is asymmetric or the game is zero-sum. Both of these extensions induce a dependency on the marginal distribution of (X_1,X_2). A specific example system is provided in [12] as well as a specific game which demonstrates unique information even when there is none available from the decision theoretic perspective.

The second conceptual difference is the way in which shared information is actually measured, within the constraints imposed by the respective operational definitions. In the Ibroja PID, shared information is measured as the maximum co-information over the optimization space Δ_p. I_{ccs} also relies on co-information, but breaks down the pointwise contributions and includes only those terms that unambiguously correspond to redundant information between the inputs about the output. This is important because co-information conflates redundant and synergistic effects [8,12] so cannot itself be expected to fully separate them. I_{ccs} is calculated using the distribution with maximum entropy subject to the game theoretic operational constraints (equality of all pairwise marginals). However, note that maximizing co-information subject to the extended game theoretic constraints is equivalent to maximizing entropy.

A decomposition of mutual information can be obtained using I_{ccs} following the partial information decomposition framework [8].

$$
\begin{aligned}
\text{Unq}X_1 &= I[Y; X_1] - I_{ccs}[Y; (X_1, X_2)], \\
\text{Unq}X_2 &= I[Y; X_2] - I_{ccs}[Y; (X_1, X_2)], \\
\text{Shar}_{S+M} &= I_{ccs}[Y; (X_1, X_2)], \\
\text{Syn} &= I[Y; (X_1, X_2)] - I[Y; X_1] - I[Y; X_2] + I_{ccs}[Y; (X_1, X_2)],
\end{aligned}
$$

The inclusion of $p(x_1, x_2)$ in the constraints for $\tilde{q}$ means that the measured shared and unique information is not invariant to the predictor-predictor marginal dependence. With I_{ccs} this affects the decomposition in an intuitive way: negative or no correlation between predictors results in more unique information, while when correlation between the predictors increases, shared information increases (driven by increased source shared information) and unique information decreases; see Figure 7 in [12]. However, the PID computed with I_{ccs} is not non-negative. In particular, the unique information terms can take negative values, which can be challenging to interpret.

In [13], it was recently suggested that the PID formalism could be applied to decompose multivariate entropy directly. The concepts of redundancy and synergy can apply just as naturally to entropy, resulting in a Partial Entropy Decomposition (PED) which can separate a bivariate entropy into four terms representing shared uncertainty, unique uncertainty in each variable, and synergistic uncertainty which arises only from the system as a whole. This approach shows that mutual information is actually the difference between redundant and synergistic entropy:

$$
I[Y; X] = H_{\text{shd}}[(Y, X)] - H_{\text{syn}}[(Y, X)]
$$

and this relationship holds for any measure of shared entropy which satisfies the PED axioms. This shows that mutual information does not only quantify common, shared or overlapping entropy, but is also affected by synergistic effects between the variables. At the global level since joint entropy is maximised when the two variables are independent (alternatively mutual information is non-negative), this implies that $H_{\text{shd}}[(Y, X)] \geq H_{\text{syn}}[(Y, X)]$. Mutual information is the expectation over local information terms that can themselves be positive, representing an decrease in the surprisal of event y when event x is observed, or negative, representing an increase in the surprisal of y when x is observed. Negative local information terms, which have been called "misinformation" [26], arise for symbols where $h(x, y) > h(x) + h(y)$; that is, those symbols provide a synergistic contribution to the joint entropy expectation sum. The existence of such locally synergistic entropy terms suggest that synergistic entropy is a reasonable thing to quantify within the PED framework. A shared entropy measure (H_{cs}) can be defined in a manner consistent with I_{ccs} as [13]:

$$
H_{cs}(Y, X_1, X_2) = \sum_{y, x_1, x_2} \tilde{q}(y, x_1, x_2) h_{cs}(y, x_1, x_2)
$$

$$
h_{cs}(y, x_1, x_2) = \max\left[-i_{\tilde{q}}(y, x_1, x_2), 0\right]
$$

This entropy perspective can give some insight into the meaning of negative terms within the I_{ccs} PID. With I_{ccs}, shared information is calculated as shared entropy with the target that is common to both inputs (positive local co-information terms in I_{ccs}) minus synergistic entropy with the target that is common to both inputs (negative local co-information terms in I_{ccs}). Negative unique information terms can therefore arise when there is more unique synergistic entropy between a target and the predictor than there is unique shared entropy between the target and the predictor. Unique synergistic entropy means there is synergistic entropy between say X_2 and Y which is not shared with X_1. This can arise for example, whenever the calculation of $I[Y;X_2]$ includes negative local terms in the expectation (for some values of y, x_2), but $I[Y;X_1]$ does not. In such cases, these negative local contributions to the mutual information must be unique; they do not appear in $I[Y;X_1]$ since that calculation has no negative terms.

The PED of our three variables also provides a way to separate the I_{ccs} shared information into mechanistic and source shared terms. The source shared information can be obtained from the three way partial entropy term, $H_{shd}[(Y,X_1,X_2)]$. This term represents the entropy that is common to all three variables, therefore it is included in the calculation of both $I[Y;X_1]$ and $I[Y;X_2]$ and so is shared information. However, it is possible that this quantity also includes some mechanistic shared information. This can only happen if $H_{cs}[(Y,X_1,X_2)] > H_{cs}[(X_1,X_2)]$—i.e., the two inputs share more entropy in the context of the full system then they do when ignoring (by marginalising away) the output. This corresponds to a negative partial entropy term $H_{shd}[(X_1,X_2)]$. Therefore we calculate source and mechanistic shared information, from Equation (32) in [13], as:

$$I_{shdS}[Y;(X_1,X_2)] = \min\left(H_{cs}[(Y,X_1,X_2)], H_{cs}[(X_1,X_2)]\right),$$
$$I_{shdM}[Y;(X_1,X_2)] = I_{ccs}[Y;(X_1,X_2)] - I_{shdS}[Y;(X_1,X_2)]$$

The first expression quantifies the source shared entropy: it is the three-way shared entropy with any mechanistic shared entropy removed. Since I_{ccs} quantifies source and mechanistic shared information together, we obtain the mechanistic shared information by subtracting off the calculated source shared information. Source shared information defined in this way is always positive, but mechanistic shared information can be negative. Negative mechanistic shared information can arise when, for example, both $I[Y;X_1]$ and $I[Y;X_2]$ contain negative local information terms, and those local information terms are common, reflected in a negative local co-information term. Alternatively, there is synergistic entropy between Y and X_1 that overlaps with synergistic entropy between Y and X_2. Synergistic entropy between the target and a predictor is by definition a mechanistic effect, since it is uncertainty that does not arise in the predictor alone, but is only obtained when the output (i.e., the mechanism) is considered. Please see [13] for further details. Since this approach relies on terms from the partial entropy decomposition as well as the partial information decomposition using I_{ccs}, we refer to it here as an Entropic Information Decomposition (EID).

5. Information Decomposition (ID) Spectra

We now describe a simple visual display [29] in which all the transmitted mutual information components appear, together with the residual output entropy. These displays are referred to as "spectra" because different colours are used for different components. Here the spectra are shown as stacked bar charts, which facilitates presentation of many spectra in a single figure. These spectra convey a simple but important message when applied to the goal of distinguishing between modulatory and additive interactions, whether in real or artificial neural systems. The important message is that modulatory and additive forms of interaction can have similar or even identical effects under some conditions, but very different effects under others. Such plots can also be used to compare the information processing performed in a system under different parameter regimes. They can also be used to compare the kinds of information processing performed by individual subjects or groups of subjects when completing psychophysical tasks; see Section 8.

5.1. Definition and Illustrations

The first five components are the partial information measures considered in Section 4: unique informations, shared source and mechanistic information and synergy. To this is added the residual output entropy.

The residual output entropy is $H(Y)_{res} = H(Y|X_1, X_2)$, which appears in the following decomposition, from Equation (6) in [21],

$$H(Y) = I_{unq}[Y; X_1|X_2] + I_{unq}[Y; X_2|X_1] + I_{shdS+M}[Y; (X_1, X_2)] + I_{syn}[Y; (X_1, X_2)] + H(Y|X_1, X_2) \quad (27)$$

and here we also use the decomposition

$$I_{shdS+M}[Y; (X_1, X_2)] = I_{shdS}[Y; (X_1, X_2)] + I_{shdM}[Y; (X_1, X_2)].$$

In our discussion, we consider four different spectra as an illustrative test set. First, we take $s_1 = 10.0$ and $s_2 = 0.05$ to represent the situation where the RF input is strong and the CF input is extremely weak. Secondly, in the case where $s_1 = 0.05, s_2 = 10.0$, the RF input is extremely weak while the CF input is strong. Thirdly, when $s_1 = 1.0, s_2 = 0.05$ the RF input is weak and the CF input is extremely weak. Finally, when $s_1 = 1.0, s_2 = 5.0$, the RF input is weak and the CF input is of moderate strength.

5.2. Ibroja Spectra

It is useful to bear in mind when interpreting these spectra that the information components are not independent quantities since they satisfy the constraints (18)–(21) and (27); so these non-negative components are negatively correlated. Figure 4a,b show PID decompositions when the two inputs have a correlation of either 0.78 or 0. In both cases modulatory and additive transfer functions lead to very similar decompositions when the RF input is strong (charts M1 and A1), or of moderate strength (charts M3 and A3), and the CF input is very weak, since there is little or no difference between charts M1 and A1 and between M3 and A3. Thus, when context is absent or very weak the modulatory transfer function becomes effectively equivalent to an additive function.

When the RF input is either very weak (charts M2 and A2) or less weak but with strong CF input (charts M4 and A4), modulatory and additive transfer functions have very different effects. Consider the case where the RF input is very weak and the CF is strong. The modulatory function transmits little or no input information (chart M2), implying that RF input is necessary to information transmission. In contrast, the additive transfer function in that case transmits information unique to the CF input with shared information if the two inputs are correlated (chart A2). Cases where RF input is present but weak show the modulatory effect of the CF input. Consider transmission in the case of weak RF input with extremely weak CF input (charts M3 and A3). The output residuals are then high, showing that little information is transmitted. What is transmitted is a combination of shared information and information unique to the RF input. If the RF input is weak but the CF input is strong, however, then the modulatory function transmits more unique information about the RF than when the CF input is weak, together with some synergy, some mechanistic shared, and some source shared if the inputs are correlated (chart M4). In contrast, the additive transfer function transmits no information unique to the RF but only information unique to the CF and shared information if the inputs are correlated (chart A4).

(a) PID spectra, correlation of 0.78

(b) PID spectra, correlation of zero

(c) EID spectra, correlation of 0.78

(d) EID spectra, correlation of zero

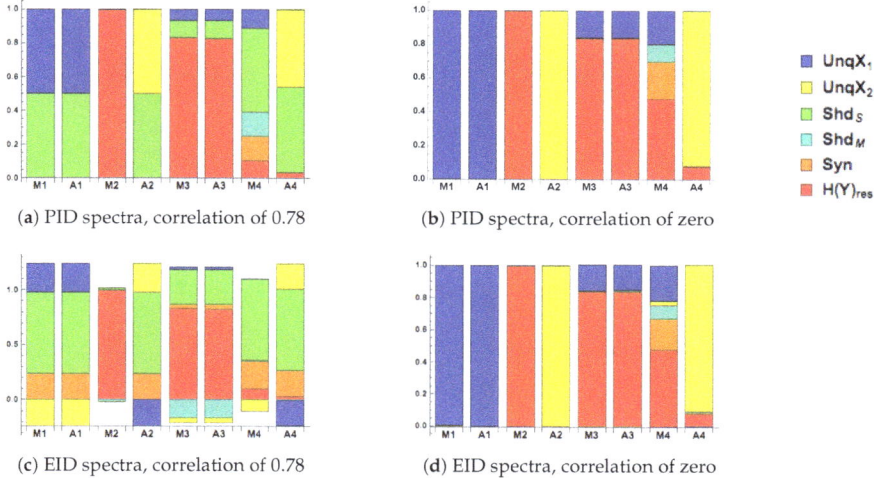

Figure 4. Partial information decomposition (PID) and entropic information decomposition (EID) spectra (in bits), based on additive (A) and modulatory (M) transfer functions for four combinations of signal strengths: 1. ($s_1 = 10.0, s_2 = 0.05$), 2. ($s_1 = 0.05, s_2 = 10.0$), 3. ($s_1 = 1.0, s_2 = 0.05$), 4. ($s_1 = 1.0, s_2 = 5.0$), and two values of the correlation between inputs: 0.78 and zero.

5.3. EID Spectra

The EID spectra can have negative partial information measures, and so when interpreting them it is useful to bear in mind the constraints (18)–(21). Therefore, for example, if the UnqX$_1$ component is negative then, since the classical Shannon measures are fixed, it would follow from (18) and (20) that the components Shar$_{S+M}$ and Syn would be larger than if the UnqX$_1$ component were equal to zero; of course the component Shar$_{S+M}$ is split further into *Source* and *Mechanistic* terms, as discussed in Section 4. In particular, if it were the case that $I[Y; X_1|X_2]$ were equal to zero then the synergy component would be positive and equal in magnitude to the UnqX$_1$ component. Therefore, when a negative component is present this is likely to make the relative magnitudes of the partial information components appear different than in the corresponding Ibroja spectra, even though the same essential message might be being expressed.

Consider Figure 4c. We note that the use of the modulatory and the additive transfer functions leads to very similar spectra in charts M1 and A1, and M3 and A3. In charts M1 and A1, we see that when the RF input is strong the residual output is zero and the information is transmitted mainly via the source-shared component, but with some synergy and some unique information about the RF, as well as some unique misinformation from the CF. Charts M2 and A2 reveal a marked difference in the spectra due to the transfer functions. When the modulatory transfer function is employed and the RF input is extremely weak then almost no information is transmitted. In contrast, the use of the additive transfer function leads to all the information being transmitted, mostly in the form of source shared information, with some synergy, some unique information about the CF and some misinformation from the RF. In charts M3 and A3, the output residual is very high and so very little information is transmitted when the RF input is weak and the CF input is extremely weak, and what is transmitted is a combination of positive source shared information and negative mechanistic shared information. Chart M4, where the CF input is moderate but the RF input is weak, indicates that more information about the RF is transmitted than was the case in chart M3, since the output residual is smaller. This information is transmitted mainly via source shared information and synergy, with some unique misinformation from the CF.

We now briefly consider Figure 4d. Charts M1 and A1 show that all the information is transmitted in a form unique to the RF. We see a striking difference between charts M2 and A2, with no information being transmitted in M1 and all the information unique to the CF being transmitted in A2. Charts M3 and A3 appear to be identical, with some information unique to the RF being transmitted and a high output residual. Chart M4 shows that about one-half of the information is transmitted, mainly due to that unique to the RF and synergy but also with some mechanistic shared and a little unique to the CF. Much more information is transmitted in A4, predominantly in a form unique to the CF. A pleasing feature of Figure 4d is that the source shared information component is zero in all the charts, while the mechanistic shared component in chart M4 is positive; this is exactly what would be expected when the inputs are uncorrelated, and here there are no negative mechanistic shared components unlike in Figure 4c where the inputs are strongly correlated.

5.4. Contextual Modulation and Information Decompositions

In Section 3, the conditions M1–M3 express the notion of contextual modulation. Here, we translate these conditions using (18)–(21) into corresponding expressions of contextual modulation for ID measures, denoted by S1–S3 for non-negative decompositions, with amended conditions S1′–S2′ for the EID when it has negative components.

S1: If the *RF* signal is strong enough, and the CF input is extremely weak, then both UnqX2 and Syn are close to zero, UnqX1 can have its maximum value, and the sum of UnqX1 and Shar$_{S+M}$ can equal the total output entropy. This shows that the RF input is sufficient, thus allowing the information in the *RF* to be transmitted, and that the CF input is not necessary.

S2: All five partial information components are close to zero when the RF input is extremely weak no matter how strong the CF input. This shows that the RF input is necessary for information to be transmitted, and that the CF input is not sufficient to transmit the information in the RF input.

S3: When $s_1 < s_2$ and when the RF input is weak, then the sum of UnqX1 and Syn is larger when the CF input is moderate than it is when the CF input is weak. The same is true of the sum of UnqX1 and Shar$_{S+M}$. Thus the CF input modulates the transmission of information about the RF input.

The following conditions provide amendments to S1-S2 when the EID has negative components:

S1′: When UnqX2 < 0, UnqX2 and Syn are approximately of the same magnitude, the sum of UnqX1 and Syn can have its maximum value, and the sum of UnqX1 and Shar$_{S+M}$ can equal the total output entropy.

S2′: If at least one component is negative, then we can set the left-hand sides of (18)–(21) to zero and use the rule that the sum of the magnitudes of the negative components is approximately equal to the sum of the magnitudes of the positive components. If in any of (18)–(21) there is no negative term then all terms on the right-hand side are close to zero.

We now discuss the spectra in relation to the these conditions. First we discuss the PID charts in Figure 4a. In charts M1 and A1, we see that Syn and UnqX2 are apparently equal to zero and that the sum of UnqX1 and Shar$_{S+M}$ is equal to 1, the value of the total output entropy; UnqX1 is equal to 0.5 which is presumably the maximum value it can take. Therefore Condition S1 is satisfied for the modulatory and the additive transfer function. For charts M2 and A2, we see in M2 that all five of the components are apparently zero, and hence condition S2 holds for the modulatory transfer function, but this is not the case with the additive transfer function in A2 since the values of UnqX2 and Shar$_S$ are appreciable. Inspection of charts M3 and M4 shows that the sum of Syn and UnqX1 and the sum of UnqX1 and Shar$_{S+M}$ are larger in M4 than in M3, thus supporting condition S3. In charts A3 and A4 we see the same for the sum of UnqX1 and Shar$_{S+M}$, but the opposite for the sum of Syn and UnqX1, and so S3 is not fully supported in the additive case.

We now consider the EID charts in Figure 4c. In charts M1 and A1, UnqX2 is negative and UnqX2 and Syn have approximately the same magnitude. Therefore, the sum of UnqX1 and Shar$_{S+M}$

is equal to 1, the value of the total output entropy. Also, UnqX1 is just larger than 0.2, presumably the largest value it can take. Therefore, the conditions of S1′ are satisfied in both the modulatory and additive cases. For charts M2 and A2, we see in M2 that the residual output entropy is almost equal to 1, that UnqX1, UnqX2 and Syn are apparently zero and that the little negative mechanistic shared information is counterbalanced by a similar amount of positive source shared information, thus supporting condition S2′, since all the right-hand sides in (18)–(21) are close to zero. This condition is, however, not supported in the additive case since the values of UnqX2, Shar$_S$, Syn and UnqX1 (negative) are all appreciable. Considering charts M3 and M4, we notice that the sum of Syn and UnqX1 and also the sum of UnqX1 and Shar$_{S+M}$ are larger in M4 than in M3, thus supporting condition S3. In charts A3 and A4 we see the same for the sum of UnqX1 and Shar$_{S+M}$, but the opposite for the sum of Syn and UnqX1, and so S3 is not fully supported in the additive case. Hence, when the correlation between inputs is strong, we find that the conclusions for both PID and EID are the same with regard to the use of modulatory and additive transfer functions.

In Figure 4b,d, the respective PID and EID spectra are virtually identical, and so the same conclusions will hold for both decompositions. In charts M1 and A1, UnqX2 and Syn are apparently zero, the sum of UnqX1 and Shar$_{S+M}$ is equal to the total output entropy and this time UnqX1 is fully maximized. Therefore condition S1 is supported in both charts. In chart M2 the residual output entropy is close to 1 and so all five information components are close to zero, thus supporting condition S2. We notice that the sum of Syn and UnqX1 and also the sum of UnqX1 and Shar$_{S+M}$ are larger in M4 than in M3, thus supporting condition S3. In charts A3 and A4 we see that both these sums are smaller in A4 than in A3, and so S3 is not supported in the additive case.

5.5. Comparison of PID and EID

Close comparison of the EID and PID spectra sheds light on both the information processing properties of the form of modulation considered here, and on relations between PID and EID. Most importantly for the purposes of this paper both PID and EID show the distinctive properties of the modulatory interaction, in which the modulatory transfer function is employed. First, no information dependent on the inputs is transmitted when the RF input is very weak whatever the value of the CF input. This shows that the RF input is necessary for this transfer function to transmit information about the input and that the CF input is not sufficient. Second, information is transmitted about the RF input for all states of the CF input including those in which it is absent or very weak. This shows that the RF input is sufficient for this transfer function to transmit information about the input and that the CF input is not necessary. Third, when the RF input is strong no information dependent on the CF input is transmitted by the output, but when the RF input is present but weak then the output transmits less information dependent on the the RF input when context is very weak.

This shows the modulatory effect of the CF input. Fourth, modulatory interactions produce the same components as additive interactions when the CF input is very weak, but very different components when the CF input is stronger and the RF input is present but weak. This shows conditions that distinguish these two forms of interaction. In general, the two inputs have equivalent opportunities to effect the output for additive interactions, whereas the effects of the CF input are conditional upon the RF input for the modulatory interaction. Fifth, when the two inputs are uncorrelated there is little difference between the EID and PID decompositions other than the splitting of shared into source and mechanistic by EID.

The spectra displayed may also shed some light on the negative components of EID, which still await a clear and widely accepted interpretation. First, negative components are zero or tiny when the two inputs are uncorrelated. Second, synergy and source shared were never negative in the conditions studied. Third, negative unique components seem to be compensated for by positive synergistic components. Fourth, source shared is never negative and positive only when the two inputs are correlated. Whether these observations will aid interpretation of the negative components remains to be seen.

The spectra shown here are all for specific values of the two input strengths, so to see whether the observations listed in the two preceding paragraphs hold for other values of those strengths the following section presents surfaces showing each of the output components that depend on input as a function of the two input strengths.

6. Analysis of the Transfer Functions Using the Ibroja PID over a Wide Range of Input Strengths

The five Ibroja surfaces were constructed as a function of the RF and CF signal strengths, s_1 and s_2. In Figure 5, we notice the striking differences in the surfaces for each measure between the use of the modulatory and the additive transfer function. In Figure 5b,d, there is a clear asymmetry that mimics that shown in Figure 2d,f.

We notice, in particular, that it appears that $UnqX_1$ is zero when $s_2 > s_1$ while $UnqX_2$ is zero for $s_1 > s_2$. In Figure 5a, $UnqX_1$ rises towards its maximum as s_1 increases, and the rise is similar for $s_2 > 2$. For $s_1 > 2$ the shape of this plot matches that in Figure 2c. In Figure 5c, we note that $UnqX_2$ appears to be zero for all values of s_1 and s_2. In Figure 5f,h,j plots of $Shar_S$, $Shar_M$ and Synergy are symmetric about the line $s_1 = s_2$ when based on the additive transfer function, and the maximum values of $Shar_M$ and Synergy happen along the line $s_1 = s_2$, while $Shar_S$ flattens quite quickly onto a plateau for most values of s_1 and s_2. On the other hand, there is no symmetry in Figure 5e,g,i, where the surfaces of $Shar_M$ and Synergy rise and fall as s_1 increases and the pattern is similar for $s_2 > 2$, while the $Shar_S$ surface rises quickly onto a plateau. The plot of synergy in Figure 5g appears to match exactly the plot of $I[Y; X_2|X_1]$ in Figure 2e, as expected, since it appears from Figure 5c that $UnqX_2 = 0$.

In Figure 6, the surfaces for $UnqX_1$ and $UnqX_2$ are similar to the corresponding plots in Figure 5. In particular, we note that again it appears from Figure 6c that $UnqX_2 = 0$. Again, Figure 6g appears to match the corresponding plot of $I[Y; X_2|X_1]$ in Figure 2e. In Figure 6e,f, the $Shar_S$ surface is zero for all values of s_1 and s_2; this is expected since the source shared information should be zero when the inputs are uncorrelated. By inspecting the surfaces in Figures 5e,f and 6e,f, we notice (as expected) that the source shared information is much larger when the inputs are strongly correlated than when they are uncorrelated. The plots of mechanistic shared information in Figures 5g,h and 6g,h indicate that the presence of strong correlation does not have much effect. In Figure 6h,j, symmetry is again apparent, with the maximum values occurring along the line $s_1 = s_2$.

Of special interest is the finding that $UnqX_2$ appears to be zero. This suggests that X_2 can modify the transmission of information from the receptive field input X_1 to the output Y without transmitting any unique information about itself. This conclusion would be much stronger if it were possible to prove mathematically that $UnqX_2 = 0$, given the system defined in Sections 2 and 3. We now state some formal results which indicate that this is indeed the case. We also define a class of transfer functions, that includes our modulatory transfer function T_M, for which $UnqX_2 = 0$.

We saw also in the surfaces of $UnqX_1$ and $UnqX_2$, produced by the additive transfer function, that $UnqX_2$ appears to be zero when $s_1 > s_2$, and also that $UnqX_1$ appears to be zero when $s_1 < s_2$. We also state some mathematical results to confirm these impressions, as well as proving that when $s_1 = s_2$ both uniques are zero. Then, using (18)–(21), the exact Ibroja decomposition is derived. Proofs are given in the appendix. We now state the results.

(**a**) Modulatory, UnqX_1

(**b**) Additive, UnqX_1

(**c**) Modulatory, UnqX_2

(**d**) Additive, UnqX_2

(**e**) Modulatory, Shar$_S$

(**f**) Additive, Shar$_S$

(**g**) Modulatory, Shar$_M$

(**h**) Additive, Shar$_M$

(**i**) Modulatory, Synergy

(**j**) Additive, Synergy

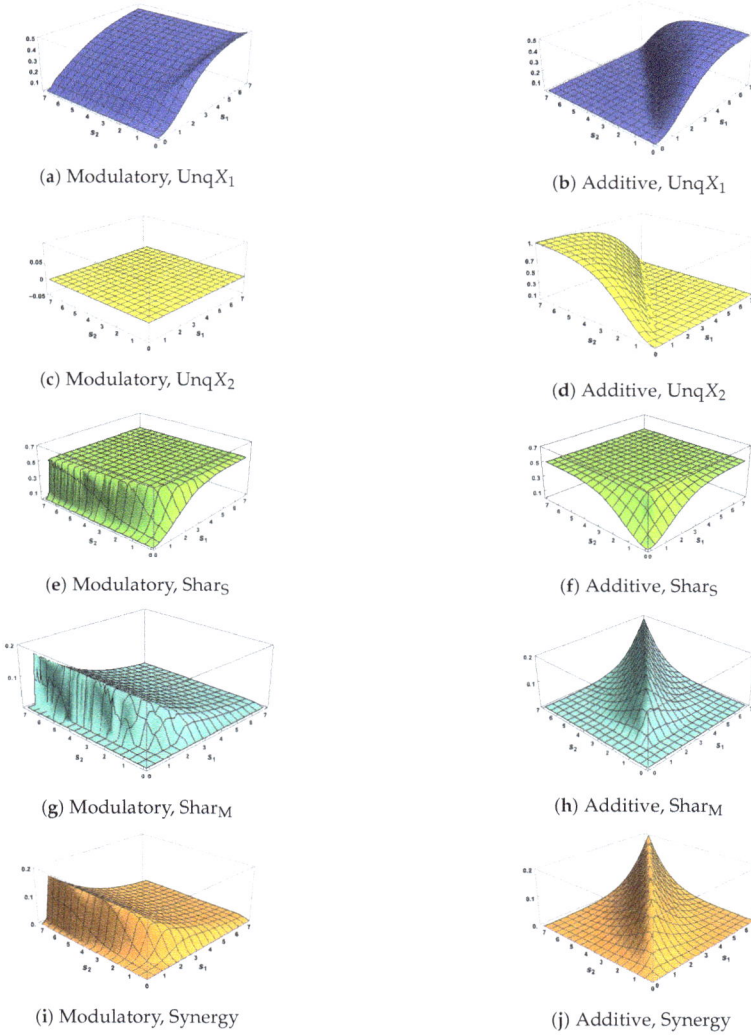

Figure 5. Ibroja surfaces, based on additive and modulatory transfer functions, and a correlation between inputs of 0.78.

(a) Modulatory, UnqX$_1$

(b) Additive, UnqX$_1$

(c) Modulatory, UnqX$_2$

(d) Additive, UnqX$_2$

(e) Modulatory, Shar$_S$

(f) Additive, Shar$_S$

(g) Modulatory, Shar$_M$

(h) Additive, Shar$_M$

(i) Modulatory, Synergy

(j) Additive, Synergy

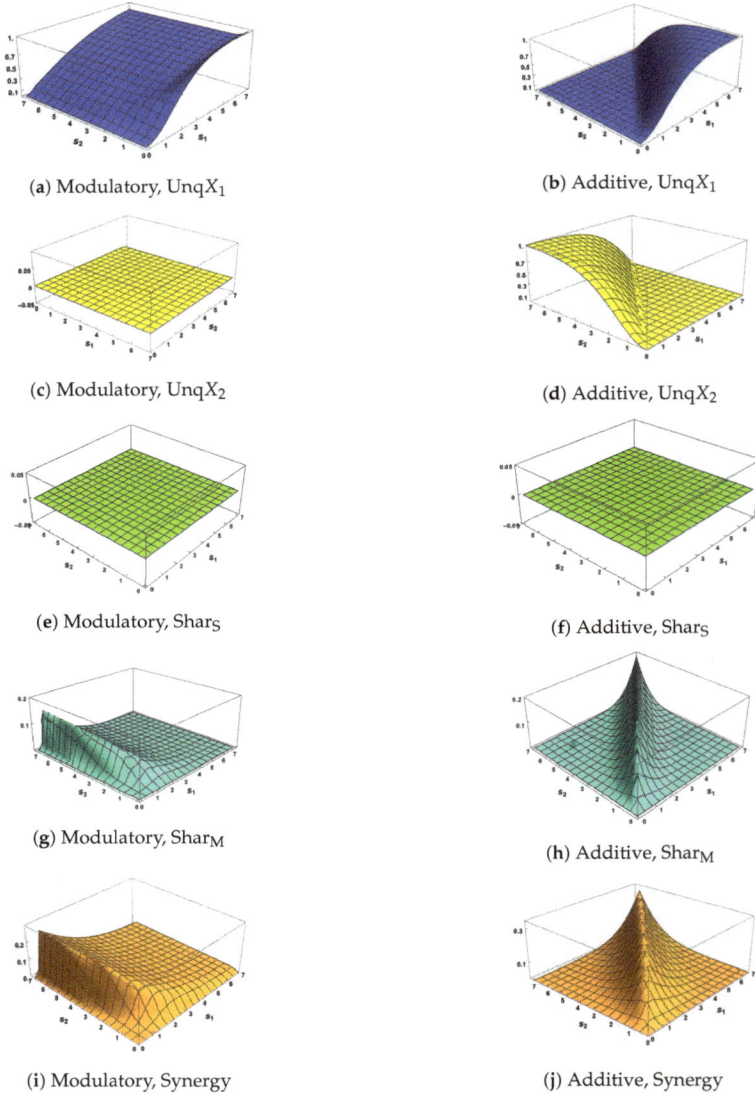

Figure 6. Ibroja surfaces, based on additive and modulatory transfer functions, and zero correlation between inputs.

Let F be a function of two real variables, x, y, which has the property that

$$F(-x,-y) = -F(x,y) \quad \text{and} \quad F(-x,y) = -F(x,-y), \text{ for } x > 0, y > 0. \tag{28}$$

We consider $F(r,c)$ as a transfer function, for integrated RF input r and integrated CF input c, and, as in Section 3, we pass the value of F through a logistic nonlinearity to obtain output conditional probabilities of the form, with $r = s_1 x_1$ and $c = s_2 x_2$,

$$Pr(Y = 1 | X_1 = x_1, X_2 = x_2) = 1/(1 + \exp[-F(s_1 x_1, s_2, x_2)]). \qquad (29)$$

We also assume that the joint p.m.f. for (X_1, X_2) has the form given in (1)–(3).

Theorem 2. *For the trivariate probability distribution defined in (1)–(3), (29) and a transfer function as defined in (28), suppose that $g \geq \frac{1}{2}$ and $h \geq \frac{1}{2}$ but g and h are not both equal to $\frac{1}{2}$, where g and h are defined by*

$$g = Pr(Y = 1 | X_1 = 1, X_2 = 1) \quad and \quad h = Pr(Y = 1 | X_1 = 1, X_2 = -1). \qquad (30)$$

Suppose also that $\lambda \neq 0, \mu \neq 0, s_1 > 0, s_2 > 0$. Then, for such a system, $UnqX_2 = 0$ in the Ibroja PID.

The conclusion of Theorem 2 also holds when the conditions on g, h are: $g \leq \frac{1}{2}, h \leq \frac{1}{2}$ but both g, h are not equal to $\frac{1}{2}$. The conclusion also holds when $g = \frac{1}{2}, h = \frac{1}{2}$, although in this case all of the information components are zero since the total mutual information $I[Y; (X_1, X_2)] = 0$, because Y is independent from (X_1, X_2).

We now state the results for the two transfer functions used in this study.

Corollary 1. *If the modulatory transfer function T_M is used in the system described in Theorem 2, and under the conditions stated there, then $UnqX_2 = 0$ in the Ibroja PID.*

Corollary 2. *If the additive transfer function T_A is used in the system described in Theorem 2, and under the conditions stated there, then $UnqX_2 = 0$ in the Ibroja PID when $s_1 \geq s_2$.*

It is shown by Theorem 2 that there is a general class of transfer functions which, when used in the system described in Sections 2 and 3, and which satisfy the conditions of the Theorem 2, have the property of not transmitting any unique information about the modulator. The modulatory transfer function used in this work is a member of this class. The additive transfer function T_A is also a member of this class but it does not satisfy the conditions required in Theorem 2 for all values of s_1 and s_2.

We now present a result regarding $UnqX_1$ and $UnqX_2$ when the additive transfer function is used in the system considered in Sections 2 and 3.

Theorem 3. *For the trivariate probability distribution defined in Sections 2 and 3, with the additive transfer function T_A, suppose that $\lambda \neq 0, \mu \neq 0, s_1 > 0, s_2 > 0$. Then, for such a system, $UnqX_1 = 0$ in the Ibroja PID when $s_1 \leq s_2$. When $s_1 = s_2$ then both $UnqX_1$ and $UnqX_2$ are zero in the Ibroja PID.*

Given the results of Theorems 2 and 3, and since the Ibroja PID is a non-negative decomposition, we can now state the following exact results.

Theorem 4. *For the trivariate probability distribution defined in (1)–(4), suppose that $\lambda \neq 0, \mu \neq 0, s_1 > 0, s_2 > 0$. Then, with u_M, v_M, u_A, v_A defined in (15)–(16), we have*

(a)   When transfer function T_M is employed then

   (i) $Syn = I(Y; X_2 | X_1) = h(z_M) - 2\lambda h(u_M) - 2\mu h(v_M);$

   (ii) $Shar_{S+M} = I(Y; X_2) = 1 - h(w_M);$

   (iii) $UnqX_1 = I(Y; X_1 | X_2) - I(Y; X_2 | X_1) = h(w_M) - h(z_M), \quad and \quad UnqX_2 = 0.$

(b)   When the transfer function T_A is used and $s_1 = s_2$ then

    (i) $Syn = I(Y; X_2|X_1) = h(z_A) - 2\lambda h(u_A) - 2\mu;$

    (ii) $Shar_{S+M} = I(Y; X_1) = 1 - h(z_A);$

    (iii) $UnqX_1 = UnqX_2 = 0;$

(c)   When the transfer function T_A is used and $s_1 < s_2$ then

    (i) $Syn = I(Y; X_1|X_2) = h(w_A) - 2\lambda h(u_A) - 2\mu h(v_A);$

    (ii) $Shar_{S+M} = I(Y; X_1) = 1 - h(z_A);$

    (iii) $UnqX_2 = I(Y; X_2|X_1) - I(Y; X_1|X_2) = h(z_A) - h(w_A)$ *and* $UnqX_1 = 0.$

(d)   When the transfer function T_A is used and $s_1 > s_2$ then

    (i) $Syn = I(Y; X_2|X_1) = h(z_A) - 2\lambda h(u_A) - 2\mu h(v_A);$

    (ii) $Shar_{S+M} = I(Y; X_2) = 1 - h(w_A);$

    (iii) $UnqX_1 = I(Y; X_1|X_2) - I(Y; X_2|X_1) = h(w_A) - h(z_A),$ *and* $UnqX_2 = 0.$

For the trivariate binary system considered in Sections 2 and 3, these results show that the Ibroja PID is a minimum mutual information PID, as was found in [30,31] for the trivariate Gaussian system. Finally, we give the PID for any non-negative decomposition in the case where $\lambda = 0$ or $\mu = 0$, so that the correlation between inputs is -1 or $+1$, respectively.

Theorem 5. *Consider the probability distribution defined in (1)–(4). When the correlation between the inputs, X_1, X_2, is $+1$, we have that*

(a) $UnqX_1 = UnqX_2 = Syn = 0$, and $Shar_{S+M} = 1 - h(u)$.

when the correlation between the inputs, X_1, X_2, is -1, we have that

(b) $UnqX_1 = UnqX_2 = Syn = 0$, and $Shar_{S+M} = 1 - h(v)$,

where, from (15)–(16), $u = u_M$, $v = v_M$ when the transfer function T_M is employed and $u = u_A$, $v = v_A$ when the transfer function T_A is used.

7. Analysis of the Transfer Functions Using EID over a Wide Range of Input Strengths

As in the previous section, five EID surfaces were constructed as a function of the RF and CF signal strengths, s_1 and s_2, in the definition of the trivariate binary system. Many of the properties of the resulting surfaces are common with the Ibroja PID surfaces: the opposite asymmeteries of the unique information terms for the additive system (Figures 7b,d and 8b,d), the symmetry in s_1 and s_2 of the other terms for the additive transfer function, and the asymmetries for the modulatory transfer function where the surfaces are relatively constant along the s_2 axis. However, there are also some differences, most noticeably the presence of negative terms.

Figure 7c shows that for the modulatory transfer function, the EID shows negative unique information about X2.

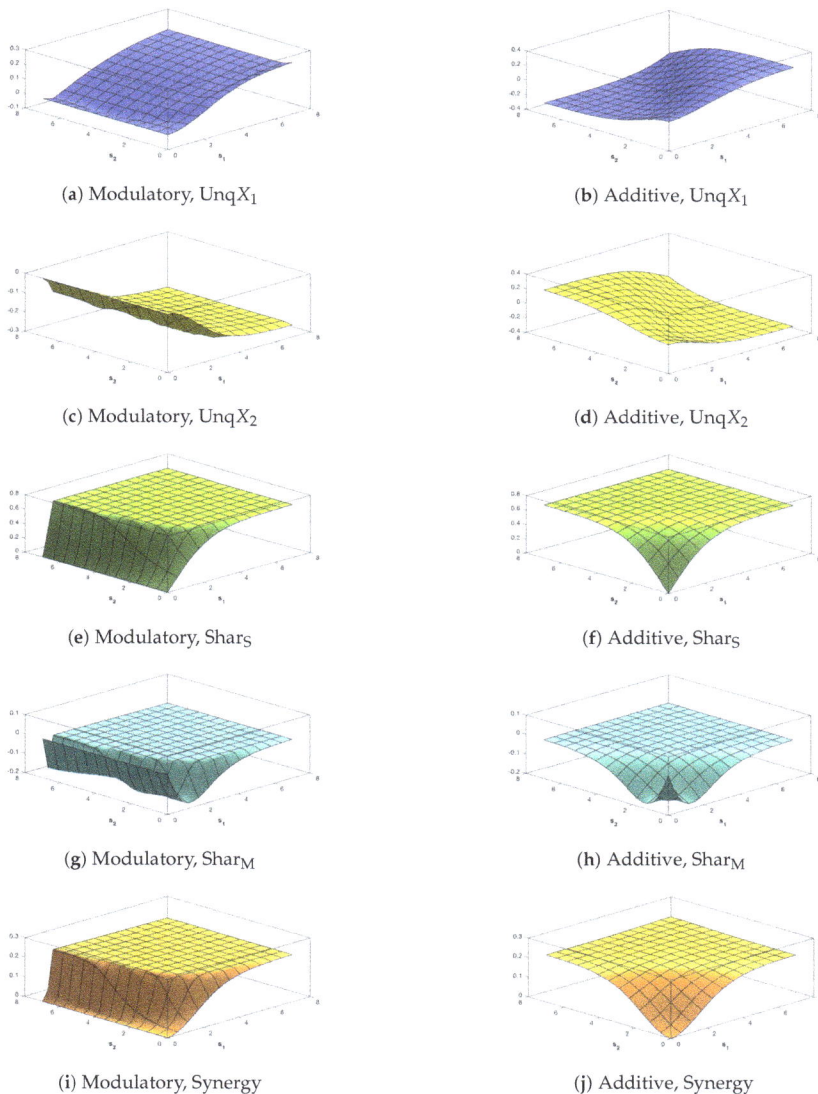

(**a**) Modulatory, UnqX_1

(**b**) Additive, UnqX_1

(**c**) Modulatory, UnqX_2

(**d**) Additive, UnqX_2

(**e**) Modulatory, Shar$_S$

(**f**) Additive, Shar$_S$

(**g**) Modulatory, Shar$_M$

(**h**) Additive, Shar$_M$

(**i**) Modulatory, Synergy

(**j**) Additive, Synergy

Figure 7. EID surfaces, based on additive and modulatory transfer functions, and a correlation between inputs of 0.78.

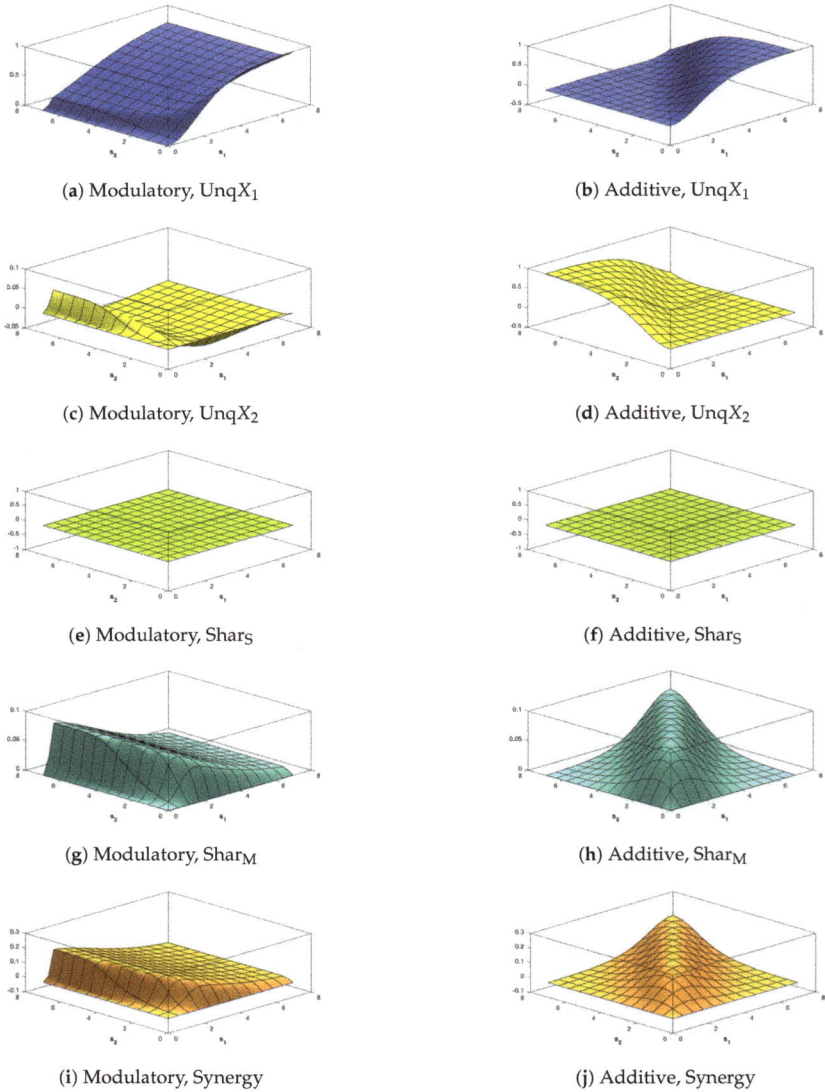

(**a**) Modulatory, UnqX$_1$

(**b**) Additive, UnqX$_1$

(**c**) Modulatory, UnqX$_2$

(**d**) Additive, UnqX$_2$

(**e**) Modulatory, Shar$_S$

(**f**) Additive, Shar$_S$

(**g**) Modulatory, Shar$_M$

(**h**) Additive, Shar$_M$

(**i**) Modulatory, Synergy

(**j**) Additive, Synergy

Figure 8. EID surfaces, based on additive and modulatory transfer functions, and a zero correlation between inputs.

This is relatively constant irrespective of the strength of the CF signal, and increases in magnitude with stronger RF signals. Shar$_{S+M}$ is here split into separate source and mechanistic components. The source shared information for the modulatory transfer function plateaus for $s_1, s_2 > 2$ (Figure 7e). In this case, there is a very strong correlation between the two inputs, which is reflected in the shared source information. The source shared information is fixed due to the high correlation between the inputs, however, the univariate information in the CF decreases as a function of s_1. Therefore the unique X_2 information is negative. Similarly, as $I[Y; X_2|X_1]$ is UnqX$_2$ plus synergy in (21), the negative unique interacts with the plateau of positive synergy to result in the $I[Y; X_2|X_1]$ surface (Figure 2e).

In Figure 7g, we note that the mechanistic shared component is negative for small values of s_1, while in Figure 7h it is negative for some small values of s_1, s_2. In contrast, Figure 8g,h show that the mechanistic component is non-negative when the correlation between the inputs is zero.

In general, the univariate mutual information $I[Y; X_2]$ is a sum of positive and negative terms, representing shared and synergistic entropy respectively between the two variables in the calculation. Since mutual information is non-negative, the positive terms always outweigh the negative terms in the mutual information expectation summation. However, if some of the positive terms in the calculation of $I[Y; X_2]$ are shared, or overlapping, with corresponding positive local information terms of $I[Y; X_1]$, those terms will contribute to the shared information term of the decomposition, and not be counted in the unique information terms. If enough of the shared entropy between X_2 and Y is overlapping with that shared between X_1 and Y, and the negative synergistic entropy terms in $I[Y; X_2]$ are not shared with X_1, then the unique synergistic entropy between Y and X_2 can be larger than the unique redundant entropy between Y and X_2, resulting in a net negative $UnqX_2$ information term.

To illustrate this consider a specific example, when $s_1 = s_2 = 2$, with correlation between inputs of 0.78. We can consider the local contributions to the univariate mutual information $I[Y; X_1]$. As $I[Y; X_1]$ is an expectation computed with a summation we can consider each local term in the summation which we denote $e(y, x_1) = p(y, x_1) i(y, x_1)$:

$$e(-1, -1) = e(1, 1) \quad = \quad 0.46$$
$$e(-1, 1) = e(1, -1) = -0.06$$

and similarly for $I(Y; X_2)$, the $e(y, x_2)$ are:

$$e(-1, -1) = e(1, 1) \quad = \quad 0.40$$
$$e(-1, 1) = e(1, -1) = -0.11$$

Note that here the strong similarity in the profile of the local information terms results from the high correlation between the two inputs. Local co-information values when $x_1 = x_2 = y = -1$ and when $x_1 = x_2 = y = 1$ show that the terms are largely, but not completely, overlapping (0.37 bits). There are no other local contributions to the I_{ccs} shared information measure.

Further consideration of these pointwise terms reveals that there are some positive and some negative local unique contributions to the univariate information for both predictors. The shared local information for the state $(y, x_1, x_2) = (-1, -1, -1)$ is 0.37 bits. The corresponding $(y, x_1) = (-1, -1)$ term in the calculation of $I(Y; X_1)$ gives 0.46 bits of information. Since 0.37 bits of that is shared with X_2, $0.46 - 0.37 = 0.09$ bits are unique to X_1 for that local contribution. Similarly there is a contribution of 0.09 bits of unique X_1 information when $(y, x_1) = (1, 1)$. Considering the same local terms for X_2 there are again 0.37 bits shared with X_1 and now $0.40 - 0.37 = 0.03$ bits of unique X_2 information. So in total, when the output matches the RF X_1 input, those states contribute 0.18 bits to the unique X_1 information and 0.06 bits to the unique X_2 information.

Moving to the cross-terms, since there is no corresponding local shared information these contributions to the univariate mutual information are entirely unique. So for X_1 the unique information is $2 \times -0.06 = -0.12$ bits, and X_2 has $2 \times -0.11 = -0.22$ bits of unique information. So the total net unique information in X_1 is $0.18 - 0.12 = 0.06$ bits, and for X_2 there are $0.06 - 0.22 = -0.16$ bits of unique information. This shows that in this system both variables have both positive and negative contributions to unique information, and that a negative value results when the negative contributions are larger.

In this case, when the sign of either input matches the sign of the output, they have locally redundant entropy, some of which is shared with the other input, but a small fraction of which is unique to that variable (i.e., related to the residual variance over that determined by the correlations between the variables). Instead, when the sign of the input does not match the sign of the output, there is local synergistic entropy between the variables. In other words, that particular local value of the input variable is misleading about the corresponding local output value, in the following sense.

Imagine a gambler was trying to predict the output of the system, starting with knowledge of the marginal distribution of the output $p(Y)$. They would determine a gambling strategy to optimise payout based on that distribution of Y. Observing the value of an input variable, combined with knowledge of the function of the system, would allow the gambler to form a new distribution of the output, $p(Y|X_2 = x_2)$. In this updated conditional distribution some specific values of the output would have higher probability than under $p(Y)$, and some would have lower probability. In the alternate sign cross terms in this example, the actual outcome is one of those that had lower probability under the conditional distribution obtained after observing the input. The particular (local) evidence provided by the value of the input on that trial moved the conditional distribution in the wrong direction for that output value—i.e., it was misleading about that particular output value, because it suggested it was less likely to happen, but then it did happen anyway. The fact that negative local values correspond to misleading evidence from the perspective of prediction explains why they have been termed misleading information or "misinformation" [26].

Therefore for both variables there are some unique information contributions that are both positive and negative (positive when the sign of the input is preserved in the output, and negative when the sign is changed in the output). Because a change in the sign of the output is rare, as a consequence of the design of the transfer function, that joint event is less likely to happen than would be predicted from the independent marginal local probability of the two events. The surprisal of the joint event is greater than the sum of the surprisal of the individual events. In conditional probability terms, $p(y|x_1) < p(y)$, the likelihood of seeing that value of y is decreased by conditioning on that value of x_1.

While in Figure 7c, the unique X_2 information is always negative, as shown in the example above there can be both positive and negative components. It would be possible to further split I_{ccs} to consider positive and negative terms separately, and so keep these shared vs. synergistic entropy effects separate throughout the decomposition. However here we focus on the net unique information effects to present a simpler decomposition and one that can be directly compared with the I_{broja} PID. Note that in Figure 8c the balance is different. Here the two inputs are independent. Without the strong correlation between the inputs the positive local information terms are smaller, and the balance between positive and negative contributions to unique information is closer. Therefore, there is a narrow parameter region, when $s_1 < 2$ in which there is net positive unique information about X_2. In Figure 8, which shows all the surfaces for independent inputs, the surfaces for the modulatory transfer function do not plateau so much. They remain mostly constant along s_2 axis, and along the s_1 axis UnqX$_1$ increases while Shar$_M$ and Synergy decrease (Shar$_S$ is always zero here due to the fact the inputs are independent.)

8. Applications of ID Measures to Psychophysical Data

We now turn our attention to demonstrating the practicality of using PID and EID to decompose spectra from real-world data. We use the example of a behavioural lateral masking paradigm whereby the driving RF input is a centrally presented gabor patch (a sinusoidal grating combined with a gaussian function) of varying contrast. CF input takes the form of high-contrast gabor patches that flank the central target in the upper and lower visual fields; see Figure 9 for example stimuli. Neurophysiological studies have demonstrated that, in this experimental setup, when flankers are presented concurrently with targets but placed outside the classical receptive field, the cell's response to the target is modulated [32,33]. Furthermore, due to the size of stimuli, orientation, contrast, and their wavelength, CF input can suppress detection of the centrally presented target gabor [32,33]. This paradigm is a suitable testbed for PID measures since it measures the influence of a modulatory input (CF), surrounding flanker stimuli, on performance, in this instance a contrast detection task on a centrally presented gabor (RF). Furthermore, the paradigm can be manipulated to conform to the predictions outlined in Section 3.

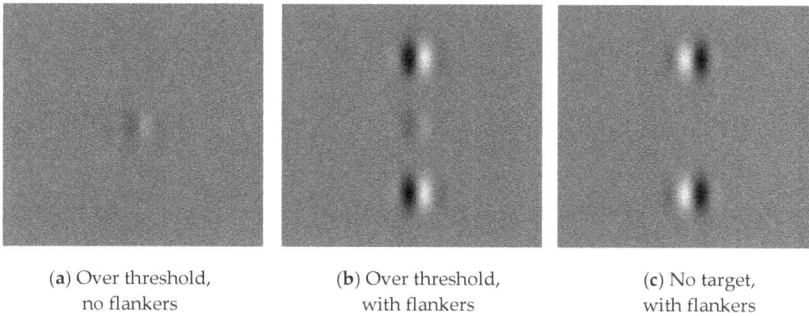

(a) Over threshold,   (b) Over threshold,   (c) No target,
no flankers   with flankers   with flankers

Figure 9. Examples of gabor patch stimuli used in the psychophysical experiment. In all conditions, the task was to detect the presence of a centrally presented target gabor.

We tested 21 participants from the University of Stirling's undergraduate psychology programme (Mean age = 19.1 years, SD = 1.3), who all had normal or corrected to normal vision. Ethical approval for the study was obtained from the University of Stirling's research ethics committee. Participants first completed a two-alternative forced choice staircase experiment, in which individual contrast sensitivity thresholds were established. Participants were asked to report whether a Gabor patch appeared to the left or right of a central fixation cross; the Gabor patch steadily decreased in contrast over the course of the experiment until a threshold of 60% accuracy was determined. This procedure was run twice with participants, and the average contrast threshold was used. After thresholds were established, participants completed the main experiment in which they were tasked with detecting a central target gabor in three conditions: (1) Over threshold target; (2) At threshold target; (3) No target present. In all three conditions, flankers were either present or not with equal occurrence; see Figure 9 for example stimuli.

Participants completed 100 trials per condition (except in the "No target" conditions, where they viewed 25 trials per condition, giving 450 trials in total), and all stimuli were presented for 500 ms, with a 2000 ms inter-stimulus interval for participants to respond.

Gabor patch stimuli for both the staircase and the main experimental paradigms were viewed on a gamma corrected CRT monitor (Tatung C7BBR, 60 Hz refresh rate, Taipei, Taiwan) at a distance of 80 cm, had a spatial frequency of 0.5 cycles per degree, and subtended a visual angle of no more than $1.93°$ in horizontal and vertical dimensions. From upper to lower flanker, the whole image subtended no more than $8.22°$ of vertical visual angle. All stimuli were presented on a medium grey background (RGB, 128,128,128). Gabors were phase shifted by $\pm90°$ to present equal weightings of black/white. Flanker gabors in the main experiment were presented at 0.85 Michelson contrast across all trials, whereas central target gabor contrast varied by individual (Mean = 0.012, SD = 0.003).

Table 1. Estimated accuracy, with estimated standard error, for each combination of the three conditions and the absence or presence of flankers.

	No Target	At Threshold	Over Threshold
Without Flankers	0.9096 (0.0273)	0.8797 (0.0289)	0.9824 (0.0037)
With Flankers	0.9629 (0.0150)	0.3766 (0.0532)	0.9849 (0.0039)

Summary statistics for the accuracy data are shown in Table 1. Of particular note is the suppression of contrast detection accuracy in the "At Threshold" condition when flankers are present. We found, using a 3 (Threshold: Over, At, No target) by 2 (Flankers: With vs. Without) repeated measures ANOVA model (Huynh-Feldt corrections reported where appropriate), that accuracy for detection of the central gabor patch was lower in "at threshold" conditions in comparison to "over threshold" [$F(1.147, 22.937)$

= 66.401, $p < 0.001, \eta^2 = 0.769$]; post hoc comparison, Mean difference = 0.356, $p < 0.001$]. Furthermore, the presence of flankers further reduced the contrast detection accuracy [F(1, 20) = 55.508, $p < 0.001$, $\eta^2 = 0.735$], however this was a consequence of flanker stimuli suppressing contrast detection when target was at threshold, but not when the target was over threshold [F(1.334, 26.678) = 85.042, $p < 0.001, \eta^2 = 0.81$]. These results indicate that the CF input in these conditions served to suppress contrast detection; however the nature of the suppressive effect found could be additive/subtractive or modulatory.

Group ID spectra for the analysis of this experiment show that in conditions where the central target gabor was presented over threshold, i.e., in a case of near certainty, the majority of information transmitted in Y is unique to X_1, the driving RF input. The influence of CF flanker stimuli in this condition makes very little contribution to the output (Figure 10). In contrast, in conditions of uncertainty, i.e., at threshold, the unique contributions of X_1 driving RF input is, by definition, much reduced, and the effect of the X_2 modulatory CF input is much increased via its contribution to the synergistic component. This latter effect occurs even though the unique contribution of the CF input at threshold is small. The pattern of decompositions observed when the target driving RF input is weak is similar to that of the modulatory transfer function examined in Section 5, except for the occurrence of a small amount of unique information from the X_2 modulatory CF input.

Figure 10 shows group decomposition spectra, however the decomposition may vary across subjects. Fortunately, enough data was collected for analysis of the individual data to be possible. We show Ibroja spectra for individual subjects of interest also in Figure 10. When the RF input is over threshold (i.e., strong), information transmitted is again unique to the RF in both subjects 10 and 18. However, at threshold (i.e., weak RF input) interactions that meet the criteria for modulation do occur for many subjects. Subject 10 is a clear example of a subject for whom the flanking context did indeed seem to function as a modulator. Information unique to the target stimulus was transmitted, but information unique to the flanking context, X_2, was at or near zero. X_2 must have contributed to output, however, because there is a substantial synergistic component. Such subjects therefore display a decomposition that is remarkably similar to that for the modulatory function studied in previous sections.

A few subjects performed very differently at threshold. Subject 18's responses at threshold conveyed no unique information about the target; unique information to CF input dominates, but again with substantial shared information and synergy between RF and CF. Therefore, the target, X_1, input contributed to the synergy, but the subject's response conveyed unique information only about X_2. Thus, under these conditions for these subjects, the central target, X_1, modulated transmission of information about the flankers, X_2, not the other way round. This demonstrates the value of using ID spectra to analyze such data.

Accuracy data for subject 18 suggests a very strong suppressive effect of CF input on contrast detection when the central target was presented at threshold (Accuracy in at threshold condition with flankers is 3%). The presence of some information unique to X_2 in the group data is therefore largely due to a few subjects whose performance at threshold was mainly transmitting information about the flankers. It may be that there were subjects for whom the threshold was underestimated. Overall, the decompositions of these psychophysical data confirm the rich expressive power of the decomposition spectra, and we expect to see far more use of them for such purposes in the near future.

To summarise, the nature of the modulation presented above is uncovered through use of decomposition measures. The suppression of contrast detection accuracy observed here when the RF input is weak coincides with less unique information transmitted about the RF in the output, and in addition, shared information and a synergistic relationship between RF and CF inputs. EID spectra suggest that the shared information is not mechanistic (see Section 4). Differing PID spectra between individual participants highlights the efficacy of PID for disambiguating modulatory interactions at the single subject level. The empirically observed spectra shown in this section may also cast some light on relations between PID and EID. Overall, these two forms of decomposition are mostly in agreement. With respect to the negative EID components they again show that where negative unique components occur they seem to be compensated for by equivalent positive increases in the synergy. In addition, these results show that most EID components are positive, with negative components being the exception rather than the rule.

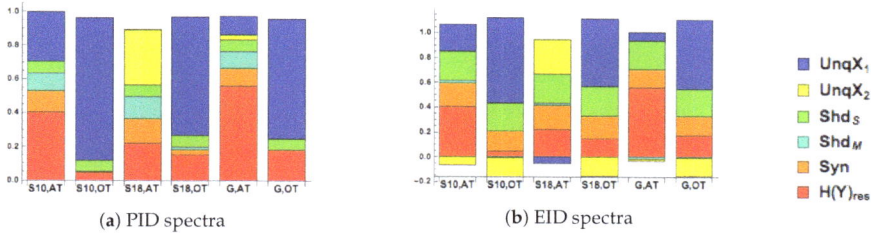

(a) PID spectra   (b) EID spectra

Figure 10. Partial information decomposition (PID) and EID spectra (in bits) calculated for subject 10 (S10), subject 18 (S18) and the whole group of subjects (G) in the contrast detection experiment calculated at threshold (AT) and over threshold (OT).

9. Conclusions and Discussion

9.1. Implications of These Findings for Conceptions of 'Modulation' in the Cognitive and Neurosciences

Intuition suggests that any variable that affects output must transmit information specifically about itself in that output. That is clearly incorrect because output can be pure synergy. Furthermore, as shown in Figure 2e, conventional information theoretic analysis weakens that view by showing that the conditional mutual information transmitted about the modulator is at or near zero unless the primary drive is present but weak. The PID and EID analyses reported in Sections 5–7 now show that the conditional mutual information transmitted about the modulator was greater than zero when the primary input was present but weak because the synergistic component is then greater than zero, not because the modulator transmits unique information about itself. Thus, the intuitive view that to have any effect modulators must transmit specific information about themselves is shown to be seriously misleading.

Signals can have a kind of dual "semantics", one concerned with the message being transmitted, and one being concerned with the strength, salience, confidence, or precision with which that message is conveyed. The notion of contextual modulation requires a distinction between signal strength and signal semantics because it implies that the signal's strength can be modulated without changing its semantic content. A set of criteria to be met by what we call a modulatory transfer function were stated in Section 3. The surfaces given in Sections 6 and 7 for PID and EID analyses respectively show that our modulatory transfer functions meet these criteria. Section 5.2 showed a set of four ID spectra that together would imply that a transfer function is modulatory. ID spectra have substantial expressive power so it is possible that, when applied to empirical data from the cognitive and neurosciences, they may reveal that modulatory interactions take various and unexpected forms.

Another perspective from which to view our distinction between drive and modulation is that of the receiver of the output signal. Such a receiver can confidently infer the sign of the driving input from the output alone when the driving input is sufficiently strong. This is true whatever the strength of the modulatory input. Nothing can be confidently inferred from the output alone about the sign of the modulatory input, however, no matter what the strength of that modulatory input. This again supports our claim that modulatory inputs do not contribute to the message being conveyed by the semantics of output.

9.2. Comparisons between PID and EID

The most important outcome of the findings reported above is that they show that both EID and PID support all the main conclusions made above with respect to the defining properties and functions of modulatory interactions. Important strengths of EID shown here are that it distinguishes between source and mechanistic forms of shared information, and it relates them appropriately to the correlation between the two inputs. This is also the case with PID when the separation of shared information from [23] is included in the Ibroja decomposition.

9.3. Using EID and PID to Analyze and Interpret Psychophysical Data

The application of PID spectra to psychophysical data is useful in distinguishing ways in which two distinct inputs can contribute to a single measure of output. The methods outlined here can establish the underlying nature of statistical interactions in real world systems that cannot be studied with traditional multi-variate statistics alone. Future studies will apply these measures to continuous data streams to elucidate the strength of modulatory effects in complex neuroimaging data for example.

9.4. Using ID Spectra to Analyze and Interpret Empirical Data in General

The spectra and surfaces shown here were computed from a known transfer function, but the inverse problem may also arise. That is, to what extent can a transfer function, or properties of it, be inferred from an ID spectra, or set of spectra? For example, ID spectra could be computed from neurobiological observations, from psychophysical observations, from the activities of local processing elements in deep learning architectures, or from the input-output activity of a system as a whole. Work on information decomposition has so far focussed on the forward problem, i.e., on computing the spectra given a known transfer function. When ID spectra are computed from empirical data, however, then issues concerning the inverse problem will become more prominent and the application of formal statistical modelling will be required. Future studies will apply these measures to continuous data streams to elucidate the strength of modulatory effects in complex neuroimaging data for example. I_{ccs} and the EID can be easily computed for continuous Gaussian variables, which together with a semi-parametric Gaussian copula assumption results in a promising approach for robustly estimating these quantities from experimental data [34]. Further study of the statistical properties of these methods when applied to experimental data, for example in terms of limited sampling bias [35] and optimal permutation tests for valid statistical inference [36] are important areas for future work. For some recent work with fMRI data, see [37].

Empirical studies will rarely provide enough data to compute the equivalent of the surfaces shown above, so it is spectra that empirical studies will usually provide. The studies above show that the conditions under which the spectra are measured must be carefully chosen if modulatory and additive functions are to be distinguishable. We assume that transfer functions cannot be rigorously inferred from observed spectra, but they can be examined to see whether or not they meet the requirements for a modulatory interaction as described above. This will not fully constrain the unknown transfer function producing the observed output because those requirements can be met in many different ways. If an observed spectrum does meet our criteria for a modulatory interaction, then further experiments might be designed to distinguish between different ways in which those criteria can be met.

9.5. Modulatory Regulation of Activity as a Crucial and Non-Trivial Aspect of Information Processing

Though the topics dealt with in this Special Issue have implications for many disciplines they have special implications for the computational, cognitive, and neurosciences. This new perspective on multivariate information decomposition substantially enhances our notions of what "information processing" can be, and that is at the heart of all of those disciplines. Information processing is more than simply transferring information from one time or place to another. As others have argued it also includes creating new information via synergetic interactions between separate inputs; see [26,38]. Our argument here is that in addition to "enhancing computational capabilities via synergy" information processing also includes distinguishing between currently relevant and currently irrelevant inputs. That is far from trivial, and though we have not considered the various criteria by which relevance can be assessed, we have done so elsewhere; see e.g., [19,39]. Here we have shown that it is possible to use any such assessment to amplify relevant and disamplify irrelevant signals without corrupting their semantic content. ID spectra can now be used as a way of exploring information processing within biological systems. It will be of particular interest to see whether interactions similar to those produced by our modulatory transfer function can be observed at the cellular level. We have shown that it is possible to use any such assessment to amplify relevant and disamplify irrelevant signals without corrupting their semantic content. Whether biology uses such modulatory interactions can now be explored by applying the ID spectra that we have proposed to biological data. The ID spectra could also be used to enhance our understanding of the information processing performed by local processors within various machine-learning architectures. It will also be possible to build new architectures designed to exploit the computational capabilities made possible by modulatory interactions such as those analysed here.

Acknowledgments: We thank Elena Gheorghiu for help with design of the Gabor patch stimuli, and Eva Kriechbaum & Aimee Lord for assistance with data collection and analysis. William A. Phillips is partially supported by a European Human Brain Project (EU grant 604102) to Lars Muckli. We also thank anonymous reviewers for helpful comments which have resulted in an improved version of the paper.

Author Contributions: William A. Phillips conceived the investigation, designed the computational studies, introduced the concept of an ID spectra, wrote Sections 1 and 9, and contributed to Sections 3, 5 and 8. Benjamin Dering wrote Section 8. Robin A. A. Ince produced all the EID outputs, wrote Sections 4.2 and 7 and contributed to Section 3. Jim W. Kay produced all the PID outputs, wrote Sections 2, 3.1, 4.1 and 6, wrote the appendix and contributed to Sections 3 and 5. All the authors have read and approved the final manuscript.

Conflicts of Interest: The authors declare no conflict of interest.

Appendix A. Preliminary Results

Consider the logistic function L, from $\mathbb{R}$ to the open interval $(0, 1)$, which is strictly increasing and has the following properties

$$L(x) = 1/(1 + \exp(-x)), \quad L(-x) = 1 - L(x), \quad 0 < L(x) < 1,$$
$$L(x) > \tfrac{1}{2}, \ L(x) = \tfrac{1}{2}, \ L(x) < \tfrac{1}{2} \iff x > 0, \ x = 0, \ x < 0, \quad \text{respectively.} \tag{A1}$$

From (14)–(16), we may use (A1) to write the values of u, v in the form

$$u_M = L[T_M(1,1)], \ v_M = L[T_M(1,-1)], \ u_A = L[T_A(1,1)], \ v_A = L[T_A(1,-1)]. \tag{A2}$$

Now, we write from (10) that

$$T_M(-1,-1) = -\tfrac{1}{2}s_1(1 + \exp(s_1 s_2)) = -T_M(1,1),$$
$$T_M(-1,1) = -\tfrac{1}{2}s_1(1 + \exp(-s_1 s_2)) = -T_M(1,-1) \tag{A3}$$

and so using (A1) it follows that

$$\Pr(Y = 1 | X_1 = -1, X_2 = -1) = L[T_M(-1,-1)] = L[-T(1,1)] = 1 - L[T_M(1,1)] = 1 - u_M$$
$$\Pr(Y = 1 | X_1 = -1, X_2 = 1) = L[T_M(-1,1)] = L[-T(1,-1)] = 1 - L[T_M(1,-1)] = 1 - v_M$$

A similar argument using the additive transfer function, T_A, shows that

$$\Pr(Y = 1 | X_1 = -1, X_2 = -1) = 1 - u_A, \quad \text{and} \quad \Pr(Y = 1 | X_1 = -1, X_2 = 1) = 1 - v_A.$$

Therefore, the conditional output probabilities are $\{1-u, 1-v, v, u\}$ when taken in the order $\{--, -+, +-, ++\}$, where (u, v) are replaced by (u_M, v_M) when using the transfer function T_M, and by (u_A, v_A) when using T_A. It follows from (1)–(4) that the joint p.m.f. $p(y, x_1, x_2)$ may be written as

$$\{\lambda u, \ \mu v, \ \mu(1-v), \ \lambda(1-u), \ \lambda(1-u), \ \mu(1-v), \ \mu v, \ \lambda u\}, \tag{A4}$$

where the probabilities are written in the order $\{p_{---}, p_{--+}, p_{-+-}, p_{-++}, p_{+--}, p_{+-+}, p_{++-}, p_{+++}\}$, so, for example, $\Pr(Y = -1, X_1 = +1, X_2 = -1) = p_{-+-}$. We find the marginal distribution of the output Y. From (A4), we have that

$$\Pr(Y = -1) = \lambda u + \mu v + \mu(1-v) + \lambda(1-u) = \lambda + \mu = \tfrac{1}{2},$$

and so Y, as well as X_1 and X_2 has a uniform binary distribution.

We now calculate the various Shannon entropy terms that will be required in the sequel. Since each of the three variables has a marginal uniform binary distribution, we can say that

$$H(Y) = 1, \quad H(X_1) = 1, \quad \text{and} \quad H(X_2) = 1. \tag{A5}$$

From (2) and (3), and noting that $\lambda + \mu = \tfrac{1}{2}$, we can write the Shannon entropy of the marginal (X_1, X_2) distribution as

$$H(X_1, X_2) = -2\lambda \log \lambda - 2\mu \log \mu = 1 - (2\lambda) \log(2\lambda) - (1 - 2\lambda) \log(1 - 2\lambda) = 1 + h(2\lambda),$$

where the function h is defined in (17). From (A4), we may write the marginal p.m.f.s of (Y, X_1) and (Y, X_2) in the order $\{--, -+, +-, ++\}$.

$$p(y, x_1): \ \{\lambda u + \mu v, \lambda(1 - u) + \mu(1 - v), \lambda(1 - u) + \mu(1 - v), \lambda u + \mu v\} = \{\tfrac{1}{2}z, \tfrac{1}{2}(1 - z), \tfrac{1}{2}(1 - z), \tfrac{1}{2}z\}, \tag{A6}$$

where, as in (17), $z = 2\lambda u + 2\mu v$.

$$p(y, x_2): \ \{\lambda u + \mu(1 - v), \lambda(1 - u) + \mu v, \lambda(1 - u) + \mu v, \lambda u + \mu(1 - v)\} = \{\tfrac{1}{2}w, \tfrac{1}{2}(1 - w), \tfrac{1}{2}(1 - w), \tfrac{1}{2}w\}, \tag{A7}$$

where, as in (17), $w = 2\lambda u + 2\mu(1 - v)$.

We now calculate the Shannon entropies of the marginal (Y, X_1) and (Y, X_2) distributions.

$$H(Y, X_1) = -z \log(\tfrac{1}{2}z) - (1 - z) \log(\tfrac{1}{2}(1 - z)) = 1 + h(z). \tag{A8}$$

$$H(Y, X_2) = -w \log(\tfrac{1}{2}w) - (1 - w) \log(\tfrac{1}{2}(1 - w)) = 1 + h(w). \tag{A9}$$

Finally, from (A4), we find the Shannon entropy of the joint distribution of (Y, X_1, X_2).

$$
\begin{aligned}
H(Y, X_1, X_2) &= -2\lambda u \log(\lambda u) - 2\mu v \log(\mu v) - 2\lambda(1 - u) \log[(\lambda(1 - u)] - 2\mu(1 - v) \log[\mu(1 - v)], \\
&= -2\lambda \log \lambda - 2\mu \log \mu - 2\lambda[-u \log u - (1 - u) \log(1 - u)] \\
&\qquad - 2\mu[-v \log v - (1 - v) \log(1 - v)], \\
&= H(X_1, X_2) + 2\lambda h(u) + 2\mu h(v). \tag{A10}
\end{aligned}
$$

Appendix B. Proof of Theorem 1

(a) From (6), and using (A5), (A6), (A9) and (A10), we have that

$$
\begin{aligned}
I[Y; X_1 | X_2] &= H(Y, X_2) + H(X_1, X_2) - H(X_2) - H(Y, X_1, X_2) \\
&= 1 + h(w) + H(X_1, X_2) - 1 - H(X_1, X_2) - 2\lambda h(u) - 2\mu h(v) \\
&= h(w) - 2\lambda h(u) - 2\mu h(v).
\end{aligned}
$$

(b) From (7), and using (A5), (A6), (A8) and (A10), we have that

$$
\begin{aligned}
I[Y; X_2 | X_1] &= H(Y, X_1) + H(X_1, X_2) - H(X_1) - H(Y, X_1, X_2) \\
&= 1 + h(z) + H(X_1, X_2) - 1 - H(X_1, X_2) - 2\lambda h(u) - 2\mu h(v) \\
&= h(z) - 2\lambda h(u) - 2\mu h(v).
\end{aligned}
$$

(c) From (9), and using (A5) and (A8), we have

$$
I[Y; X_1] = H(Y) + H(X_1) - H(Y, X_1) = 2 - 1 - h(z) = 1 - h(z).
$$

(d) From (9), and using (A5) and (A9), we have

$$
I[Y; X_2] = H(Y) + H(X_2) - H(Y, X_2) = 2 - 1 - h(w) = 1 - h(w).
$$

(e) From (8) and parts (a) and (b), we have that

$$
I[Y; X_1; X_2) = 1 - h(z) - (h(w) - 2\lambda h(u) - 2\mu h(v)) = 1 - h(z) - h(w) + 2\lambda h(u) + 2\mu h(v).
$$

(f) From (5) and using (A5), (A6) and (A10), we have

$$
I[Y; (X_1, X_2)] = H(Y) + H(X_1, X_2) - H(Y, X_1, X_2) = 1 - 2\lambda h(u) - 2\mu h(v).
$$

Appendix C. Proof of Theorem 2

From Lemma 6 in [9] a necessary and sufficient condition for $\mathrm{Unq}X_2$ to vanish is that there exists a row stochastic matrix $S = [\sigma(x_1; x_2)]$ such that

$$
\Pr(Y = y, X_2 = x_2) = \sum_{x_1 \in B} \Pr(Y = y, X_1 = x_1)\sigma(x_1; x_2). \tag{A11}
$$

We first find expressions for the joint p.m.f. in this more general case, but the work involved is very similar to that leading to (A4) above. From (29), (30) and (A1) we note that

$$
g = L[F(s_1, s_2)], \quad h = L[F(s_1, -s_2)] \text{ and } 0 < g, h < 1.
$$

Also, since $\lambda \neq 0$ and $\mu \neq 0$ and $\lambda + \mu = \frac{1}{2}$, we have that $0 < \lambda < \frac{1}{2}$ and $0 < \mu < \frac{1}{2}$. From (28), (29) and (A1) we have that

$$
\begin{aligned}
\Pr(Y = 1 | X_1 = -1, X_2 = 1) &= L[F(-s_1, s_2)] = L[-F(s_1, -s_2)] = 1 - L[F(s_1, -s_2)] = 1 - h \\
\Pr(Y = 1 | X_1 = -1, X_2 = -1) &= L[F(-s_1, -s_2)] = L[-F(s_1, s_2)] = 1 - L[F(s_1, s_2)] = 1 - g
\end{aligned}
$$

It follows that the joint p.m.f. of (Y, X_1, X_2) is

$$
\{\lambda g, \ \mu h, \ \mu(1-h), \ \lambda(1-g), \ \lambda(1-g), \ \mu(1-h), \ \mu h, \ \lambda g\}, \tag{A12}
$$

and that the p.m.f.s for (Y, X_1) and (Y, X_2), in the order $\{--, -+, +-, ++\}$, are

$$p(y, x_1): \quad \{\lambda g + \mu h, \lambda(1-g) + \mu(1-h), \lambda(1-g) + \mu(1-h), \lambda g + \mu v h\}, \qquad \text{(A13)}$$
$$p(y, x_2): \quad \{\lambda g + \mu(1-h), \lambda(1-g) + \mu h, \lambda(1-g) + \mu h, \lambda g + \mu(1-h)\}. \qquad \text{(A14)}$$

Note that, since $\lambda + \mu = \frac{1}{2}$, we can write

$$\lambda(1-g) + \mu(1-h) = \tfrac{1}{2} - \lambda g - \mu h, \text{ and } \lambda(1-g) + \mu h = \tfrac{1}{2} - \lambda g - \mu(1-h).$$

From (A11), we now write out the system of equations that we will use to find a stochastic matrix.

$$\lambda g + \mu(1-h) = (\lambda g + \mu h)\sigma_{--} + (\tfrac{1}{2} - \lambda g - \mu h)\sigma_{+-} \qquad \text{(A15)}$$
$$\tfrac{1}{2} - \lambda g - \mu(1-h) = (\lambda g + \mu h)\sigma_{-+} + (\tfrac{1}{2} - \lambda g - \mu h)\sigma_{++} \qquad \text{(A16)}$$
$$\tfrac{1}{2} - \lambda g - \mu(1-h) = (\tfrac{1}{2} - \lambda g - \mu h)\sigma_{--} + (\lambda g + \mu h)\sigma_{+-} \qquad \text{(A17)}$$
$$\lambda g + \mu(1-h) = (\tfrac{1}{2} - \lambda g - \mu h)\sigma_{-+} + (\lambda g + \mu h)\sigma_{++} \qquad \text{(A18)}$$

Using (A15) and (A17), we first solve for σ_{--} and σ_{+-} and obtain

$$\begin{bmatrix} \lambda g + \mu(1-h) \\ \tfrac{1}{2} - \lambda g - \mu(1-h) \end{bmatrix} = \begin{bmatrix} \lambda g + \mu h & \tfrac{1}{2} - \lambda g - \mu h \\ \tfrac{1}{2} - \lambda g - \mu h & \lambda g + \mu h \end{bmatrix} \begin{bmatrix} \sigma_{--} \\ \sigma_{+-} \end{bmatrix}$$

Hence, inverting the matrix, we can write

$$\begin{bmatrix} \sigma_{--} \\ \sigma_{+-} \end{bmatrix} = \frac{1}{\Delta} \begin{bmatrix} \lambda g + \mu h & \lambda g + \mu h - \tfrac{1}{2} \\ \lambda g + \mu h - \tfrac{1}{2} & \lambda g + \mu h \end{bmatrix} \begin{bmatrix} \lambda g + \mu(1-h) \\ \tfrac{1}{2} - \lambda g - \mu(1-h) \end{bmatrix},$$

where the determinant $\Delta = \lambda g + \mu h - \tfrac{1}{4}$. Now, $\Delta > 0$ provided that $g \geq \tfrac{1}{2}, h \geq \tfrac{1}{2}$ and g, h are not both equal to $\tfrac{1}{2}$. After some manipulation we obtain

$$\begin{bmatrix} \sigma_{--} \\ \sigma_{+-} \end{bmatrix} = \frac{1}{\Delta} \begin{bmatrix} \lambda g + \tfrac{1}{2}\mu - \tfrac{1}{4} \\ \mu(h - \tfrac{1}{2}) \end{bmatrix}$$

and so when $g \geq \tfrac{1}{2}$ and $h \geq \tfrac{1}{2}$, but both are not equal to $\tfrac{1}{2}$ then σ_{--} and σ_{+-} are both non-negative and they sum to 1.

Very similar calculations for solving (A16) and (A18) give that

$$\begin{bmatrix} \sigma_{-+} \\ \sigma_{++} \end{bmatrix} = \frac{1}{\Delta} \begin{bmatrix} \mu(h - \tfrac{1}{2}) \\ \lambda g + \tfrac{1}{2}\mu - \tfrac{1}{4} \end{bmatrix}$$

and the same reasoning as above shows that σ_{-+} and σ_{++} are both non-negative and they also sum to 1. Hence we have found a row stochastic matrix

$$S = \begin{bmatrix} \sigma_{--} & \sigma_{+-} \\ \sigma_{-+} & \sigma_{++} \end{bmatrix}$$

which satisfies (A11), and we conclude that $\text{Unq}X_2 = 0$.

Appendix D. Proof of Corollary 1

It follows from (A3) that T_M satisfies the properties of F in (28). We now show that $u_M > \tfrac{1}{2}$ and that $v_M > \tfrac{1}{2}$. From (A2) and (11) we have that

$$u_M = L[\tfrac{1}{2}s_1(1 + \exp(s_1 s_2))] \text{ and } v_M = L[\tfrac{1}{2}s_1(1 + \exp(-s_1 s_2))].$$

Since $s_1 > 0, s_2 > 0$, we conclude from (A1) that u_M and v_M are both greater than $\frac{1}{2}$. Hence, from Theorem 2, $\mathrm{Unq}X_2 = 0$.

Appendix E. Proof of Corollary 2

Using (11) we know that

$$T_A(-1,-1) = -s_1 - s_2 = -(s_1 + s_2) = -T_A(1,1),$$
$$T_A(-1,1) = -s_1 + s_2 = -(s_1 - s_2) = -T_A(1,-1)$$

and so T_A has the properties of F defined in (28). Also

$$u_A = L[s_1 + s_2] \text{ and } v_A = L[s_1 - s_2],$$

and so from (A1), and the assumption that $s_1 > 0, s_2 > 0$, we have that $u_A > \frac{1}{2}$ and also that $v_A \geq \frac{1}{2}$ if and only if $s_1 \geq s_2$. Hence, from Theorem 2 it follows that $\mathrm{Unq}X_2 = 0$.

Appendix F. Proof of Theorem 3

From Lemma 6 in [9], a necessary and sufficient condition for $\mathrm{Unq}X_1$ to vanish is that there exists a row stochastic matrix $T = [\tau(x_2; x_1)]$ such that

$$\Pr(Y = y, X_1 = x_1) = \sum_{x_2 \in B} \Pr(Y = y, X_2 = x_2)\tau(x_2; x_1). \tag{A19}$$

Since we are using T_A, here $g = u_A$ and $h = v_A$. From (A19), we now write out the system of equations that we will use to find a stochastic matrix.

$$\lambda g + \mu h = (\lambda g + \mu(1-h))\tau_{--} + (\tfrac{1}{2} - \lambda g - \mu(1-h))\tau_{+-}$$
$$\tfrac{1}{2} - \lambda g - \mu h = (\lambda g + \mu(1-h))\tau_{-+} + (\tfrac{1}{2} - \lambda g - \mu(1-h))\tau_{++}$$
$$\tfrac{1}{2} - \lambda g - \mu h = (\tfrac{1}{2} - \lambda g - \mu(1-h))\tau_{--} + (\lambda g + \mu(1-h))\tau_{+-}$$
$$\lambda g + \mu h = (\tfrac{1}{2} - \lambda g - \mu(1-h))\tau_{-+} + (\lambda g + \mu(1-h))\tau_{++}$$

We note that the only difference in this system of equations, as compared with (A15)–(A18) is that h has been replaced by $1 - h$, and so one would expect that the result will hold when $u_A > \frac{1}{2}$ and $v_A \leq \frac{1}{2}$, and this turns out to be the case.

Following the same argument used in the proof of Theorem 2 it turns out that

$$T = \frac{1}{\lambda g + \mu(1-h) - \frac{1}{4}} \begin{bmatrix} \lambda g + \frac{1}{2}\mu - \frac{1}{4} & \mu(\frac{1}{2} - h) \\ \mu(\frac{1}{2} - h) & \lambda g + \frac{1}{2}\mu - \frac{1}{4} \end{bmatrix},$$

and we see that T is a row stochastic matrix provided that $g \geq \frac{1}{2}$ and $h \leq \frac{1}{2}$ and g and h cannot both be equal to $\frac{1}{2}$. From the proof of Corollary 2, we know that $u_A > \frac{1}{2}$ and from (A1) we know that $v_A \leq \frac{1}{2}$ if and only if $s_1 \leq s_2$.

For the last part, we know from (A1) that, when $s_1 = s_2$,

$$v_A = L[s_1 - s_2] = L[0] = \tfrac{1}{2}.$$

From (A14), with $g = u_A$ and $h = v_A = \frac{1}{2}$, we have that the marginal distributions of (Y, X_1) and (Y, X_2) are identical. Hence since both marginals have the same range space, B^2, it follows from [9] (Corollary 8) that $\mathrm{Unq}X_1 = 0$ and $\mathrm{Unq}X_2 = 0$. This completes the proof.

Appendix G. Proof of Theorem 4

(a) We saw in Theorem 2, Corollary 1, that $\text{Unq}X_2 = 0$. It follows from (21) and Theorem 1(b) that

$$Syn = I[Y; X_2|X_1] = h(z_M) - 2\lambda h(u_M) - 2\mu h(v_M).$$

From Theorem 1(a) and (20) we have that

$$UnqX_1 = I(Y; X_1|X_2) - I(Y; X_2|X_1) = h(w_M) - h(z_M),$$

and from (19) we deduce that

$$Shar_{S+M} = I(Y; X_2) = 1 - h(w_M).$$

(b) In this case, $v_A = \frac{1}{2}$, so $h(v_A) = 1$, and $z_A = w_A$. From Theorem 3, we know that $\text{Unq}X_1 = 0$ and $\text{Unq}X_2 = 0$. From (20), (21) we have

$$Syn = I[Y; X_1|X_2] = I[Y; X_2|X_1] = h(z_A) - 2\lambda h(u_A) - 2\mu.$$

and from (18) and (19) it follows that

$$Shar_{S+M} = I[Y; X_1] = I[Y; X_2] = 1 - h(z_A).$$

(c) From Theorem 3, $\text{Unq}X_1 = 0$, and using (18), (20) and (21) we obtain

$$Syn = I[Y; X_1|X_2] = h(w_A) - 2\lambda h(u_A) - 2\mu h(v_A),$$

$$UnqX_2 = I[Y; X_2|X_1] - I[Y; X_1|X_2] = h(z_A) - h(w_A),$$

and

$$Shar_{S+M} = I[Y; X_1] = 1 - h(z_A).$$

(d) From Theorem 2, Corollary 2, $\text{Unq}X_2 = 0$. Using the same deductions as in part (a), we find that

$$Syn = I[Y; X_2|X_1] = h(z_A) - 2\lambda h(u_A) - 2\mu h(v_A),$$
$$UnqX_1 = I[Y; X_1|X_2] - I[Y; X_2|X_1] = h(w_A) - h(z_A),$$
$$Shar_{S+M} = I[Y; X_2] = 1 - h(w_A).$$

Appendix H. Proof of Theorem 5

For part (a), the correlation between inputs is $+1$, and so we know from (2), (3) and (17) that

$$\lambda = \tfrac{1}{2}, \mu = 0, z = v, w = v.$$

Hence, from Theorem 1(a, b)

$$I[Y : X_1|X_2] = h(v) - h(v) = 0, \quad \text{and} \quad I[Y; X_2|X_1] = h(u) - h(u) = 0.$$

From (20) and (21) it follows that $\text{Unq}X_1 = \text{Unq}X_2 = Syn = 0$. Then from Theorem 1 and (18) it follows that $Shar_{S+M} = I[Y; X_1] = 1 - h(u)$.

In (b), the correlation between inputs is -1, and so we know from (2), (3) and (17) that

$$\lambda = 0, \mu = \tfrac{1}{2}, z = v, w = 1 - v.$$

Hence, from Theorem 1(a,b), and noting that $h(1-v) = h(v)$,

$$I[Y:X_1|X_2] = h(1-v) - h(v) = 0, \quad \text{and} \quad I[Y;X_2|X_1] = h(v) - h(v) = 0.$$

From (20) and (21) it follows that $\mathrm{Unq}X_1 = \mathrm{Unq}X_2 = \mathrm{Syn} = 0$. Then from Theorem 1 and (18) it follows that $\mathrm{Shar}_{S+M} = I[Y;X_1] = 1 - h(v)$.

References

1. Gilbert, C.D.; Sigman, M. Brain States: Top-Down Influences in Sensory Processing. *Neuron* **2007**, *54*, 677–696.
2. Phillips, W.A.; Singer, W. In search of common foundations for cortical computation. *Behav. Brain Sci.* **1997**, *20*, 657–722.
3. Phillips, W.A.; Silverstein, S.M. Convergence of biological and psychological perspectives on cognitive coordination in schizophrenia. *Behav. Brain Sci.* **2003**, *26*, 65–138.
4. Lamme, V.A.F. Beyond the classical receptive field: Contextual modulation of V1 responses. In *The Visual Neurosciences*; Werner, J.S., Chalupa, L.M., Eds.; MIT Press: Cambridge, MA, USA, 2004; pp. 720–732.
5. Kay, J.; Floreano, D.; Phillips, W.A. Contextually guided unsupervised learning using local multivariate binary processors. *Neural Netw.* **1998**, *11*, 117–140.
6. Larkum, M. A cellular mechanism for cortical associations: An organizing principle for the cerebral cortex. *Trends Neurosci.* **2013**, *36*, 141–151.
7. Phillips, W.A.; Larkum, M.E.; Harley, C.W.; Silverstein, S.M. The effects of arousal on apical amplification and conscious state. *Neurosci. Conscious.* **2016**, 1–13, doi:10.1093/nc/niw015.
8. Williams, P.L.; Beer, R.D. Nonnegative Decomposition of Multivariate Information. *arXiv* **2010**, arXiv:1004.2515.
9. Bertschinger, N.; Rauh, J.; Olbrich, E.; Jost, J.; Ay, N. Quantifying Unique Information. *Entropy* **2014**, *16*, 2161–2183.
10. Griffith, V.; Koch, C.; Griffith, V. Quantifying synergistic mutual information. In *Guided Self-Organization: Inception. Emergence, Complexity and Computation*; Springer: Berlin/Heidelberg, Germany, 2014; Volume 9, pp. 159–190.
11. James, R.G.; Emenheiser, J.; Crutchfield, J.P. Unique Information via Dependency Constraints. *arXiv* **2017**, arXiv:1709.06653.
12. Ince, R.A.A. Measuring multivariate redundant information with pointwise common change in surprisal. *Entropy* **2017**, *19*, 318.
13. Ince, R.A.A. The Partial Entropy Decomposition: Decomposing multivariate entropy and mutual information via pointwise common surprisal. *arXiv* **2017**, arXiv:1702.01591.
14. Phillips, W.A.; Kay, J.; Smyth, D. The discovery of structure by multi-stream networks of local processors with contextual guidance. *Netw. Comput. Neural Syst.* **1995**, *6*, 225–246.
15. Cover, T.M.; Thomas, J.A. *Elements of Information Theory*; Wiley-Interscience: New York, NY, USA, 1991.
16. Schneidman, E.; Bialek, W.; Berry, M.J. Synergy, Redundancy, and Population Codes. *J. Neurosci.* **2003**, *23*, 11539–11553.
17. Kay, J. Neural networks for unsupervised learning based on information theory. In *Statistics and Neural Networks: Advances at the Interface*; Kay, J.W., Titterington, D.M., Eds.; Oxford University Press: Oxford, UK, 1999; pp. 25–63.
18. Kay, J.; Phillips, W.A. Activation functions, computational goals and learning rules for local processors with contextual guidance. *Neural Comput.* **1997**, *9*, 895–910.
19. Kay, J.W.; Phillips, W.A. Coherent infomax as a computational goal for neural systems. *Bull. Math. Biol.* **2011**, *73*, 344–372.
20. James, R.G.; Crutchfield, J.P. Multivariate Dependence beyond Shannon Information. *Entropy* **2017**, *19*, 530.
21. Wibral, M.; Priesemann, V.; Kay, J.W.; Lizier, J.T.; Phillips, W.A. Partial information decomposition as a unified approach to the specification of neural goal functions. *Brain Cognit.* **2017**, *112*, 25–38.
22. Harder, M.; Salge, C.; Polani, D. Bivariate measure of redundant information. *Phys. Rev. E* **2013**, *87*, doi:10.1103/PhysRevE.87.012130.
23. Pica, G.; Piasini, E.; Chicharro, D.; Panzeri, S. Invariant components of synergy, redundancy, and unique information. *Entropy* **2017**, *19*, 451, doi:10.3390/e19090451.

24. Wibral, M.; Lizier, J.T.; Vögler, S.; Priesemann, V.; Galuske, R. Local active information storage as a tool to understand distributed neural information processing. *Front. Neuroinf.* **2014**, *8*, doi:10.3389/fninf.2014.00001.

25. Lizier, J.T.; Prokopenko, M.; Zomaya, A. Local information transfer as a spatiotemporal filter for complex systems. *Phys. Rev. E* **2008**, *77*, doi:10.1103/PhysRevE.77.026110.

26. Wibral, M.; Lizier, J.T.; Priesemann, V. Bits from brains for biologically inspired computing. *Front. Robot. AI* **2015**, doi:10.3389/frobt.2015.00005.

27. Van de Cruys, T. Two Multivariate Generalizations of Pointwise Mutual Information. In Proceedings of the Workshop on Distributional Semantics and Compositionality, Portland, Oregon, 24 June 2011; pp. 16–20.

28. Church, K.W.; Hanks, P. Word Association Norms, Mutual Information, and Lexicography. *Comput. Linguist.* **1990**, *16*, 22–29.

29. James, R.G.; Ellison, C.J.; Crutchfield, J.P. Anatomy of a bit: Information in a time series observation. *Chaos* **2011**, 037109, doi:10.1063/1.3637494

30. Olbrich, E.; Bertschinger, N.; Rauh, J. Information decomposition and synergy. *Entropy* **2015**, *17*, 3501–3517.

31. Barrett, A.B. An exploration of synergistic and redundant information sharing in static and dynamical Gaussian systems. *Phys. Rev. E* **2015**, *91*, doi.org/10.1103/PhysRevE.91.052802

32. Chen, C.C.; Kasamatsu, T.; Polat, U.; Norcia, A.M. Contrast response characteristics of long-range lateral interactions in cat striate cortex. *Neuroreport* **2001**, *12*, 655–661.

33. Polat, U.; Mizobe, K.; Pettet, M.W.; Kasamatsu, T.; Norcia, A.M. Collinear stimuli regulate visual responses depending on cell's contrast threshold. *Nature* **1998**, *391*, 580–584.

34. Ince, R.A.A.; Giordano, B.L.; Kayser, C.; Rousselet, G.A.; Gross, J.; Schyns, P.G. A Statistical Framework for Neuroimaging Data Analysis Based on Mutual Information Estimated via a Gaussian Copula. *Hum. Brain Mapp.* **2017**, *38*, 1541–1573.

35. Panzeri, S.; Senatore, R.; Montemurro, M.A.; Petersen, R.S. Correcting for the Sampling Bias Problem in Spike Train Information Measures. *J. Neurophys.* **2007**, *98*, 1064–1072.

36. Ince, R.A.A.; Mazzoni, A.; Bartels, A.; Logothetis, N.K.; Panzeri, S. A Novel Test to Determine the Significance of Neural Selectivity to Single and Multiple Potentially Correlated Stimulus Features. *J. Neurosci. Methods* **2012**, *210*, 49–65.

37. Stramaglia, S.; Angelini, L.; Wu, G.; Cortes, J.; Faes, L.; Marinazzo, D. Synergistic and redundant information flow detected by unnormalized Granger causality: Application to resting state fMRI. *IEEE Trans. Biomed. Eng.* **2016**, *63*, 2518–2524.

38. Timme, N.M.; Ito, S.; Myroshnychenko, M.; Nigam, S.; Shimono, M.; Yeh, F.-C. High-Degree Neurons Feed Cortical Computations. *PLoS Comput. Biol.* **2016**, *12*, e1004858, doi:10.1371/journal. pcbi.1004858.

39. Phillips, W.A.; Clark, A.; Silverstein, S.M. On the functions, mechanisms, and malfunctions of intracortical contextual modulation. *Neurosci. Biobehav. Rev.* **2015**, *52*, 1–20.

Article

Quantifying Information Modification in Developing Neural Networks via Partial Information Decomposition

Michael Wibral [1,2,*], Conor Finn [3,4], Patricia Wollstadt [1], Joseph T. Lizier [3] and Viola Priesemann [2,5]

[1] MEG Unit, Brain Imaging Center, Goethe University, 60528 Frankfurt, Germany; patricia.wollstadt@gmx.de
[2] Max Planck Institute for Dynamics and Self-Organization, 37077 Göttingen, Germany; vp@ds.mpg.de
[3] Complex Systems Research Group and Centre for Complex Systems, Faculty of Engineering & IT,
     The University of Sydney, Sydney, NSW 2006, Australia; conor.finn@sydney.edu.au (C.F.);
     joseph.lizier@sydney.edu.au (J.T.L.)
[4] CSIRO Data61, Marsfield, NSW 2122, Australia
[5] Bernstein Center for Computational Neuroscience, 37077 Göttingen, Germany
[*] Correspondence: wibral@em.uni-frankfurt.de

Received: 13 July 2017; Accepted: 12 September 2017; Published: 14 September 2017

Abstract: Information processing performed by any system can be conceptually decomposed into the transfer, storage and modification of information—an idea dating all the way back to the work of Alan Turing. However, formal information theoretic definitions until very recently were only available for information transfer and storage, not for modification. This has changed with the extension of Shannon information theory via the decomposition of the mutual information between inputs to and the output of a process into unique, shared and synergistic contributions from the inputs, called a partial information decomposition (PID). The synergistic contribution in particular has been identified as the basis for a definition of information modification. We here review the requirements for a functional definition of information modification in neuroscience, and apply a recently proposed measure of information modification to investigate the developmental trajectory of information modification in a culture of neurons vitro, using partial information decomposition. We found that modification rose with maturation, but ultimately collapsed when redundant information among neurons took over. This indicates that this particular developing neural system initially developed intricate processing capabilities, but ultimately displayed information processing that was highly similar across neurons, possibly due to a lack of external inputs. We close by pointing out the enormous promise PID and the analysis of information modification hold for the understanding of neural systems.

Keywords: information theory; partial information decomposition; neural computation; neural development; self-organisation

1. Introduction

Shannon's quantitative description of information and its transmission through a communication channel via the entropy and the channel capacity, respectively, has drawn considerable interest from the field of neuroscience from the very beginning. This is because information processing in neural systems is typically performed in a highly distributed way by many communicating processing elements, the neurons.

However, in contrast to a channel in Shannon's sense, the purpose of dendritic connections in a neural system is not to simply relay information for the sake of reliable communication.

Instead, communication between neurons serves the purpose of collecting multiple streams of information at a neural processing element that *modifies* this information, i.e., that synthesizes the incoming streams into output information that is not available from any of these streams in isolation. This becomes immediately clear when looking at the meshed structure of nervous systems, where multiple communication streams converge on single neurons, and where neural output signals are sent in a divergent manner to many different receiving neurons. This structure differs dramatically from a structure solely focused on the reliable transmission of information where many parallel, but non-interacting streams would suffice. Thus, the meshed architecture seems to have evolved to "fuse" information from different input sources (including a neuron's recent spiking history and its current state) in a nontrivial way, e.g., other than simply multiplexing it in the output. In other words, the distributed computation in neural systems may heavily rely on information modification [1].

Attempts at formally defining information modification have presented a considerable challenge, however, in contrast to the well established measures of information transfer [2–6] and active information storage [7–9]. This is because identifying the "modified" information in the output of a processing element amounts to distinguishing it from the information from any input that survives the passage through the processor in unmodified form. These unmodified parts, in turn, may either come uniquely from one of the inputs, uniquely from another input (unique mutual information between an input and the output), or may be provided by multiple inputs simultaneously (shared mutual information between several inputs and the output). A decomposition of the mutual information between the inputs and the output of this kind is called a partial information decomposition (PID) [10] (Some authors prefer the simpler term "information decomposition", also in this special issue.).

In a PID, the problem of identifying modified information is equivalent to identifying the part of the (joint) mutual information that *is not* unique mutual information from one or another input, and that *is also not* shared mutual information from multiple inputs. This remaining part has been termed synergistic mutual information in the work of Williams and Beer [10], and has been identified with information modification in [11] for the reasons given above.

The recent pioneering work by Williams and Beer [10] revealed that the standard axioms of information theory do not uniquely define the unique, shared and synergistic contributions to the mutual information, and that additional axioms must be chosen for its meaningful decomposition. Among several possible choices of additional axioms or assumptions available at the time of this study (see Section 4.1) we here adhere to the definition given independently by Bertschinger, Rauh, Olbrich, Jost, and Ay ("BROJA-measure", [12]) and Griffith and Koch [13]. Our decision is based on two properties of the BROJA measure that seem necessary for an application to the problem of information modification as described above: first, in their definition, the presence of non-zero synergistic mutual information for the case of two inputs and one output cannot be deduced from the (two) marginal distributions of one input and one output variable. This property distinguishes the BROJA measure from the others available at the time; Bertschinger et al. [12] referred to it as Assumption (∗∗), and showed that the BROJA measure is the only measure that satisfies both Assumption (∗) and (∗∗), where Assumption (∗) indicates that the existence of unique information only depends on the pairwise marginal distributions between the individual inputs and the output. The measures from Williams and Beer [10] and Harder et al. [14] only satisfy Assumption (∗), but not (∗∗)—see [12] for details. Second, the BROJA measure is placed on a rigorous mathematical footing, being derived directly from the aforementioned Assumption (∗) rather than postulated ad hoc; furthermore, it has an operational interpretation in terms of expected utilities from the output based on knowledge of only each input, and many mathematical properties proven.

In our proof-of-principle study, we apply the BROJA decomposition of mutual information to the analysis of the emergent information processing in self-organizing neural cultures, and show that these novel information theoretic concepts indeed provide a meaningful contribution to our understanding of neural computation in this system.

2. Methods

In the following, we consider the neural data produced by two neurons as coming from two stationary random processes $\mathcal{X}_1$, $\mathcal{X}_2$, composed of random variables $X_1(i)$ and $X_2(i)$, $i = 1 \ldots n$, with realizations $x_1(i)$, $x_2(i)$. The corresponding embedding or state space vectors are given in bold font, e.g., $\mathbf{x}^l(i) = \{x(i), x(i-1) \ldots, x(i-l+1)\}$. The state space vector $\mathbf{x}^l(i)$ is constructed such that it renders the variable $x(i+1)$ conditionally independent of all random variables $x(j)$ with $j < i - l + 1$, i.e., $p(x(i+1)|\mathbf{x}^l(i), x(j)) = p(x(i+1)|\mathbf{x}^l(i))$.

2.1. Definition and Estimation of Unique, Shared and Synergistic Mutual Information

For two input random variables X_1, X_2 and an output random variable Y (Figure 1), Bertschinger et al. [12] defined the four unique, shared and synergistic contributions to the joint mutual information $I(Y : X_1, X_2)$ as:

$$\tilde{I}_{unq}(Y : X_1 \setminus X_2) := \min_{Q \in \Delta_P} I_Q(Y : X_1 | X_2) \tag{1}$$

$$\tilde{I}_{unq}(Y : X_2 \setminus X_1) := \min_{Q \in \Delta_P} I_Q(Y : X_2 | X_1) \tag{2}$$

$$\tilde{I}_{shd}(Y : X_1; X_2) := \max_{Q \in \Delta_P} (I(Y : X_1) - I(Y : X_1 | X_2)) \tag{3}$$

$$\tilde{I}_{syn}(Y : X_1; X_2) := I(Y : (X_1, X_2)) \\ - \min_{Q \in \Delta_P} I_Q(Y : (X_1, X_2)), \tag{4}$$

where I is the standard mutual or conditional mutual information [15,16], $\tilde{I}_{unq}$ is the unique, $\tilde{I}_{shd}$ the shared, and $\tilde{I}_{syn}$ the synergistic mutual information. In our notation, the comma separates variables within a set that are considered jointly, the colon separates the (sets of) random variables between which the mutual information is computed, while the semicolon or backslash separates sets of random variables that we are decomposing such mutual information across. For the latter, the semicolon is used for measures where the sets of random variables are considered symmetrically (i.e., shared and synergistic information), while the backslash is used for asymmetric cases (i.e., unique information in one but not the other). Δ_Q in the above definitions is the space of probability distributions Q that have the same pairwise marginal distributions between each input and the output as the original joint distribution P of X_1, X_2, Y, i.e.:

$$\Delta_P = \{Q \in \Delta : Q(X_1 = x_1, Y = y) = P(X_1 = x_1, Y = y) \\ \text{and } Q(X_2 = x_2, Y = y) = P(X_2 = x_2, Y = y)\}. \tag{5}$$

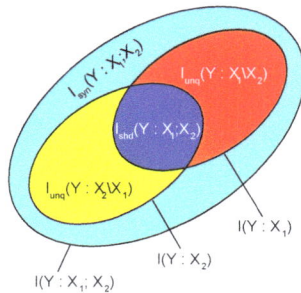

Figure 1. Decomposition of the joint mutual information between two input variables X_1, X_2 and the output variable Y. Modified from [17], CC-BY license.

2.2. Mapping of Neural Recordings to Input and Output Variables for PID, and Definition of Information Modification

In our application to developing neural cultures, we always consider two spike trains (A,B) at a time: the past *state* of the spike train A, $\mathbf{X_A^-}$, is one of the input variables and the past *state* of a spike train B, $\mathbf{X_B^-}$, is considered as the other input variable. Empirically, these states are usually constructed using the aforementioned embedding or state space vectors $\mathbf{x}^l(\mathbf{i})$ of length l. The output variable X_A^+ is simply spike train A's current spiking behavior (spiking or not).

This output variable is computed from external inputs ($\mathbf{X_B^-}$) as well as the output variable's own history $\mathbf{X_A^-}$. When analyzing this computation, one wishes to focus on the operations of information storage, transfer and modification, in alignment with established views of distributed information processing in complex systems [18,19]. In this study, specifically, we will focus on information modification, yet we first need to decompose the output variable in terms of information storage and information transfer, where the latter will also contain the information modification (see [11] and Figure 2):

$$I(X_A^+ : \mathbf{X_A^-}, \mathbf{X_B^-}) = I(X_A^+ : \mathbf{X_A^-}) + I(X_A^+ : \mathbf{X_B^-}|\mathbf{X_A^-}). \tag{6}$$

Here, $I(X_A^+ : \mathbf{X_A^-})$ is the active information storage [7], the predictive information from the past state of the variable to its next value. Then, $I(X_A^+ : \mathbf{X_B^-}|\mathbf{X_A^-})$ is the transfer entropy [2], the predictive information from the past of the other source B to the next value of A, in the context of the past of A.

In order to identify information modification, we need to take this decomposition further to reveal two sub-components of each of these information storage and transfer terms. These sub-components result from a partial information decomposition of $I(X_A^+ : \mathbf{X_A^-}, \mathbf{X_B^-})$ into four parts (see Figure 2):

1. The unique mutual information of the output spike train's own past $\tilde{I}_{unq}(X_A^+ : \mathbf{X_A^-} \setminus \mathbf{X_B^-})$—this can be considered as information uniquely stored in the past output of the spike train that reappears at the present sample.
2. The unique information from the other spike train $\tilde{I}_{unq}(X_A^+ : \mathbf{X_B^-} \setminus \mathbf{X_A^-})$—this is the information that is transferred unmodified from the input to the output of the receiving spike train (also known as the state independent transfer entropy [20]).
3. The shared mutual information about the output of spike train A that can be obtained both from the past states of spike train A and of spike train B, $\tilde{I}_{shd}(X_A^+ : \mathbf{X_B^-}; \mathbf{X_A^-})$—this is information that is redundantly stored in the past of both spike trains and that reappears at the present sample.
4. The synergistic mutual information $\tilde{I}_{syn}(X_A^+ : \mathbf{X_B^-}; \mathbf{X_A^-})$, i.e., the information in the output of spike train A, X_A^+, that can only be obtained when having knowledge about both the past state of the external input, $\mathbf{X_B^-}$, and the past state of the receiving spike train, $\mathbf{X_A^-}$. (This is also known as the state dependent transfer entropy [20]).

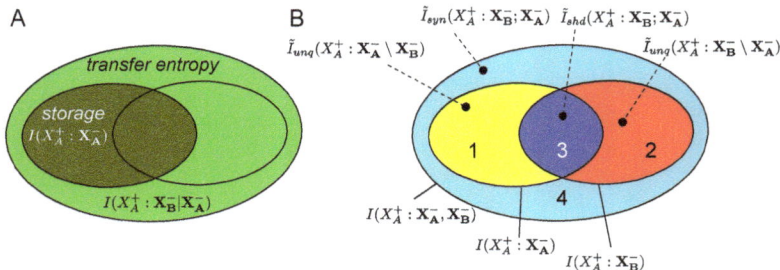

Figure 2. Mapping between the decomposition into storage and transfer (**A**) and individual or joint mutual information terms, and PID components (**B**). Numbers in (**B**) refer to the enumeration of components given in Section 2.2. Number "4" is the modified information.

We see that Components 1 and 3 above form the active information storage in Equation (6), while Components 2 and 4 form the transfer entropy term. Component 4, as the synergistic mutual information contributed by the storage and the transfer source, is what we consider to be the *modified information* (following [11]). The same underlying definition of information modification from [11] was used by Timme et al. [21] in an earlier study of dynamics of spiking activity of neural cultures, yet with another PID measure and considering multiple external inputs to a neuron (see discussion for further details).

2.3. PID Estimation

PID terms were estimated by minimizing the conditional mutual information as indicated in the first equation of the system 1. To perform the minimization, we used a stochastic approach where alternative trial distributions in Δ_P are created by swapping probability mass δ_p between the symbols of the current distribution such that the constraints defining Δ_P are satisfied. If this swap of probability mass leads to a reduction in the conditional mutual information, the trial distribution is made the current distribution, and a new trial distribution is created. If the trial distribution fails to reduce the conditional mutual information, then a new trial distribution is created from the current distribution. This latter process is repeated for a maximum of n unsuccessful swaps *in a row* (here $n = 20,000$), with a reset of the counter in case of a successful swap. If after these trials no reduction is reached, then we assume that we have found the optimum possible with the current increment in probability mass δ_p and that a finer resolution is needed. Hence, the increment is halved: $\delta_p \leftarrow \delta_p/2$. This process starts with an initial δ_p equal to the largest probability mass assigned to any symbol in the distribution P, and is repeated until the numerical precision of the machine or programming language is exhausted (here, we performed 63 divisions of the original δ_p by a factor of 2, using Numpy 1.11.2 under Python 3.4.3 and 128-bit floating point numbers). The algorithm is available in the open source toolbox IDT^{xl} [22]. We note that better solutions based on convex optimization exist (see Makkeh et al. [23] in this special issue) and that these are implemented in newer versions of IDT^{xl}; at the time of performing this study, however, these implementations were not available to us yet.

2.4. Statistical Testing

Results obtained for the joint mutual information, and for the four PID measures *normalized* by the joint mutual information, were subjected to pairwise statistical tests for differences in the median between recordings days (see Section 2.5) by means of permutation tests. An uncorrected p-value of $p < 5 \times 10^{-4}$, corresponding to p-value of $p < 0.05$ with Bonferroni correction for multiple comparisons across five measures, and 20 pairs of recording days, was considered significant.

We normalized the PID values to remove influences from changes in the overall activity of the culture (that change the entropy of the inputs) and to abstract from changes in the overall joint mutual information. Note that we did not test these normalized PID values for significance against surrogate data, as the focus here was on changes with development of the culture. Moreover, the four normalized PID terms analyzed here are not independent from each other, but instead sum up to a value of 1, making the construction of a meaningful statistical test difficult.

2.5. Electrophysiological Data—Acquisition and Preprocessing

The spike recordings were obtained by Wagenaar et al. [24] from a single in vitro culture of $M \approx 50,000$ cortical neurons. The data are available online at [25]; of the data provided in this repository, we used culture/experiment "2-1", days 7, 14, 21, 28, and 34. Details on the preparation, maintenance and recording setting can be found in the original publication. In brief, cultures were prepared from embryonic E18 rat cortical tissue. Recordings lasted more than 30 min. The recording system comprised an 8×8 array of 59 titanium nitride electrodes with 30 μm diameter and 200 μm inter-electrode spacing, manufactured by Multichannel Systems (Reutlingen, Germany). As described in the original publication, spikes were detected online using a threshold based detector as upward or downward excursions beyond 4.5 times the estimated root mean squared (RMS) noise [26]. Spike waveforms were stored, and used to remove duplicate detections of multiphasic spikes. Spike sorting was not employed, and thus spike data represent multi-unit activity. To obtain a tractable amount of data, we randomly picked spike time series from the dataset, and of these only selected those that developed at least a moderate level of activity with maturation of the culture (channels 01, 02, 03, 04, 05, 07, 11, 13, 14, 16, 19, 50, 53, 57, 58, 60), to guarantee a certain level of (Shannon) information to be present. In total, our analyses comprise 16 spike time series, i.e., 240 pairs of spike time series.

From these spike time series, the realizations $x_A^+(i)$ of the random variable $X_A^+(i)$ were constructed by applying bins of 8 ms length; empty bins were denoted by zeros, whereas bins that contained at least one spike were denoted with ones. The corresponding approximate past state vectors $\mathbf{x}_{A/B}^{-1}(i)$ were constructed with finite past length $l = 3$, and to balance the need for low dimensionality for an unbiased estimate and a coverage of as much past history as possible, three past bins of size 8 ms, 32 ms and 32 ms were defined, where the shortest bin was the bin closest to i, and where both the 8 ms and the two 32 ms bins were set to one or zero depending on whether or not a spike occurred anywhere within these bins. This approach to cover a longer history at a low dimensionality amounts to a compressing of the information in the history of the process, aiming to retain what we perceive to be the most relevant information. This approach is similar to the one used by Timme et al. [27], except for the use of nonuniform binwidths in our case. Alternative approaches to large bin widths exist that are either based (i) on nonuniform embedding, picking the most informative past samples (or bins with a small width on the order of the inverse sampling rate) from a collection of candidates (e.g., [28–30]), and the IDT^{xl} toolbox [22]; or (ii) on varying the lag between an a vector of evenly spaced past bins and the current sample [4,31,32], but both of these approaches might be less suitable for relatively sparse binary data, such as spike trains.

3. Results

3.1. PID of Information Processing in Neural Cultures

From the original report by Wagenaar et al. [26], the following aspects of the development of the "dense" culture analyzed here can be observed: (i) by preparation, neurons were unconnected and mostly silent at first; then (ii) show spontaneous activity and begin to be connected (compare the increase of mutual information between inputs and outputs in Figure 3); later, they (iii) become densely connected and thereby strongly responsive to each other's spiking activity (Quote from [26]: *"We ... found that functional projections grew rapidly during the first week in vitro in dense cultures, reaching across the entire array within 15 days (Figure 9)"*); while, in a last stage, (iv) connectivity often leads to activity pattern where all neurons become simultaneously active in large, culture spanning bursts of activity. This can, for example, be seen for data used here in the development of large, system spanning neural avalanches with maturation of the culture (see Figure 4 in [33] and Figure 13 in the corresponding preprint [34]; for the definition of neural avalanches as used here, see [35]). The number of such system spanning avalanches was [0, 0, 7, 50, 73] for the five recording weeks. At the same time, the mean avalanche sizes (defined in [35]) also increased as (1.05, 1.31, 1.81, 4.39, 3.42)—note the jump from week 2 to 3 in both measures, and compare to the normalized shared information in Figure 4.

From the viewpoint of partial information decomposition, we hypothesized that stages (i) and (ii) should be characterized by a high fraction of unique information from a neuron's own history because neurons that do not yet receive sufficient input to trigger their firing can only have unique mutual information with their own history.

Unique information from other neuron's inputs, and also synergy between both neurons' past states should be visible in stage (iii) because we assume that neurons, even in vitro are wired to fuse information from multiple sources with their the information of their own state. Thus, we expected non-trivial computation in the form of synergy to be visible as long as the input distributions are sufficiently different from a neuron's own history.

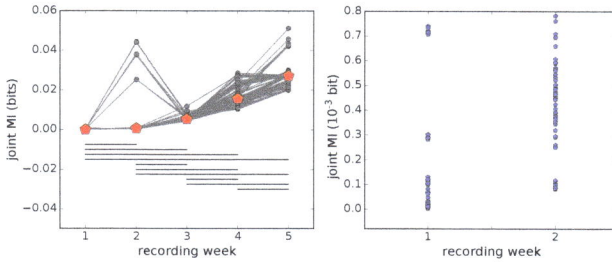

Figure 3. Left: development of the joint mutual information with network maturation. Grey symbols and lines—joint mutual information (MI) from individual pairs of spike time series, red symbols—median over all pairs. Horizontal black lines connect significantly different pairs of median values ($p < 0.05$, permutation test, Bonferroni corrected for multiple comparisons); **Right:** magnification of the joint mutual information estimates in the first two recording weeks. Note that the three large outliers from week 2 have been omitted from the plot. These tiny, but non-zero, values form the basis for the *normalized* non-zero PID terms presented in Figure 4—also leading to considerable variance there.

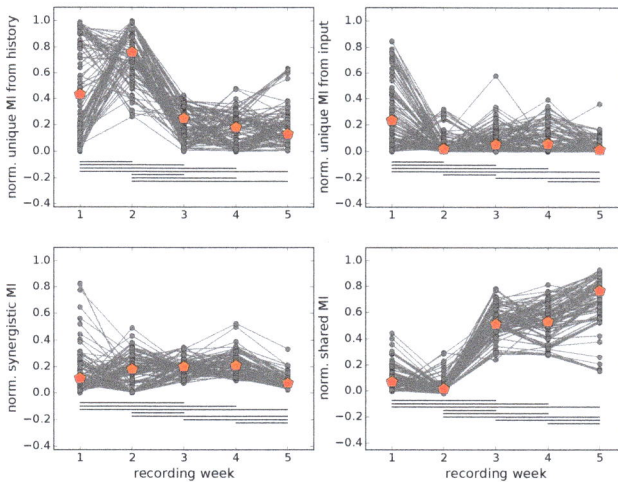

Figure 4. Development of *normalized* PID contributions (i.e., PID terms normalized by the joint mutual information) with network maturation. Grey symbols and lines—PID values from individual pairs of spike time series, red symbols—median over all pairs. Horizontal black lines connect significantly different pairs of median values ($p < 0.05$, permutation test, Bonferroni corrected for multiple comparisons). On the lower right, note the sudden increase in normalized shared mutual information from week 2 to 3 that coincides with the onset of system spanning neural avalanches (see text).

In the last stage (iv), partial information decomposition should then be dominated by shared mutual information because when all input distributions are more or less identical and highly correlated, then there can only be shared information (at least when using the BROJA measure).

From preliminary investigations [36], we also expected the joint mutual information between both inputs and the output to rise. Given the caveat that we analyzed multi-unit activity here, instead of single units (i.e., single neurons) obtained by spike sorting, our results comply with these hypotheses: the initial two recording weeks were dominated by unique information from a spike time series' own history, while, in intermediate recording weeks, synergistic and shared information were dominant, and shared information finally prevailed in the last two recordings (Figure 3 shows the joint mutual information, with the normalized PID contributions to this shown in Figure 4 and raw PID contributions in Figure 5).

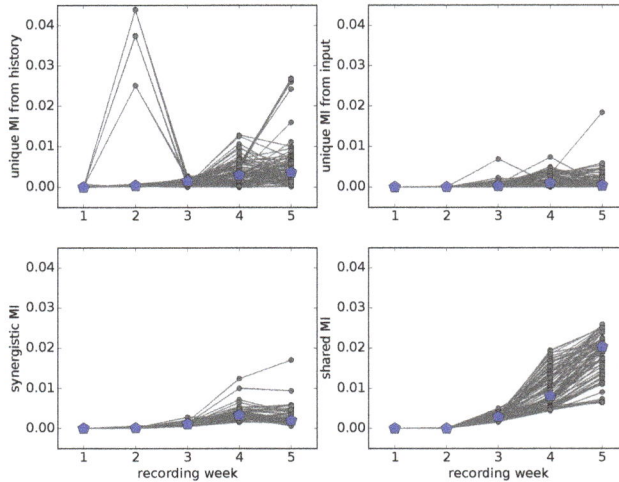

Figure 5. Development of *raw* PID contributions with network maturation. Grey symbols and lines—PID values from individual pairs of spike time series, blue symbols—median over all pairs. Note that we do not provide statistical tests here as the visible differences are heavily influenced by the corresponding differences in the joint mutual information (see Figure 3).

4. Discussion

We here applied PID to neural spike recordings with the objective to compute a measure of information modification, and, for the first time, to assess its face validity given what is already known about information processing in developing neural cultures. Our analyses of the synergistic part of the mutual information between information storage and transfer sources, which we see as a promising candidate measure of information modification, complied with our intuition on how information modification should rise with development as neurons get connected and their synaptic weights adapt to the environment of the culture (i.e., with a lack of external input to the culture). The end of this rise in (relative) information modification and a final drop caused by a jump in the (relative) shared part in the mutual information was also expected given that a computation must always be understood as the composition of a mechanism and an input distribution. This input distribution is well known to get more and more similar over neurons as the culture approaches the typical bursting behavior that synchronizes all activity. With all input distributions being similar in this way, there is reduced scope for modifying information—hence the observed drop in the last recording week. In summary, the partial information decomposition used here and the results for its synergistic part capture well our intuition of what should happen in this simple neural system. This increases confidence in the usefulness and interpretability of PID measures in the analysis of neural data from more complex neural systems.

Two additional aspects seem important here: first, our analyses underline one of the key theoretical advances of PID, that all four PID terms, and especially shared and synergistic ones, can coexist simultaneously—a fact overlooked in early attempts to define shared (or 'redundant') and synergistic contributions to the mutual information (see references in [10]); second, no knowledge on the typical development of neural cultures was necessary to arrive at our PID results; in other words, the development of computation in the culture could have been derived from our PID analysis alone.

This makes PID of neural activity a useful first step when investigating the computational architecture of a neural system.

In the sections below, we discuss some caveats to consider with this relatively young analysis technique, where several competing definitions of a PID coexist, not all of them equally suitable for computing information modification [11]. We also expand on the aforementioned relation between measures of information transfer and modification. Moreover, we would like to highlight and expand on the fact that a computation is a composition of an input distribution and a mechanism working on these inputs. Neglecting the importance of the input distribution and understanding a PID as directly describing a computational mechanism is a frequent misunderstanding that we would like to clarify here. We close by highlighting potential uses of PID in neuroscience.

4.1. Which Definition of Synergistic Mutual Information to Use?

In contrast to earlier studies of synergy or information modification in neural data [21,37], we here used the definition of unique, shared and synergistic mutual information as given by Bertschinger et al. [12] and by Griffith and Koch [13] (BROJA-decomposition). As initially outlined, in our view, this definition was the only published one at the time of experiment that had the properties necessary for a mapping between information modification and the synergistic part of the mutual information, and is sufficiently easy to compute because of the convexity of the underlying optimization.

However, the BROJA definition has also been criticized because a decomposition is only possible for the case of two inputs (although these inputs themselves can be arbitrarily large groups of variables). We consider this an acceptable restriction for some purposes in neuroscience as it seems to map well to the properties of cortical neurons; for example, the pyramidal neurons of the neocortex keep exactly two classes of inputs separate via their apical and basal dendrites (see [17,38,39] and references therein). In addition, as long as one is only interested in the computations performed by a neuron based on its own history and all its inputs considered together as *one* (vector-valued) input variable, this framework is sufficient (see, for example, the treatment of this case in the theoretical study presented in [17]).

In contrast to the BROJA-decomposition, Williams and Beer suggested in their original work [10] an alternative definition that allowed the decomposition of the mutual information between multiple inputs (considered separately, not as a group) and an output into a partial order (a mathematical "lattice") of shared information terms. While this decomposition into a lattice of terms clearly is desirable, the measure of shared information given by Williams and Beer [10] (known as I_{min}) also has several properties that have been questioned. First, it does not respect the locality of information, i.e., point-wise interpretations of this shared information are not continuous with respect to the underlying probability distribution functions [11]. Second, it suggests the presence of shared information in situations where in each realization only a single source ever holds non-zero information about a target [40]. We note that the latter is an issue for the BROJA-decomposition as well.

Third, several authors have questioned the presence of non-zero shared mutual information under I_{min} when there is no pairwise mutual information between the inputs themselves while the output is a simple collection of these inputs (known as "two bit copy"). A desire for zero shared mutual information in this case was formalized in a so-called identity axiom by Harder et al. [14]. This axiom suggests that if two inputs X_1, X_2 with no mutual information between them ($I(X_1 : X_2) = 0$) are combined into an output that is simply their collection, i.e., $Y = \{X_1, X_2\}$ then the shared part of the joint mutual information $I(Y : X_1, X_2)$ must be zero. However there are significant arguments against the inclusion of such an axiom, and in support of the presence of shared information in the two bit copy problem; see, e.g., Bertschinger et al. [41], and, in this issue, by Ince [42] and Finn et al. [40]. For example, there can be no measure of redundant information that simultaneously satisfies the original three PID axioms, has non-negative PI atoms, and possesses the identity property [43].

Debates continue on this aspect, and, in the future, it will be interesting to check the consistency of results reported here with respect to alternative decompositions, such as those presented by Finn et al. [40] or Ince [42] in this issue.

In summary, the BROJA measure used in this study has several appealing properties, yet it lacks the ability to decompose the information of more than two input variables into a lattice. Several contributions to this special issue present progress on lattice-compatible distributions [40,42] and also investigate the consequences of the symmetrical, or asymmetrical treatment of information sources and targets [44] (also see the work of Olbrich et al. [45] on this topic).

4.2. Previous Studies of Information Modification in Neural Data

Timme et al. [21] studied information modification in the dynamics of spiking activity of neural cultures with a focus on the relation between information modification at a neuron and its position in the underling (effective) network structure. They report, for example, that neurons which modify "large amounts of information tended to receive connections from high out-degree neurons". Both their study and ours have in common the same underlying definition of information modification [11]. Their study differs slightly from ours in examining synergy between two external inputs to a neuron, conditional on that neuron's past, whereas we examine synergy between one external input and the receiving neuron's past. A more important difference between their study and ours, however, is the choice of PID measure (see above). Specifically, they used the Williams and Beer [10] I_{min} measure, in contrast to the BROJA measure used here—see Section 4.1 for details on the consequences of these choices.

Another important difference is the use of multi unit activity in our study, while Timme et al. [21] used spike sorted data that represents the activity of single neurons. However, for the data-set we used, spike extraction was relatively conservative, using a high threshold and removing events with spurious waveforms [26]. This resulted in a relatively low average multi-unit activity of less than 3.5 Hz. This is comparable to the mean rate of 2.1 Hz reported by Timme et al. [21]. From this, we estimate that only one or two close by neurons typically contribute to the recorded multi-unit activity. Thus, this difference may be relatively minor in practice. Conceptually, however, the information contained in single and multi-unit activity clearly differs in interpretation—see the next section for the more details.

We also note that there are earlier applications of the concept of synergy (meant as synergistic mutual information) to neural data (e.g., [46–49]) that relied on the computation of interaction information. However, when interpreting these studies, it should be kept in mind that these report the *difference* between shared information and synergistic information—as detailed by Williams and Beer [10]. If both are present in the data (a possibility that may simply have been overlooked by most researchers before Williams and Beer [10]), then this view of a 'net-synergy' only gives a partial view of the coding principles involved.

4.3. Information Represented by Multi and Single Unit Data

As detailed in the methods section, we performed our analyses on multi unit activity, i.e., we considered all spiking activity picked up by a recording electrode—potentially coming from multiple neurons. Thus, the information processing analyzed here is that of a cluster of neurons close to the recording electrodes, but not that of individual neurons, limiting the direct interpretation of our results. This problem can be alleviated by using spike sorting algorithms, e.g., based on the individual waveforms to assign each spike to an individual neuron, and then analyzing only the spikes of individual neurons. This has indeed been done in the study by Timme et al. [21] and improves the interpretation of the results in terms of neural coding. Ideally, it should be included in follow-up studies on information modification via PID as well. However, as the multi-unit activity reported here most likely contained only one or two single units (see previous section), we expect very similar results for an analysis of single units.

4.4. Measured Information Modification versus the Capacity of a Mechanism for Modifying Information

To appreciate the findings of the current study, it is important to realize that any computation is a composition of (i) a mechanism and (ii) an input distribution. As an extreme example, take an "exclusive or" (XOR)-gate, which has only one bit of synergy when fed by two uniformly distributed random bit inputs. However, when we clamp one of these inputs, for example X_1, to producing just '0s, then all the information (still one bit of joint mutual information) is unique information from the other input X_2. This result must hold for all PID measures by virtue of the equations linking classic mutual information terms and PID terms (Equations (1)–(3) in [12], also consult Figure 1), and due to the fact that the mutual information of the clamped input and the output must be equal to or smaller than the entropy of that input, which is zero. Feeding the XOR gate with an alternative input distribution $p_a(x_1, x_2)$ of the form $p_a(0,1) = 3/8$, $p_a(1,1) = 3/8$, $p_a(0,0) = p_a(1,0) = 1/8$ yields 0.811 bits of synergy and 0.188 of unique information from x_1, using the BROJA PID.

Another simple example would be a logical conjunction (AND)–gate fed by two different input distributions: when fed by two independent streams of input bits with uniform probabilities of zeros and ones, the BROJA PID results in 0.5 bit of synergy and 0.311 bit of shared information [12]. Feeding the same mechanism with an alternative input distribution $p_b(x_1, x_2)$ of the following form: $p_b(0,0) = 3/8$, $p_b(1,1) = 3/8$, $p_b(0,1) = p_b(1,0) = 1/8$ results in approximately 0.406 bit of synergy and 0.549 bit of shared information as measured by the BROJA PID.

This dependence of the PID on the input distribution means that describing a computation in terms of information modification via the synergistic information describes the joint operation of input distribution and mechanism (with the consequences related to bursting activity in neural cultures that were noted above). Indeed, this is the correct information theoretic description of how the system modified information in the specific computation reflected in the data. This description does not, however, inform us about how much information modification the mechanism performing the computation is capable of in principle. This is analogous to the situation of a communication channel in Shannon's theory where the mutual information $I_{P_X}(X : Y)$ between the input X and the output Y of a channel informs us about how much information is actually communicated across the channel when it is fed by the input distribution P_X. However, $I_{P_X}(X : Y)$ will not inform us about how much information we could in principle communicate across the channel, i.e., the capacity of the channel defined by:

$$\mathcal{C} = \underset{P_X}{\text{argmax}}\ I_{P_X}(X : Y). \tag{7}$$

Thus, for describing the potential of a mechanism to modify information, we must define an information modification capacity in analogy to the definition of an information transmission capacity (Equation (7)) by maximizing the synergistic mutual information over all input distributions as:

$$\mathcal{C}_{\text{mod}} = \underset{P_{Y^-,X^-}}{\text{argmax}}\ \tilde{I}_{syn}(Y^+ : \mathbf{Y}^- ; \mathbf{X}^-), \tag{8}$$

$$= \underset{P_{Y^-,X^-}}{\text{argmax}} \left[I(Y^+ : (\mathbf{Y}^-, \mathbf{X}^-)) - \underset{Q \in \Delta_{P_{Y^+,Y^-,X^-}}}{\min} I_Q(Y^+ : (\mathbf{Y}^-, \mathbf{X}^-)) \right]. \tag{9}$$

How tractable the double optimization process implied in Equation (9) is in practice and whether analytical simplifications can be derived remains the topic of future work. However, other measures of PID that do not rely on an optimization over the space of probability distributions (such as the one by Finn et al. [40] in this special issue) may allow for the computation of a capacity for information modification—given the mechanism is known.

We would like to emphasize that maximizing synergy, or any other PID term, over possible input distributions is different from maximizing the same PID term via changes to the mechanism that yields the output, while keeping the input distributions fixed. This latter approach is considered in detail by the contribution of Rauh et al. [50] in this special issue.

4.5. On the Distinction between Information Modification and Noise

We emphasize that the definition of information modification used here (and first put forward in [11]) will not count information that is created de novo in an information processing element and then appears in its output. This is because modified information in the output has to be explained ultimately by the input to a processing element and the state (or history) of that element, taken together. This clearly does not hold for information just created independently of the processor's history. In other words, the information created de novo is counted as output noise instead of as modified information by our definition of information modification—a property that we consider desirable for any measure of information modification.

4.6. On the Relation between Transfer Entropy and Information Modification

As introduced in Section 2.2, the transfer entropy between two processes $\mathcal{X}$, $\mathcal{Y}$, where the variables Y_t, X_t carry the current values of the processes and the variables $\mathbf{X}^-$, $\mathbf{Y}^-$ carry the past state information is defined as [2]:

$$TE(\mathcal{X} \rightarrow \mathcal{Y}) = I(Y_t : \mathbf{X}^- | \mathbf{Y}^-) \,. \tag{10}$$

As first noted by Williams and Beer [20], the (conditional) mutual information on the right-hand side can be decomposed using a PID as well. As shown in Section 2.2, this conditional mutual information is composed of both a unique contribution from the source, and a synergistic contribution where the current value y_t is determined jointly—and not explainable in any simpler way—by the combination of past states $\mathbf{x}^-$ and $\mathbf{y}^-$, i.e., the input from $\mathcal{X}$ to $\mathcal{Y}$ and $\mathcal{Y}$'s own history. (Of course, either component could be zero for a given distribution). Williams and Beer [10] suggested to call this synergistic part of the transfer entropy the (receiver) state dependent transfer entropy (SDTE) to highlight the interplay between sender and receiver in modifying the information. Obviously, such a subdivision of transfer is highly useful where computable. Naturally, the overlaps between the concepts of information modification and (multivariate) transfer entropy become more involved if $\mathcal{Y}$ receives more than one external input. What we label as information modification in this case would comprise the SDTE above, but perhaps not all of it, and also have additional contributions (see below and Figure 6).

This is a special case of the general effect that the synergistic components of a PID may change if additional inputs are considered, e.g., when the additional input on its own brings in information that is itself redundant with the information seen as synergistic between the other inputs. See, for example, the component labeled with $\{X_2\}\{MX_1\}$ in Figure 6, which is synergistic when not considering X_2, but redundantly also provided by X_2 alone. In more detail, a PID may decompose the information provided by a larger set of sources into many different shared (redundant), unique and synergistic components between subsets of these inputs. These components are placed onto a lattice (a partial order) by some variants of PID measures (see Section 4.1).

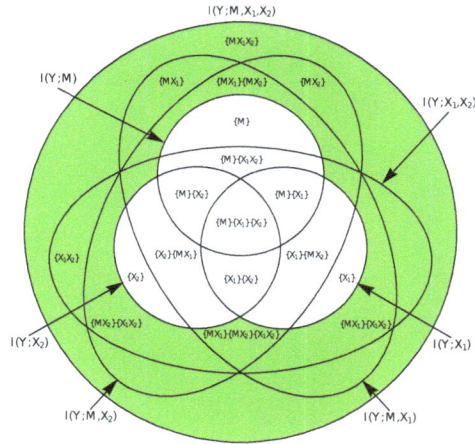

Figure 6. PID diagram for three input variables—two of them external inputs (X_1, X_2), and one representing the past state of the receiving system ($M = Y^-$). The parts of the diagram highlighted in green would be considered information modification. These parts represent the information in the receiver that can only be explained by two or more input variables considered jointly.

4.7. New Research Perspectives in Neuroscience Based on PID and Information Modification

In closing, we would like to highlight the vast potential that PID and the analysis of information modification have both in understanding biological neural systems, and in designing artificial ones.

As detailed in [1], the comparison of shared vs. synergistic mutual information in the output of a neuron or neural network allows us to address directly issues of robust coding vs. maximizing coding capacity, and thereby helps us to understand fundamental design principles of biological networks.

Conversely, PID can also be used to define information theoretic goal functions and to derive local learning rules for neurons in artificial neural networks with unprecedented detail and precision as explicated in [17], extending popular information theoretic goal functions like infomax [51], or coherent infomax (see [52] and references therein). In particular, the formulation of novel PID estimators that no longer rely on an optimization step (see the work of Finn et al. [40] in this special issue) has seemingly removed remaining difficulties with an analytical treatment of this approach.

Moreover, the PID formalism lends itself easily to the analysis of both neural and behavioral data, enabling a direct comparison of the two. This will take our understanding of the relationship between neural activity and behavior beyond the level of an analysis of mere representations, i.e., beyond decoding representations of objects and intentions, to finding the loci of particular aspects of neural computation. For example, in a human performing a task requiring an XOR computation, one may look for hot-spots of synergistic mutual information in the system.

Ultimately, the ability to obtain a complete fingerprint of a neural computation in terms of active information storage, information transfer and, now, information modification makes it possible to identify algorithms implemented by a neural system—or at least strongly confines the search space. This finally allows to fully address the algorithmic level of understanding neural systems as formulated more than 30 years ago by David Marr ([53], also see [1]).

5. Conclusions

We used a recent extension of information theory here to measure where and when in a neural network information is not simply communicated through a channel but modified. The definition of information modification here builds on the concept of synergistic mutual information as introduced by Williams and Beer [10], and the measure defined by Bertschinger et al. [12]. We show that, in the developing neural culture analyzed here, the contribution of synergistic mutual information rose as the network became more connected with development but ultimately dropped again as the activity became largely synchronized in bursts across the whole neural culture such that most mutual information was shared mutual information.

Acknowledgments: This project was supported through a Universities Australia/German Academic Exchange Service (DAAD) Australia–Germany Joint Research Cooperation Scheme grant (2016-17): "Measuring neural information synthesis and its impairment". Joseph T. Lizier was also supported through the Australian Research Council DECRA grant DE160100630, and a Faculty of Engineering and IT Early Career Researcher and Newly Appointed Staff Development Scheme 2016 grant. Viola Priesemann received financial support from the Max Planck Society, and from the German Ministry for Education and Research (BMBF) via the Bernstein Center for Computational Neuroscience (BCCN) Göttingen under Grant No. 01GQ1005B. The authors would like to thank all participants of the 2016 Workshop on Partial Information Decomposition held by the Frankfurt Institute for Advanced Studies (FIAS) and the Goethe University for their inspiring discussions on PID.

Author Contributions: Michael Wibral and Viola Priesemann conceived of the study; Michael Wibral and Viola Priesemann analyzed the data; Conor Finn, Joseph T. Lizier, Patricia Wollstadt and Michael Wibral wrote the code for PID computation in the IDT^{xl} toolbox; Michael Wibral, Viola Priesemann, and Joseph T. Lizier wrote the manuscript.

Conflicts of Interest: The authors declare no conflict of interest.

References

1. Wibral, M.; Lizier, J.T.; Priesemann, V. Bits from Brains for Biologically Inspired Computing. *Front. Robot. AI* **2015**, *2*, 5.
2. Schreiber, T. Measuring information transfer. *Phys. Rev. Lett.* **2000**, *85*, 461–464.
3. Vicente, R.; Wibral, M.; Lindner, M.; Pipa, G. Transfer entropy—A model-free measure of effective connectivity for the neurosciences. *J. Comput. Neurosci.* **2011**, *30*, 45–67.
4. Wibral, M.; Pampu, N.; Priesemann, V.; Siebenhühner, F.; Seiwert, H.; Lindner, M.; Lizier, J.T.; Vicente, R. Measuring information-transfer delays. *PLoS ONE* **2013**, *8*, e55809.
5. Wollstadt, P.; Martínez-Zarzuela, M.; Vicente, R.; Díaz-Pernas, F.J.; Wibral, M. Efficient transfer entropy analysis of non-stationary neural time series. *PLoS ONE* **2014**, *9*, e102833.
6. Lizier, J.T.; Prokopenko, M.; Zomaya, A.Y. Local information transfer as a spatiotemporal filter for complex systems. *Phys. Rev. E* **2008**, *77*, 026110.
7. Lizier, J.T.; Prokopenko, M.; Zomaya, A.Y. Local measures of information storage in complex distributed computation. *Inf. Sci.* **2012**, *208*, 39–54.
8. Wibral, M.; Lizier, J.T.; Vögler, S.; Priesemann, V.; Galuske, R. Local active information storage as a tool to understand distributed neural information processing. *Front. Neuroinf.* **2014**, *8*, doi:10.3389/fninf.2014.00001.
9. Gomez, C.; Lizier, J.T.; Schaum, M.; Wollstadt, P.; Grützner, C.; Uhlhaas, P.; Freitag, C.M.; Schlitt, S.; Bölte, S.; Hornero, R.; et al. Reduced Predictable Information in Brain Signals in Autism Spectrum Disorder. *Front. Neuroinf.* **2014**, *8*, 9.
10. Williams, P.L.; Beer, R.D. Nonnegative Decomposition of Multivariate Information. *arXiv* **2010**, arXiv:1004.2515.
11. Lizier, J.T.; Flecker, B.; Williams, P.L. Towards a synergy-based approach to measuring information modification. In Proceedings of the 2013 IEEE Symposium on Artificial Life (ALIFE), Singapore, 16–19 April 2013; pp. 43–51.
12. Bertschinger, N.; Rauh, J.; Olbrich, E.; Jost, J.; Ay, N. Quantifying Unique Information. *Entropy* **2014**, *16*, 2161–2183.
13. Griffith, V.; Koch, C. Quantifying Synergistic Mutual Information. In *Guided Self-Organization: Inception*; Prokopenko, M., Ed.; Springer: Berlin, Germany, 2014; Volume 9, pp. 159–190.
14. Harder, M.; Salge, C.; Polani, D. Bivariate measure of redundant information. *Phys. Rev. E* **2013**, *87*, 012130.
15. Fano, R. *Transmission of Information*; The MIT Press: Cambridge, MA, USA, 1961.
16. Cover, T.M.; Thomas, J.A. *Elements of Information Theory*; Wiley-Interscience: New York, NY, USA, 1991.
17. Wibral, M.; Priesemann, V.; Kay, J.W.; Lizier, J.T.; Phillips, W.A. Partial information decomposition as a unified approach to the specification of neural goal functions. *Brain Cognit.* **2017**, *112*, 25–38.
18. Langton, C.G. Computation at the edge of chaos: Phase transitions and emergent computation. *Phys. D Nonlinear Phenom.* **1990**, *42*, 12–37.

19. Mitchell, M. Computation in Cellular Automata: A Selected Review. In *Non-Standard Computation*; Gramß, T., Bornholdt, S., Groß, M., Mitchell, M., Pellizzari, T., Eds.; Wiley-VCH Verlag GmbH & Co. KGaA: Weinheim, Germany, 1998; pp. 95–140.

20. Williams, P.L.; Beer, R.D. Generalized Measures of Information Transfer. *arXiv* **2011**, arXiv:1102.1507.

21. Timme, N.M.; Ito, S.; Myroshnychenko, M.; Nigam, S.; Shimono, M.; Yeh, F.C.; Hottowy, P.; Litke, A.M.; Beggs, J.M. High-Degree Neurons Feed Cortical Computations. *PLoS Comput. Biol.* **2016**, *12*, 1–31.

22. Wollstadt, P.; Lizier, J.T.; Finn, C.; Martinez-Zarzuela, M.; Vicente, R.; Lindner, M.; Martinez-Mediano, P.; Wibral, M. The Information Dynamics Toolkit, IDT^{xl}. Available online: https://github.com/pwollstadt/IDTxl (accessed on 25 August 2017).

23. Makkeh, A.; Theis, D.O.; Vicente Zafra, R. Bivariate Partial Information Decomposition: The Optimization Perspective. **2017**, under review.

24. Wagenaar, D.A.; Pine, J.; Potter, S.M. An extremely rich repertoire of bursting patterns during the development of cortical cultures. *BMC Neurosci.* **2006**, *7*, 11.

25. Wagenaar, D.A. Network Activity of Developing Cortical Cultures In Vitro. Available online: http://neurodatasharing.bme.gatech.edu/development-data/html/index.html (accessed on 25 August 2017).

26. Wagenaar, D.; DeMarse, T.B.; Potter, S.M. MeaBench: A toolset for multi-electrode data acquisition and on-line analysis. In Proceedings of the 2nd International IEEE EMBS Conference on Neural Engineering, Arlington, VA, USA, 16–20 March 2005; pp. 518–521.

27. Timme, N.; Shinya, I.; Maxym, M.; Fang-Chin, Y.; Emma, H.; Pawel, H.; Beggs, J.M. Multiplex Networks of Cortical and Hippocampal Neurons Revealed at Different Timescales. *PLoS ONE* **2014**, *9*, 1–43.

28. Faes, L.; Marinazzo, D.; Montalto, A.; Nollo, G. Lag-specific transfer entropy as a tool to assess cardiovascular and cardiorespiratory information transfer. *IEEE Trans. Biomed. Eng.* **2014**, *61*, 2556–2568.

29. Lizier, J.T.; Rubinov, M. Inferring effective computational connectivity using incrementally conditioned multivariate transfer entropy. *BMC Neurosci.* **2013**, *14*, P337.

30. Montalto, A.; Faes, L.; Marinazzo, D. MuTE: A MATLAB Toolbox to Compare Established and Novel Estimators of the Multivariate Transfer Entropy. *PLoS ONE* **2014**, *9*, e109462.

31. Lindner, M.; Vicente, R.; Priesemann, V.; Wibral, M. TRENTOOL: A Matlab open source toolbox to analyse information flow in time series data with transfer entropy. *BMC Neurosci.* **2011**, *12*, 1–22.

32. Wollstadt, P.; Meyer, U.; Wibral, M. A Graph Algorithmic Approach to Separate Direct from Indirect Neural Interactions. *PLoS ONE* **2015**, *10*, e0140530.

33. Levina A.; Priesemann V. Subsampling scaling. *Nat. Commun.* **2017**, *8*, 15140.

34. Levina, A.; Priesemann, V. Subsampling Scaling: A Theory about Inference from Partly Observed Systems. *arXiv* **2017**, arXiv:1701.04277.

35. Priesemann, V.; Wibral, M.; Valderrama, M.; Pröpper, R.; le van Quyen, M.; Geisel, T.; Triesch, J.; Nikolić, D.; Munk, M.H.J. Spike avalanches in vivo suggest a driven, slightly subcritical brain state. *Front. Syst. Neurosci.* **2014**, *8*, 108.

36. Priesemann, V.; Lizier, J.; Wibral, M.; Bullmore, E.; Paulsen, O.; Charlesworth, P.; Schröter, M. Self-organization of information processing in developing neuronal networks. *BMC Neurosci.* **2015**, *16*, P221.

37. Timme, N.; Alford, W.; Flecker, B.; Beggs, J.M. Synergy, redundancy, and multivariate information measures: an experimentalist's perspective. *J. Comput. Neurosci.* **2014**, *36*, 119–140.

38. Phillips, W.A. Cognitive functions of intracellular mechanisms for contextual amplification. *Brain Cognit.* **2017**, *112*, 39–53.

39. Larkum, M. A cellular mechanism for cortical associations: an organizing principle for the cerebral cortex. *Trends Neurosci.* **2013**, *36*, 141–151.

40. Finn, C.; Prokopenko, M.; Lizier, J.T. Pointwise Partial Information Decomposition Using the Specificity and Ambiguity Lattices. **2017**, under review.

41. Bertschinger, N.; Rauh, J.; Olbrich, E.; Jost, J. Shared Information—New Insights and Problems in Decomposing Information in Complex Systems. In *Proceedings of the European Conference on Complex Systems 2012*; Gilbert, T., Kirkilionis, M., Nicolis, G., Eds.; Springer International Publishing: Cham, Vietnam, 2013; pp. 251–269.

42. Ince, R.A.A. Measuring Multivariate Redundant Information with Pointwise Common Change in Surprisal. *Entropy* **2017**, *19*, 318.

43. Rauh, J.; Bertschinger, N.; Olbrich, E.; Jost, J. Reconsidering unique information: Towards a multivariate information decomposition. In Proceedings of the 2014 IEEE International Symposium on Information Theory, Honolulu, HI, USA, 29 June–4 July 2014; pp. 2232–2236.

44. Pica, G.; Piasini, E.; Chicharro, D.; Panzeri, S. Invariant Components of Synergy, Redundancy, and Unique Information among Three Variables. *Entropy* **2017**, *19*, 451.

45. Olbrich, E.; Bertschinger, N.; Rauh, J. Information Decomposition and Synergy. *Entropy* **2015**, *17*, 3501–3517.

46. Schneidman, E.; Puchalla, J.L.; Segev, R.; Harris, R.A.; Bialek, W.; Berry, M.J. Synergy from Silence in a Combinatorial Neural Code. *J. Neurosci.* **2011**, *31*, 15732–15741.

47. Stramaglia, S.; Wu, G.R.; Pellicoro, M.; Marinazzo, D. Expanding the transfer entropy to identify information circuits in complex systems. *Phys. Rev. E* **2012**, *86*, 066211.

48. Stramaglia, S.; Cortes, J.M.; Marinazzo, D. Synergy and redundancy in the Granger causal analysis of dynamical networks. *New J. Phys.* **2014**, *16*, 105003.

49. Stramaglia, S.; Angelini, L.; Wu, G.; Cortes, J.M.; Faes, L.; Marinazzo, D. Synergetic and redundant information flow detected by unnormalized Granger causality: Application to resting state fMRI. *IEEE Trans. Biomed. Eng.* **2016**, *63*, 2518–2524.

50. Rauh, J.; Banerjee, P.; Olbrich, E.; Jost, J.; Bertschinger, N. On Extractable Shared Information. *Entropy* **2017**, *19*, 328.

51. Linsker, R. Self-organisation in a perceptual network. *IEEE Comput.* **1988**, *21*, 105–117.

52. Kay, J.W.; Phillips, W.A. Coherent Infomax as a computational goal for neural systems. *Bull. Math. Biol.* **2011**, *73*, 344–372.

53. Marr, D. *Vision: A Computational Investigation into the Human Representation and Processing of Visual Information*; Henry Holt and Co., Inc.: New York, NY, USA, 1982.

Article

The Partial Information Decomposition of Generative Neural Network Models

Tycho M.S. Tax [1,*,†], Pedro A.M. Mediano [2,*,†] and Murray Shanahan [2]

[1]   Corti, Nørrebrogade 45E 2, 2200 Copenhagen N, Denmark
[2]   Department of Computing, Imperial College London, London SW7 2RH, UK; m.shanahan@imperial.ac.uk
*   Correspondence: tt@cortilabs.com (T.M.S.T.); pmediano@imperial.ac.uk (P.A.M.M.);
      Tel.: +31-643-92-93-33 (T.M.S.T.); +44-20-759-48445 (P.A.M.M.)
†   These authors contributed equally to this work.

Received: 8 July 2017; Accepted: 1 September 2017; Published: 6 September 2017

Abstract: In this work we study the distributed representations learnt by generative neural network models. In particular, we investigate the properties of redundant and synergistic information that groups of hidden neurons contain about the target variable. To this end, we use an emerging branch of information theory called partial information decomposition (PID) and track the informational properties of the neurons through training. We find two differentiated phases during the training process: a first short phase in which the neurons learn redundant information about the target, and a second phase in which neurons start specialising and each of them learns unique information about the target. We also find that in smaller networks individual neurons learn more specific information about certain features of the input, suggesting that learning pressure can encourage disentangled representations.

Keywords: partial information decomposition; neural networks; information theory

1. Introduction

Neural networks are famously known for their excellent performance, yet are infamously known for their thin theoretical grounding. While common deep learning "tricks" that are empirically proven successful tend to be later discovered to have a theoretical justification (e.g., the Bayesian interpretation of dropout [1,2]), deep learning research still operates "in the dark" and is guided almost exclusively by empirical performance.

One common topic in learning theory is the study of data representations, and in the case of deep learning it is the hierarchy of such representations that is often hailed as the key to neural networks' success [3]. More specifically, disentangled representations have received increased attention recently [4–6], and are particularly interesting given their reusability and their potential for transfer learning [7,8]. A representation can be said to be disentangled if it has factorisable or compositional structure, and has consistent semantics associated to different generating factors of the underlying data generation process.

In this article we explore the evolution of learnt representations in the hidden layer of a restricted Boltzmann machine as it is being trained. Are groups of neurons correlated or independent? To what extent do neurons learn the same information or specialise during training? If they do so, when? To answer these questions, we need to know how multiple sources of information (the neurons) contribute to the correct prediction of a target variable—which is known as a multivariate information problem.

To this end, the partial information decomposition (PID) framework by Williams and Beer [9], which seeks a rigorous mathematical generalisation of mutual information to the multivariate setting, provides an excellent foundation for this study [9]. In PID, the information that multiple sources contain about a target is decomposed into unique non-negative *information atoms*, the distribution of which gives insight into the interactions between the sources.

1.1. Why Information Theory?

Information theory was developed to optimise communication through noisy channels, and it quickly found other areas of application in mathematical and computer sciences. Nevertheless, it is not commonly linked to machine learning, and it is not part of the standard deep learning engineer's toolkit or training. So why, then, is information theory the right tool to study neural networks?

To answer that question, we must first consider some of the outstanding theoretical problems in deep learning: what kind of stimuli do certain neurons *encode*; how do different layers *compress* certain features of an input image; or how can we *transfer* learnt information from one dataset to another?

These problems (encodings, compression, transfer) are precisely among the problems information theory was made to solve. Casting these questions within the established framework of information theory gives us a solid language to reason about these systems and a comprehensive set of quantitative methods to study them.

We can also motivate this choice from a different perspective: in the same way as neuroscientists have been using information theory to study computation in biological brains, here we try to understand an artificially developed *neural code* [10]. Although the code used by artificial neural networks is most likely much simpler than the one used by biological brains, deep learning researchers can benefit from the neuroscientists' set of tools.

1.2. Related Work

When it comes to representations, the conventional way of obtaining insights about a network has typically been through visualisation. Famously, Le et al. [11] trained a neural network on web-scraped images and reported finding neural receptive fields consisting mostly of human faces, human bodies and cat faces [11]. Later, Zeiler and Fergus [12] devised a technique to visualise the features learnt by neurons in hidden layers, and provided good qualitative evidence to support the long-standing claim that deeper layers learn increasingly abstract features of the input [12].

While visualisation is a great exploration tool, it provides only qualitative insights and is therefore unable to make strong statements about the learning dynamics. Furthermore, as later work showed, the specific values of weights are highly sensitive to the details of the optimisation algorithm used, and therefore cannot be used to make definite judgements about the network's behaviour [13,14].

More recently, there is a small line of emerging work investigating the behaviour of neural networks from an information-theoretic perspective [15–20], with some work going as far back as [21]. The most relevant of these is the work by Schwartz-Ziv and Tishby [16], who show that feed-forward deep neural networks undergo a dynamic transition between drift- and diffusion-like regimes during training.

The main contribution of this article is to show how PID can be used for the analysis of learning algorithms, and its application to neural generative models. The results of our PID analysis show two distinct learning phases during the optimisation of the network, and a decrease in the specialisation of single neurons in bigger networks.

2. Methods

2.1. Restricted Boltzmann Machines

We deal with the problem of multiclass classification, in which we have a dataset $\mathcal{D}$ of $(\mathbf{x}, y)$ tuples, where y is a discrete *label* (also called the *target* variable) and $\mathbf{x}$ is a vector of predictor variables. The goal is to learn an approximation to the joint distribution of the predictors and the labels, $p(\mathbf{x}, y)$. We will use a class of neural generative models known as Boltzmann machines.

Boltzmann machines (BMs) are energy-based probabilistic graphical models, the origin of which goes as far back as Paul Smolensky's Harmonium [22]. Of particular interest are the so-called restricted Boltzmann machines (RBMs). These are called *restricted* because all the nodes in the model are separated in two layers, and intra-layer connections are prohibited. These typically receive the names of *visible* and *hidden* layers.

In this article we follow [23] and perform classification with a *discriminative* RBM (DRBM). To do this, we introduce the vector of target classes y as part of the visible layer, such that the DRBM represents the joint distribution over the hidden, visible, and target class variables. The distribution parametrized by the DRBM is:

$$p(y, \mathbf{x}, \mathbf{h}) = \frac{1}{Z} e^{-E(y, \mathbf{x}, \mathbf{h})} \, , \tag{1}$$

where $E(y, \mathbf{x}, \mathbf{h})$ is the DRBM *energy function*, given by

$$E(y, \mathbf{x}, \mathbf{h}) = -\mathbf{h}^T \mathbf{W} \mathbf{x} - \mathbf{b}^T \mathbf{x} - \mathbf{c}^T \mathbf{h} - \mathbf{h}^T \mathbf{U} \vec{y} - \mathbf{d}^T \vec{y} \, , \tag{2}$$

where $\vec{y} = (1_{y=i})_{i=1}^C$ for the C different classes. For comparison, the energy function for a standard RBM is the same but with the last two terms removed. Figure 1 shows a schematic diagram of a DRBM and the variables involved.

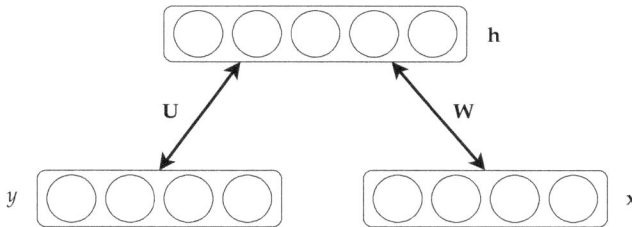

Figure 1. Graphical representation of the discriminative restricted Boltzmann machine (DRBM) and its components. Vectors $\mathbf{x}$ and $\mathbf{y}$ correspond to the training input and label, respectively, $\mathbf{h}$ is the activation of the hidden neurons, and $\mathbf{U}$ and $\mathbf{W}$ are the weight matrices to be learned. (Adapted from [23]).

Now that the model is specified, we calculate the predictive posterior density $p(y|\mathbf{x}, \theta)$ given DRBM parameters $\theta = \{\mathbf{W}, \mathbf{b}, \mathbf{c}, \mathbf{U}, \mathbf{d}\}$. At this point, the restricted connectivity of RBMs comes into play—this connectivity induces conditional independence between all neurons in one layer given the other layer. This resulting intra-layer conditional independence allows us to factorise $p(y_i, x_i|\theta)$ and to write the following conditional distributions [24]:

$$p(X_i = 1|\mathbf{h}) = \sigma(b_i + \sum_j W_{ji} h_j)$$

$$p(H_j = 1|y, \mathbf{x}) = \sigma(c_j + U_{jy} + \sum_i W_{ji} x_i) \tag{3}$$

$$p(y|\mathbf{h}) = \frac{e^{d_y + \sum_j U_{jy} h_j}}{\sum_{y^*} e^{d_{y^*} + \sum_j U_{jy^*} h_j}} \, ,$$

where $\sigma(t) = (1 + e^{-t})^{-1}$ is the standard sigmoid function. With these equations at hand, we can classify a new input vector $\mathbf{x}^*$ by sampling from the predictive posterior $p(y|\mathbf{x}^*, \theta)$, or we can sample from the joint distribution $p(y, \mathbf{x}|\theta)$ via Gibbs sampling.

Finally, the network needs to be trained to find the right parameters θ that approximate the distribution of the data. We use a standard maximum likelihood objective function,

$$\mathcal{L}(\theta) = -\sum_i \log p(y_i, x_i | \theta) . \tag{4}$$

Gradients of this objective cannot be obtained in closed form, and we must resort to contrastive sampling techniques. In particular, we use the constrastive divergence (CD) algorithm [24] to estimate the gradient, and we apply fixed-step size stochastic gradient updates to all parameters in the network. The technical details of CD and other contrastive sampling estimators are outside the scope of this paper, and the interested reader is referred to the original publications for more information [24,25].

2.2. Information Theory

In this section we introduce a few relevant tools from information theory (IT) that we will use to analyse the networks trained as explained in the previous section. For a broader introduction to IT and more rigorous mathematical detail, we refer the reader to [26].

We focus on systems of discrete variables with a finite number of states. Throughout the paper we will deal with the scenario in which we have one *target* variable and a number of *source* variables. We refer to the target variable as Y (matching the nomenclature in Section 2.1), to the source variables as Z_i, and let Z denote generically any nonempty subset of the set of all sources. Summations always run over all possible states of the variables considered.

Mutual information (MI) is a fundamental quantity in IT that quantifies how much information is shared between two variables Z and Y, and is given by

$$I(Y; Z) = \sum_{y,z} p(y, z) \log \left(\frac{p(y, z)}{p(y) p(z)} \right) . \tag{5}$$

MI can be thought of as a generalised (non-linear) correlation, which is higher the more a given value of Z constrains possible values of Y. Note that this is an average measure—it quantifies the information about Y gained when observing Z *on average*. In a similar fashion, *specific information* [27] quantifies the information contained in Z associated with a particular outcome y of Y, and is given by

$$I(Y = y; Z) = \sum_z p(z|y) \log \left(\frac{p(y|z)}{p(y)} \right) . \tag{6}$$

Specific information quantifies to what extent the observation of Z makes outcome y more likely than otherwise expected based on the prior $p(y)$. Conveniently, MI can easily be written in terms of specific information as

$$I(Y; Z) = \sum_y p(y) I(Y = y; Z) .$$

2.2.1. Non-Negative Decomposition of Multivariate Information

In this section we discuss the main principles of the PID framework proposed by Williams and Beer [9]. Technical details will not be covered, and the interested reader is referred to the original paper [9].

The goal of PID is to decompose the joint mutual information that two or more sources have about the target, $I(Y; \{Z_1, Z_2, \ldots, Z_n\})$, into interpretable non-negative terms. For simplicity, we present the two-variable case here, although the framework applies to an arbitrary number of sources. In the two-variable PID (or PI-2), there are three types of contributions to the total information of $\{Z_1, Z_2\}$ about Y which form the basic atoms of multivariate information:

- *Unique* information U one of the sources provides and the other does not.

- *Redundant* information R both sources provide.
- *Synergistic* information S the sources provide jointly, which is not known when either of them is considered separately.

There is a very intuitive analogy between this decomposition and set theory—in fact, the decomposition for any number of variables can be shown to have a formal lattice structure if R is mapped to the set intersection operation. This mapping corresponds to the intuitive notion that R should quantify the *overlapping information* of Z_1 and Z_2. Consequently, these quantities can be represented in a Venn diagram called the *PI diagram*, shown in Figure 2.

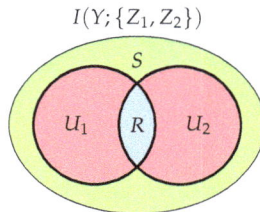

Figure 2. Partial information (PI) diagram for two source variables and a target. The outer ellipse corresponds to the mutual information (MI) between both sources and the target, $I(Y; \{Z_1, Z_2\})$, and both inner circles (highlighted in black) to the MI between each source and the target, $I(Y; Z_i)$. Coloured areas represent the PI terms described in the text.

Mathematically, the relation between MI and S, R, and U (which we refer to jointly as *PI terms*) can be written as follows:

$$
\begin{aligned}
I(Y; Z_1) &= R(Y; \{Z_1, Z_2\}) + U(Y; Z_1), \\
I(Y; Z_2) &= R(Y; \{Z_1, Z_2\}) + U(Y; Z_2), \\
I(Y; \{Z_1, Z_2\}) &= R(Y; \{Z_1, Z_2\}) + U(Y; Z_1) + U(Y; Z_2) + S(Y; \{Z_1, Z_2\}).
\end{aligned}
\tag{7}
$$

This is an underdetermined system of three equations with four unknowns, which means the PI decomposition in itself does not provide a method to calculate the PI terms. To do that, we need to specify one of the four variables in the system, typically by providing an expression to calculate either R or S. There are a number of proposals in the literature [28–31], but at the time of writing there is no consensus on any one candidate.

In this study we follow the original proposal by Williams and Beer [9] and use $I_{\min}$ as a measure of redundancy, defined as

$$
R(Y; \{Z_1, Z_2\}) = I_{\min}(Y; \{Z_1, Z_2\}) = \sum_y p(y) \min\{I(Y = y; Z_1), I(Y = y, Z_2)\},
\tag{8}
$$

where $I(Y = y; Z_i)$ is the specific information in Equation (6). (In fact, in their original article DeWeese and Meister [27] propose two quantities to measure "the information gained from one symbol": specific information and specific surprise. Confusingly, Williams and Beer's specific information is actually DeWeese and Meister's specific surprise.) Despite the known flaws of $I_{\min}$, we chose it for its tractability and inclusion-exclusion properties. With this definition of redundancy and the standard MI expression in Equation (5), we can go back to system (7) and calculate the rest of the terms.

While all the terms in PI-2 can be readily calculated with the procedure above, with more sources, the number of terms explodes very quickly—to the point that the computation of all PI terms is intractable even for very small networks. Conveniently, with $I_{\min}$, we can compute the overall redundancy, synergy, and unique information terms for arbitrarily many sources—restricted only by the computational cost and amount of data necessary to construct large joint probability tables.

We write here the overall redundancy, synergy, and unique information equations for completeness, but the interested reader is referred to [32] for a full derivation:

$$R(Y; \{Z_1, \ldots, Z_n\}) = I_{\min}(Y; \{Z_1, \ldots, Z_n\})$$
$$= \sum_y p(y) \min\{I(Y = y; Z_1), \ldots, I(Y = y; Z_n)\}$$

$$S(Y; \{Z_1, \ldots, Z_n\}) = I(Y; \{Z_1, \ldots, Z_n\}) - I_{\max}(Y; \{\mathbf{A} \in \{Z_1, \ldots, Z_n\} : |\mathbf{A}| = n - 1\})$$
$$U(Y; Z_i) = I(Y; Z_i) - I_{\min}(Y; \{Z_i, \{Z_1, \ldots, Z_n\} \backslash Z_i\}),$$

(9)

where $I_{\max}$ is defined exactly the same as $I_{\min}$, except substituting max for min [32].

3. Results

Instead of generating a synthetic dataset, we used the MNIST dataset of hand-written digits. We used a stochastic binarised version of MNIST—every time an image was fed as input to the network, the value of each pixel was sampled from a binomial distribution with a probability equal to the normalised intensity of that pixel. Then, we used Equations (3) to sample the state of the network, and repeated this process to build the probability distributions of interest.

For training, the gradients were estimated with contrastive divergence [24] and the weights were optimised with vanilla stochastic gradient descent with fixed learning rate (0.01). We did not make strong efforts to optimise the hyperparameters used during training.

To produce the results below, we trained an ensemble of 100 RBMs and took snapshots of these networks during training. Each RBM in the ensemble was initialised and trained separately using a different random number generator seed. All information-theoretic measures are reported in bits and debiased with random permutation tests. To debias the estimation of any measure on a given set of data, we generated many surrogate data sets by randomly permuting the original data, calculating the mean of the measure across all surrogates, and subtracting this from the original estimation on the unshuffled data [33].

3.1. Classification Error and Mutual Information

First, we trained a small RBM with 20 hidden neurons and inspected its learning curve during training. In Figure 3, we show the classification error and the mutual information between the predicted labels $\hat{Y}$ and the real labels Y during training, averaged for the ensemble of 100 RBMs.

As expected, classification error decreased and MI increased during training, the relationship between the two being an almost perfect line. This gives us an intuitive correspondence between a relatively abstract measure like bits and a more easily interpretable measure like error rate. We note that a perfect classifier with 0 error rate would have $I(\hat{Y}, Y) = H(Y) = \log_2(10) \approx 3.32$ bit.

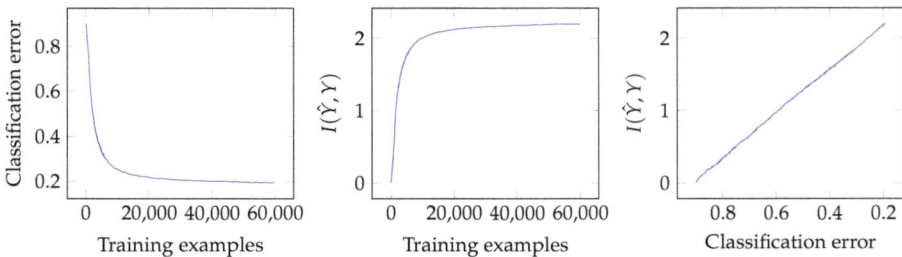

Figure 3. Classification error and mutual information between real and predicted labels, $I(\hat{Y}, Y)$, calculated through training. Note: X-axis in the rightmost plot is reversed for illustration pusposes, so that training time goes from left to right.

As should be apparent to any occasional reader of the machine learning literature, the classification error presented in Figure 3 is worse than the authors reported originally in [23], and significantly worse than the state of the art on this dataset. The main reason for this is that we are restricting our network to a very small size to obtain a better resolution of the phenomena of interest.

3.2. Phases of Learning

In this section we investigate the evolution of the network through training, and show three complementary pieces of evidence for the existence of two separate learning phases. We describe the main results illustrated in Figures 4 and 5.

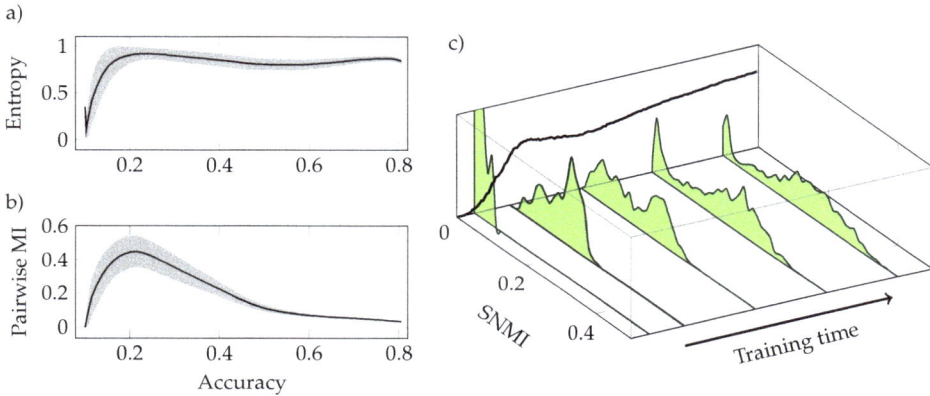

Figure 4. Single-neuron entropy and mutual information follow non-trivial patterns during training. (**a**) Entropy quickly rises up to close to its maximum value of 1 bit. (**b**) Inter-neuron correlation as measured by pairwise MI peaks midway through training. (**c**) Histograms of single-neuron MI (SNMI) split midway through training, implying that some neurons actually lose information. Average SNMI is shown in black projected on the frame box.

First, in Figure 4a, we show the evolution of the average entropy of single neurons in the hidden layer, where the average is taken over all neurons in the same network. Entropy increases rapidly at the start of training until it settles around the 0.8 to 0.9 bit range, relatively close to its maximum possible value of 1 bit. This means that throughout most of the training (including the final state), most of the informational capacity of the neurons was being actively used—if this were not the case, in a network with low entropy in which most neurons do not change their states much, the encoding capability of the network would be heavily reduced.

As a measure of inter-neuron correlation, we calculated the average pairwise mutual information (PWMI) between hidden neurons H_i, defined as

$$\text{PWMI} = \langle I(H_i; H_j) \rangle_{ij} \ .$$

PWMI is shown in Figure 4b, and is the first sign of the transition mentioned above—it increases rapidly at the start, it reaches a peak at an intermediate point during training, and then decays back to near zero.

Next, we calculate the average MI between a single hidden neuron and the target, $I(Y; H_i)$, which we refer to as single-neuron mutual information (SNMI), and show the results in Figure 4c. As expected, at first neurons barely have any information about the target, and early in training we see a quick uniform increase in SNMI.

Remarkably, at the transition point there is a split in the SNMI histogram, with around half of the neurons reverting back to low values of SNMI and the other half continuing to increase. At the whole network level, we do not find any sign of this split, as shown by the monotonically decreasing error rate in Figure 3. This is a seemingly counterintuitive finding—some neurons actually get *worse* at predicting the target as the network learns. We currently do not have a solid explanation for this phenomenon, although we believe it could be due to the effects of local minima or to the neurons relying more on synergistic interactions at the cost of SNMI, as suggested by the results below.

After exploring the behaviour of individual neurons, we now turn to PID and study the interactions between them when predicting the target. Since a full PID analysis of the whole network is intractable, we follow a procedure inspired by [34] to estimate the PI terms of the learnt representation: we sample pairs of neurons, calculate the PI terms for each of them, apply random permutation correction to each pair separately, and finally compute averages over all pairs. We present results obtained with I_{min} following Section 2.2.1, but qualitatively identical results are obtained if the more modern measures in [28,35] are used.

We calculate synergy S, redundancy R, and total unique information $U = U(Y; Z_1) + U(Y; Z_2)$, as well as their normalised versions calculated by dividing S, R, or U by the joint mutual information $I(Y; \{H_1, H_2\})$. Results are depicted in Figure 5, and error intervals shown correspond to two standard deviations across pairs.

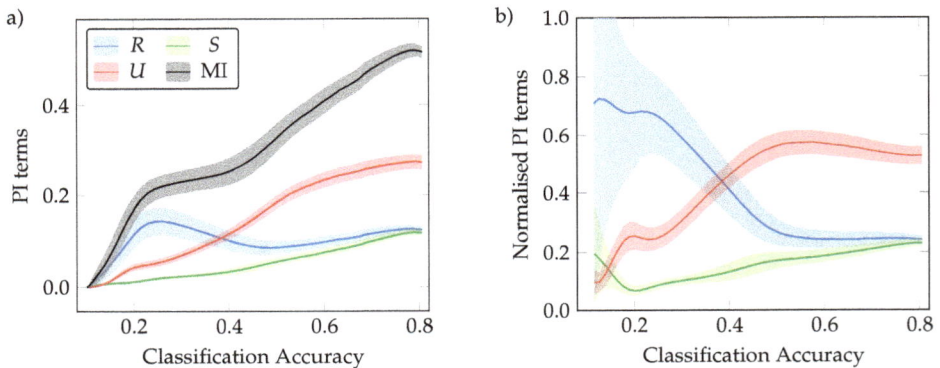

Figure 5. PI terms (**a**) and PI terms normalised by joint mutual information (**b**). Mutual information (MI) in black, redundancy (R) in blue, synergy (S) in green, and unique information (U) in red. MI increases consistently during training, but the PI terms reveal a transition between a redundancy-dominated phase and a unique-information-dominated phase.

Here we see again a transition between two phases of learning. Although synergy and joint MI increase steadily at all times, there is a clear distinction between a first phase dominated by redundancy and a second one dominated by unique information. It is at this point that neurons specialise, suggesting that this is when disentangled representations emerge.

These three phenomena (peak in PWMI, split SNMI histogram, and redundant-unique information transition) do not happen at the same time. In the figures shown, the peak in PWMI marks the onset of the decline of redundancy, and the split in SNMI happens between then and the point when redundancy is overtaken by unique information. However, this is a consistent pattern we have observed in networks of multiple sizes, and in bigger networks these three events tend to come closer in time (results not shown).

We note that there is a relation between PWMI and R and between SNMI and U. As indicated in Equation (7), SNMI is an upper bound on that neuron's unique information; and usually, higher PWMI comes with higher redundancy between the neurons. However, although they follow similar shapes, these magnitudes do not quantify the same thing. Take the OR logical gate as an example—if we feed it a uniform distribution of all possible inputs (00, 01, 10, 11), both input bits will be perfectly uncorrelated, yet their redundancy (according to I_{min}) will be nonzero.

These findings are in line with those of Schwartz-Ziv and Tishby [16], who observe a similar transition in a feed-forward neural network classifier. One of the pieces of evidence for Schwartz-Ziv and Tishby's [16] claim is in the change of gradient signal dynamics from a drift to a diffusion regime. We did not analyse gradient dynamics as part of this study, but investigating the relationship between informational and dynamical accounts of learning is certainly a promising topic.

3.3. Neural Interactions

In this last set of experiments, we examine the representations learnt jointly by larger groups of neurons. Due to computational constraints, we run the analyses only on fully trained networks instead of at multiple points during training. We train networks of different sizes, ranging from 20 to 500 hidden neurons (using the same algorithm, but allowing each network to train for more epochs until convergence), and consider larger groups of neurons for the PID analysis. We use a procedure similar to the one used in the previous section, but this time sampling tuples of K neurons, and calculating their overall synergy following Equation (9). We refer to this as the PI-K synergy. Results are shown in Figure 6.

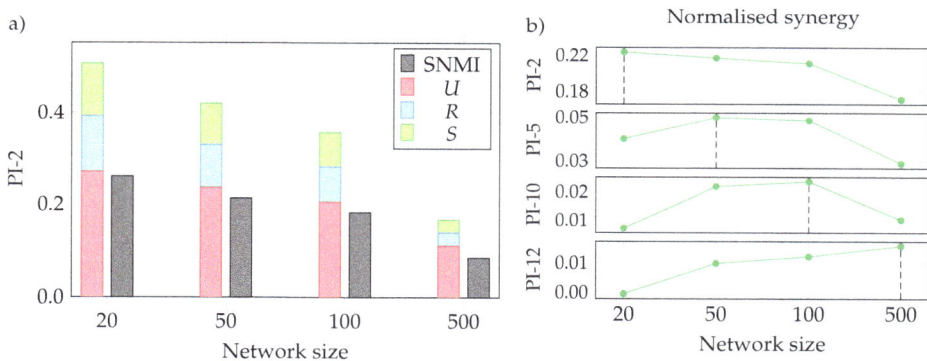

Figure 6. Partial information decomposition (PID) analysis of larger groups of neurons in networks of different sizes. (**a**) Single-neuron MI is consistently smaller in bigger networks, indicating that, although the network as a whole is a better classifier, each individual neuron has a less-efficient encoding; (**b**) Normalised PI-K synergy, with network size increasing from left to right and K from top to bottom. Network with maximum synergy for each PI-K highlighted with a vertical dashed line. The PI group size with the highest synergy becomes larger in larger networks, indicating that in bigger networks one needs to consider larger groups to capture strong cooperative interactions.

The first result in Figure 6a is that SNMI decreases consistently with network size. This represents reduced efficiency in the neurons' compression—despite the overall accuracy of the network being significantly higher for bigger networks (~20% error rate for a network with 20 hidden neurons vs. ~5% for a network with 500), each individual neuron contains less information about the target. This suggests that the representation is more distributed in bigger networks, as emphasised below.

What is somewhat counterintuitive is that normalised unique information actually grows in bigger networks, which is apparently in contradiction with more distributed representations. However, these two are perfectly compatible—bigger networks have more and less correlated neurons, and despite U growing relative to S and R, it still decreases significantly in absolute terms. Note that the U term plotted in Figure 6 is the sum of the unique information of both neurons in the pair; naturally, for one neuron $U(Y; H_i) \leq I(Y; H_i)$, Equation (7).

Interestingly, Figure 6b shows that the network size that achieves maximum normalised synergy shifts to the right as we inspect larger groups of neurons. For bigger networks, bigger groups carry more synergistic information, meaning that representations also become more distributed. There is a consistent pattern that in bigger networks we need to explore increasingly high neural groups to see any meaningful PI values, which means that perhaps part of the success of bigger networks is that they make better use of higher-order correlations between hidden neurons. This can be seen as a signature of bigger networks achieving richer and more complex representations [36].

3.4. Limitations

The main limitation of the vanilla PID formulation is that the number of PI terms scales very rapidly with bigger group sizes—the number of terms in the PI decomposition of a system with n sources is the $(n-1)$-st Dedekind number, which is 7579 terms for a five-variable system and has not been computed yet for systems of more than eight variables. For this reason, we have restricted our analyses mostly to pairs of neurons, although in practice we expect larger groups of neurons to have strong effects on the prediction. Potentially some approximation to the whole PID or a reasonable grouping of PI terms could help scale this type of analysis to larger systems.

On a separate topic, some of the phenomena of interest we have described in this article (two phases of learning, peak in correlation between the neurons) happen very early on during training, in practice. In a real-world ML setting, most of the time is spent in the last phase where error decreases very slowly; and so far, we have not seen any unusual behaviour in that region. Future work should focus on this second phase and try to characterise it in more detail, with the aim of improving performance or speeding convergence.

4. Conclusions

In this article we have used information theory—in particular the partial information decomposition framework—to explore the latent representations learned by a restricted Boltzmann machine. We have found that the learning process of neural generative models has two distinct phases: a first phase dominated by redundant information about the target, and another phase in which neurons specialise and each of them learns unique information about the target and synergy. This is in line with the findings of Schwartz-Ziv and Tishby [16] in feed-forward networks, and we believe further research should explore the differences between generative and discriminative models in this regard.

Additionally, we found that representations learned by bigger networks are more distributed, yet significantly less efficient at the single-neuron level. This suggests that the learning pressure of having fewer neurons encourages those neurons to specialise more, and therefore yields more disentangled representations. The interesting challenge is to find a principled way of encouraging networks towards disentangled representations while preserving performance.

An interesting piece of follow-up work would be to investigate whether these findings generalise to other deep generative models—most notably, variational autoencoders [37]. We believe that further theoretical study of these learning systems is necessary to help us understand, interpret, and improve them.

Acknowledgments: The authors would like to thank Raúl Vicente for insightful discussions, and Pietro Marchesi for valuable feedback in the early stages of this project.

Author Contributions: Tycho M.S. Tax and Pedro A.M. Mediano designed the experiments; Tycho M.S. Tax performed the experiments; Tycho M.S. Tax and Pedro A.M. Mediano analysed the results; and Pedro A.M. Mediano, Tycho M.S. Tax and Murray Shanahan wrote the paper. All authors have read and approved the final manuscript.

Conflicts of Interest: The authors declare no conflict of interest.

References

1. Srivastava, N.; Hinton, G.; Krizhevsky, A.; Sutskever, I.; Salakhutdinov, R. Dropout: A Simple Way to Prevent Neural Networks from Overfitting. *J. Mach. Learn. Res.* **2014**, *15*, 1929–1958.

2. Gal, Y.; Ghahramani, Z. Dropout as a Bayesian Approximation: Representing Model Uncertainty in Deep Learning. *arXiv* **2015**, arXiv:1206.5538.

3. Bengio, Y.; Courville, A.; Vincent, P. Representation Learning: A Review and New Perspectives. *arXiv* **2012**, arXiv:1206.5538.

4. Higgins, I.; Matthey, L.; Glorot, X.; Pal, A.; Uria, B.; Blundell, C.; Mohamed, S.; Lerchner, A. Early Visual Concept Learning with Unsupervised Deep Learning. *arXiv* **2016**, arXiv:1606.05579.

5. Mathieu, M.; Zhao, J.; Sprechmann, P.; Ramesh, A.; LeCun, Y. Disentangling Factors of Variation in Deep Representations Using Adversarial Training. *arXiv* **2016**, arXiv:1611.03383.

6. Siddharth, N.; Paige, B.; Van de Meent, J.W.; Desmaison, A.; Wood, F.; Goodman, N.D.; Kohli, P.; Torr, P.H.S. Learning Disentangled Representations with Semi-Supervised Deep Generative Models. *arXiv* **2017**, arXiv:1706.00400.

7. Lake, B.M.; Ullman, T.D.; Tenenbaum, J.B.; Gershman, S.J. Building Machines That Learn and Think Like People. *arXiv* **2016**, arXiv:1604.00289.

8. Garnelo, M.; Arulkumaran, K.; Shanahan, M. Towards Deep Symbolic Reinforcement Learning. *arXiv* **2016**, arXiv:1609.05518.

9. Williams, P.L.; Beer, R.D. Nonnegative Decomposition of Multivariate Information. *arXiv* **2010**, arXiv:1004.2515.

10. Rieke, F.; Bialek, W.; Warland, D.; de Ruyter van Steveninck, R. *Spikes: Exploring the Neural Code*; MIT Press: Cambridge, MA, USA, 1997; p. 395.

11. Le, Q.V.; Ranzato, M.; Monga, R.; Devin, M.; Chen, K.; Corrado, G.S.; Dean, J.; Ng, A.Y. Building High-Level Features Using Large Scale Unsupervised Learning. *arXiv* **2011**, arXiv:1112.6209.

12. Zeiler, M.D.; Fergus, R. Visualizing and Understanding Convolutional Networks. In Proceedings of the European Conference on Computer Vision, Zurich, Switzerland, 6–12 September 2014; Springer: Berlin, Germany, 2014; pp. 818–833.

13. Choromanska, A.; Henaff, M.; Mathieu, M.; Arous, G.B.; LeCun, Y. The Loss Surfaces of Multilayer Networks. *arXiv* **2014**, arXiv:1412.0233.

14. Kawaguchi, K. Deep Learning Without Poor Local Minima. *arXiv* **2016**, arXiv:1605.07110.

15. Sørngård, B. Information Theory for Analyzing Neural Networks. Master's Thesis, Norwegian University of Science and Technology, Trondheim, Norway, 2014.

16. Schwartz-Ziv, R.; Tishby, N. Opening the Black Box of Deep Neural Networks via Information. *arXiv* **2017**, arXiv:1703.00810.

17. Achille, A.; Soatto, S. On the Emergence of Invariance and Disentangling in Deep Representations. *arXiv* **2017**, arXiv:1706.01350.

18. Tishby, N.; Zaslavsky, N. Deep Learning and the Information Bottleneck Principle. *arXiv* **2015**, arXiv:1503.02406.

19. Berglund, M.; Raiko, T.; Cho, K. Measuring the Usefulness of Hidden Units in Boltzmann Machines with Mutual Information. *Neural Netw.* **2015**, *64*, 12–18.

20. Balduzzi, D.; Frean, M.; Leary, L.; Lewis, J.; Ma, K.W.D.; McWilliams, B. The Shattered Gradients Problem: If Resnets are the Answer, Then What is the Question? *arXiv* **2017**, arXiv:1702.08591.

21. Hinton, G.E.; van Camp, D. Keeping the Neural Networks Simple by Minimizing the Description Length of the Weights. In Proceedings of the Sixth Annual Conference on Computational Learning Theory (COLT), Santa Cruz, CA, USA, 26–28 July 1993; ACM: New York, NY, USA, 1993; pp. 5–13.

22. Smolensky, P. *Information Processing in Dynamical Systems: Foundations of Harmony Theory*; Technical Report, DTIC Document; MIT Press: Cambridge, MA, USA, 1986.

23. Larochelle, H.; Bengio, Y. Classification Using Discriminative Restricted Boltzmann Machines. In Proceedings of the 25th International Conference on Machine Learning, Helsinki, Finland, 5–9 July 2008; pp. 536–543.

24. Hinton, G.E.; Osindero, S.; Teh, Y.W. A Fast Learning Algorithm for Deep Belief Nets. *Neural Comput.* **2006**, *18*, 1527–1554.

25. Tieleman, T. Training Restricted Boltzmann Machines Using Approximations to the Likelihood Gradient. In Proceedings of the 25th International Conference on Machine Learning, Helsinki, Finland, 5–9 July 2008; ACM Press: New York, NY, USA, 2008; pp. 1064–1071.

Entropy **2017**, *19*, 474

26. Cover, T.M.; Thomas, J.A. *Elements of Information Theory*; Wiley: Hoboken, NJ, USA, 2006.
27. DeWeese, M.R.; Meister, M. How to Measure the Information Gained from one Symbol. *Netw. Comput. Neural Syst.* **1999**, *12*, 325–340.
28. Ince, R.A.A. Measuring Multivariate Redundant Information with Pointwise Common Change in Surprisal. *Entropy* **2017**, *19*, doi:10.3390/e19070318.
29. Griffith, V.; Ho, T. Quantifying Redundant Information in Predicting a Target Random Variable. *Entropy* **2015**, *17*, 4644–4653.
30. Harder, M.; Salge, C.; Polani, D. Bivariate Measure of Redundant Information. *Phys. Rev. E* **2013**, *87*, doi:10.1103/PhysRevE.87.012130.
31. Bertschinger, N.; Rauh, J.; Olbrich, E.; Jost, J. Shared Information—New Insights and Problems in Decomposing Information in Complex Systems. In *Proceedings of the European Conference on Complex Systems 2012*; Gilbert, T., Kirkilionis, M., Nicolis, G., Eds.; Springer: Berlin, Germany, 2013; pp. 251–269.
32. Williams, P.L. Information Dynamics: Its Theory and Application to EmbodiedCognitive Systems. Ph.D. Thesis, Indiana University, Bloomington, IN, USA, 2011.
33. Lizier, J.T. *The Local Information Dynamics of Distributed Computation in Complex Systems*; Springer: Berlin/Heidelberg, Germany, 2010.
34. Timme, N.; Alford, W.; Flecker, B.; Beggs, J.M. Synergy, Redundancy, and Multivariate Information Measures: An Experimentalist's Perspective. *J. Comput. Neurosci.* **2014**, *36*, 119–140.
35. Bertschinger, N.; Rauh, J.; Olbrich, E.; Jost, J.; Ay, N. Quantifying Unique Information. *Entropy* **2014**, *16*, 2161–2183.
36. Montúfar, G.; Ay, N.; Ghazi-Zahedi, K. Geometry and Expressive Power of Conditional Restricted Boltzmann Machines. *J. Mach. Learn. Res.* **2015**, *16*, 2405–2436.
37. Kingma, D.P.; Welling, M. Auto-Encoding Variational Bayes. *arXiv* **2013**, arXiv:1312.6114.

Article

Morphological Computation: Synergy of Body and Brain

Keyan Ghazi-Zahedi [1,2,*], Carlotta Langer [1] and Nihat Ay [1,2]

[1] Max Planck Institute for Mathematics in the Sciences, 04103 Leipzig, Germany;
    carlotta.langer@mis.mpg.de (C.L.); nay@mis.mpg.de (N.A.)

[2] Santa Fe Institute, Santa Fe, NM 87501, USA

*   Correspondence: zahedi@mis.mpg.de; Tel.: +49-341-9959-545

Received: 9 July 2017; Accepted: 25 August 2017; Published: 31 August 2017

Abstract: There are numerous examples that show how the exploitation of the body's physical properties can lift the burden of the brain. Examples include grasping, swimming, locomotion, and motion detection. The term Morphological Computation was originally coined to describe processes in the body that would otherwise have to be conducted by the brain. In this paper, we argue for a synergistic perspective, and by that we mean that Morphological Computation is a process which requires a close interaction of body and brain. Based on a model of the sensorimotor loop, we study a new measure of synergistic information and show that it is more reliable in cases in which there is no synergistic information, compared to previous results. Furthermore, we discuss an algorithm that allows the calculation of the measure in non-trivial (non-binary) systems.

Keywords: embodied artificial intelligence; synergistic information; information theory; morphological computation; complexity; information integration

1. Introduction

There are numerous examples that show how the exploitation of the body's physical properties can lift the burden of the brain. Examples range from grasping [1–3], swimming [4–6], locomotion [7–10], to motion detection [11]. Probably the most prominent example in this field is the Passive Dynamic Walker [9], which is a purely mechanical system that mimics human walking. It has carefully chosen length and weight proportions of the leg segments, as well as carefully designed feet. If placed on a slope, it will show a natural, appealing walking behaviour, which is a strong indication that human walking does not have to be fully controlled, but that part of it can result from physical interactions of the legs (weight, friction, etc.) with their environment (slope, gravity, etc.). Another impressive example is human grasping which exploits at least two different physical interactions. First, as a result of the skin's softness and friction, even fragile objects can be hold securely with some variation of grip posture and grip pressure. This means that the brain does not have to carefully control the position of the fingers, the tightness of the grasp, and, in particular, does not have to precisely estimate the shape of the object. This leads to a significant reduction of the computational burden for the brain in grasping. The second effect that is used is the friction in the hand's tendon network, which has been shown to perform logic computation and affect torque production capabilities [2].

The term *Morphological Computation* [12] was originally coined to describe processes in the body that would otherwise have to be conducted by the brain [13]. One of the main questions that arises, and is so far unsolved, regards the distinction between the Passive Dynamic Walker and a ball rolling downhill. Both cases are purely mechanical systems, but one would assign morphological computation to the Passive Walker, whereas one would generally have difficulties in stating that the ball is performing computation or reducing the computational complexity for a brain. There are three possible solutions to this problem. Their relation to synergy will be addressed below. First, as argued in [14,15], the Passive Walker itself is not performing computation, but it shows that morphological computation can be present in human walking. Second, the definition of physical computation [16] offers a possibility to distinguish between pure physics and physical computation. Physical computation requires four ingredients. First, a function that encodes the data of the user into an initial state of the system, second, a physical process that transforms the initial state into some target state, third, a decoding function that transforms the target state back into something that the user can process, and, finally, a theory about how the system works. The implications of this theory of physical computation, in particular with respect to morphological computation, are discussed in [17]. With respect to the ball and the Passive Walker, the theory of [16] would lead to the conclusion that both are computing if there is a user that translates their states, e.g., to measure the slope. This is in accordance with the initial definition given by [12] and is also used as a basis for [18,19], which are discussed below. The third possibility can be summarised in the following way (cited from [20]): "nonneural body parts could be described as parts of a computational system, but they do not realise computation autonomously, only in connection with some kind of [...] central control system." If we now compare the three different approaches, they don't seem to be entirely different. All three cases argue that morphological computation requires the interaction of a brain with the body. In the context of this work, this is understood as a synergistic perspective on morphological computation. This will be explained in more detail, after related work on formalising morphological computation is presented next.

Pfeifer and Iida [21] state that "One problem with the concept of morphological computation is that while intuitively plausible, it has defied serious quantification efforts." Since then, there are basically two different streams of formalising morphological computation, which can be divided into a dynamical systems approach and an information theoretic approach. The two approaches do not stand in opposition but should rather be seen as complimentary [22]. The first approach [18,19] models processes in the body in the context of reservoir computing [23,24]. This means that the body is understood as a type of physical reservoir computer and the controller or brain harnesses the body dynamics to produce a behaviour. Examples are the spine-driven robot [7], which uses the spine dynamics as part of its controller and the dynamics of an octopus arm that can be used for computation [25]. Within this first approach, there are also several works that discuss the importance of a tight body–brain–environment coupling, of which the following are just a few examples [13,26–31]. Although very intuitive and compelling, this approach does not allow to quantify how the body reduces the computational burden of the brain. This is the motivation for the second, information theoretic approach [15]. The guiding idea is to model the sensorimotor loop as a causal graph [32] (details will follow below) and, based on that, ask how much internal processes (with respect to the agent's perspective) contributed to an observed behaviour, as opposed to body–environment interactions. The information theoretic measures have been successfully applied to quantify morphological computation in muscle models [14] and soft robotics [1] and relations have been drawn to unique and synergistic information [33] based on the work by [34]. Unfortunately, at that time, the synergistic information could only be calculated for simple binary models of the sensorimotor loop, which prohibited a further investigation in non-trivial systems or even real data. This is where this work is targeted at. Based on the complexity measure by Ay [35] and Perrone and Ay [36], we investigate synergistic information in binary and non-binary models of the sensorimotor loop and compare the results to our previous work in [33].

This work is organised in the following way. Section two discusses in detail the relation between synergistic information and morphological computation, based on previous work and the causal model of the sensorimotor loop. The third section presents the parametrised model of the sensorimotor loop, which is used to analyse and compare the new measure with previous work. The fourth section presents numerical results which are discussed in the final section.

2. A Synergistic Perspective on Morphological Computation

The introduction gave a motivation to understand morphological computation as a process that occurs as the result of some type of control that exploits physical properties of the body. This is one way to distinguish morphological computation from purely physical processes and can be understood as a synergistic coupling of brain and body that is required for morphological computation. This section will give a formal motivation to quantifying morphological computation as synergistic information, which is based on our previous work [15,33]. For a derivation of an information theoretic quantification, it is helpful to have a causal model of the sensorimotor loop [32], which is presented first.

2.1. Causal Model of the Sensorimotor Loop

We assume that there is a canonical way to separate a cognitive system into four parts, namely brain, sensors, actuators, and body. We are fully aware that the system–environment separation is a very difficult and yet unsolved question for biological systems (see e.g., [37] for a discussion). This holds even more in the case of the distinction between the body and brain. Yet, in order to derive a quantification, we have to assume that there is such a distinction.

In our conceptual model of the sensorimotor loop, which is derived from [22], the brain or controller sends signals to the actuators that influence the environment (see Figure 1). We prefer the notion of the system's *Umwelt* [31,38,39], which is the part of the system's environment that can be affected by the system and itself affects the system. The state of the actuators and the *Umwelt* are not directly accessible to the cognitive system, but the loop is closed as information about both the *Umwelt* and the actuators are provided to the controller by the system's sensors. In addition to this general concept, which is widely used in the embodied artificial intelligence community (see e.g., [22]), we introduce the notion of *world* to the sensorimotor loop, that is, the system's morphology and the system's *Umwelt*. This differentiation between body and world is analogous to the agent–environment distinction made in the context of reinforcement learning [40], where the environment is defined as everything that cannot be changed arbitrarily by the agent.

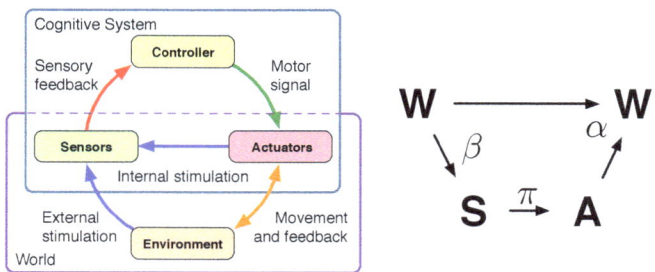

Figure 1. Sensorimotor Loop. Left-hand side: schematics of the sensorimotor loop (redrawn from [22]), Right-hand side: causal diagram of a reactive system.

Revisiting the list of examples given in the introduction (e.g., locomotion, grasping), it is seen that most behaviours that are interesting in the context of morphological computation can be modelled sufficiently as reactive behaviours. Hence, for the remainder of this work, we will omit the controller and assume that the sensors are directly connected to the actuators. For a discussion of the causal diagram for non-reactive systems, the reader is referred to [32]. Quantifications of morphological computation for non-reactive systems are discussed in [15].

The causal diagram of the sensorimotor loop is shown on the right-hand side of Figure 1. The random variables A, S, W, and W' refer to actuator signals, sensor signals, and the current and next world state. Directed edges reflect causal dependencies between the random variables. The random variables

S and *A* are not to be mistaken with the sensors and actuators. The variable *S* is the output of the sensors, which is available to the controller or brain, and the action *A* is the input that the actuators take. Consider an artificial robotic system as an example. Then the sensor state *S* could be the pixel matrix delivered by a camera sensor and the action *A* could be a numerical value that is taken by a motor controller and converted in currents to drive a motor.

Throughout this work, we use capital letters ($X, Y, \ldots$) to denote random variables, non-capital letters ($x, y, \ldots$) to denote a specific value that a random variable can take, and calligraphic letters ($\mathcal{X}, \mathcal{Y}, \ldots$) to denote the alphabet for the random variables. This means that x_t is the specific value that the random variable X can take a time $t \in \mathbb{N}$, and it is from the set $x_t \in \mathcal{X}$. Greek letters refer to generative kernels, i.e., kernels that describe an actual underlying mechanism or a causal relation between two random variables. In the causal graphs throughout this paper, these kernels are represented by direct connections between corresponding nodes. This notation is used to distinguish generative kernels from others, such as the conditional probability of *a* given that *w* was previously seen, denoted by $p(a|w)$, which can be calculated or sampled but that does not reflect a direct causal relation between the two random variables *A* and *W* (see Figure 1).

We abbreviate the random variables for better comprehension in the remainder of this work, as all measures consider random variables of consecutive time indices. Therefore, we use the following notation. Random variables without any time index refer to time index *t* and hyphened variables to time index $t + 1$. The two variables *W* and *W'* refer to W_t and W_{t+1}, respectively.

2.2. Quantifying Morphological Computation as Synergistic Information

We can now restate the two original concepts of quantifying morphological computation [15] (see Figure 2) and also discuss their relationship to synergistic information [33] as defined by [34]. This will build the foundation for the comparison with quantifying synergistic information based on the measure proposed by [35].

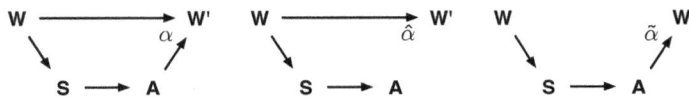

Figure 2. Visualisation of the two concepts MC$_A$ and MC$_W$. Left-hand side: causal diagram for a reactive system. Centre: causal diagram assuming no effect of the action *A* on the next world state *W'*. Right-hand side: causal diagram assuming no effect of the previous world state *W* on the next world state *W'*.

The basis for both original concepts MC$_A$ and MC$_W$ is the world dynamics kernel $\alpha(w'|w, a)$, which describes how the next world states *W'* depends on the current world state *W* and the current action *A* (see Figure 1, right-hand side, and Figure 2, left-hand side, respectively). For the first concept MC$_A$, let us assume that there is no dependence of the next world state *W'* on the current action *A*. In this case, the world dynamics kernel $\alpha(w'|w, a)$ reduces to $\hat{\alpha}(w'|w)$ (which is given by $\hat{\alpha}(w'|w) = \sum_a p(w', w, a)/p(w)$, see also Figure 2, centre). As a result, we would state that we have maximal morphological computation, as the system's behaviour is not controlled by the brain at all. An example of such a system is the Passive Dynamic Walker that was discussed in the introduction of this work. We can measure how much the observed behaviour differs from this assumption with the Kullback-Leibler divergence. This leads to the following formalisation:

$$\mathrm{MC}_A := \sum_{w',w,a} p(w', w, a) \log_2 \frac{\alpha(w'|w, a)}{\hat{\alpha}(w'|w)} \tag{1}$$

$$= I(W'; A|W) \tag{2}$$

Unfortunately, Equation (1) is zero for maximal morphological computation, which is why we initially chose to normalise and invert it, leading to the following definition:

$$\text{MC}_A := 1 - \frac{1}{\log_2 |\mathcal{W}|} \sum_{w',w,a} p(w',w,a) \log_2 \frac{\alpha(w'|w,a)}{\hat{\alpha}(w'|w)} \tag{3}$$

The second concept, MC_W starts with the opposite assumption, namely, that the current world state W does not have any influence on the next world state W' (see Figure 2, right-hand side). In this case, the world dynamics kernel $\alpha(w'|w,a)$ reduces to $\tilde{\alpha}(w'|a)$ (which is given by $\tilde{\alpha}(w'|a) = \sum_w p(w',w,a)/p(a)$, see also Figure 2) and analogously to the following definition for MC_W:

$$\text{MC}_W := \sum_{w',w,a} p(w',w,a) \log_2 \frac{\alpha(w'|w,a)}{\tilde{\alpha}(w'|a)} \tag{4}$$

$$= I(W';W|A) \tag{5}$$

The relation of the measures to transfer entropy [41,42] and the information bottleneck [43] are discussed in [15]. In the context of this work, we focus on their relation to synergistic information as defined by [34] and [35,36].

Next, we briefly restate the information decomposition by [34,44] that was used in the context of morphological computation in [33].

2.3. Synergistic Information Based on the Decomposition of the Multivariate Mutual Information

Consider three random variables X, Y, and Z. Suppose that a system wants to predict the value of the random variable X, but it can only access the information in Y or Z. The question is, how is the information that Y and Z carry about X distributed over Y and Z? In general, there may be *redundant* or *shared* information (information contained in both Y and Z), but there may also be *unique* information (information contained in either Y or Z). Finally, there is also the possibility of *synergistic* or *complementary* information, i.e., information that is only available when Y and Z are taken together. The classical example for synergy is the XOR function: if Y and Z are binary random variables and if $X = Y \text{ XOR } Z$, then neither Y nor Z contain any information about X (in fact, X is independent of Y and X is independent of Z), but when Y and Z are taken together, they completely determine X.

The total information that (Y, Z) contains about X can be quantified by the mutual information $I(X; Y, Z)$. However, there is no canonical way to separate these four kinds of informations. Different variations have been proposed (see e.g., [34,36,44–47]), but a final definition has yet to be found.

Mathematically, one would like to have four functions, namely shared information ($SI(X : Y; Z)$), unique information of Y ($UI(X : Y \setminus Z)$), unique information of Z ($UI(X : Z \setminus Y)$), and finally synergistic information (also named complementary information in [34], $CI(X : Y; Z)$) that satisfy

$$I(X : (Y, Z)) = SI(X : Y; Z) + UI(X : Y \setminus Z) \tag{6}$$

$$+ UI(X : Z \setminus Y) + CI(X : Y; Z). \tag{7}$$

It follows from the defining equations [34] and the chain rule of mutual information that an information decomposition always satisfies

$$I(X : Y|Z) = UI(X : Y \setminus Z) + CI(X : Y; Z). \tag{8}$$

Several candidates have been proposed for SI, UI, and CI so far (see e.g., [45,46]). A new candidate will be presented below (see Section 4.2). In this section, we will describe the decomposition of [34] that is defined in the following way.

Let Σ be the set of all possible joint distributions of X, Y, and Z. Fix an element $P \in \Sigma$ (the "true" joint distribution of X, Y, and Z).

Define

$$\Lambda_P = \Big\{ Q \in \Sigma : Q(X = x, Y = y) = P(X = x, Y = y)$$

$$\text{and } Q(X = x, Z = z) = P(X = x, Z = z)$$

$$\text{for all } x \in \mathcal{X}, y \in \mathcal{Y}, z \in \mathcal{Z} \Big\} \tag{9}$$

as the set of all joint distributions that have the same marginal distributions on the pairs (X, Y) and (X, Z). Then

$$UI(X : Y \setminus Z) = \min_{Q \in \Lambda_P} I_Q(X : Y|Z), \tag{10}$$

$$SI(X : Y; Z) = \max_{Q \in \Lambda_P} CoI_Q(X; Y; Z), \tag{11}$$

$$CI(X : Y; Z) = I(X : (Y, Z)) - \min_{Q \in \Lambda_P} I_Q(X : (Y, Z)), \tag{12}$$

where CoI denotes the co-information as defined in [48]. Here, a subscript Q in an information quantity means that the quantity is computed with respect to Q as the joint distribution.

In [34], the formulas for UI, CI, and SI are derived from considerations about decision problems in which the objective is to predict the outcome of X. In the context of morphological computation, we want to apply the information decomposition in the following way. We will set $X = W'$, $Y = W$, and $Z = A$. In the context of the sensorimotor loop, W and A not only have information about W' but they actually *control* W'. However, from an abstract point of view, the situation is similar: in the sensorimotor loop, we also expect to find aspects of redundant, unique, and complementary influence of W and A on W'. Formally, since everything is defined probabilistically, we can still use the same functions UI, CI, and SI. We believe that the arguments behind the definition of UI, CI, and SI remain valid in the setting of the sensorimotor loop where we need it.

The reason for investigating the unique and synergistic information is indicated in Equation (8), which we will rewrite here in terms of the sensorimotor loop in the following way:

$$MC_W = I(W'; W|A) \tag{13}$$

$$= UI(W' : W \setminus A) + CI(W' : W; A) \tag{14}$$

Equation (14) shows that MC_W can be decomposed into unique and synergistic information. This also shows the advantage of the information decomposition approach as defined by [34]. Synergistic and unique information can be computed from the other, if the conditional mutual information $I(W'; W|A)$ is known. The conditional mutual information can be easily derived from observation, given that there are enough samples with respect to the dimensionality of W', W, and A. We are not only interested in mathematically rigorous definitions but also in applicability to real data. This is where the decomposition by [34] currently has a disadvantage. There is no algorithm known to us to compute synergistic and unique information for non-trivial systems, i.e., non-binary systems. For the binary model of the sensorimotor loop (see below), we used an approximation in our previous work [33] that was already described in the original paper [34]. We will compare our previous results with a new measure that can be computed also for non-trivial systems as below. However, there is a more important problem with the definitions given by [34]. They do not incorporate the probability distributions over the inputs, in this case, the random variables W and A, which means that the synergistic measure $CI(W' : W; A)$ can falsely detect correlations in the input as synergistic information (this will be shown in the results section and discussed at the end of this work).

Applicability to non-trivial systems and potential false positives are the two reasons why we introduce a measure for synergistic information based on [35,36] in the remainder of this section.

2.4. Synergistic Information as the Difference between the Whole and the Sum of Its Parts

The basic idea of the complexity measure defined by [35] is summarised in the paper in the following way: "The whole is more than the sum of its elementary parts." The underlying information theoretic idea is best explained along the schematics shown in Figure 3.

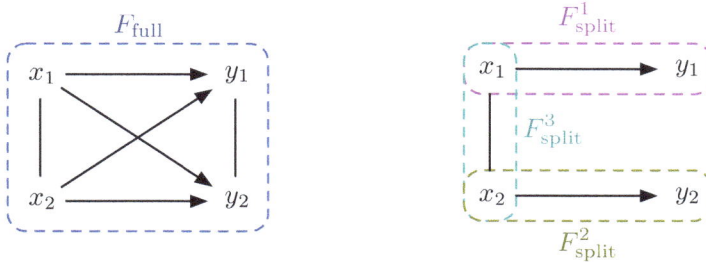

Figure 3. Quantifying complexity. Left-hand side: full model of two input and two output variables. Right-hand side: split model, as proposed by [35].

The left-hand side of Figure 3 shows the "whole", while the right-hand side shows "the elementary parts" of a stochastic system with two input variables, X_1 and X_2, and two output variables, Y_1 and Y_2. We refer to the graphical model on the left-hand side of Figure 3 as the *full model*, whereas the model on the right-hand side will be referred to as the *split model*. The full model assumes that every output node is connected to every input node. The split model assumes every input node only affects one output node. The undirected connection between the input nodes indicates that the input distribution is taken into account in both models. Both models are defined by their feature sets F. In the example given above, the feature set for the full model is given by $F_{full} = \{\{X_1, Y_1, X_2, Y_2\}\}$ and the feature set for the split model is given by $F_{split} = \{\{X_1, Y_1\}, \{X_2, Y_2\}, \{X_1, X_2\}\}$. Note that we have explicitly included the feature corresponding to the input distribution $\{X_1, X_2\}$. The divergence between the full and split models is defined as a measure for complexity in [35]. Variations of this measure have been proposed in [49,50] and compared in [51].

As discussed earlier in this work, we are primarily interested in the relation of the three random variables W, A, and W', which represent the current world state, the current action, and the next world state, respectively. Hence, we translate the quantification for complexity [35] that was originally formulated for four random variables (see Figure 3) to three random variables (see Figure 4). The resulting quantification is also known as synergistic information and was first discussed in [36].

Given three random variables X, Y, and Z (see Figure 4), the synergistic information is defined as the averaged Kullback-Leibler divergence

$$\sum_{y,z} p(y,z) D(p_{full}(x|y,z)||p_{split}(x|y,z)), \tag{15}$$

where the full and split models are defined by the feature sets depicted in Figure 4.

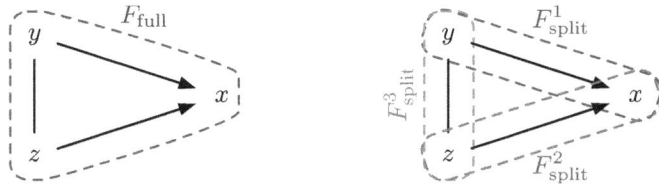

Figure 4. Quantifying synergy. Left-hand side: full model of two input and two output variables. Right-hand side: split model, as proposed by [36].

In their work, the authors showed that the main difference between the synergistic measure defined by [34] and the synergistic measure defined by [36] is that the former does not take the input distribution into account. This is shown by the following Equation (cited from [36]):

$$CI(X:Y;Z) = I(X,Y:Z) - \min_{q \in \Delta_P} I_q(X,Y:Z) \tag{16}$$

$$\Delta_P = \Big\{ Q \in \Sigma : Q(X=x, Y=y) = P(X=x, Y=y)$$

$$\text{and } Q(X=x, Z=z) = P(X=x, Z=z)$$

$$\text{and } Q(Y=y, Z=z) = P(Y=y, Z=z)$$

$$\text{for all } x \in \mathcal{X}, y \in \mathcal{Y}, z \in \mathcal{Z} \Big\} \tag{17}$$

The difference between the measure proposed by [34,36] is seen in line 3 of Equation (17) (compare with Equation (9)). Another difference is that the measure by [34] can so far not be calculated for non-trivial systems, e.g., with the iterative scaling algorithm discussed in the next section.

Applying the measure by [36] to the sensorimotor loop and in the context of morphological computation, we are asking how much the observation of our embodied agent's behaviour differs from the assumption that there is no synergistic term. In other words, we are measuring the averaged Kullback-Leibler divergence

$$\text{MC}_{SY} = \sum_{w,a} p(w,a) D(p_{\text{full}}(w'|w,a) || p_{\text{split}}(w'|w,a)) \tag{18}$$

where the feature set of the full model is defined as $F_{\text{full}} = \{\{W', W, A\}\}$ and the feature set of the split model is defined as $F_{\text{split}} = \{\{W', W\}, \{W', A\}, \{W, A\}\}$ (see Figure 4, where $X = W'$, $Y = W$, and $Z = A$).

2.5. Maximum Entropy Estimation with the Iterative Scaling Algorithm

There is a standard method to calculate the maximum entropy estimation of a probability distribution based on features, known as iterative scaling, which is well-established in this field and goes back to the work of Darroch and Ratcliff [52] and Csiszár [53]. The algorithm can be summarised in the following form (for joint distributions). Let $\hat{p}$ be the target distribution (in our case $\hat{p}(w', w, a) = \sum_s p(w)\beta(s|w)\pi(a|s)\alpha(w'|w,a)$, see Section 3.1), V be the set of random variables (in the context of this work $V = \{W', W, A\}$), and F be the feature set (e.g., $F_{\text{split}} = \{\{W', W\}, \{W', A\}, \{W, A\}\}$). For the sake of simplicity in presentation, we use the following abbreviations: $p(V) = p(w', w, a)$, $p(F_i)$ is either $p(w, a)$, $p(w', a)$, or $p(w', w)$ (depending on the index i) and $p(V \backslash F_i | F_i)$ is either $p(w'|w, a)$, $p(w|w', a)$, or $p(a|w', w)$ depending on the selected feature. The target distribution (that is approximated with the maximum entropy method) is denoted by $\hat{p}(V)$. As an example, the target distribution for the first feature is given by $\hat{p}(F_1) = \sum_a \hat{p}(w', w, a)$. Iterative scaling is then defined in the following way:

$$p^{(0)}(V) = \frac{1}{|V|} \tag{19}$$

$$p^{(n+1)}(V) = \hat{p}(F_{n \bmod |F|}) \cdot p^{(n)}(V \backslash F_{n \bmod |F|} | F_{n \bmod |F|}). \tag{20}$$

In words, we initialise the joint distribution $p^{(0)}$ to be the uniform distribution. In each iteration, we pick one feature from the set of features. We then multiply the marginal distribution of this feature in the target distribution ($\hat{p}(F_{n \bmod |F|})$) with the conditional distribution of the remaining variables conditioned on the chosen feature ($p^{(n)}(V \backslash F_{n \bmod |F|} | F_{n \bmod |F|})$). In the notion above, $F_{n \bmod |F|}$ refers to the iterative selection of features based on the current iteration step n, modulo the number $|F|$ of defined features. This algorithm is proved to converge [52,53] and is used in the numerical simulations below (source code is available at [54]). Next, we discuss the parametrised model of the sensorimotor loop which is used in this work to evaluate and compare the different measures.

3. Parametrised Model of the Sensorimotor Loop

The previous section discussed five different measures, namely MC_W, MC_A, $UI(W' : W \backslash A)$ $CI(W' : W; A)$, and MC_{SY}. The first four were evaluated in previous publications [14,15] based on a binary model of the sensorimotor loop. MC_W was also evaluated on data from hopping models [14] and in the context of soft robotics [1]. This is why we concentrate on $CI(W' : W; A)$ and MC_{SY} in the context of this work. A new proposal for the unique information will follow below (see Section 4.2).

In our previous work [33], we used an approximation to calculate the synergistic information and were only able to apply it to binary systems. As stated earlier, we are interested in applying synergistic information to real-world applications, hence, this section has two goals. The first goal is to compare how the previous results to the new measure. The second goal is to investigate how the new measure perform on non-binary systems in a fully controlled setting, i.e., model setting. This is a necessary step, before applying the measures to real data, as we need to understand if the results comply with the intuitive understanding of morphological computation.

The next section introduces the binary and non-binary model of the sensorimotor loop. The binary model is used to compare the results of our new measure MC_{SY} with our previous results on $CI(X : Y; Z)$ [33].

3.1. Binary Model of the Sensorimotor Loop

The causal diagram of the sensorimotor loop (see Figure 1) implies that we need to define four different maps, namely the world dynamics kernel $\alpha(w'|w, a)$, the policy $\pi(a|s)$, the sensor map $\beta(s|w)$, and finally, the input distribution $p(w)$. Note that in this section we operate on binary random variables, i.e., $w', w, s, a \in \Omega = \{-1, 1\}$. The binary model of the sensorimotor loop is then defined by the following set of equations (see also [1,15]):

$$\alpha_{\phi, \psi, \chi}(w'|w, a) = \frac{e^{\phi w' w + \psi w' a + \chi w' w a}}{\sum_{w'' \in \Omega} e^{\phi w'' w + \psi w'' a + \chi w'' w a}} \tag{21}$$

$$\beta_\zeta(s|w) = \frac{e^{\zeta s w}}{\sum_{s'' \in \Omega} e^{\zeta s'' w}} \tag{22}$$

$$\pi_\mu(a|s) = \frac{e^{\mu a s}}{\sum_{a' \in \Omega} e^{\mu a' s}} \tag{23}$$

$$p_\tau(w) = \frac{e^{\tau w}}{\sum_{w'' \in \Omega} e^{\tau w''}} \tag{24}$$

In the context of this work, we will only vary the parameters ϕ, ψ, and χ, which means that we will change the causal dependence of $W \rightarrow W'$ (parameter ϕ), the causal dependence of $A \rightarrow W'$ (parameter ψ), and finally, the causal dependence of $(W, A) \rightarrow W'$ (parameter χ). The other parameters are set such that they result in a uniform distribution of the corresponding kernels, i.e., $\tau, \mu, \zeta = 0$.

Next, we will first present the modification that allows us to model the non-binary sensorimotor loop, before the results are discussed in the next section (see Section 4).

3.1.1. Non-Binary Model of the Sensorimotor Loop

A generalisation from the binary model to a non-binary model requires to modify the function that operates on the state values. In case of a binary alphabet $\Omega = \{-1, 1\}$, the function was simply given by the product. We explain the approach to generalise this function used in this work based on the policy $\pi(a|s)$. For the non-binary model, the policy is now given by:

$$\pi_\mu(a|s) = \frac{e^{\mu f(a,s)}}{\sum_{a' \in \Omega} e^{\mu f(a's)}}. \tag{25}$$

There are various ways in which the function $f(a, s)$ can be chosen. Our choice of the function $f(a, s)$ is derived from the requirement to have a single parameter μ that allows us to smoothly transition from an independence to a strong dependence (we will briefly discuss a different method at the end of this paragraph). Therefore, we chose to normalise the values $a \in \mathcal{A} = \{1, 2, \ldots, N\}$ and $s \in \mathcal{S} = \{1, 2, \ldots, N\}$ such that they are mapped onto the interval $[-1, 1]$. This allows us to use define $f(a, s)$ similar to the binary case (see Equation (23)) as the product of both mapped random variables, i.e.,

$$f(a, s) = \left(2\frac{a-1}{N-1} - 1\right)\left(2\frac{s-1}{N-1} - 1\right). \tag{26}$$

As this example indicates, we chose the same number of bins for all random variables $w', w, s, a \in \Omega = \{1, 2, \ldots, N\}$.

It must be noted here that this choice of $f(a, s)$ is a projection of the full space of couplings to a subspace. In particular, there will be cases in which the effects of ψ, ϕ, and χ cannot be fully separated. One such example is the case in which ψ, ϕ, and χ are large and w', w, and a are all equal to N. The reason is that the bases spanning the space given by our choice of $f(a, s)$ are not orthogonal (as in the case of Equation (21) to Equation (24)). Walsh bases are one possible way to define the function(s) f such that they are orthogonal with respect to the L^2 inner product given by the uniform distribution.

This concludes the description of the parametrised binary and non-binary model of the sensorimotor loop. The next section presents the numerical results obtained with these two models.

4. Numerical Simulations

In this section, we plot the numerical results for two measures, $CI(W' : W; A)$ and MC_{SY}. $CI(W' : W; A)$ is plotted to compare our previous results [33] with the results obtained from MC_{SY} with the iterative scaling algorithm. The entire source code used in this work is available at [54].

4.1. Results for the Binary Sensorimotor Loop

This section begins with revisiting the point made earlier about the difference between $CI(W' : W; A)$ and MC_{SY} with respect to including the input distribution as part of the feature set. Figure 5 shows two experiments with the binary model of the sensorimotor loop. Both plots show the results for $\chi = 0.0$ and $\psi, \phi \in [0, 5.0]$, i.e., without the synergistic term $\chi w' w a$ (see Equation (21)). The first plot (left-hand side), shows the result for $CI(W' : W; A)$. Along the diagonal $\phi \approx \psi$, we see a region in which the synergistic information is non-zero. This is counter-intuitive as the higher-order interaction term is set to zero ($\chi = 0$, see Equation (21)) and the bases in the binary model of the sensorimotor loop are orthogonal. Hence, high values of ϕ and ψ should not result in non-zero values for the synergistic information in these cases. The second plot shows the results for MC_{SY} with the feature set $F = \{\{W, A\}, \{W, W'\}, \{A, W'\}\}$. The plot shows that this measure results in zero synergistic information for $\chi = 0.0$ and any choice of $\phi, \psi \in [0, 5]$, which is what we expect.

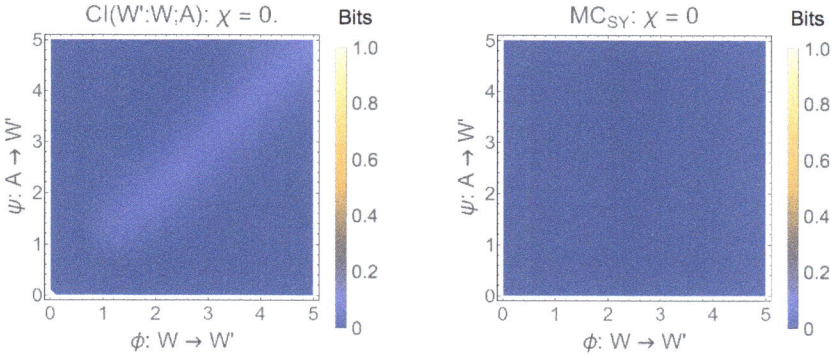

Figure 5. $CI(W' : W; A)$ (left-hand side) and MC_{SY} (right-hand side) without synergistic information present in the model, i.e., $\chi = 0$. The comparison of the plots reveal that $CI(W' : W; A)$ has regions with non-zero values, although no synergistic information is present. By that we mean that the higher order interaction term $\chi w'wa$ is set to zero (see Equation (21)). No false positives are found for MC_{SY} in this case.

The difference between these two approaches becomes more evident if we increase the synergistic coupling factor χ. Figure 6 shows the two measures for varying values of $\chi \in \{0, 1.25, 2.5, 3.75, 5.0\}$. We see that for increasing values of χ, the amount of detected synergistic information increases for both measures, however in different ways. $CI(W' : W; A)$ shows increasing regions with high values along the diagonal, whereas MC_{SY} shows areas in which the synergistic information is close to zero for values of $\chi \neq 0$. The latter is surprising because, in the binary case, the three basis are orthogonal and hence the synergistic information should be distinguishable from the pair-wise interactions, also for high values of ψ and ϕ. This issue will be addressed again in the discussion. The plots for the non-binary case, in particular for 4 and 8 bins, are shown below (see Figure 7). Note, that $CI(W' : W; A)$ was omitted for those cases, as it was not computable for non-trivial systems at this time.

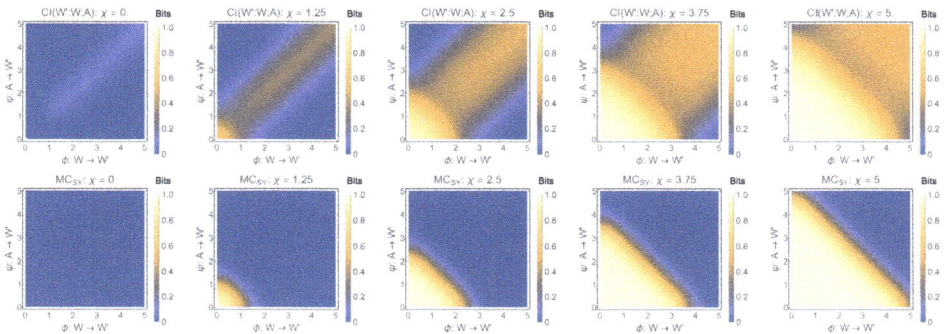

Figure 6. $CI(W' : W; A)$ (**top**) and MC_{SY} (**bottom**) for with $w', w, s, a \in \Omega = \{-1, 1\}$.

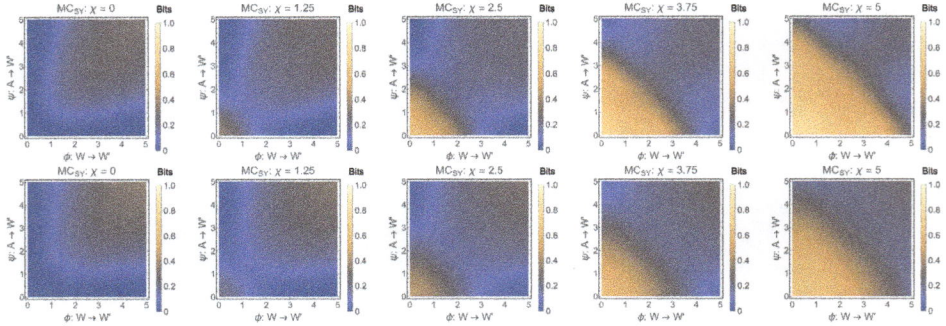

Figure 7. MC_{SY} with $w', w, s, a \in \Omega = \{1, 2, \ldots, 4\}$ (top) and $w', w, s, a \in \Omega = \{1, 2, \ldots, 8\}$ (bottom).

4.2. New Measure for Unique Information

Equation (14) shows that the conditional mutual information $I(X; Y|Z)$ is given by the sum of the unique information $UI(X : Y \backslash Z)$ and the synergistic information $CI(X : Y; Z)$. Given that MC_{SY} is a new measure for synergy (first introduced in [36]), we can now also give a new definition for the unique information $UI(X : Y \backslash Z)$ in terms of MC_W and MC_{SY}, denoted by MC_P, because it captures the part of MC_W that results from uncontrolled physical interactions:

$$MC_P = MC_W - MC_{SY}. \tag{27}$$

Note, that MC_P is not equivalent to the definition of $UI(X : Y \backslash Z)$ given above, because $CI(W' : W, A) \neq MC_{SY}$. Figure 8 shows MC_W, MC_P, and MC_{SY} for 2, 4, and 8 bins and varying values of the synergistic parameter χ.

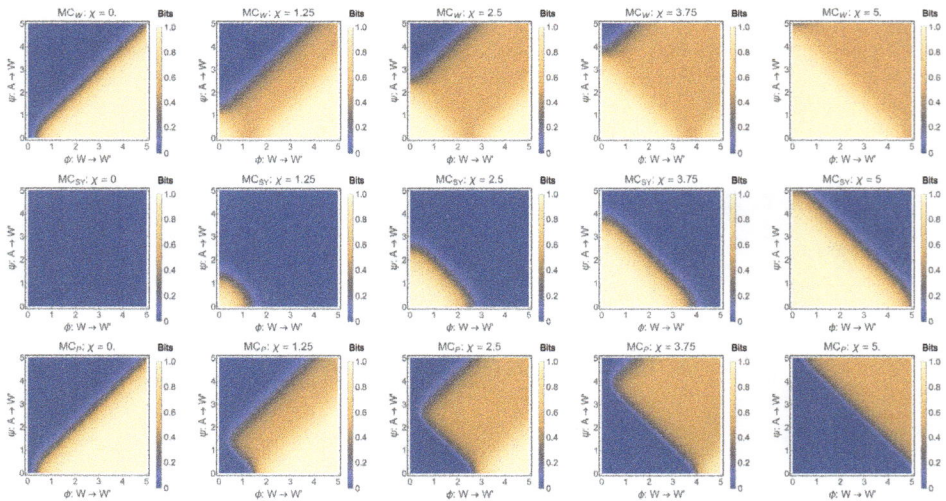

Figure 8. From left to right $\chi \in \{0, 1.25, 2.5, 3.75, 5.0\}$. From top to bottom: MC_W, MC_{SY}, MC_P, for binary system, i.e., $w', w, s, a \in \Omega = \{-1, 1\}$. Note that each plot in the upper row (MC_W) is the sum the corresponding plot in the second row (MC_{SY}) and the final row (MC_P).

Entropy **2017**, *19*, 456

5. Discussion

This work is a continuation of our previous work on the quantification of morphological computation. Initially, we proposed two measures MC_W and MC_A that are based on calculating the conditional mutual information in the sensorimotor loop [15]. In a later work, we investigated the relation of the conditional mutual information to unique $UI(W' : W \backslash A)$ and synergistic information $CI(W' : W; A)$ [54], while primarily focussing on unique information at that time. In this work, we investigated synergistic information and compared a measure based on [35] with the synergistic information that was independently discovered in [34,44]. The main difference between the two measures is that the previously utilised measure does not take the input distribution $p(w,a)$ into account. We have shown in this work that omitting the input distribution can lead to positive synergistic information in cases in which it should be zero. Furthermore, we showed that the new measure MC_{SY} can be calculated for non-trivial systems with the iterative scaling method.

Although the new measure MC_{SY} has significantly better properties (no false positives), it does show false negative results. This means that the measure MC_{SY} is close to zero for high values of χ, i.e., for a high synergistic term in our parametrised model, if two other couplings $W \to W'$ (parameter ϕ) and $A \to W'$ (parameter ψ) are large. It seems that in the case in which ϕ or ψ are large, it is increasingly difficult to detect synergistic information. At this point it is not quite clear if this is a general problem or something that is specific to the measure MC_{SY}. This is the work of currently ongoing investigations.

Acknowledgments: This work was partly funded by the German Priority Program DFG-SPP 1527 "Autonomous Learning." This publication was also made possible through the support of a grant from the John Templeton Foundation. The opinions expressed in this publication are those of the author(s) and do not necessarily reflect the views of the John Templeton Foundation.

Author Contributions: K.G.-Z. and N.A. proposed and discussed the research. K.G.-Z. carried out most of the research and took the main responsibility for writing the article. C.L. wrote the Iterative Scaling code and contributed to the writing of the paper. All authors read and approved the final manuscript.

Conflicts of Interest: The authors declare no conflict of interest.

References

1. Ghazi-Zahedi, K.; Deimel, R.; Montúfar, G.; Wall, V.; Brock, O. Morphological Computation: The Good, the Bad, and the Ugly. In Proceedings of the IEEE/RSJ International Conference on Intelligent Robots and Systems, Vancouver, BC, Canada, 24–28 September 2017.
2. Valero-Cuevas, F.J.; Yi, J.W.; Brown, D.; McNamara, R.V.; Paul, C.; Lipson, H. The tendon network of the fingers performs anatomical computation at a macroscopic scale. *IEEE Trans. Biomed. Eng.* **2007**, *54*, 1161–1166.
3. Brown, E.; Rodenberg, N.; Amend, J.; Mozeika, A.; Steltz, E.; Zakin, M.R.; Lipson, H.; Jaeger, H.M. Universal robotic gripper based on the jamming of granular material. *Proc. Natl. Acad. Sci. USA* **2010**, *107*, 18809–18814.
4. Beal, D.N.; Hover, F.S.; Triantafyllou, M.S.; Liao, J.C.; Lauder, G.V. Passive propulsion in vortex wakes. *J. Fluid Mech.* **2006**, *549*, 385–402.
5. Liao, J.C.; Beal, D.N.; Lauder, G.V.; Triantafyllou, M.S. Fish Exploiting Vortices Decrease Muscle Activity. *Science* **2003**, *302*, 1566–1569.
6. Ziegler, M.; Iida, F.; Pfeifer, R. "Cheap" underwater locomotion: Roles of morphological properties and behavioral diversity. In Proceedings of the 9th International Conference on Climbing and Walking Robots, Brussels, Belgium, 12–14 September 2006.
7. Zhao, Q.; Nakajima, K.; Sumioka, H.; Hauser, H.; Pfeifer, R. Spine dynamics as a computational resource in spine-driven quadruped locomotion. In Proceedings of the 2013 IEEE/RSJ International Conference on Intelligent Robots and Systems, Tokyo, Japan, 3–7 November 2013; pp. 1445–1451.
8. Iida, F.; Pfeifer, R. "Cheap" rapid locomotion of a quadruped robot: Self-stabilization of bounding gait. In Proceedings of the International Conference on Intelligent Autonomous Systems, Tokyo, Japan, 7–9 March 2004; pp. 642–649.
9. McGeer, T. Passive walking with knees. In Proceedings of the International Conference on Robotics and Automation, Cincinnati, OH, USA, 13–18 May 1990; pp. 1640–1645.

10. Rieffel, J.A.; Valero-Cuevas, F.J.; Lipson, H. Morphological communication: Exploiting coupled dynamics in a complex mechanical structure to achieve locomotion. *J. R. Soc. Interface* **2010**, *7*, 613–621.

11. Franceschini, N.; Pichon, J.M.; Blanes, C.; Brady, J. From insect vision to robot vision. *Philos. Trans. R. Soc. B* **1992**, *337*, 283–294.

12. Paul, C. Morphological computation: A basis for the analysis of morphology and control requirements. *Robot. Auton. Syst.* **2006**, *54*, 619–630.

13. Pfeifer, R.; Bongard, J.C. *How the Body Shapes the Way We Think: A New View of Intelligence*; The MIT Press (Bradford Books): Cambridge, MA, USA, 2006.

14. Ghazi-Zahedi, K.; Haeufle, D.F.; Montufar, G.F.; Schmitt, S.; Ay, N. Evaluating Morphological Computation in Muscle and DC-motor Driven Models of Hopping Movements. *Front. Robot. AI* **2016**, *3*. doi:10.3389/frobt.2016.00042.

15. Zahedi, K.; Ay, N. Quantifying Morphological Computation. *Entropy* **2013**, *15*, 1887–1915.

16. Horsman, C.; Stepney, S.; Wagner, R.C.; Kendon, V. When does a physical system compute? *Proc. R. Soc. A* **2014**, *470*, doi:10.1098/rspa.2014.0182 .

17. Müller, V.C.; Hoffmann, M. What is Morphological Computation? On How the Body Contributes to Cognition and Control. *Artif. Life* **2017**, *23*, 1–24.

18. Füchslin, R.M.; Dzyakanchuk, A.; Flumini, D.; Hauser, H.; Hunt, K.J.; Luchsinger, R.H.; Reller, B.; Scheidegger, S.; Walker, R. Morphological Computation and Morphological Control: Steps toward a Formal Theory and Applications. *Artif. Life* **2012**, *19*, 9–34.

19. Hauser, H.; Ijspeert, A.; Füchslin, R.M.; Pfeifer, R.; Maass, W. Towards a theoretical foundation for morphological computation with compliant bodies. *Biol. Cybern.* **2011**, *105*, 355–370.

20. Nowakowski, P.R. Bodily Processing: The Role of Morphological Computation. *Entropy* **2017**, *19*. doi:10.3390/e19070295.

21. Pfeifer, R.; Iida, F. Morphological computation: Connecting body, brain and environment. *Jap. Sci. Mon.* **2005**, *58*, 48–54.

22. Pfeifer, R.; Lungarella, M.; Iida, F. Self-Organization, Embodiment, and Biologically Inspired Robotics. *Science* **2007**, *318*, 1088–1093.

23. Jaeger, H.; Haas, H. Harnessing Nonlinearity: Predicting Chaotic Systems and Saving Energy in Wireless Communication. *Science* **2004**, *304*, 78–80.

24. Maass, W.; Natschläger, T.; Markram, H. Real-Time Computing Without Stable States: A New Framework for Neural Computation Based on Perturbations. *Neural Comput.* **2002**, *14*, 2531–2560.

25. Nakajima, K.; Hauser, H.; Kang, R.; Guglielmino, E.; Caldwell, D.G.; Pfeifer, R. Computing with a muscular-hydrostat system. In Proceedings of the 2013 IEEE International Conference on Robotics and Automation, Karlsruhe, Germany, 6–10 May 2013; pp. 1504–1511.

26. Hauser, H.; Corucci, F. Morphosis—Taking Morphological Computation to the Next Level. In *Soft Robotics: Trends, Applications and Challenges: Proceedings of the Soft Robotics Week, April 25-30, 2016, Livorno, Italy*; Laschi, C., Rossiter, J., Iida, F., Cianchetti, M., Margheri, L., Eds.; Springer International Publishing: Cham, Switzerland, 2017; pp. 117–122.

27. Nurzaman, S.G.; Yu, X.; Kim, Y.; Iida, F. Guided Self-Organization in a Dynamic Embodied System Based on Attractor Selection Mechanism. *Entropy* **2014**, *16*, 2592–2610.

28. Nurzaman, S.G.; Yu, X.; Kim, Y.; Iida, F. Goal-directed multimodal locomotion through coupling between mechanical and attractor selection dynamics. *Bioinspir. Biomim.* **2015**, *10*, 025004.

29. Pfeifer, R.; Iida, F.; Gòmez, G. Morphological computation for adaptive behavior and cognition. *Int. Congr. Ser.* **2006**, *1291*, 22–29.

30. Pfeifer, R.; Gómez, G. *Morphological Computation—Connecting Brain, Body, and Environment*; Springer: Berlin/Heidelberg, Germany, 2009; pp. 66–83.

31. Clark, A. *Being There: Putting Brain, Body, and World Together Again*; MIT Press: Cambridge, MA, USA, 1996.

32. Ay, N.; Zahedi, K. On the causal structure of the sensorimotor loop. In *Guided Self-Organization: Inception*; Prokopenko, M., Ed.; Springer: Berlin, Germany, 2014; Volume 9.

33. Ghazi-Zahedi, K.; Rauh, J. Quantifying Morphological Computation based on an Information Decomposition of the Sensorimotor Loop. In Proceedings of the 13th European Conference on Artificial Life (ECAL 2015), York, UK, 20–24 July 2015; pp. 70–77.

34. Bertschinger, N.; Rauh, J.; Olbrich, E.; Jost, J.; Ay, N. Quantifying Unique Information. *Entropy* **2014**, *16*, 2161–2183.

35. Ay, N. Information Geometry on Complexity and Stochastic Interaction. *Entropy* **2015**, *17*, 2432–2458.

36. Perrone, P.; Ay, N. Hierarchical Quantification of Synergy in Channels. *Front. Robot. AI* **2016**, *2*, 35.

37. Von Foerster, H. On Self-Organizing Systems and Their Environments. In *Understanding Understanding*; Springer: New York, NY, USA, 2003; pp. 1–19.

38. Von Uexkuell, J. A Stroll Through the Worlds of Animals and Men. In *Instinctive Behavior*; Schiller, C.H., Ed.; International Universities Press: New York, NY, USA, 1957; pp. 5–80.

39. Ay, N.; Löhr, W. The Umwelt of an embodied agent—A measure-theoretic definition. *Theory Biosci.* **2015**, *134*, 105–116.

40. Sutton, R.S.; Barto, A.G. *Reinforcement Learning: An Introduction*; MIT Press: Cambridge, CA, USA, 1998.

41. Schreiber, T. Measuring Information Transfer. *Phys. Rev. Lett.* **2000**, *85*, 461–464.

42. Bossomaier, T.; Barnett, L.; Harré, M.; Lizier, J.T. *An Introduction to Transfer Entropy*; Springer: Berlin, Germany, 2016.

43. Tishby, N.; Pereira, F.C.; Bialek, W. The information bottleneck method. In Proceedings of the 37th Annual Allerton Conference on Communication, Control and Computing, Monticello, IL, USA, 29 September–1 October 1999; pp. 368–377.

44. Griffith, V.; Koch, C. *Quantifying Synergistic Mutual Information*; Springer: Berlin/Heidelberg, Germany, 2014; pp. 159–190.

45. Williams, P.L.; Beer, R.D. Nonnegative Decomposition of Multivariate Information. *arXiv* **2010**, arXiv10042515.

46. Harder, M.; Salge, C.; Polani, D. Bivariate measure of redundant information. *Phys. Rev. E* **2013**, *87*, 012130.

47. Griffith, V.; Chong, E.K.P.; James, R.G.; Ellison, C.J.; Crutchfield, J.P. Intersection Information Based on Common Randomness. *Entropy* **2014**, *16*, 1985–2000.

48. Bell, A.J. The co-information lattice. In Proceedings of the Fifth International Workshop on Independent Component Analysis and Blind Signal Separation: ICA 2003, Nara, Japan, 1–4 April 2003.

49. Amari, S.I. *Information Geometry and Its Applications*; Springer: Berlin, Germany, 2016.

50. Oizumi, M.; Tsuchiya, N.; Amari, S.I. Unified framework for information integration based on information geometry. *Proc. Natl. Acad. Sci. USA* **2016**, *113*, 14817–14822.

51. Kanwal, M.S.; Grochow, J.A.; Ay, N. Comparing Information-Theoretic Measures of Complexity in Boltzmann Machines. *Entropy* **2017**, *19*, 16.

52. Darroch, J.N.; Ratcliff, D. Generalized Iterative Scaling for Log-Linear Models. *Ann. Math. Stat.* **1972**, *43*, 1470–1480.

53. Csiszár, I. *I*-Divergence Geometry of Probability Distributions and Minimization Problems. *Ann. Probab.* **1975**, *3*, 146–158.

54. Ghazi-Zahedi, K. Entropy++ GitHub Repository. Available online: http://github.com/kzahedi/entropy (accessed on 25 August 2017).

entropy

MDPI

Article

Information Theoretical Study of Cross-Talk Mediated Signal Transduction in MAPK Pathways

Alok Kumar Maity [1,†], **Pinaki Chaudhury** [1] and **Suman K. Banik** [2,*]

1   Department of Chemistry, University of Calcutta, 92 Acharya Prafulla Chandra Road, Kolkata 700009, India; amaity@ucla.edu (A.K.M.); pinakc@rediffmail.com (P.C.)

2   Department of Chemistry, Bose Institute, 93/1 Acharya Prafulla Chandra Road, Kolkata 700009, India

*   Correspondence: skbanik@jcbose.ac.in; Tel.: +91-33-2303-1147

†   Current Address: Department of Chemistry and Biochemistry, University of California, Los Angeles, CA 90095, USA.

Received: 28 June 2017; Accepted: 1 September 2017; Published: 5 September 2017

Abstract: Biochemical networks having similar functional pathways are often correlated due to cross-talk among the homologous proteins in the different networks. Using a stochastic framework, we address the functional significance of the cross-talk between two pathways. A theoretical analysis on generic MAPK pathways reveals cross-talk is responsible for developing coordinated fluctuations between the pathways. The extent of correlation evaluated in terms of the information theoretic measure provides directionality to net information propagation. Stochastic time series suggest that the cross-talk generates synchronisation in a cell. In addition, the cross-interaction develops correlation between two different phosphorylated kinases expressed in each of the cells in a population of genetically identical cells. Depending on the number of inputs and outputs, we identify signal integration and signal bifurcation motif that arise due to inter-pathway connectivity in the composite network. Analysis using partial information decomposition, an extended formalism of multivariate information calculation, also quantifies the net synergy in the information propagation through the branched pathways. Under this formalism, signature of synergy or redundancy is observed due to the architectural difference in the branched pathways.

Keywords: mutual information; partial information decomposition; net synergy; Langevin equation

1. Introduction

The decision making processes at the cellular level are initiated by some specialised signalling networks [1,2]. These networks play a pivotal role in making robust and precise cellular response towards endogenic and exogenic perturbations. In addition, the process of decision making resolves cellular fate as well as survival strategies in diverse peripheral conditions. Although both prokaryotic and eukaryotic cells are comprised of several common signalling networks, few signalling networks are incorporated mostly in the eukaryotes [3].

As an evolutionary outcome, cells have developed an optimal protein-protein (cognate and noncognate) interaction within the signalling pathway to transduce extra-cellular information efficiently [4]. One such signalling network is the mitogen-activated protein kinase (MAPK) pathway that plays the central role to attune with extra-cellular signal in eukaryotic cells [5–9]. Although different MAPK pathways with diverse inputs and outputs belong to a higher living species, they are sometimes interconnected through overlapping sets of signalling components. Depending on the interconnections, MAPK pathways can be classified into different groups that use one or more than one common signalling components. Moreover, as a result of cross-interaction, a single regulon regulates multiple targets in addition to its cognate target. Such type of signal association is defined as cross-talk. Cross-coupling in the signalling network can modify the functionality of a network topology and can subsume errors compared to the uncoupled one. Cross-interactions have been identified not only in eukaryotes but also in prokaryotes, as observed in the bacterial two-component system [10–14]. In the eukaryotic system, cross-talk has been identified in numerous situations [15–17]. Furthermore, cross-talk and several of its variants have also been identified at different stages of gene regulation [18–22].

Since cross-talk is observed in a broad range of biological processes, one may interrogate the functional utility of such network coordination. The cross-coupling mechanism is conveyed through generations and observed in a significant number of evolutionary descent. This signature remains prominent in the course of evolution in spite of other modifications that are taking place at the cellular level. Such character indicates that cross-talk might have a definite functional role to build up synchronised cellular regulations by spending the stored energy. If this is true, how does a cell balance the trade-off between network association and potential cost? Few comprehensive experiments on the network connectivity suggest that networks of a well-delineated cluster are correlated with each other but are uncorrelated to the rest of the network [8]. Synchronisation is necessary to attain natural activity but needs to maintain a threshold value. Otherwise, too much synchronisation may lead to a physiological disorder like epilepsy [23]. Inter-pathway cross-talk becomes prominent due to the limitation of common resources, defined as the overloaded condition. However, cross-talk effect becomes faint in the underloaded condition, where the level of available resources are present in sufficient amount [19,20,22,24,25]. A key source of survival strategy under diverse environmental conditions is the generation of fluctuations which induces non-genetic variability in a cellular population. In such a situation, cells readjust to cope with the limited resources by introducing cross-correlation among a set of genes and thus implementing a successful bet-hedging program [22]. Cross-talk also facilitates synchronisation in different organs such as cardio-respiratory interaction, brain and tissues [23].

To address the functionality of cross-talk, we undertake a representative network comprised of well-characterized features of signal transduction. To be specific, we focus on MAPK pathway, a well-studied eukaryotic signalling machinery, conserved with three kinase cascades. In *S. cerevisiae*, five MAPK signalling pathways are present, out of which only three (pheromone response, filamentous growth response and osmostress adaptation) use a common kinase protein Ste11 [5,8]. In fact, pheromone response and filamentous growth pathways also use the same kinase Ste7. Pheromone MAPK cascade (Ste11 → Ste7 → Fus3) is activated by mating pheromone. Under low nutrient condition, filamentous growth MAPK cascade (Ste11 → Ste7 → Kss1) becomes active whereas high external osmolarity activates the osmoadaption cascade (Ste11 → Pbs2 → Hog1) [5,8]. Due to inter-cascade correlation among the three signalling pathways, one pathway can be activated by the signal of another pathway in the absence of its own signal. Several experimental results suggest that such cross-talk is filtered out by cross-pathway inhibition, kinetic insulation and formation of scaffold protein [5,8,26–29]. Although activation through inter-pathway cross-talk and cross-pathway inhibition compensates each other, information is exchanged among the pathways during these interactions. This leads to distinct queries (i) is it possible for an individual signalling pathway to convey its input signal reliably downstream without experiencing any influence from the other channels of signal? (ii) Since the inter-pathway connectivity is known not to allow the uniqueness of transduced signals - what are the physiological advantages of cross-talk? (iii) Is there any participation of pathway output in the cooperative regulation of a downstream target in a synchronised manner? (iv) How is it possible for correlated pathways to keep up static as well as dynamic synchronisation in a single cell environment that is prevalently stochastic in nature? (v) Do correlated fluctuations have any capability to control the variability in the correlation between two different proteins?

In the present manuscript, we study generic *S. cerevisiae* MAPK pathways to address the potential functionality of inter-pathway cross-talk within a stochastic framework. We consider two equivalent interacting MAPK pathways, each one consisting of a linear chain of three MAPK cascade proteins [5,8]. Both pathways get activated by their corresponding external signals propagating downstream through phosphorylation (activation) and dephosphorylation (deactivation) of the cascade proteins. In addition, due to cross-talk, phosphorylation of the intermediate components of the two pathways is influenced by the activated kinase of the other pathway along with the cognate one. As the population of each cascade protein is not sufficiently high within a single cell and experiences a fluctuating environment, we express all associated chemical reactions in terms of the stochastic differential equation. We solve the coupled set of nonlinear Langevin equations using linear noise approximation (LNA) [30,31] and calculate the variance of each kinase and covariance between two different kinases (see Sec. II and Appendix). Recent theoretical development [32] shows that LNA is not only limited to high copy number but also exact up to second moments of any chemical species involved in a second-order reaction. The fluctuations associated with at least one of the species participating in each of the second-order reaction are Poissonian and uncorrelated with the fluctuations of other species. Also, LNA remains valid for faster activation and deactivation (or synthesis and degradation) rates of the corresponding components compared to the coarse-grained (steady state) time scale [30–40].

To classify the signal transduction efficacy through two pathways in the presence of cross-association, we quantify two as well as multivariate mutual information. Distributions of all kinase proteins are approximately considered Gaussian, allowing us to adopt a reduced expression of mutual information [41,42]. The reduced equation mainly depends on the variance and the covariance of the corresponding kinase. We validate our analytical calculation by exact stochastic simulation [43]. In the first subsection, we quantify two variable mutual information under the influence of cross-talk parameter. We also investigate the mutual information between two non-cognate kinases and find causality of this coordination. Since causality leads to synchronisation [23,44], it is important to measure causality relation between the pathways, i.e., who regulates whom and to which extent. If both pathways interact with each other and transduce information of the corresponding input signal with different degrees, then it is challenging to characterize the magnitude and direction of signal propagation. To overcome such difficulty, we define a new measure, net information transduction, using the expressions of two cross mutual information, which satisfactorily quantifies the amount of net signal propagation. In the connection of measuring directionality, it is important to mention the concept of transfer entropy [45] which has been applied in several systems [46–49]. Transfer entropy quantifies the directed information transfer within a system. In the calculation of transfer entropy, one considers the time series of the random variable and makes use of the time lagged data. In our study, however, we make use of mutual information evaluated at steady state to define a relative dimensionless measure, net information transduction D (see Equation (1)). D is a normalized quantity and is bounded within a range ± 1. It helps to diagnose the direction of net information flow between the two parallel cross interacting signalling pathways. We also verify inter-pathway synchronisation with the help of coordinated fluctuations of stochastic trajectories of two parallel kinases. This result implies how two kinases are synchronised within a cell. To understand this phenomenon further, we investigate how much correlation develops between the steady state levels of two different kinases in a hypothetical population of genetically identical cells.

In the second subsection, we quantify three variable mutual information (i.e., two sources to one target, or one source to two targets) when both the channels of information flow work separately. We make use of three variable mutual information along with two variable mutual information to define the measure of net synergy [50]. In the present work, we quantify net synergy using the theory of partial information decomposition (PID) [51] as the formalism deals with the calculation of multivariate mutual information. Furthermore, it defines unique, synergistic and redundant information among the variables of interest. Following the initial development by Williams and Beer [51], several modifications of the formalism have been proposed till date [52–54]. In the context of dynamical systems obeying Gaussian noise processes the formalism has been further extended by Barrett [55]. We note here that the measure of interaction information is the negative of the measure of net synergy [56]. Considering the interactions between two parallel MAPK pathway, we identify signal integration and signal bifurcation motif. In signal integration motif, two MAPKKK proteins transmit signal into a single MAPKK protein which amounts to a sub-motif with two inputs and one output. On the other hand, in signal bifurcation motif, a single MAPKKK protein transmits signal to two MAPKK protein thus leading into a sub-motif with one input and two outputs. Quantification of information transmission in these two sub-motifs requires calculation of both two variable and multivariate mutual information which in turn provides the metric of net synergy. We observe the sign of net synergy value changes depending on the signal integration as well as signal bifurcation and is mainly controlled by pathway architecture.

2. Results and Discussion

2.1. Two Variable Mutual Information

The parameters ε_2 and ε_1 control the signalling channel X and Y, respectively. In Figure 1, we show the mutual information profile as a function of ε_2 for two different sets of parameters (see Tables 1 and 2) while keeping ε_1 constant. Figure 1A shows that mutual information between x_{p1} and x_{p2} kinases decays with the increment of ε_2. Augmentation of ε_2 includes a competition between x_{p1} and y_{p1} to phosphorylate the x_2 kinase. During phosphorylation, mutual association is originated, and signal transduction is ensued. Thus, for the low value of ε_2, maximum level of mutual information is attained due to minimal phosphorylation competition. On the other hand, minimum level of mutual information is propagated at high ε_2 value due to maximum phosphorylation contribution of y_{p1}. In Figure 1B, mutual information between x_{p1} and y_{p2} is plotted, which shows a constant value as a function of ε_2. This happens as ε_2 has no influence in the alteration of mutual information. The same logic is applicable to the mutual information between y_{p1} and y_{p2} shown in Figure 1C. In Figure 1D, mutual information between y_{p1} and x_{p2} increases as a function of ε_2, as ε_2 is only responsible for establishing the cross-talk between y_{p1} and x_{p2}. This result implies that with the enhancement of cross-talk the process of signal integration through y_{p1} increases. The same profiles can be generated as a function of ε_1, while keeping ε_2 fixed. These results together indicate that $\mathcal{I}(x_{p1}; x_{p2})$ and $\mathcal{I}(y_{p1}; x_{p2})$ depend on ε_2, whereas $\mathcal{I}(x_{p1}; y_{p2})$ and $\mathcal{I}(y_{p1}; y_{p2})$ depend on ε_1. In Figure 1A–D, the green lines drawn for slower relaxation rate (see Table 2) always maintains a lower mutual information value compared to the red lines drawn for faster relaxation rate (see Table 1). Relaxation rates of the corresponding kinases i.e., $x_{p1}, x_{p2}, x_{p3}, y_{p1}, y_{p2}$ and y_{p3} are $-J_{x1x1} = (\alpha_1 + k_x s_x)$, $-J_{x2x2} = (\alpha_2 + k_{12x}\langle x_{p1}\rangle + \varepsilon_2\langle y_{p1}\rangle)$, $-J_{x3x3} = (\alpha_3 + k_{23x}\langle x_{p2}\rangle)$, $-J_{y1y1} = (\beta_1 + k_y s_y)$, $-J_{y2y2} = (\beta_2 + k_{12y}\langle y_{p1}\rangle + \varepsilon_1\langle x_{p1}\rangle)$ and $-J_{y3y3} = (\beta_3 + k_{23y}\langle y_{p2}\rangle)$, respectively, where the angular bracket $\langle \cdots \rangle$ indicates the deterministic copy number at long time limit (see Appendix). An input signal can reliably flow downstream if relaxation rate (or degradation rate) of a cascade protein is higher than that of its upstream cascade proteins [35]. For red line, we consider higher degradation rate for x_{p2} and x_{p3} (y_{p2} and y_{p3}) compared to x_{p1} (y_{p1}). Thus, faster relaxation rates are attained under this condition with high information propagation capacity.

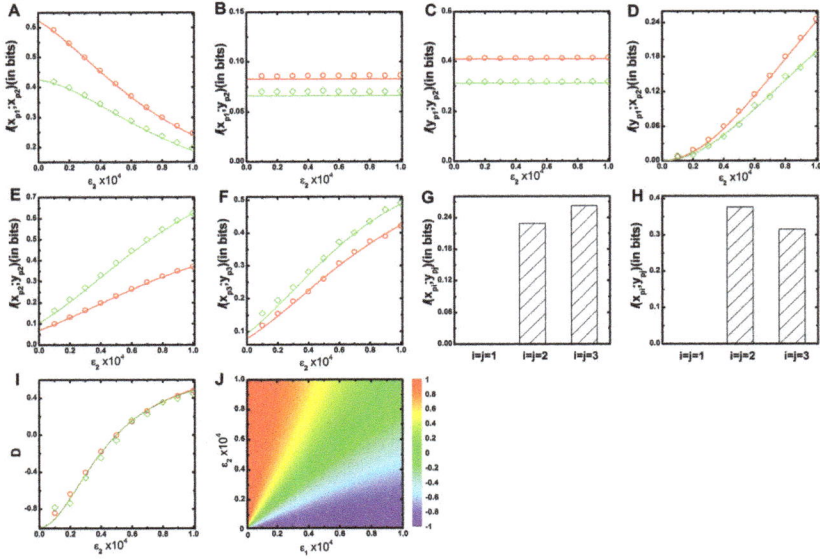

Figure 1. (color online) Two variable mutual information and net information transduction as a function of cross-talk parameter. (**A–F**) Two variable mutual information profiles $\mathcal{I}(x_{p1}; x_{p2})$, $\mathcal{I}(x_{p1}; y_{p2})$, $\mathcal{I}(y_{p1}; y_{p2})$, $\mathcal{I}(y_{p1}; x_{p2})$, $\mathcal{I}(x_{p2}; y_{p2})$ and $\mathcal{I}(x_{p3}; y_{p3})$ as a function of cross-interaction parameter ε_2 for a fixed value of $\varepsilon_1 = 0.5 \times 10^{-4}$. In all figures, red (with open circle) and green (with open diamond) lines are generated using faster (Table 1) and slower (Table 2) relaxation rate parameters, respectively. The symbols are generated using stochastic simulation algorithm [43] and the lines are due to theoretical calculation; (**G,H**) Bar diagram of two variable mutual information of three parallel cascade kinases under an equivalent cross-talk condition ($\varepsilon_1 = \varepsilon_2 = 0.5 \times 10^{-4}$) for faster (Table 1) and slower (Table 2) relaxation rate parameters, respectively; (**I**) Net information transduction D as a function of cross-interaction parameter ε_2 for a fixed value of $\varepsilon_1 = 0.5 \times 10^{-4}$. The red (with open circle) and the green (with open diamond) lines are due to faster (Table 1) and slower (Table 2) relaxation rate parameters, respectively. The figure indicates data collapse for two relaxation rate parameters. The symbols are generated using stochastic simulation algorithm [43] and the lines are obtained from theoretical calculation. All the simulation data (open circles and open diamonds) are ensemble average of 10^7 independent trajectories; (**J**) 2d-surface plot of net information transduction D as a function of two cross-talk parameters ε_1 and ε_2 for faster (Table 1) relaxation rate parameters.

Table 1. Reactions and corresponding parameter values for the MAPK network motif of *S. cerevisiae* [6,9,66], related to faster relaxation rate. Other Parameters are $s_x = s_y = 10$ molecules/cell, $x_{T1} = x_1 + x_{p1} = 250$ molecules/cell, $x_{T2} = x_2 + x_{p2} = 1700$ molecules/cell, $x_{T3} = x_3 + x_{p3} = 5000$ molecules/cell, $y_{T1} = y_1 + y_{p1} = 250$ molecules/cell, $y_{T2} = y_2 + y_{p2} = 1700$ molecules/cell and $y_{T3} = y_3 + y_{p3} = 5000$ molecules/cell. The kinetic schemes adopted in the present work follows the model of Heinrich et al. [63].

Description	Reaction	Propensity Function	Rate Constant
Activation of x_1	$x_1 + s_x \xrightarrow{k_x} x_{p1} + s_x$	$k_x s_x x_1$	$k_x = 10^{-4}$ molecules^{-1} s^{-1}
Deactivation of x_{p1}	$x_{p1} \xrightarrow{\alpha_1} x_1$	$\alpha_1 x_{p1}$	$\alpha_1 = 0.01$ s^{-1}
Activation of y_1	$y_1 + s_y \xrightarrow{k_y} y_{p1} + s_y$	$k_y s_y y_1$	$k_y = 10^{-4}$ molecules^{-1} s^{-1}
Deactivation of y_{p1}	$y_{p1} \xrightarrow{\beta_1} y_1$	$\beta_1 y_{p1}$	$\beta_1 = 0.01$ s^{-1}
Activation of x_2	$x_2 + x_{p1} \xrightarrow{k_{12x}} x_{p2} + x_{p1}$	$k_{12x} x_{p1} x_2$	$k_{12x} = 10^{-4}$ molecules^{-1} s^{-1}
Activation of x_2	$x_2 + y_{p1} \xrightarrow{\varepsilon_2} x_{p2} + y_{p1}$	$\varepsilon_2 y_{p1} x_2$	$\varepsilon_2 = (0-1) \times 10^{-4}$ molecules^{-1} s^{-1}
Deactivation of x_{p2}	$x_{p2} \xrightarrow{\alpha_2} x_2$	$\alpha_2 x_{p2}$	$\alpha_2 = 0.05$ s^{-1}
Activation of y_2	$y_2 + y_{p1} \xrightarrow{k_{12y}} y_{p2} + y_{p1}$	$k_{12y} y_{p1} y_2$	$k_{12y} = 10^{-4}$ molecules^{-1} s^{-1}
Activation of y_2	$y_2 + x_{p1} \xrightarrow{\varepsilon_1} y_{p2} + x_{p1}$	$\varepsilon_1 x_{p1} y_2$	$\varepsilon_1 = (0-1) \times 10^{-4}$ molecules^{-1} s^{-1}
Deactivation of y_{p2}	$y_{p2} \xrightarrow{\beta_2} y_2$	$\beta_2 y_{p2}$	$\beta_2 = 0.05$ s^{-1}
Activation of x_3	$x_3 + x_{p2} \xrightarrow{k_{23x}} x_{p3} + x_{p2}$	$k_{23x} x_{p2} x_3$	$k_{23x} = 5 \times 10^{-5}$ molecules^{-1} s^{-1}
Deactivation of x_{p3}	$x_{p3} \xrightarrow{\alpha_3} x_3$	$\alpha_3 x_{p3}$	$\alpha_3 = 0.05$ s^{-1}
Activation of y_3	$y_3 + y_{p2} \xrightarrow{k_{23y}} y_{p3} + y_{p2}$	$k_{23y} y_{p2} y_3$	$k_{23y} = 5 \times 10^{-5}$ molecules^{-1} s^{-1}
Deactivation of y_{p3}	$y_{p3} \xrightarrow{\beta_3} y_3$	$\beta_3 y_{p3}$	$\beta_3 = 0.05$ s^{-1}

Table 2. Reactions and corresponding parameter values for the MAPK network motif of *S. cerevisiae* [6,9,66], related to slower relaxation rate. Other Parameters are $s_x = s_y = 10$ molecules/cell, $x_{T1} = x_1 + x_{p1} = 250$ molecules/cell, $x_{T2} = x_2 + x_{p2} = 1700$ molecules/cell, $x_{T3} = x_3 + x_{p3} = 5000$ molecules/cell, $y_{T1} = y_1 + y_{p1} = 250$ molecules/cell, $y_{T2} = y_2 + y_{p2} = 1700$ molecules/cell and $y_{T3} = y_3 + y_{p3} = 5000$ molecules/cell. The kinetic schemes adopted in the present work follows the model of Heinrich et al. [63].

Description	Reaction	Propensity Function	Rate Constant
Activation of x_1	$x_1 + s_x \xrightarrow{k_x} x_{p1} + s_x$	$k_x s_x x_1$	$k_x = 10^{-4}$ molecules^{-1} s^{-1}
Deactivation of x_{p1}	$x_{p1} \xrightarrow{\alpha_1} x_1$	$\alpha_1 x_{p1}$	$\alpha_1 = 0.01$ s^{-1}
Activation of y_1	$y_1 + s_y \xrightarrow{k_y} y_{p1} + s_y$	$k_y s_y y_1$	$k_y = 10^{-4}$ molecules^{-1} s^{-1}
Deactivation of y_{p1}	$y_{p1} \xrightarrow{\beta_1} y_1$	$\beta_1 y_{p1}$	$\beta_1 = 0.01$ s^{-1}
Activation of x_2	$x_2 + x_{p1} \xrightarrow{k_{12x}} x_{p2} + x_{p1}$	$k_{12x} x_{p1} x_2$	$k_{12x} = 10^{-4}$ molecules^{-1} s^{-1}
Activation of x_2	$x_2 + y_{p1} \xrightarrow{\varepsilon_2} x_{p2} + y_{p1}$	$\varepsilon_2 y_{p1} x_2$	$\varepsilon_2 = (0-1) \times 10^{-4}$ molecules^{-1} s^{-1}
Deactivation of x_{p2}	$x_{p2} \xrightarrow{\alpha_2} x_2$	$\alpha_2 x_{p2}$	$\alpha_2 = 0.01$ s^{-1}
Activation of y_2	$y_2 + y_{p1} \xrightarrow{k_{12y}} y_{p2} + y_{p1}$	$k_{12y} y_{p1} y_2$	$k_{12y} = 10^{-4}$ molecules^{-1} s^{-1}
Activation of y_2	$y_2 + x_{p1} \xrightarrow{\varepsilon_1} y_{p2} + x_{p1}$	$\varepsilon_1 x_{p1} y_2$	$\varepsilon_1 = (0-1) \times 10^{-4}$ molecules^{-1} s^{-1}
Deactivation of y_{p2}	$y_{p2} \xrightarrow{\beta_2} y_2$	$\beta_2 y_{p2}$	$\beta_2 = 0.01$ s^{-1}
Activation of x_3	$x_3 + x_{p2} \xrightarrow{k_{23x}} x_{p3} + x_{p2}$	$k_{23x} x_{p2} x_3$	$k_{23x} = 10^{-5}$ molecules^{-1} s^{-1}
Deactivation of x_{p3}	$x_{p3} \xrightarrow{\alpha_3} x_3$	$\alpha_3 x_{p3}$	$\alpha_3 = 0.01$ s^{-1}
Activation of y_3	$y_3 + y_{p2} \xrightarrow{k_{23y}} y_{p3} + y_{p2}$	$k_{23y} y_{p2} y_3$	$k_{23y} = 10^{-5}$ molecules^{-1} s^{-1}
Deactivation of y_{p3}	$y_{p3} \xrightarrow{\beta_3} y_3$	$\beta_3 y_{p3}$	$\beta_3 = 0.01$ s^{-1}

Next, we quantify mutual information between two parallel kinases (x_{pi} and y_{pj}, with $i = j$) of the two equivalent interacting MAPK pathways. The inter pathway coupling is unidirectional when either ε_1 or ε_2 is zero but is bidirectional when both are non-zero. In this situation, both variables

(x_{pi} and y_{pj}) do not interact with each other but are regulated by a common kinase regulon incorporating coordinated fluctuations into these variables. In other words, quantification of mutual information actually evaluates the extent of cross-correlation between these two variables. We observe zero mutual information value between x_{p1} and y_{p1}, as these are uncorrelated. In Figure 1E, we show mutual information between x_{p2} and y_{p2} as a function of ε_2 keeping ε_1 fixed. The profile shows an increasing trend as cross-talk parameter ε_2 increases. Similarly, in Figure 1F, mutual information between x_{p3} and y_{p3} is shown with a similar trend as in Figure 1E. Interestingly, for faster relaxation time scale, mutual information between similar cascade kinases increases while moving from second (x_{p2} and y_{p2}) to third (x_{p3} and y_{p3}) cascade. On the other hand, an opposite trend is observed for slower relaxation time scale. This characteristic trend is further shown in Figure 1G,H using bar diagram. These results together suggest that fluctuations due to faster relaxation rate transduce correlated fluctuations in a better way compared to the slower one. In Figure 1E, mutual information is high for slower relaxation rate than the faster one, as slower rate parameters yield high level of x_{p2} and y_{p2} which in turn incorporate extra fluctuations that help to increase mutual association. A similar result is also observed in Figure 1F. Identical mutual information profiles of $\mathcal{I}(x_{p2}; y_{p2})$ and $\mathcal{I}(x_{p3}; y_{p3})$ can be generated as function of ε_1 keeping ε_2 fixed. These results suggest that both the cross-talk parameters ε_1 and ε_2 contribute equally to the development of an association between two parallel pathways.

Both the mutual information between x_{p2} and y_{p2}, x_{p3} and y_{p3} are capable of providing a satisfactory explanation of enhancement of cross-talk with the increment of inter pathway interaction parameters (ε_1 and ε_2). Under equivalent interactions condition ($\varepsilon_1 = \varepsilon_2$), each pathway shares its information with other to an equal extent and is quantified not only by $\mathcal{I}(x_{p2}; y_{p2})$ and $\mathcal{I}(x_{p3}; y_{p3})$ but also by $\mathcal{I}(x_{p1}; y_{p2})$ and $\mathcal{I}(y_{p1}; x_{p2})$. However, characterization of the direction of information transduction is difficult under unequal condition ($\varepsilon_1 \neq \varepsilon_2$). Except the equivalent condition ($\varepsilon_1 = \varepsilon_2$) where the net information ($\mathcal{I}(y_{p1}; x_{p2}) - \mathcal{I}(x_{p1}; y_{p2})$) flow is zero, it has a definite value with directionality (positive or negative value) at all other conditions. Since the definition of mutual information is symmetric in nature and usage of the same is difficult to provide directionality of information propagation, we define a dimensionless quantity, *net information transduction* (D) using $\mathcal{I}(x_{p1}; y_{p2})$ and $\mathcal{I}(y_{p1}; x_{p2})$ as

$$D = \frac{\mathcal{I}(y_{p1}; x_{p2}) - \mathcal{I}(x_{p1}; y_{p2})}{\mathcal{I}(y_{p1}; x_{p2}) + \mathcal{I}(x_{p1}; y_{p2})}. \tag{1}$$

The above expression implies that it is maximal ($D = 1$) when $\mathcal{I}(x_{p1}; y_{p2})$ is zero, i.e., no information propagation from x_{p1} to y_{p2} ($\varepsilon_1 = 0$). It is minimal ($D = -1$) when $\mathcal{I}(y_{p1}; x_{p2})$ is zero, which specifies zero information propagation from y_{p1} to x_{p2} ($\varepsilon_2 = 0$). In Figure 1I, we show the profile of D as a function of ε_2 while keeping ε_1 fixed, where the value of D changes from negative to positive as ε_2 increases. It suggests that at low ε_2, information flowing from X to Y pathway dominates over the flow from Y to X. In other words, in this regime, the net information flow is accounted for by X $\rightarrow$ Y, leading to a negative value of D. On the other hand, at high ε_2, the direction of net information propagation is from Y to X due to reverse situation and generates a positive D value. The opposite scenario can be observed if one generates the profile of D as a function of ε_1 for fixed ε_2 In this connection, it is important to mention that both the relaxation time scale limits generate a similar profile of D. As a result, both the profiles of D exhibit data collapse when depicted as a function of ε_2 for fixed ε_1 (Figure 1I) or vice versa. This observation indicates that normalised profiles of D are independent of relaxation time scales. In Figure 1J, we also show a 2d-surface plot of D as a function of both ε_1 and ε_2 for faster relaxation time scale (Table 1). The surface plot indicates zero (or near to zero) value of D along the diagonal region ($\varepsilon_1 \approx \varepsilon_2$). However, the off diagonal region is positive for $\varepsilon_1 < \varepsilon_2$ and negative for $\varepsilon_1 > \varepsilon_2$.

In Figure 2A,B, we show two 2d-surface plots of mutual information between x_{p2} and y_{p2}, x_{p3} and y_{p3} kinases, respectively, as a function of two cross-interaction parameters ε_1 and ε_2 under faster relaxation time scale (Table 1). Both figures show maximum mutual information at high values of the two parameters. Since, ε_1 and ε_2 are equally responsible for developing the cross-correlation between two pathways, one can check the effect of maximisation of mutual information by increasing any one of these two parameters. Although we can quantify the influence of cross-talk with the help of two variable mutual information, $\mathcal{I}(x_{p2}; y_{p2})$ and $\mathcal{I}(x_{p3}; y_{p3})$, it is difficult to get an insight how the static and dynamic populations of the phosphorylated kinases are correlated. To this end, we have checked such correlation as shown in Figure 2C. Dynamic correlation can be measured using stochastic trajectories of the two variables in a single cell. If sufficient association between two trajectories exist, then correlated fluctuations are observed i.e., one trajectory closely follows the other. Otherwise, uncorrelated fluctuations (trajectories do not follow each other) are seen in the absence of cross-talk. In Figure 2C, we show stochastic time series of different phosphorylated kinases under varied conditions. Here, four different sets of ε_1 and ε_2 parameters have been used - mentioned as I, II, III and IV in Figure 2A,B. The stochastic time series exhibit correlated fluctuations at high ε_1 and ε_2. However, uncorrelated time series are seen at low ε_1 and ε_2. Each of these time series are generated from a single run of stochastic simulation and represents the dynamics in a single cell.

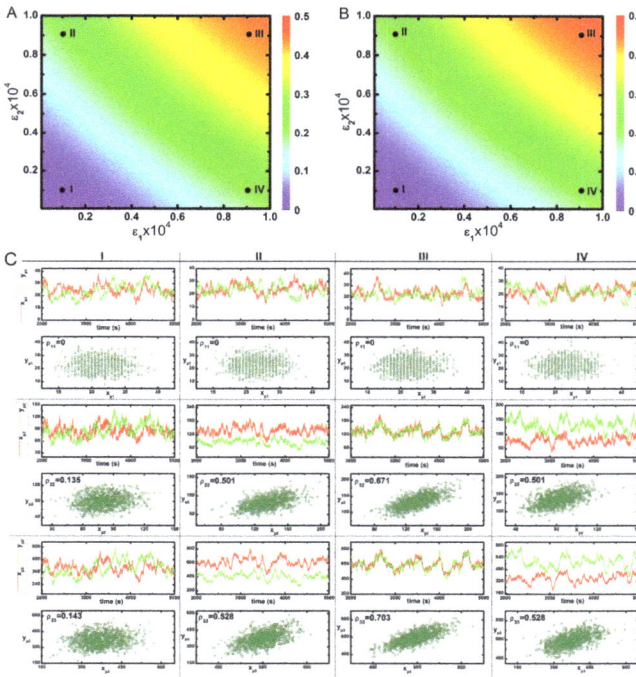

Figure 2. (color online) 2d-surface plots of two variable mutual information, stochastic time trajectories and scatter plots. (**A,B**) 2d-surface plot of two variable mutual information $\mathcal{I}(x_{p2}; y_{p2})$ and $\mathcal{I}(x_{p3}; y_{p3})$ as a function of two cross-talk parameters ε_1 and ε_2 for faster relaxation rate parameters (Table 1). In both figures, I, II, III and IV correspond to four different values of ε_1 and ε_2; (**C**) Stochastic time trajectories and steady state population of two parallel kinases for four different sets of ε_1 and ε_2. For CI, CII, CIII and CIV we have used $\varepsilon_1 = \varepsilon_2 = 0.1 \times 10^{-4}$, $\varepsilon_1 = 0.1 \times 10^{-4}$ and $\varepsilon_2 = 0.9 \times 10^{-4}$, $\varepsilon_1 = \varepsilon_2 = 0.9 \times 10^{-4}$ and $\varepsilon_1 = 0.9 \times 10^{-4}$ and $\varepsilon_2 = 0.1 \times 10^{-4}$, respectively. In each scatter plot, $\rho_{ij}(i = j)$ represents analytical value of Pearson's correlation coefficient. The stochastic trajectories and the scatter plots are generated using stochastic simulation algorithm [43] and the surface plots are due to theoretical calculation.

Next, we have executed 1000 independent stochastic simulations and measured the steady state phosphorylated kinase levels from each run to draw a scatter plot. Here, the collection of symbols mimics the behaviour of a hypothetical population of 1000 genetically identical cells. Each symbol in the scatter plots is due to a single stochastic run and represents copies of phosphorylated kinases (x_{pi} and y_{pj}, $i, j = 1, 2, 3$) in a single cell expressed at steady state. In other words, each symbol signifies the steady state phosphorylated protein levels in an individual cell within a hypothetical population. Using this concept, we measure static correlation among the pairs of phosphorylated kinases ((x_{p1}, y_{p1}), etc.) produced in different cells within the population. The motivation behind creating a scatter plot is to measure the static correlation between two phosphorylated protein levels expressed in different cells [57–59]. In Figure 2C, we show static correlation between different phosphorylated kinases expressed in different cells. For plots with high ε_1 and ε_2 values, most of the symbols are aligned diagonally in a narrow strip but for low ε_1 and ε_2, symbols are distributed in a much larger space.

Correlated variation of these kinases in the population is observed along the diagonal direction. However, an uncorrelated variation along the off-diagonal direction reflects an increase in variability between two phosphorylated kinases with respect to each other in a population. Therefore, enhancement of cross-talk decreases variability between two kinases in a population. These results imply that cross-talk develops correlated fluctuations between kinases in a population, thereby assisting in the successful development of a robust adaptation machinery as observed in the bet-hedging program under diverse environmental conditions [22]. Here, an increase in the correlation between two different protein pool (with high Pearson's correlation coefficient) is observed for maximal cross-talk. On the other hand, a decrease in correlation is seen for minimal cross-interaction. Similar behaviour can be seen under slower relaxation time scale (Table 2) as shown in Figure 3. The primary difference between the nature of correlation between (x_{p2}, y_{p2}) and (x_{p3}, y_{p3}) are visible from Figure 2 and Figure 3. In Figure 2, the correlation between (x_{p3}, y_{p3}) is always higher than (x_{p2}, y_{p2}) for all four conditions. On the other hand, in Figure 3, it shows an opposite trend. We note that coordinated fluctuations of different MAPK proteins are generated due to cross-talk between two parallel pathways that ultimately lead to similar cellular function with high reliability. Fluctuations in the transcription factor are much faster than the transcript dynamics. On the other hand, fluctuations in the transcription factor (known as extrinsic fluctuations) are comparable with the gene switching (on-off) time scale. Such time scale agreement gets reflected in the dynamics of the transcript [36].

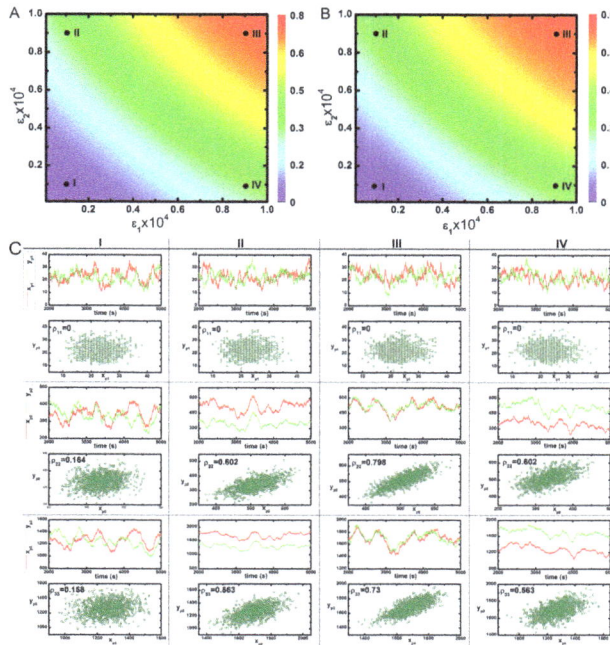

Figure 3. (color online) 2d-surface plots of two variable mutual information, stochastic time trajectories and scatter plots. (**A,B**) 2d-surface plot of two variable mutual information $\mathcal{I}(x_{p2}; y_{p2})$ and $\mathcal{I}(x_{p3}; y_{p3})$ as a function of two cross-talk parameters ε_1 and ε_2 for slower relaxation rate parameters (Table 2). In both figures, I, II, III and IV correspond to four different values of ε_1 and ε_2; (**C**) Stochastic time trajectories and steady state population of two parallel kinases for four different sets of ε_1 and ε_2. For CI, CII, CIII and CIV we have used $\varepsilon_1 = \varepsilon_2 = 0.1 \times 10^{-4}$, $\varepsilon_1 = 0.1 \times 10^{-4}$ and $\varepsilon_2 = 0.9 \times 10^{-4}$, $\varepsilon_1 = \varepsilon_2 = 0.9 \times 10^{-4}$ and $\varepsilon_1 = 0.9 \times 10^{-4}$ and $\varepsilon_2 = 0.1 \times 10^{-4}$, respectively. In each scatter plot, $\rho_{ij}(i = j)$ represents analytical value of Pearson's correlation coefficient. The stochastic trajectories and the scatter plots are generated using stochastic simulation algorithm [43] and the surface plots are due to theoretical calculation.

2.2. Three Variable Mutual Information

In the foregoing discussion, we have shown the effect of cross-talk in terms of conventional two variable mutual information. However, as cross-interaction between two pathways develops a complex network, a comprehensive study of three variable mutual information provides an extra insight. In the present study, three variable mutual information is defined as $\mathcal{I}(x_{p1}, y_{p1}; x_{p2})$ and $\mathcal{I}(x_{p1}; y_{p2}, x_{p2})$ for two types of branched pathways (see Section 3.2). One of the branched pathways is two inputs (x_{p1} and y_{p1}) and one output (x_{p2} or y_{p2}) motif where two input signals are integrated into a single output. The other is one input (x_{p1} or y_{p1}) and two outputs (x_{p2} and y_{p2}) motif where the input signal is bifurcated into two outputs. In this subsection, we investigate the efficacy of such signal integration as well as signal bifurcation. Since marginal and joint distributions of all cascade proteins are considered approximately Gaussian, we adopt multivariate mutual information theory [49–51] to analytically estimate three variable mutual information [55]. Each branched motif consists of two signal propagating channels that work together. It is thus interesting to investigate whether these signalling channels perform separately and what significant change arises in the estimation of three variable mutual information. The change in the magnitude of mutual information is defined by net synergy [50] and is evaluated using the theory of partial information decomposition [49,51,55] in terms of the difference between three variable mutual information and two corresponding two variable mutual information. The value of net synergy is either positive or negative; a positive value indicates synergy (extra information) whereas negative value measures redundancy (deficit of information) [50,55].

In Figure 4A, we show mutual information, $\mathcal{I}(x_{p1}, y_{p1}; x_{p2})$ of two inputs and one output model as a function of ε_2 for a fixed value of ε_1. The profile shows a bifunctional behaviour with the increment of ε_2; initially it decreases up to a certain value of ε_2, and then it increases. At low ε_2, a minimal amount of signal is propagated from y_{p1} to x_{p2}. Consequently, the motif reduces to a single input-output motif and the motif regains its native form due to the significant contribution of ε_2. In Figure 1A, we show that two variable mutual information between x_{p1} and x_{p2} decreases with the increment of ε_2. Similar situation arises in Figure 4A for low value of ε_2. On the contrary, $\mathcal{I}(y_{p1}; x_{p2})$ increases with the increment of ε_2 (Figure 1D). Thus, two opposing effects work together to generate the convex profile. In Figure 4B, we plot net synergy of the motif as a function of ε_2 for a fixed value of ε_1 and it is seen to increase monotonically. It is pertinent to mention here that for this motif, one always gets a positive net synergy value as a function of ε_2. This result implies that an integrating signalling motif transduces more information compared to the summation of two isolated channels. The extra information, i.e., synergy facilitates fidelity of the output kinase. Intuitively, the sum of the reduction in the uncertainty (cross-correlation) of the output kinase contributed by each input signal is lower than the reduction in the uncertainty of the output provided by both signals together. This phenomenon implicates the aspect of integration of multiple signals in cellular signalling network motif as observed in *V. harveyi* quorum-sensing circuit [60,61].

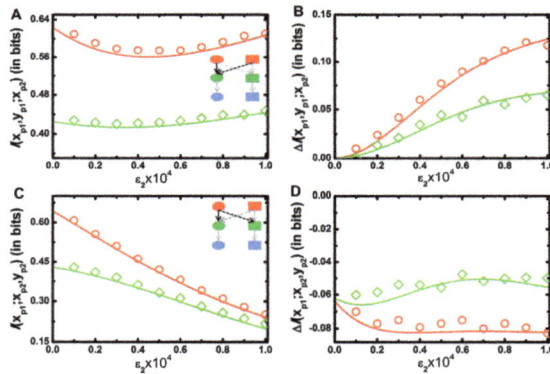

Figure 4. (color online) Three variable mutual information as a function of cross-talk parameter. (**A,B**) Three variable mutual information $\mathcal{I}(x_{p1}, y_{p1}; x_{p2})$ (**A**) and net synergy $\Delta\mathcal{I}(x_{p1}, y_{p1}; x_{p2})$ ((**B**) for signal integration motif. Schematic diagram of signal integration motif in composite MAPK network (see inset in (**A**). (**C,D**) - Three variable mutual information $\mathcal{I}(x_{p1}; x_{p2}, y_{p2})$ (**C**) and net synergy $\Delta\mathcal{I}(x_{p1}; x_{p2}, y_{p2})$ (**D**) for signal bifurcation motif. Schematic diagram of signal bifurcation motif in composite MAPK network (see inset in (**C**). All the figures are drawn as a function of cross-interaction parameter ε_2 for a fixed value of $\varepsilon_1 = 0.5 \times 10^{-4}$. Here red (with open circle) and green (with open diamond) lines are drawn for faster (Table 1) and slower (Table 2) relaxation rate parameters, respectively. The symbols are generated using stochastic simulation algorithm [43] and the lines are obtained from theoretical calculation. All the simulation data (open circles and open diamonds) are ensemble average of 10^7 independent trajectories.

In Figure 4C, we show mutual information $\mathcal{I}(x_{p1}; x_{p2}, y_{p2})$ of one input and two outputs motif with the increment of ε_2 for a fixed value of ε_1. The mutual information value decreases with ε_2 since propagation of information from x_{p1} to x_{p2} is only inhibited by the cross-interaction. However, ε_2 does not have any influence in information propagation from x_{p1} to y_{p2} and remains unaltered. Thus, three variable mutual information profile follows a decreasing trend. Figure 4D shows decreasing trend of net synergy profile as a function of ε_2 for a fixed value of ε_1. Importantly, for this motif negative values of net synergy are observed irrespective of the value of ε_2. This indicates redundancy in the information transmission in this composite motif compared to the sum of the individual one. Naturally, predictability about the output kinases decreases when two isolated signal propagation channels work together to form a bifurcated signal transduction motif. This result implies that although bifurcated signalling model reduces mutual information, it has a biological significance of the activation of multiple signalling channels in the presence of a single input as identified in the chemotaxis system of *E. coli* [62]. In all figures (Figure 4A–D), the red lines are plotted for faster relaxation rate constants (Table 1) of x_{p2}, x_{p3}, y_{p2} and y_{p3} and the green lines are for slower relaxation rate constants (Table 2).

3. Materials and Methods

In Figure 5, we show a schematic diagram of two interacting parallel MAPK pathways (named as X and Y). Each MAPK pathway consists of three kinase components, i.e., x_1, x_2, x_3 (X pathway) and y_1, y_2, y_3 (Y pathway) [7,28,63–65]. x_i and x_{pi} represent dephosphorylated and phosphorylated form of a kinase proteins, respectively, and the same applies to y_j and y_{pj} (here $i, j = 1, 2, 3$). The first cascade protein of a MAPK pathway gets phosphorylated with an exposure to the external stimulus. While phosphorylated, it positively regulates the phosphorylation of its own downstream kinase along with the kinase of the other pathway. The phosphorylated intermediate kinase regulates phosphorylation of the last kinase. To maintain the pool of phosphorylated kinase within a cell, a dephosphorylation process is in action with the help of phosphatase molecules. The cross-pathway interactions between two parallel MAPK pathways are denoted by the dashed lines in Figure 5 along with the cross-interaction rate parameters ε_1 and ε_2. S_x and S_y are the two extra-cellular signals acting on the X and Y pathway, respectively.

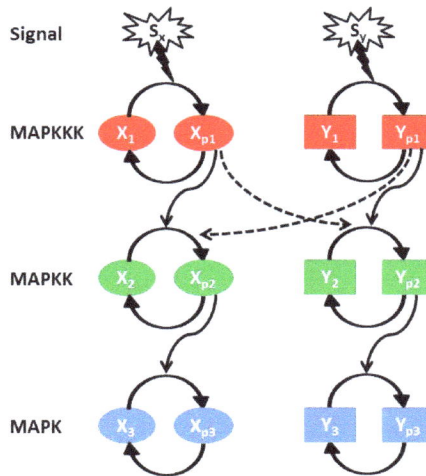

Figure 5. (color online) Schematic diagram of two parallel MAPK (equivalent and identical) signalling pathways (X and Y). Each pathway consists of three successively connected cascade kinases, MAPKKK (red), MAPKK (green) and MAPK (blue). The first activated kinase facilitates the activation of the second one and then the second kinase regulates the activation of the last one. Both signalling pathways are exposed to two different signals (S_x and S_y). Cross-talk is developed due to inter-pathway interactions. ε_1 and ε_2 are the cross-interaction parameters and the directionality of these interactions are $x_{p1} \rightarrow y_{p2}$ and $y_{p1} \rightarrow x_{p2}$, respectively.

Both pathways get causally correlated through cross-interactions, and a cross-talk develops as a consequence. Causal relationships are frequently examined in various circumstances that are subjected to stochastic fluctuations [23,44,50,55]. In the present manuscript, we quantify the causal relationship in terms of mutual information. Here, the two cross-interaction parameters ε_1 and ε_2 play a significant role in establishing different levels of cross-talk. The parameter ε_1 controls information flow from X to Y pathway ($x_{p1} \rightarrow y_{p2}$), but the parameter ε_2 is responsible for Y to X pathway ($y_{p1} \rightarrow x_{p2}$) information flow. In this connection, it is important to mention that during mating process, both pheromone and filamentous growth pathways are activated to a roughly equal extent, whereas during invasive growth process, only filamentous growth pathway is activated [26]. These observations corroborate with our model development. In our calculation, we only consider the post-translationally modified forms of all MAPK proteins. Thus, in the model, the total population of a MAPK protein is the sum of the phosphorylated and the unphosphorylated form of the protein and is considered to be constant ($(x_i + x_{pi}) = x_{Ti} = (y_j + y_{pj}) = y_{Tj} = $ constant, here $i = j$). In addition, we consider a physiologically relevant parameter set for our calculation [6,9,66].

3.1. Two Variable Mutual Information

Adopting Shannon's information theory [41,42], we have calculated two variable mutual information between two phosphorylated kinases,

$$\mathcal{I}(x_{pi}; y_{pj}) = \sum_{x_{pi}} \sum_{y_{pj}} p(x_{pi}, y_{pj}) \log_2 \left[\frac{p(x_{pi}, y_{pj})}{p(x_{pi})p(y_{pj})} \right]. \tag{2}$$

A generalised index x_{pi} and y_{pj} have been considered to represent the copy number of two different phosphorylated kinases. Similarly, $p(x_{pi})$ and $p(y_{pj})$ are the marginal and $p(x_{pi}, y_{pj})$ is the joint probability distributions associated with the corresponding kinases. For the calculation of mutual information between two kinases of X signalling pathway, we have replaced y_{pj} by x_{pi} (where $i \neq j$) and the reverse replacement has been followed for Y signalling pathway. For the estimation of mutual information between two equivalent kinases (x_{pi} and y_{pj}) of the respective pathways, we have used the same formula for $i = j$ condition. Mutual information can also be written in the form of the entropy function. Hence, Equation (2) can be redefined as

$$\mathcal{I}(x_{pi}; y_{pj}) = H(x_{pi}) + H(y_{pj}) - H(x_{pi}, y_{pj}). \tag{3}$$

Here, $H(x_{pi})$ and $H(y_{pj})$ are individual and $H(x_{pi}, y_{pj})$ is total entropy of the respective kinases. In the present study, both probability distribution functions (marginal and joint) are approximately considered to be Gaussian as experimental studies on MAPK pathway show Gaussian dynamics [67]. Thus, using Gaussian channel approximation [41,42,55], Equation (3) takes the reduced form

$$\mathcal{I}(x_{pi}; y_{pj}) = \frac{1}{2} \log_2 \left[\frac{\sigma_{x_{pi}}^2 \sigma_{y_{pi}}^2}{\sigma_{x_{pi}}^2 \sigma_{y_{pi}}^2 - \sigma_{x_{pi}y_{pi}}^4} \right], \tag{4}$$

where $\sigma_{x_{pi}}^2$ and $\sigma_{y_{pj}}^2$ are variances and $\sigma_{x_{pi}y_{pj}}^2$ is covariance of the corresponding kinases (for detailed calculation see Appendix). At this point it is important to mention that no prior knowledge is required about the nature of probability distribution function for evaluating mutual information using Equation (2). For exact or approximate Gaussian distribution, one can reduce Equation (2) to Equation (4) applying Gaussian channel approximation. However, for systems with non-Gaussian distribution, one can still use Equation (2) with proper analytical expressions of probability distribution functions that may contribute expressions of higher moments in Equation (4).

In the present work, all expressions of two variable mutual information are calculated using Equation (4). The analytical results are then validated by evaluating probability distribution functions (Equation (2)) using exact numerical simulation [43]. The two variable mutual information value is bounded within a scale $0 \le \mathcal{I}(x_{pi}; y_{pj}) \le \min(H(x_{pi}), H(y_{pj}))$. To quantify the association between two equivalent kinases, we have used Pearson's correlation coefficient ($\rho_{ij}, i = j$) [68]

$$\rho_{ij} = \frac{\sigma^2_{x_{pi}y_{pj}}}{\sigma_{x_{pi}}\sigma_{y_{pj}}}. \tag{5}$$

3.2. Three Variable Mutual Information

The three variable mutual information are calculated for both signal integration and signal bifurcation motif. In the first motif, two phosphorylated input kinases interact with one output kinase. Hence the complete description of mutual information is given by

$$\mathcal{I}(x_{p1}, y_{p1}; x_{p2}) = \sum_{x_{p1}, y_{p1}} \sum_{x_{p2}} p(x_{p1}, y_{p1}, x_{p2}) \log_2 \left[\frac{p(x_{p1}, y_{p1}, x_{p2})}{p(x_{p1}, y_{p1})p(x_{p2})} \right], \tag{6}$$

where $p(x_{p1}, y_{p1}, x_{p2})$ and $p(x_{p1}, y_{p1})$ are the joint distribution functions of the corresponding components. On the other hand, $p(x_{p2})$ is the marginal distribution of phosphorylated x_2 kinase. One can also write Equation (6) in terms of the respective entropy

$$\mathcal{I}(x_{p1}, y_{p1}; x_{p2}) = H(x_{p1}, y_{p1}) + H(x_{p2}) - H(x_{p1}, y_{p1}, x_{p2}). \tag{7}$$

Similarly, using Gaussian approximation [41,42,55], one can reduce Equation (7) into the following form

$$\mathcal{I}(x_{p1}, y_{p1}; x_{p2}) = \frac{1}{2}\log_2 \left[\frac{\sigma^2_{x_{p2}}(\sigma^2_{x_{p1}}\sigma^2_{y_{p1}} - \sigma^4_{x_{p1}y_{p1}})}{|\Delta_1|} \right], \tag{8}$$

with

$$|\Delta_1| = \begin{pmatrix} \sigma^2_{x_{p1}} & \sigma^2_{x_{p1}y_{p1}} & \sigma^2_{x_{p1}x_{p2}} \\ \sigma^2_{y_{p1}x_{p1}} & \sigma^2_{y_{p1}} & \sigma^2_{y_{p1}x_{p2}} \\ \sigma^2_{x_{p2}x_{p1}} & \sigma^2_{x_{p2}y_{p1}} & \sigma^2_{x_{p2}} \end{pmatrix}.$$

Here, the magnitude of three variable mutual information is bounded within a scale $0 \le \mathcal{I}(x_{p1}, y_{p1}; x_{p2}) \le \min(H(x_{p1}, y_{p1}), H(x_{p2}))$. Using PID formalism, the three variable mutual information can be decomposed into two parts [50,55]. As a result, the net synergy expression becomes

$$\begin{aligned} \Delta\mathcal{I}(x_{p1}, y_{p1}; x_{p2}) &= \mathcal{I}(x_{p1}, y_{p1}; x_{p2}) - \mathcal{I}(x_{p1}; x_{p2}) - \mathcal{I}(y_{p1}; x_{p2}) \\ &= \frac{1}{2}\log_2 \left[\frac{(\sigma^2_{x_{p1}}\sigma^2_{y_{p1}} - \sigma^4_{x_{p1}y_{p1}})(\sigma^2_{x_{p1}}\sigma^2_{x_{p2}} - \sigma^4_{x_{p1}x_{p2}})(\sigma^2_{y_{p1}}\sigma^2_{x_{p2}} - \sigma^4_{y_{p1}x_{p2}})}{|\Delta_1|\sigma^2_{x_{p1}}\sigma^2_{y_{p1}}\sigma^2_{x_{p2}}} \right]. \end{aligned} \tag{9}$$

Furthermore, one can calculate mutual information for the signal bifurcating motif with the help of associated distribution functions

$$\mathcal{I}(x_{p1}; x_{p2}, y_{p2}) = \sum_{x_{p1}} \sum_{x_{p2}, y_{p2}} p(x_{p1}, x_{p2}, y_{p2}) \log_2 \left[\frac{p(x_{p1}, x_{p2}, y_{p2})}{p(x_{p1})p(x_{p2}, y_{p2})} \right]. \tag{10}$$

and the entropy representation of Equation (10) is

$$\mathcal{I}(x_{p1}; x_{p2}, y_{p2}) = H(x_{p1}) + H(x_{p2}, y_{p2}) - H(x_{p1}, x_{p2}, y_{p2}). \tag{11}$$

Using Gaussian approximation [41,42,55] Equation (11) becomes

$$\mathcal{I}(x_{p1}; x_{p2}, y_{p2}) = \frac{1}{2} \log_2 \left[\frac{\sigma_{x_{p1}}^2 (\sigma_{x_{p2}}^2 \sigma_{y_{p2}}^2 - \sigma_{x_{p2}y_{p2}}^4)}{|\Delta_2|} \right], \tag{12}$$

with

$$|\Delta_2| = \begin{pmatrix} \sigma_{x_{p1}}^2 & \sigma_{x_{p1}x_{p2}}^2 & \sigma_{x_{p1}y_{p2}}^2 \\ \sigma_{x_{p2}x_{p1}}^2 & \sigma_{x_{p2}}^2 & \sigma_{x_{p2}y_{p2}}^2 \\ \sigma_{y_{p2}x_{p1}}^2 & \sigma_{y_{p2}x_{p2}}^2 & \sigma_{y_{p2}}^2 \end{pmatrix}.$$

In addition, the three variable mutual information value is bounded within a range $0 \leq \mathcal{I}(x_{p1}; x_{p2}, y_{p2}) \leq \min(H(x_{p1}), H(x_{p2}, y_{p2}))$. In this case, one can also use the theory of PID to decompose the three variable mutual information into two parts and calculate the net synergy [50]

$$\begin{aligned} \Delta\mathcal{I}(x_{p1}; x_{p2}, y_{p2}) &= \mathcal{I}(x_{p1}; x_{p2}, y_{p2}) - \mathcal{I}(x_{p1}; x_{p2}) - \mathcal{I}(x_{p1}; x_{p2}) \\ &= \frac{1}{2} \log_2 \left[\frac{(\sigma_{x_{p2}}^2 \sigma_{y_{p2}}^2 - \sigma_{x_{p2}y_{p2}}^4)(\sigma_{x_{p1}}^2 \sigma_{x_{p2}}^2 - \sigma_{x_{p1}x_{p2}}^4)(\sigma_{x_{p1}}^2 \sigma_{y_{p2}}^2 - \sigma_{x_{p1}y_{p2}}^4)}{|\Delta_2| \sigma_{x_{p1}}^2 \sigma_{x_{p2}}^2 \sigma_{y_{p2}}^2} \right]. \tag{13} \end{aligned}$$

For analytical calculation, we have adopted Equations (4), (5), (8), (9), (12) and (13) which contain only variance and covariance expressions, whereas we adopt numerical simulation for evaluation of the expressions given in Equations (2), (6) and (10).

At this point it is important to mention that we validate our analytical calculation by exact stochastic simulation, commonly known as stochastic simulation algorithm or Gillespie algorithm [43]. The validation signifies how much closer the system dynamics with the Gaussian statistics. Corroboration of analytical and simulation results indicate a valid consideration of linear noise approximation. In our numerical simulation, we have used 10^7 trajectories. We note that the dynamics of different MAPK pathways exhibit different temporal evolution [8]. However, the average physiological time scale for the activation of MAPK pathways is ~ 1 h [15]. Keeping this information in mind, each stochastic simulation was executed for 5000 s (~ 1.38 h) to ensure that the system dynamics reaches steady state. Using the final value of the variables of each run we carry out the calculation of different variances and covariances to evaluate different expressions of mutual information.

4. Conclusions

To summarize, we have investigated evolutionarily conserved yeast MAPK signalling pathway. In our phenomenological model, we study two parallel MAPK signalling pathways where one signalling pathway in addition to its cognate pathway activates the non-cognate pathway through cross-talk with an emphasis to understand the change in the dynamical behaviour of the system in the presence of cross-talk at the single cell level. The model nonlinear Langevin equations have been solved under the purview of LNA to quantify the variance and the covariance associated with the different phosphorylated kinase. These quantities assist in the evaluation of mutual information (two variable and multivariate) under Gaussian channel approximation. Quantification of mutual information has been carried out with the variation of two cross-talk parameters ε_1 and ε_2. The two variable mutual information shows that cross-talk establishes correlation in the signal propagation among the two pathways. To represent a better insight into the directionality of the net information flow, we have defined a new dimensionless parameter (net information transduction D), which varies on a scale of

−1 to +1. Depending on the sign of D, we have deciphered the fidelity of one pathway compared to the other.

We show that cross-talk generates correlated fluctuations at the population level. A minimum and a maximum degree of coordination are observed at the low and high level of cross-talk, respectively. Our analysis thus suggests that coordinated fluctuations are the causal effect of cross-talk in MAPK signalling pathways. Furthermore, we demonstrate the impact of correlated fluctuations in the reduction of variability between two different kinases using scatter plots. At the high degree of cross-talk, scatter plots show high correlation coefficient compared to the lower level of cross talk. These results together imply that cross-talk not only develops synchronisation in a cell but also reduces variability due to the development of correlation between two different phosphorylated kinase levels in a population. Depending on the number of inputs and outputs, we have identified two types of signalling motifs from the composite network. Also, quantification of multivariate mutual information allows us to calculate the net synergy associated with these two different motifs. The signal integration motif (two inputs and one output) reveals high fidelity, whereas the signal bifurcation motif (one input and two outputs) shows redundancy in information propagation.

Based on the aforesaid theoretical discussion, we suggest a satisfactory explanation about the synchronisation in the outputs - a causal effect of cross-talk in parallel MAPK signalling pathways. Nevertheless, one question apparently arises - what is the importance of such synchronisation in cellular physiology? Such functional correlation is possibly required for both the outputs to perform in a combined way to regulate several essential downstream genes. Several experimental results on MAPK cross-talk in *S. cerevisiae* provide interesting evidence that corroborates with our theoretical analysis. Phosphorylated Fus3 and Kss1 are both responsible for the activation of transcription factor Ste12 that regulates different downstream genes [64]. Additionally, both activated Fus3 and Hog1 assist in arresting the cell cycle in G1 phase temporarily [5,8]. Cross-talk is also highly significant for the eukaryotic cells where the promoter of TATA binding proteins is solely controlled by MAPK signalling pathways [15,69], whereas these binding proteins are essential for the expression of most nuclear genes. Also, they act as a potential vehicle for developing coordination among the multiple disparate classes of genes. Thus, coordinated signalling of MAPK pathways paves the way for TATA binding proteins to establish the association among large-scale nuclear genes. Gene regulation in *S. cerevisiae* is known to be controlled by more than one transcription factors that bind cooperatively at many promoter sites. This phenomenon suggests that coordinated fluctuations between the outputs of MAPK signalling pathways are necessary to express the gene product in a controlled way. It is also noticed that coordinated fluctuations among gene products are developed through transcriptional as well as translational cross-talk [25,57,58,70]. We propose that it could be more convenient for a cell to establish a functional connection among all intracellular processes if the correlation is initiated in the signalling pathway, not solely in the gene regulation stage. In fact, one interesting signature which was observed in different experiments is that cross-talk is prominent at the low concentration level that is manifested in diverse environmental cues [22,24]. Thus, in these situations, fluctuations in the cellular components are very high, and it is improbable for cells to adopt a constructive decision for survival [71]. Our results indicate that such decision making program becomes easy when correlated fluctuations among the essential proteins are successfully implemented through the bet-hedging program [22].

Overall, we suggest that synchronisation between MAPK signalling pathways is a result of cross-talk. Our analytical calculation supplemented by exact numerical simulation is a general approach and can be applied to other cross-talk pathways to quantify the strength of cross-interactions. In future, we plan to address the influence and physiological relevance of cross-talk in other network motifs. Our theoretical observations in the present work could be verified upon the quantification of phosphorylated kinase protein in a single cell using flow cytometry and time lapse microscopy [72–74]. These experimental approaches can be implemented to measure the amount of intra-cellular phosphorylated kinases by treating cells with external stimuli, fixing and permeabilizing cells with appropriate chemicals, and then staining with phospho-specific antibodies for different kinases. After that, one can quantify the intensity of phosphorylated kinases in individual cells of a colony. Using these data, distribution profiles of the concentration of phosphorylated kinases could be developed. These quantifiable distribution profiles could be used to quantify the mutual information.

Acknowledgments: We thank Debi Banerjee, Ayan Biswas, Sandip Kar and Jayanta Mukhopadhyay for critical reading and for making constructive suggestions. Alok Kumar Maity acknowledges University Grants Commission for a research fellowship (UGC/776/JRF(Sc)). Suman K. Banik acknowledges financial support from Council of Scientific and Industrial Research, Government of India (01(2771)/14/EMR-II) and Bose Institute (Institutional Programme VI - Systems Biology), Kolkata.

Author Contributions: Alok Kumar Maity conceived and designed the experiments; Alok Kumar Maity performed the experiments; Alok Kumar Maity, Pinaki Chaudhury and Suman K. Banik analyzed the data; Pinaki Chaudhury and Suman K. Banik contributed reagents/materials/analysis tools; Alok Kumar Maity, Pinaki Chaudhury and Suman K. Banik wrote the paper.

Conflicts of Interest: The authors declare no conflict of interest.

Abbreviations

The following abbreviations are used in this manuscript:

MAPK   Mitogen activated protein kinase
LNA   Linear noise approximation
PID   Partial information decomposition

Appendix A. Calculation of variance and covariance

The MAPK network motif shown in Figure 5 is explicated through stochastic Langevin equations. Each pathway (X or Y) is activated by the initiation of an extra cellular signal (S_x or S_y). When the first cascade kinase is activated, it regulates the activation of downstream kinases of the same as well as the parallel pathway through cross-interaction. Once activated, the second kinase regulates the activation of the last kinase. The active and inactive states can be identified in terms of phosphorylated (x_{pi} and y_{pj}) and dephosphorylated (x_i and y_j) forms of each kinase ($i, j = 1, 2, 3$), respectively. To construct the theoretical model of the composite MAPK network motif, we have considered the total population (phosphorylated and dephosphorylated form) of all kinases to be a constant $((x_i + x_{pi}) = x_{Ti} = (y_j + y_{pj}) = y_{Tj} = \text{constant}; i = j)$. Thus, for X pathway the stochastic differential equations for x_{p1}, x_{p2} and x_{p3} are [63]

$$\frac{dx_{p1}}{dt} = k_x s_x (x_{T1} - x_{p1}) - \alpha_1 x_{p1} + \xi_1(t), \tag{A1a}$$

$$\frac{dx_{p2}}{dt} = k_{12x} x_{p1}(x_{T2} - x_{p2}) + \varepsilon_2 y_{p1}(x_{T2} - x_{p2})$$
$$-\alpha_2 x_{p2} + \xi_2(t), \tag{A1b}$$

$$\frac{dx_{p3}}{dt} = k_{23x} x_{p2}(x_{T3} - x_{p3}) - \alpha_3 x_{p3} + \xi_3(t). \tag{A1c}$$

The first and the second terms on the right hand side of Equation (A.1) denote phosphorylation and dephosphorylation rate of the corresponding kinase. Here, k_x, k_{12x} and k_{23x} are activation and α_1, α_2 and α_3 are deactivation rate constants of x_{p1}, x_{p2} and x_{p3}, respectively. ε_2 is the cross-interaction parameter that controls signal propagation from Y to X pathway ($y_{p1} \to x_{p2}$). The ξ_i-s ($i = 1, 2, 3$) are Gaussian white noise terms with zero mean and finite noise strength. While writing Equation (A.1) we have used the conservation relation $x_i = x_{Ti} - x_{pi}$. Similarly, the stochastic Langevin equations associated with the components of the Y pathway can be written as [63]

$$\frac{dy_{p1}}{dt} = k_y s_y (y_{T1} - y_{p1}) - \beta_1 y_{p1} + \eta_1(t), \tag{A2a}$$

$$\frac{dy_{p2}}{dt} = k_{12y} y_{p1} (y_{T2} - y_{p2}) + \varepsilon_1 x_{p1} (y_{T2} - y_{p2})$$
$$-\beta_2 y_{p2} + \eta_2(t), \tag{A2b}$$

$$\frac{dy_{p3}}{dt} = k_{23y} y_{p2} (y_{T3} - y_{p3}) - \beta_3 y_{p3} + \eta_3(t). \tag{A2c}$$

In Equation (A.2), the first and the second terms stand for phosphorylation and dephosphorylation rate. Here, k_y, k_{12y} and k_{23y} are activation and β_1, β_2 and β_3 are deactivation rate constants of y_{p1}, y_{p2} and y_{p3}, respectively. The cross-interaction parameter is ε_1 that controls signal transduction from X to Y pathway ($x_{p1} \to y_{p2}$). The noise terms η_i-s ($i = 1, 2, 3$) are considered to be Gaussian white noise with zero mean and finite noise strength. For Y pathway, constant constraint $y_i = y_{Ti} - y_{pi}$ is also valid. The statistical properties of ξ_i-s and η_j-s ($i, j = 1, 2, 3$) are

$$\langle \xi_1 \rangle = \langle \xi_2 \rangle = \langle \xi_3 \rangle = \langle \eta_1 \rangle = \langle \eta_2 \rangle = \langle \eta_3 \rangle = 0,$$
$$\langle \xi_i(t) \xi_j(t') \rangle = \langle |\xi_i|^2 \rangle \delta_{ij} \delta(t - t'),$$
$$\langle \eta_i(t) \eta_j(t') \rangle = \langle |\eta_i|^2 \rangle \delta_{ij} \delta(t - t'),$$
$$\langle \xi_i(t) \eta_j(t') \rangle = \langle |\xi_i \eta_j| \rangle \delta_{ij} \delta(t - t'),$$
$$\langle |\xi_1|^2 \rangle = k_x s_x (x_{T1} - \langle x_{p1} \rangle) + \alpha_1 \langle x_{p1} \rangle = 2\alpha_1 \langle x_{p1} \rangle,$$
$$\langle |\xi_2|^2 \rangle = (k_{12x} \langle x_{p1} \rangle (x_{T2} - \langle x_{p2} \rangle)$$
$$+ \varepsilon_2 \langle y_{p1} \rangle (x_{T2} - \langle x_{p2} \rangle) + \alpha_2 \langle x_{p2} \rangle$$
$$= 2\alpha_2 \langle x_{p2} \rangle,$$
$$\langle |\xi_3|^2 \rangle = k_{23x} \langle x_{p2} \rangle (x_{T3} - \langle x_{p3} \rangle) + \alpha_3 \langle x_{p3} \rangle$$
$$= 2\alpha_3 \langle x_{p3} \rangle,$$
$$\langle |\eta_1|^2 \rangle = k_y s_y (y_{T1} - \langle y_{p1} \rangle) + \beta_1 \langle y_{p1} \rangle$$
$$= 2\beta_1 \langle y_{p1} \rangle,$$
$$\langle |\eta_2|^2 \rangle = k_{12y} \langle y_{p1} \rangle (y_{T2} - \langle y_{p2} \rangle)$$
$$+ \varepsilon_1 \langle x_{p1} \rangle (y_{T2} - \langle y_{p2} \rangle) + \beta_2 \langle y_{p2} \rangle$$
$$= 2\beta_2 \langle y_{p2} \rangle,$$
$$\langle |\eta_3|^2 \rangle = k_{23y} \langle y_{p2} \rangle (y_{T3} - \langle y_{p3} \rangle) + \beta_3 \langle y_{p3} \rangle$$
$$= 2\beta_3 \langle y_{p3} \rangle,$$
$$\langle |\xi_1 \eta_1| \rangle = \langle |\xi_1 \eta_2| \rangle = \langle |\xi_1 \eta_3| \rangle = \langle |\xi_2 \eta_1| \rangle = \langle |\xi_2 \eta_2| \rangle$$
$$= \langle |\xi_2 \eta_3| \rangle = \langle |\xi_3 \eta_1| \rangle = \langle |\xi_3 \eta_1| \rangle = \langle |\xi_3 \eta_3| \rangle$$
$$= 0.$$

To solve the nonlinear Equations (A.1–A.2), we adopt LNA [30–34,36–40,75–77]. Linearizing Equations (A.1–A.2) around steady state $\delta z(t) = z(t) - \langle z \rangle$, where $\langle z \rangle$ is the average population of z at long time limit, one arrives at

$$
\frac{d}{dt}\begin{pmatrix} \delta x_{p1} \\ \delta x_{p2} \\ \delta x_{p3} \\ \delta y_{p1} \\ \delta y_{p2} \\ \delta y_{p3} \end{pmatrix} = \begin{pmatrix} J_{x1x1} & J_{x1x2} & J_{x1x3} & J_{x1y1} & J_{x1y2} & J_{x1y3} \\ J_{x2x1} & J_{x2x2} & J_{x2x3} & J_{x2y1} & J_{x2y2} & J_{x2y3} \\ J_{x3x1} & J_{x3x2} & J_{x3x3} & J_{x3y1} & J_{x3y2} & J_{x3y3} \\ J_{y1x1} & J_{y1x2} & J_{y1x3} & J_{y1y1} & J_{y1y2} & J_{y1y3} \\ J_{y2x1} & J_{y2x2} & J_{y2x3} & J_{y2y1} & J_{y2y2} & J_{y2y3} \\ J_{y3x1} & J_{y3x2} & J_{y3x3} & J_{y3y1} & J_{y3y2} & J_{y3y3} \end{pmatrix} \begin{pmatrix} \delta x_{p1} \\ \delta x_{p2} \\ \delta x_{p3} \\ \delta y_{p1} \\ \delta y_{p2} \\ \delta y_{p3} \end{pmatrix} + \begin{pmatrix} \zeta_1 \\ \zeta_2 \\ \zeta_3 \\ \eta_1 \\ \eta_2 \\ \eta_3 \end{pmatrix}. \quad (A3)
$$

Here

$$
\begin{aligned}
J_{x1x1} &= -(k_x s_x + \alpha_1), \\
J_{x1x2} &= J_{x1x3} = J_{x1y1} = J_{x1y2} = J_{x1y3} = 0, \\
J_{x2x1} &= k_{12x}(x_{T2} - \langle x_{p2} \rangle), \\
J_{x2x2} &= -(k_{12x}\langle x_{p1} \rangle + \varepsilon_2 \langle y_{p1} \rangle + \alpha_2), \\
J_{x2y1} &= \varepsilon_2 (x_{T2} - \langle x_{p2} \rangle), \\
J_{x2x3} &= J_{x2y2} = J_{x2y3} = 0, \\
J_{x3x1} &= J_{x3y1} = J_{x3y2} = J_{x3y3} = 0, \\
J_{x3x2} &= k_{23x}(x_{T3} - \langle x_{p3} \rangle), \\
J_{x3x3} &= -(k_{23x}\langle x_{p2} \rangle + \alpha_3), \\
J_{y1y1} &= -(k_y s_y + \beta_1), \\
J_{y1x1} &= J_{y1x2} = J_{y1x3} = J_{y1y2} = J_{y1y3} = 0, \\
J_{y2y1} &= k_{12y}(y_{T2} - \langle y_{p2} \rangle), \\
J_{y2y2} &= -(k_{12y}\langle y_{p1} \rangle + \varepsilon_1 \langle x_{p1} \rangle + \beta_2), \\
J_{y2x1} &= \varepsilon_1 (y_{T2} - \langle y_{p2} \rangle), \\
J_{y2y3} &= J_{y2x2} = J_{y2x3} = 0, \\
J_{y3y1} &= J_{y3x1} = J_{y3x2} = J_{y3x3} = 0, \\
J_{y3y2} &= k_{23y}(y_{T3} - \langle y_{p3} \rangle), \\
J_{y3y3} &= -(k_{23y}\langle y_{p2} \rangle + \beta_3).
\end{aligned}
$$

The generalised matrix form of Equation (A.3) is

$$
\frac{d\delta \mathbf{A}}{dt} = \mathbf{J}_{A=\langle A \rangle} \delta \mathbf{A}(t) + \mathbf{\Theta}(t), \quad (A4)
$$

where $\mathbf{J}$ is the Jacobian matrix evaluated at steady state. The diagonal elements of $\mathbf{J}$ matrix define the relaxation rate of each kinase and the off-diagonal elements represent the interaction rate between two different kinases [75–77]. Moreover, $\delta \mathbf{A}$ and $\mathbf{\Theta}$ are the fluctuations matrix and the noise matrix of the kinases, respectively. To calculate the different variances and covariances in the stationary state we make use of the Lyapunov matrix equation [33,36,37]

$$
\mathbf{J}\sigma + \sigma \mathbf{J}^T + \mathbf{D} = 0, \quad (A5)
$$

where σ is the covariance matrix and $\mathbf{D} = \langle \mathbf{\Theta}\mathbf{\Theta}^T \rangle$ is the diffusion matrix that depends on different noise strengths. Here, $\langle \cdots \rangle$ represents ensemble average and T stands for transpose of a matrix. Solution of Equation (A.5) provides the expressions of variance and covariance of the kinases

$$\sigma^2_{x_{p1}} = \frac{\alpha_1 \langle x_{p1} \rangle}{J_{x1x1}}. \tag{A6a}$$

$$\sigma^2_{x_{p2}} = \frac{\alpha_2 \langle x_{p2} \rangle}{J_{x2x2}} + \frac{\alpha_1 \langle x_{p1} \rangle J^2_{x2x1}}{J_{x1x1}J_{x2x2}(J_{x1x1} + J_{x2x2})} + \frac{\beta_1 \langle y_{p1} \rangle J^2_{x2y1}}{J_{y1y1}J_{x2x2}(J_{y1y1} + J_{x2x2})}. \tag{A6b}$$

$$\sigma^2_{y_{p1}} = \frac{\beta_1 \langle y_{p1} \rangle}{J_{y1y1}}. \tag{A6c}$$

$$\sigma^2_{y_{p2}} = \frac{\beta_2 \langle y_{p2} \rangle}{J_{y2y2}} + \frac{\beta_1 \langle y_{p1} \rangle J^2_{y2y1}}{J_{y1y1}J_{y2y2}(J_{y1y1} + J_{y2y2})} + \frac{\alpha_1 \langle x_{p1} \rangle J^2_{y2x1}}{J_{x1x1}J_{y2y2}(J_{x1x1} + J_{y2y2})}. \tag{A6d}$$

$$\sigma^2_{x_{p1}x_{p2}} = \sigma^2_{x_{p2}x_{p1}} = \frac{\alpha_1 \langle x_{p1} \rangle J_{x2x1}J_{x2x2}}{J_{x1x1}J_{x2x2}(J_{x1x1} + J_{x2x2})}. \tag{A6e}$$

$$\sigma^2_{x_{p1}y_{p2}} = \sigma^2_{y_{p2}x_{p1}} = \frac{\alpha_1 \langle x_{p1} \rangle J_{y2x1}J_{y2y2}}{J_{x1x1}J_{y2y2}(J_{x1x1} + J_{y2y2})}. \tag{A6f}$$

$$\sigma^2_{y_{p1}y_{p2}} = \sigma^2_{y_{p2}y_{p1}} = \frac{\beta_1 \langle y_{p1} \rangle J_{y2y1}J_{y2y2}}{J_{y1y1}J_{y2y2}(J_{y1y1} + J_{y2y2})}. \tag{A6g}$$

$$\sigma^2_{y_{p1}x_{p2}} = \sigma^2_{x_{p2}y_{p1}} = \frac{\beta_1 \langle y_{p1} \rangle J_{x2y1}J_{x2x2}}{J_{y1y1}J_{x2x2}(J_{y1y1} + J_{x2x2})}. \tag{A6h}$$

$$\sigma^2_{x_{p1}y_{p1}} = \sigma^2_{y_{p1}x_{p1}} = 0. \tag{A6i}$$

$$\sigma^2_{x_{p3}} = \frac{\alpha_3 \langle x_{p3} \rangle}{J_{x3x3}} + \frac{\alpha_2 \langle x_{p2} \rangle J^2_{x3x2}}{J_{x2x2}J_{x3x3}(J_{x2x2} + J_{x3x3})} + \frac{\alpha_1 \langle x_{p1} \rangle J^2_{x2x1}J^2_{x3x2}(J_{x1x1} + J_{x2x2} + J_{x3x3})}{J_{x1x1}J_{x2x2}J_{x3x3}(J_{x1x1} + J_{x2x2})(J_{x1x1} + J_{x3x3})(J_{x2x2} + J_{x3x3})}$$
$$+ \frac{\beta_1 \langle y_{p1} \rangle J^2_{x2y1}J^2_{x3x2}(J_{y1y1} + J_{x2x2} + J_{x3x3})}{J_{y1y1}J_{x2x2}J_{x3x3}(J_{y1y1} + J_{x2x2})(J_{y1y1} + J_{x3x3})(J_{x2x2} + J_{x3x3})}. \tag{A6j}$$

$$\sigma^2_{y_{p3}} = \frac{\beta_3 \langle y_{p3} \rangle}{J_{y3y3}} + \frac{\beta_2 \langle y_{p2} \rangle J^2_{y3y2}}{J_{y2y2}J_{y3y3}(J_{y2y2} + J_{y3y3})} + \frac{\beta_1 \langle y_{p1} \rangle J^2_{y2y1}J^2_{y3y2}(J_{y1y1} + J_{y2y2} + J_{y3y3})}{J_{y1y1}J_{y2y2}J_{y3y3}(J_{y1y1} + J_{y2y2})(J_{y1y1} + J_{y3y3})(J_{y2y2} + J_{y3y3})}$$
$$+ \frac{\alpha_1 \langle x_{p1} \rangle J^2_{x2x1}J^2_{y3y2}(J_{x1x1} + J_{y2y2} + J_{y3y3})}{J_{x1x1}J_{y2y2}J_{y3y3}(J_{x1x1} + J_{y2y2})(J_{x1x1} + J_{y3y3})(J_{y2y2} + J_{y3y3})}. \tag{A6k}$$

$$\sigma^2_{x_{p2}y_{p2}} = \sigma^2_{y_{p2}x_{p2}}$$
$$= \frac{\alpha_1 \langle x_{p1} \rangle J_{x2x1}J_{y2x1}(J_{x2x2} + J_{y2y2} + 2J_{x1x1})}{J_{x1x1}(J_{x2x2} + J_{y2y2})(J_{x2x2} + J_{x1x1})(J_{y2y2} + J_{x1x1})} + \frac{\beta_1 \langle y_{p1} \rangle J_{x2y1}J_{y2y1}(J_{x2x2} + J_{y2y2} + 2J_{y1y1})}{J_{y1y1}(J_{x2x2} + J_{y2y2})(J_{x2x2} + J_{y1y1})(J_{y2y2} + J_{y1y1})}. \tag{A6l}$$

$$\sigma^2_{x_{p3}y_{p3}} = \sigma^2_{y_{p3}x_{p3}}$$
$$= \frac{\alpha_1 \langle x_{p1} \rangle J_{x3x2}J_{y3y2}J_{x2x1}J_{y2x1}C_1}{\begin{array}{c} J_{x1x1}(J_{x2x2} + J_{y2y2})(J_{x2x2} + J_{y3y3})(J_{x3x3} + J_{y2y2})(J_{x3x3} + J_{y3y3}) \\ \times (J_{x2x2} + J_{x1x1})(J_{x3x3} + J_{x1x1})(J_{y2y2} + J_{x1x1})(J_{y3y3} + J_{x1x1}) \end{array}}$$
$$+ \frac{\beta_1 \langle y_{p1} \rangle J_{x3x2}J_{y3y2}J_{x2y1}J_{y2y1}C_2}{\begin{array}{c} J_{y1y1}(J_{x2x2} + J_{y2y2})(J_{x2x2} + J_{y3y3})(J_{x3x3} + J_{y2y2})(J_{x3x3} + J_{y3y3}) \\ \times (J_{x2x2} + J_{y1y1})(J_{x3x3} + J_{y1y1})(J_{y2y2} + J_{y1y1})(J_{y3y3} + J_{y1y1}) \end{array}}. \tag{A6m}$$

Here,

$$\begin{aligned}
C_1 &= (J_{x2x2} + J_{y2y2})(J_{x2x2} + J_{y3y3})(J_{x3x3} + J_{y2y2})(J_{x3x3} + J_{y3y3}) + 2J_{x1x1}((J_{x3x3} + J_{y2y2}) \\
&\quad (J_{x3x3} + J_{y3y3})(J_{y2y2} + J_{y3y3}) + J^2_{x2x2}(J_{x3x3} + J_{y2y2} + J_{y3y3}) + J_{x2x2}(J_{x3x3} + J_{y2y2} + J_{y3y3})^2) \\
&\quad + 2J^2_{x1x1}(J_{x2x2} + J_{x3x3} + J_{y2y2} + J_{y3y3})^2 + 2J^3_{x1x1}(J_{x2x2} + J_{x3x3} + J_{y2y2} + J_{y3y3}), \\
C_2 &= (J_{x2x2} + J_{y2y2})(J_{x2x2} + J_{y3y3})(J_{x3x3} + J_{y2y2})(J_{x3x3} + J_{y3y3}) + 2J_{y1y1}((J_{x3x3} + J_{y2y2}) \\
&\quad (J_{x3x3} + J_{y3y3})(J_{y2y2} + J_{y3y3}) + J^2_{x2x2}(J_{x3x3} + J_{y2y2} + J_{y3y3}) + J_{x2x2}(J_{x3x3} + J_{y2y2} + J_{y3y3})^2) \\
&\quad + 2J^2_{y1y1}(J_{x2x2} + J_{x3x3} + J_{y2y2} + J_{y3y3})^2 + 2J^3_{y1y1}(J_{x2x2} + J_{x3x3} + J_{y2y2} + J_{y3y3}).
\end{aligned}$$

In our calculation, we use the analytical expressions of variance and covariance for evaluating the value of mutual information and correlation coefficient.

References

1. Tyson, J.J.; Chen, K.C.; Novak, B. Sniffers, buzzers, toggles and blinkers: dynamics of regulatory and signaling pathways in the cell. *Curr. Opin. Cell Biol.* **2003**, *15*, 221–231.

2. Alon, U. *An Introduction to Systems Biology: Design Principles of Biological Circuits*; CRC Press: Boca Raton, FL, USA, 2006.

3. Lyons, S.M.; Prasad, A. Cross-Talk and Information Transfer in Mammalian and Bacterial Signaling. *PLoS ONE* **2012**, *7*, doi:10.1371/journal.pone.0034488.

4. Tareen, A.; Wingreen, N.S.; Mukhopadhyay, R. Modeling Evolution of Crosstalk in Noisy Signal Transduction Networks. *arXiv* **2017**, arXiv:physics.bio-ph/1707.01467.

5. Bardwell, L. Mechanisms of MAPK signalling specificity. *Biochem. Soc. Trans.* **2006**, *34*, 837–841.

6. Ferrell, J.E. Tripping the switch fantastic: How a protein kinase cascade can convert graded inputs into switch-like outputs. *Trends Biochem. Sci.* **1996**, *21*, 460–466.

7. Huang, C.Y.F.; Ferrell, J.E. Ultrasensitivity in the mitogen-activated protein kinase cascade. *Proc. Natl. Acad. Sci. USA* **1996**, *93*, 10078–10083.

8. Saito, H. Regulation of cross-talk in yeast MAPK signaling pathways. *Curr. Opin. Microbiol.* **2010**, *13*, 677–683.

9. Voliotis, M.; Perrett, R.M.; McWilliams, C.; McArdle, C.A.; Bowsher, C.G. Information transfer by leaky, heterogeneous, protein kinase signaling systems. *Proc. Natl. Acad. Sci. USA* **2014**, *111*, E326–E333.

10. Laub, M.T.; Goulian, M. Specificity in two-component signal transduction pathways. *Annu. Rev. Genet.* **2007**, *41*, 121–145.

11. Posas, F.; WurglerMurphy, S.M.; Maeda, T.; Witten, E.A.; Thai, T.C.; Saito, H. Yeast HOG1 MAP kinase cascade is regulated by a multistep phosphorelay mechanism in the SLN1-YPD1-SSK1 "two-component" osmosensor. *Cell* **1996**, *86*, 865–875.

12. Rowland, M.A.; Deeds, E.J. Crosstalk and the evolution of specificity in two-component signaling. *Proc. Natl. Acad. Sci. USA* **2014**, *111*, doi:10.1073/pnas.1317178111.

13. Siryaporn, A.; Perchuk, B.S.; Laub, M.T.; Goulian, M. Evolving a robust signal transduction pathway from weak cross-talk. *Mol. Syst. Biol.* **2010**, *6*, doi:10.1038/msb.2010.105.

14. Trach, K.A.; Hoch, J.A. Multisensory Activation of the Phosphorelay Initiating Sporulation in Bacillus Subtilis—Identification and Sequence of the Protein-Kinase of the Alternate Pathway. *Mol. Microbiol.* **1993**, *8*, 69–79.

15. Gustin, M.C.; Albertyn, J.; Alexander, M.; Davenport, K. MAP kinase pathways in the yeast Saccharomyces cerevisiae. *Microbiol. Mol. Biol. Rev.* **1998**, *62*, 1264–1300.

16. Kunkel, B.N.; Brooks, D.M. Cross talk between signaling pathways in pathogen defense. *Curr. Opin. Plant Biol.* **2002**, *5*, 325–331.

17. Oeckinghaus, A.; Hayden, M.S.; Ghosh, S. Crosstalk in NF-κB signaling pathways. *Nat. Immunol.* **2011**, *12*, 695–708.

18. Iborra, F.J.; Escargueil, A.E.; Kwek, K.Y.; Akoulitchev, A.; Cook, P.R. Molecular cross-talk between the transcription, translation, and nonsense-mediated decay machineries. *J. Cell Sci.* **2004**, *117*, 899–906.

19. Mather, W.H.; Hasty, J.; Tsimring, L.S.; Williams, R.J. Translational Cross Talk in Gene Networks. *Biophys. J.* **2013**, *104*, 2564–2572.

20. Mauri, M.; Klumpp, S. A Model for Sigma Factor Competition in Bacterial Cells. *PLoS Comput. Biol.* **2014**, *10*, doi:10.1371/journal.pcbi.1003845.

21. Riba, A.; Bosia, C.; El Baroudi, M.; Ollino, L.; Caselle, M. A Combination of Transcriptional and MicroRNA Regulation Improves the Stability of the Relative Concentrations of Target Genes. *PLoS Comput. Biol.* **2014**, *10*, doi:10.1371/journal.pcbi.1003490.

22. Tsimring, L.S. Noise in biology. *Rep. Prog. Phys.* **2014**, *77*, doi:10.1088/0034-4885/77/2/026601.

23. Palus, M.; Komarek, V.; Hrncir, Z.; Sterbova, K. Synchronization as adjustment of information rates: Detection from bivariate time series. *Phys. Rev. E* **2001**, *63*, doi:10.1103/PhysRevE.63.046211.

24. Cookson, N.A.; Mather, W.H.; Danino, T.; Mondragon-Palomino, O.; Williams, R.J.; Tsimring, L.S.; Hasty, J. Queueing up for enzymatic processing: correlated signaling through coupled degradation. *Mol. Syst. Biol.* **2011**, *7*, doi:10.1038/msb.2011.94.

25. Komili, S.; Silver, P.A. Coupling and coordination in gene expression processes: a systems biology view. *Nat. Rev. Genet.* **2008**, *9*, 38–48.

26. Bardwell, L.; Zou, X.F.; Nie, Q.; Komarova, N.L. Mathematical models of specifcity in cell signaling. *Biophys. J.* **2007**, *92*, 3425–3441.

27. Komarova, N.L.; Zou, X.F.; Nie, Q.; Bardwell, L. A theoretical framework for specificity in cell signaling. *Mol. Syst. Biol.* **2005**, *1*, doi:10.1038/msb4100031.

28. McClean, M.N.; Mody, A.; Broach, J.R.; Ramanathan, S. Cross-talk and decision making in MAP kinase pathways. *Nat. Genet.* **2007**, *39*, 567.

29. Ubersax, J.A.; Ferrell, J.E. Mechanisms of specificity in protein phosphorylation. *Nat. Rev. Mol. Cell Biol.* **2007**, *8*, 530–541.

30. Van Kampen, N.G. *Stochastic Processes in Physics and Chemistry*; North-Holland: Amsterdam, The Netherlands, 2011.

31. Gardiner, C.W. *Stochastic Methods*; 4th ed.; Springer: Berlin, Germany, 2009.

32. Grima, R. Linear-noise approximation and the chemical master equation agree up to second-order moments for a class of chemical systems. *Phys. Rev. E* **2015**, *92*, doi:10.1103/PhysRevE.92.042124.

33. Elf, J.; Ehrenberg, M. Fast evaluation of fluctuations in biochemical networks with the linear noise approximation. *Genome Res.* **2003**, *13*, 2475–2484.

34. Maity, A.K.; Bandyopadhyay, A.; Chaudhury, P.; Banik, S.K. Role of functionality in two-component signal transduction: a stochastic study. *Phys. Rev. E* **2014**, *89*, doi:10.1103/PhysRevE.89.032713.

35. Maity, A.K.; Chaudhury, P.; Banik, S.K. Role of relaxation time scale in noisy signal transduction. *PLoS ONE* **2015**, *10*, doi:10.1371/journal.pone.0123242.

36. Paulsson, J. Summing up the noise in gene networks. *Nature* **2004**, *427*, 415–418.

37. Paulsson, J. Models of stochastic gene expression. *Phys. Life Rev.* **2005**, *2*, 157–175.

38. Grima, R.; Thomas, P.; Straube, A.V. How accurate are the nonlinear chemical Fokker-Planck and chemical Langevin equations? *J. Chem. Phys.* **2011**, *135*, doi:10.1063/1.3625958.

39. Thomas, P.; Matuschek, H.; Grima, R. How reliable is the linear noise approximation of gene regulatory networks? *BMC Genom.* **2013**, *14* (Suppl. S4), doi:10.1186/1471-2164-14-S4-S5.

40. Thomas, P.; Straube, A.V.; Timmer, J.; Fleck, C.; Grima, R. Signatures of nonlinearity in single cell noise-induced oscillations. *J. Theor. Biol.* **2013**, *335*, 222–234.

41. Cover, T.M.; Thomas, J.A. *Elements of Information Theory*; Wiley Interscience: Hoboken, NJ, USA, 2012.

42. Shannon, C.E. A Mathematical Theory of Communication. *Bell Syst. Tech. J.* **1948**, *27*, 623–656.

43. Gillespie, D.T. A general method for numerically simulating the stochastic time evolution of coupled chemical reactions. *J. Comput. Phys.* **1976**, *22*, 403–434.

44. Granger, C.W.J. Investigating Causal Relations by Econometric Models and Cross-Spectral Methods. *Econometrica* **1969**, *37*, doi:10.2307/1912791.

45. Schreiber, T. Measuring Information Transfer. *Phys. Rev. Lett.* **2000**, *85*, 461–464.

46. Faes, L.; Kugiumtzis, D.; Nollo, G.; Jurysta, F.; Marinazzo, D. Estimating the decomposition of predictive information in multivariate systems. *Phys. Rev. E* **2015**, *91*, doi:10.1103/PhysRevE.91.032904.

47. Faes, L.; Porta, A.; Nollo, G. Information Decomposition in Bivariate Systems: Theory and Application to Cardiorespiratory Dynamics. *Entropy* **2015**, *17*, 277–303.

48. Spinney, R.E.; Lizier, J.T.; Prokopenko, M. Transfer entropy in physical systems and the arrow of time. *Phys. Rev. E* **2016**, *94*, doi:10.1103/PhysRevE.94.022135.

49. Wibral, M.; Priesemann, V.; Kay, J.W.; Lizier, J.T.; Phillips, W.A. Partial information decomposition as a unified approach to the specification of neural goal functions. *Brain Cognit.* **2017**, *112*, 25–38.

50. Schneidman, E.; Bialek, W.; Berry, M.J. Synergy, redundancy, and independence in population codes. *J. Neurosci.* **2003**, *23*, 11539–11553.

51. Williams, P.L.; Beer, R.D. Nonnegative decomposition of Multivariate Information. *arXiv* **2010**, arXiv:cs.IT/1004.2515.

52. Harder, M.; Salge, C.; Polani, D. Bivariate measure of redundant information. *Phys. Rev. E* **2013**, *87*, doi:10.1103/PhysRevE.87.012130.

53. Bertschinger, N.; Rauh, J.; Olbrich, E.; Jost, J.; Ay, N. Quantifying Unique Information. *Entropy* **2014**, *16*, 2161–2183.

54. Griffith, V.; Koch, C. Quantifying synergistic mutual information. In *Guided Self-Organization: Inception, Emergence, Complexity and Computation*; Prokopenko, M., Ed.; Springer: Berlin, Germany, 2014; Volume 9, pp. 159–190.

55. Barrett, A.B. Exploration of synergistic and redundant information sharing in static and dynamical Gaussian systems. *Phys. Rev. E* **2015**, *91*, doi:10.1103/PhysRevE.91.052802.

56. Faes, L.; Porta, A.; Nollo, G.; Javorka, M. Information Decomposition in Multivariate Systems: Definitions, Implementation and Application to Cardiovascular Networks. *Entropy* **2017**, *19*, doi:10.3390/e19010005.

57. Dunlop, M.J.; Cox, R.S.; Levine, J.H.; Murray, R.M.; Elowitz, M.B. Regulatory activity revealed by dynamic correlations in gene expression noise. *Nat. Genet.* **2008**, *40*, 1493–1498.

58. Munsky, B.; Neuert, G.; van Oudenaarden, A. Using Gene Expression Noise to Understand Gene Regulation. *Science* **2012**, *336*, 183–187.

59. Elowitz, M.B.; Levine, A.J.; Siggia, E.D.; Swain, P.S. Stochastic gene expression in a single cell. *Science* **2002**, *297*, 1183–1186.

60. Waters, C.M.; Bassler, B.L. Quorum sensing: Cell-to-cell communication in bacteria. *Ann. Rev. Cell Dev. Biol.* **2005**, *21*, 319–346.

61. Mehta, P.; Goyal, S.; Long, T.; Bassler, B.L.; Wingreen, N.S. Information processing and signal integration in bacterial quorum sensing. *Mol. Syst. Biol.* **2009**, *5*, doi:10.1038/msb.2009.79.

62. Armitage, J.P. Bacterial tactic responses. *Adv. Microb. Physiol.* **1999**, *41*, 229–289.

63. Heinrich, R.; Neel, B.G.; Rapoport, T.A. Mathematical models of protein kinase signal transduction. *Mol. Cell* **2002**, *9*, 957–970.

64. Suderman, R.; Deeds, E.J. Machines vs. Ensembles: Effective MAPK Signaling through Heterogeneous Sets of Protein Complexes. *PLoS Comput. Biol.* **2013**, *9*, doi:10.1371/journal.pcbi.1003278.

65. Tănase-Nicola, S.; Warren, P.B.; ten Wolde, P.R. Signal detection, modularity, and the correlation between extrinsic and intrinsic noise in biochemical networks. *Phys. Rev. Lett.* **2006**, *97*, doi:10.1103/PhysRevLett.97.068102.

66. Bardwell, L.; Cook, J.G.; Chang, E.C.; Cairns, B.R.; Thorner, J. Signaling in the yeast pheromone response pathway: specific and high-affinity interaction of the mitogen-activated protein (MAP) kinases Kss1 and Fus3 with the upstream MAP kinase kinase Ste7. *Mol. Cell Biol.* **1996**, *16*, 3637–3650.

67. Atay, O.; Skotheim, J.M. Spatial and temporal signal processing and decision making by MAPK pathways. *J. Cell Biol.* **2017**, *216*, 317–330.

68. Whitlock, M.C.; Schluter, D. *The Analysis of Biological Data*; Roberts and Company Publishers: Greenwood Village, CO, USA, 2009.

69. White, R.J.; Sharrocks, A.D. Coordinated control of the gene expression machinery. *Trends Genet.* **2010**, *26*, 214–220.

70. Stewart-Ornstein, J.; Weissman, J.S.; El-Samad, H. Cellular Noise Regulons Underlie Fluctuations in Saccharomyces cerevisiae. *Mol. Cell* **2012**, *45*, 483–493.

71. Junttila, M.R.; Li, S.P.; Westermarck, J. Phosphatase-mediated crosstalk between MAPK signalling pathways in the regulation of cell survival. *FASEB J.* **2008**, *22*, 954–965.

72. Gao, R.; Stock, A.M. Probing kinase and phosphatase activities of two-component systems in vivo with concentration-dependent phosphorylation profiling. *Proc. Natl. Acad. Sci. USA* **2013**, *110*, 672–677.

73. Schulz, K.R.; Danna, E.A.; Krutzik, P.O.; Nolan, G.P. Single-cell phospho-protein analysis by flow cytometry. *Curr. Protoc. Immunol.* **2012**, Chapter 8, 1–20, doi:10.1002/0471142735.im0817s78.

74. Yaginuma, H.; Kawai, S.; Tabata, K.V.; Tomiyama, K.; Kakizuka, A.; Komatsuzaki, T.; Noji, H.; Imamura, H. Diversity in ATP concentrations in a single bacterial cell population revealed by quantitative single-cell imaging. *Sci. Rep.* **2014**, *4*, doi:10.1038/srep06522.

75. De Ronde, W.H.; Tostevin, F.; ten Wolde, P.R. Feed-forward loops and diamond motifs lead to tunable transmission of information in the frequency domain. *Phys. Rev. E* **2012**, *86*, doi:10.1103/PhysRevE.86.021913.

76. Mehta, P.; Goyal, S.; Wingreen, N.S. A quantitative comparison of sRNA-based and protein-based gene regulation. *Mol. Syst. Biol.* **2008**, *4*, doi:10.1038/msb.2008.58.

77. Thattai, M.; van Oudenaarden, A. Stochastic gene expression in fluctuating environments. *Genetics* **2004**, *167*, 523–530.

entropy

MDPI

Article

Analyzing Information Distribution in Complex Systems

Sten Sootla, Dirk Oliver Theis and Raul Vicente *

Institute of Computer Science, University of Tartu, Ulikooli 17, 50090 Tartu, Estonia;
stensoootla@gmail.com (S.S.); dotheis@ut.ee (D.O.T.)
*   Correspondence: raulvicente@gmail.com

Received: 21 July 2017; Accepted: 1 November 2017; Published: 24 November 2017

Abstract: Information theory is often utilized to capture both linear as well as nonlinear relationships between any two parts of a dynamical complex system. Recently, an extension to classical information theory called partial information decomposition has been developed, which allows one to partition the information that two subsystems have about a third one into unique, redundant and synergistic contributions. Here, we apply a recent estimator of partial information decomposition to characterize the dynamics of two different complex systems. First, we analyze the distribution of information in triplets of spins in the 2D Ising model as a function of temperature. We find that while redundant information obtains a maximum at the critical point, synergistic information peaks in the disorder phase. Secondly, we characterize 1D elementary cellular automata rules based on the information distribution between neighboring cells. We describe several clusters of rules with similar partial information decomposition. These examples illustrate how the partial information decomposition provides a characterization of the emergent dynamics of complex systems in terms of the information distributed across their interacting units.

Keywords: information theory; partial information decomposition; Ising model; cellular automata

1. Introduction

The universe is full of systems that comprise a large number of interacting elements. Even if the immediate local interactions of these elements are rather simple, the global observable behaviour that they give rise to is often complex. Such systems, intuitively understood to be physical manifestations of the expression "the whole is more than the sum of its parts", are aptly called *complex systems*. Canonical examples of complex systems include the human brain, ant colonies and financial markets. Indeed, most of these systems have many relatively simple parts (e.g., neurons) interacting nonlinearly, whose collective behavior engenders complex phenomena (e.g., consciousness).

In addition to physical systems, many mathematical models have been developed that capture the essence of different complex systems. These theoretical models are particularly interesting because one has complete knowledge of how their various parts are connected together and which rules they obey while interacting with each other. Nevertheless, the emergent global structures are often so complex that their exact evolution is difficult to predict from the initial conditions and the interaction rules without actually simulating the system. Cellular automata and the Ising model are quintessential examples of such models.

One way to analyze these complex models is to treat them as information processing systems and measure the amount of information that their elements have about each other. Often, such analysis is done by using a well-known quantity from classical information theory, *mutual information*, and its various derivations, which measure statistical dependencies between a pair of random variables. These measures are particularly useful because of their sensitivity to both linear as well as nonlinear interactions between random variables. Among other things, they allow one to quantify the amount of information that is stored [1], transferred [2–5] and modified [6] in different parts of the system.

However, only measuring the information that is processed between *two* sub-components is rather restrictive. Indeed, even the simplest of logic gates has more elements, being composed of a pair of inputs and an output, which statistical dependencies we are interested to characterize. While one could consider the inputs as a single sub-component, this would not capture the intricate interactions among the inputs themselves. In particular, components in the input ensemble can provide information uniquely, redundantly, or synergistically about the output [7].

To capture this distribution of information between two inputs and a single output, an extension to classical information theory is needed [7]. Recently, several axiomatic frameworks have been developed to account for such extension and they are often referred to as *partial information decomposition* (PID) [7–11]. For a review of the uses of partial information decomposition in Neuroscience, see [12,13]. In this article, we capitalize on a recently developed numerical estimator for PID [14] for a particular version of PID [8], and use it to characterize the emergent dynamics of several complex systems (2D Ising model and 1D cellular automata) in terms of the information distribution across their interacting sub-units.

The remaining of this article is organized as follows. In the Background sections, we give a brief overview of partial information decomposition including its numerical estimation, as well as the basics of Ising and elementary cellular automata models. The Methods section details both the numerical simulation and PID analyses for both systems. The Results section describes the results of applying the PID estimator to the dynamics of neighboring cells in the Ising model and elementary cellular automata. We conclude by discussing the implications of the obtained results and related work, as well as the limitations of applying the current approach to other systems such as artificial neural networks, and provide suggestions for future work.

2. Background

2.1. Partial Information Decomposition

Mutual information measures the amount of information two random variables, or more generally, two random vectors have about each other. However, it is often worthwhile to ask how much information an ensemble of input (source) random variables carries about some output (target) variable. A trivial solution would be to measure the mutual information between the whole input ensemble considered as a single random vector and the output. However, this would not capture the interactions between the input variables themselves. Moreover, by considering the input ensemble as a single unit, knowledge about how the interactions between specific individual units and the output differ is lost.

This section briefly reviews the partial information decomposition proposed by [8]—a specific mathematical framework for decomposing mutual information between a group of input variables and single source variable.

2.1.1. Formulation

The simplest non-trivial system to analyze that has an ensemble of inputs and a single output is a system with *two* inputs. Given this setup, one can ask how much information one input variable has about the output that the other does not, how much information they share about the output, and how much information they jointly have about the output such that both inputs must be present for this information to exist.

More formally, let Y and Z be two random variables that are considered as sources to a third random variable X. The mutual information between the pair (Y, Z) and X is defined in terms of entropies as

$$MI(X; Y, Z) = H(X) - H(X|Y, Z).$$

The partial information decomposition framework aims to decompose this mutual information into *unique*, *redundant* and *complementary information* terms.

Unique information quantifies the amount of information that only one of the input variables has about the output variable. The unique information that Y has about the output X is denoted as $UI(X : Y \setminus Z)$. Similarly, $UI(X : Z \setminus Y)$ denotes the unique information that Z has about the target X.

Shared information quantifies the amount of information both inputs share about the output variable. It is also sometimes called *redundant* information because, if both inputs contain the same information about the output, it would suffice to observe only one of the input variables. The shared information is denoted as $SI(X : Y; Z)$. (To be consistent with "Elements of Information Theory", the notation used in this article for PID terms deviates a little from the one introduced by Bertschinger et al. [8]. Specifically, a colon (:) is used to partition the set of random variables to a single output (on the left-hand side) and a set of inputs (on the right-hand side). As before, a semicolon (;) is used to separate the input variables on the right-hand side, signifying that these variables are considered to be separate entities, not part of a single random vector.)

Complementary or *synergistic* information quantifies the amount of information that is only present when both inputs are considered jointly. The complementary information is denoted as $CI(X : Y; Z)$.

It is generally agreed [7–10] that mutual information can be decomposed into the four terms just described as follows:

$$MI(X; Y, Z) = SI(X : Y; Z) + UI(X : Y \setminus Z) + UI(X : Z \setminus Y) + CI(X : Y; Z). \tag{1}$$

The same sources also agree on the decomposition of information that a single variable, either Y or Z, has about the output X:

$$\begin{aligned} MI(X; Y) &= UI(X : Y \setminus Z) + SI(X : Y; Z), \\ MI(X; Z) &= UI(X : Z \setminus Y) + SI(X : Y; Z). \end{aligned} \tag{2}$$

It is important to note that thus far in this section, no formulas for actually calculating the PID terms have been given, and only several relationships that such a decomposition should satisfy have been stated. The only computable quantities so far are the mutual information terms on the left-hand side of Equations (1) and (2). The discussion of computing the specific PID terms is developed in the next section, which is heavily inspired by an intuitive overview of the paper "Quantifying Unique Information" by Bertschinger et al. [8], provided by Wibral et al. [13].

2.1.2. Calculating PID Terms

It turns out that the current tools from classical information theory—entropy and various forms of mutual information—are not enough to calculate any of the terms of the PID [7]. Indeed, there are only three Equations (1) and (2) relating to the four variables of interest, making the system undetermined. In order to make the problem tractable, a definition of at least one of the PID terms must be given [8].

Taking inspiration from decision theory, Bertschinger et al. [8] were able to provide such a definition for unique information. Their insight was that, if a variable contains unique information, there must be a way to exploit it. In other words, there must exist a situation such that an agent having access to unique information has an advantage over another agent who does not possess this knowledge. Given such a situation, the agent in possession of unique information can prove it to others by designing a bet on the output variable, such that, on average, the bet is won by the designer.

In particular, suppose there are two agents, Alice and Bob, Alice having access to the random variable Y and Bob having access to the random variable Z from Equation (1). Neither of them have access to the other player's random variable, and both of them can observe, but not directly modify, the output variable X. Alice can prove to Bob that she has unique information about X via Y by constructing a bet on the outcomes of X. Since Alice can only directly *modify* Y and *observe* the outcome X, her reward will depend only on the distribution $p(X, Y)$. Similarly, Bob's reward will depend only on the distribution $p(X, Z)$. From this, it follows that the results of the bet are *not* dependent on the full distribution $p(X, Y, Z)$, but rather only on its marginals $p(X, Y)$ and $p(X, Z)$.

Let $p \equiv p(X, Y, Z)$ be the original joint probability distribution that we are interested in computing the PID of, and let Δ be the set of *all* joint probability distributions of X, Y and Z. Under the assumption that the unique information depends only on the two marginal distributions of p, a set of probability distributions Δ_p can be defined such that the unique information stays constant for any element in this set. Such a set consist only of the probability distributions that have the same marginal distributions of the pairs (X, Y) and (X, Z) as p. It is defined as follows:

$$\Delta_p = \{q \in \Delta : q(X = x, Y = y) = p(X = x, Y = y)$$
$$\text{and } q(X = x, Z = z) = p(X = x, Z = z) \text{ for all } x \in X, y \in Y, z \in Z\}$$

Putting the observation that unique information is constant on Δ_p and Equation (2) together, it becomes apparent that shared information will also be constant on Δ_p. Thus, only complementary information varies when considering arbitrary distribution q from Δ_p. The last observation makes sense intuitively and is to be expected, since "complementary information should capture precisely the information that is carried by the joint dependencies between X, Y and Z" [8].

Using the chain rule for information as well as decompositions (1) and (2), the following identities can be derived:

$$MI(X; Y|Z) = UI(X : Y \setminus Z) + CI(X : Y; Z),$$
$$MI(X; Z|Y) = UI(X : Z \setminus Y) + CI(X : Y; Z). \tag{3}$$

Now, if a distribution $q_0 \in \Delta_p$ could be found that yields vanishing synergy, the unique information could be calculated using quantities from classical information theory. Indeed, from Equation (3), it can be seen that when synergy is 0, the mutual information and unique information terms coincide. Bertschinger et al. [8] prove that a distribution $q_0 \in \Delta_p$ with this property only exists for specific measures of unique, shared and complementary information. They define the suitable measure for unique information as follows:

$$\widetilde{UI}(X : Y \setminus Z) = \min_{q \in \Delta_p} MI_q(X; Y|Z), \tag{4}$$

$$\widetilde{UI}(X : Z \setminus Y) = \min_{q \in \Delta_p} MI_q(X; Z|Y), \tag{5}$$

where the subscript q under the mutual information symbol means that the quantity is calculated over the distribution q.

Replacing these measures with the corresponding quantities in Equations (1) and (2), measures for shared and complementary information can be defined as follows:

$$\widetilde{SI}(X : Y; Z) = \max_{q \in \Delta_p} MI_q(X; Y) - MI_q(X; Y|Z), \tag{6}$$

$$\widetilde{CI}(X : Y; Z) = MI(X; Y, Z) - \min_{q \in \Delta_p} MI_q(X; Y, Z). \tag{7}$$

These four constrained optimization problems (Equations (4)–(7)) are all equivalent in the sense that it would suffice to solve only one of these problems and the obtained optimal joint distribution q would produce the optimal value for all the remaining three measures as well.

2.1.3. Numerical Estimator

Bertschinger et al. showed that "the optimization problems involved in the definitions of $\widetilde{UI}$, $\widetilde{SI}$ and $\widetilde{CI}$ ... are convex optimization problems on convex sets" [8]. A notable property of convex functions is that their local and global minimums coincide, making the optimization problems that involve such functions relatively easy to solve. Indeed, many effective algorithms have been developed that solve even large convex problems both efficiently and reliably [15].

However, in this particular case, the convex optimization problem is not trivial because "the optimization problems ... can be very ill-conditioned, in the sense that there are directions in which the function varies fast, and other directions in which the function varies slowly [8]." This means that there exists extremely small eigenvalues in the positive definite matrix that needs to be inverted as part of the convex optimization procedure, making the method numerically unstable. To tackle this problem in [14], the optimization problem is analyzed in detail and found that the problematic issues occur mostly at the boundary of the feasible region. Hence, the authors proposed and compared several versions of interior point methods to provide a fast estimator of PID terms together with a certificate of its approximation quality.

The analyzed numerical estimator takes the approach of solving the optimization problem given in Equation (7) and then using the resulting distribution q to find the other quantities of interest. The user interface of the estimator is rather simple, abstracting away all the technical details of its inner workings: it takes as input a probability distribution $p(X, Y, Z)$ and outputs the scalars $MI(X; Y, Z)$, $UI(X : Y \setminus Z)$, $UI(X : Z \setminus Y)$, $SI(X : Y; Z)$ and $CI(X : Y; Z)$. For all of the analyses conducted, the convex program is solved in CVXOPT [16] ,using an interior point method. When the interior point method failed to converge, we refined the solution by solving iteratively the Karush–Kuhn–Tucker equations of the program until a desired level of tolerance was reached. See [14] for a detailed study of the performance of different algorithms to solve the optimization problem in Equation (7).

2.2. Ising Model

The Ising model, first conceived by Wilhelm Lenz in 1920 [17], is a mathematical model of ferromagnetism. The model abstracts away the rather complex details of atomic structures of magnets, consisting simply of a discrete lattice of cells or sites, denoted as s_i, each of which has an associated binary value of either -1 or $+1$ [18]. Conceptually, the lattice can be thought of as a physical material, where the sites roughly represent the unpaired electrons of its atoms. The binary value of each site intuitively corresponds to the direction of the electron's spin. A value of -1 means that the spin is considered to point down, otherwise it is said to be pointing up. A given set of spins, denoted as s (without the subscript), is called the *configuration* of the lattice [18].

The probability of a configuration s at thermal equilibrium is given by the Boltzmann distribution:

$$P_\beta(s) = \frac{e^{-\beta E(s)}}{\sum_s e^{-\beta E(s)}}, \tag{8}$$

where the sum in the denominator is over all possible spin configurations, $E(s)$ denotes the *energy* associated with the configuration s, and $\beta = \frac{1}{k_B T}$, where T is the temperature and k_B is the Boltzmann constant. Thus, β is proportional to the inverse temperature of the system.

The probability of a configuration s depends on two quantities: the internal energy of the configuration under discussion, and the temperature. Two observations that stem from Equation (8) are of importance. First, the lower the energy $E(s)$ of a configuration s, the higher its probability. Second, the higher the temperature T (or equivalently, the lower the parameter β), the more diffuse the distribution becomes. The latter mathematical property models the physical fact that, at high temperatures, the thermal "oscillation" of the atoms break the alignment of the spins, demagnetizing the material.

Assuming that the external magnetic field interacting with the lattice is omitted, and the interaction strength between pairs of nearest neighbors is fixed to be equal to the Boltzmann constant k_B, the energy of a spin configuration s simplifies to

$$H(s) = -\sum_{\langle ij \rangle} s_i s_j, \tag{9}$$

where the sum is over all different nearest neighboring pairs of spins (each pair counted only once). The minus sign in front of the sum accounts for a lower energy state (and, thus, with a higher probability) is achieved when neighboring spins take on the same value, as this yields a positive product. It can be intuitively thought as if the spins are intrinsically trying to align with their neighbors, while the temperature of the system quantifies the amount of prohibition that prevents them from doing so.

The Ising model in two or more dimensions exhibits a second order phase transition with a critical temperature T_c such that, for temperatures $T < T_c$, the expected magnetization (net alignment of spins) quickly rises to be different from zero. The Ising model is thus a prototypical example of many complex systems exhibiting collective order even under the constant presence of a source of disorder.

2.3. Elementary Cellular Automata

Elementary cellular automata (ECA) are discrete dynamical complex systems that consist of a one-dimensional array of cells, each of which has an associated binary value. Every automaton is uniquely defined by its rule table—a function that maps the value of a cell to a new value based on the cell's current value and the values of its two immediate neighbors. Since each rule table corresponds to a unique 8-bit binary number, there are only $2^8 = 256$ elementary cellular automata in total, each of which is associated with a unique decimal number from 0 to 255.

Elementary cellular automata can be simulated in time by simultaneously applying the update rule to each cell in the one-dimensional array, producing a two-dimensional plot where the vertical axis represents time. The result of evolving the rule 30, given an initial lattice configuration of all white cells except the center, can be seen in Figure 1. Notably, the figure shows that the evolution of the dynamics can be rather non-trivial. Indeed, cellular automata are interesting precisely because, despite their simplicity, the patterns that emerge as a function of the rule table and the initial configuration can be quite complex. For example, elementary cellular automata have been shown to be capable of generating random numbers [19], modelling city traffic [20] and simulating any Turing machine [21]. On the other hand, many rules quickly converge into an uninteresting homogeneous or repetitive state.

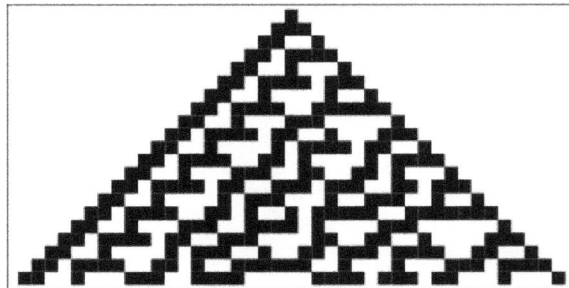

Figure 1. A space-time diagram of the evolution of rule 30 [22].

Because the set of all elementary cellular automata is rather diverse, consisting of both computationally interesting as well as uninteresting rules, it would make sense to try to group them based on the apparent complexity of their behaviour. In his seminal paper "Universality and Complexity in Cellular Automata" [23], Stephen Wolfram did just that.

After qualitatively analyzing the global structures that the different rules give rise to, given random initial states, Wolfram proposed a classification scheme that partitions all elementary cellular automata into four classes. The proposed classes are as follows:

- Class 1: Cellular automata that converge to a homogeneous state. For example, rule 0, which takes any state into a 0 state, belongs to this class.
- Class 2: Ceullar automata that converge to a repetitive or periodic state. For example, rule 184, which has been used to model traffic, belongs to this class.
- Class 3: Cellular automata that evolve chaotically. For example, rule 30, which Mathematica uses as a random number generator [24], belongs to this class.
- Class 4: Cellular automata in which persistent propagating structures are formed. For example, rule 110, which is capable of universal computation, belongs to this class. It is conjectured that other rules in this class are also universal.

3. Methods

3.1. Methodology for Analyzing the Ising Model

To estimate the PID terms in the Ising model, a two-dimensional model with Glauber dynamics [25], periodic boundary conditions and a square lattice of size 128×128 was simulated. A single simulation consisted of a burn-in period of 10^4 updates, followed by 10^5 updates from which the samples were gathered. As in the paper by Barnett et al. [26], "each update comprised L (potential) spin-flips according to Glauber transition probabilities", where L is the size of the lattice. Hence, the probability to accept a transition is given by

$$P(s \rightarrow s_n) = \frac{1}{1 + e^{\frac{\Delta E(s \rightarrow s_n)}{T}}},$$ (10)

where s and s_n denote the old and new lattice configurations, respectively, T stands for temperature and $\Delta E(s \rightarrow s_n) = E(s) - E(s_n)$ is the difference between the energies of the two successive configurations.

In other words, using Algorithm 1 as a subprocedure, the model was simulated according to Algorithm 2 with $B = 10^4$, $N = 10^5$ and $L = 128 \times 128$. This procedure was performed at 102 temperature points spaced evenly over the interval [2.0, 2.8], which encloses the theoretical phase transition at $T_c \approx 2.269$.

Algorithm 1: A single Glauber dynamics update, which consists of L spin-flip attempts

---

1 **Input:** A lattice configuration s, temperature T and lag L
2 **for** $i = 1...L$ **do**
3   Choose a random site from the lattice;
4   Flip the spin associated with the chosen site to obtain a configuration s_n;
5   Calculate $P(s \rightarrow s_n)$;
6   Generate a random number x uniformly at random within the range [0, 1];
7   **if** $x \leq P(s \rightarrow s_n)$ **then**
8         $s = s_n$;   ▷ accept the new configuration
9 **return** s;

Algorithm 2: The full Glauber dynamics algorithm

---

1 **Input:** Temperature T, burn-in period B, lag L, and the number of samples to draw N
2 Initialize a random lattice configuration s;
3 **for** $i = 1...B$ **do**
4   s = Run Algorithm 1 on inputs s, T and L;
5 set *samples* to empty list   ▷ List to save the sampled configurations to
6 **for** $i = 1...N$ **do**
7   s = Run Algorithm 1 on input s, T and L;
8   save configuration s to *samples*;
9 **return** *samples*

The obtained 10^5 lattice configurations at each temperature point were subsequently used to construct the probability distributions that the PID estimator takes as input. One-hundred sites were chosen uniformly at random at the beginning of the simulation, and they stayed the same for all temperature points. Figure 2 illustrates the 100 randomly chosen sites of the 128×128 lattice. For each site, the relative frequency of the spin configurations of its local neighborhood (the site itself along with four of its neighbors) was measured, yielding a total of 100 joint probability distributions of five random variables per temperature point. An example of one such distribution at temperature $T \approx 2.119$ is given by Table 1, where the first random variable C represents the center site, and the following four random variables represent its immediate neighbors. For example, the last row of the table illustrates that the configuration where all the spins point upwards at a specific location on the lattice has a probability of 0.776, meaning that it appears approximately $0.776 \times 10^5 = 77{,}600$ times out of a total of 10^5 configurations sampled. The high probability of "all aligned" spins is to be expected, since the samples are taken while the Ising model is in the ordered, low temperature regime.

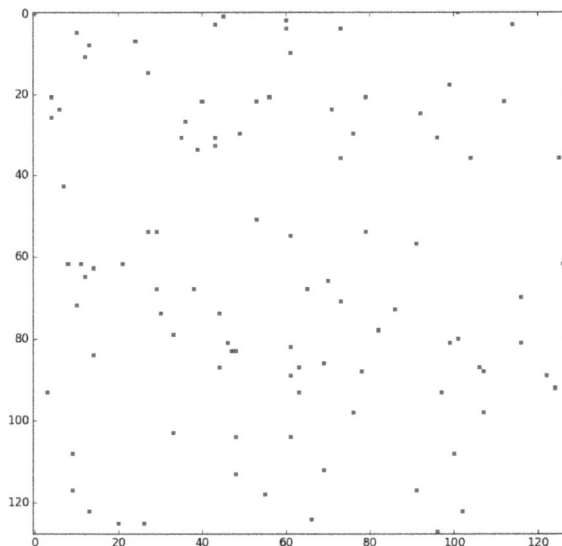

Figure 2. One-hundred randomly chosen sites (blue dots) of a 128×128 square lattice.

Having created 100 probability distributions for each of the 102 temperature points, it remains to feed the distributions into the PID estimator for analysis. However, this can not be done naively with the current setup, as the estimator works with probability distributions of 3 random vectors only, where one of them is thought of as an output and the remaining as inputs. Thus, the distributions of the same form as the one in Table 1 must be reconfigured such that they are understood by the estimator, i.e., it must be decided how neighboring sites are partitioned into 2 sets of inputs and an output. Two different setups were considered. First, the center site was taken to be the output, and only 2 neighbors were chosen without repetitions uniformly at random (out of the possible set of 4 neighbors) as inputs. Second, the center was again considered as an output, but, in this experiment, all 4 neighbors were taken into consideration as inputs: the full set of neighbors was randomly partitioned into 2 disjoint pairs, such that each pair was a two-dimensional random vector. After estimating the PID terms, an arithmetic mean across the sites was taken at each temperature point, yielding 102 average PID vectors, one for each temperature point.

Table 1. Joint probability distribution of a random site and its four neighbors at temperature $T \approx 2.119$. The column labels represent the location of the sites with respect to the neighboring center (C) site: upper (U), right (R), down (D), left (L).

C	U	R	D	L	Pr
−1	−1	−1	−1	−1	0.004
−1	−1	−1	−1	1	0.002
−1	−1	−1	1	−1	0.003
−1	−1	−1	1	1	0.003
..	..	..	..	..	..
1	1	1	−1	1	0.035
1	1	1	1	−1	0.033
1	1	1	1	1	0.776

Due to the randomness present in the Glauber dynamics and in choosing the 100 sites from the lattice for analysis, the results may vary across different runs. To gain more confidence in the results, the whole experiment described above (simulating the Ising model, choosing 100 random sites for analysis, estimating the PID of the local neighborhood of the sites) was repeated 8 times and the results averaged. In the very first run, each initial spin configuration was initialized randomly at each temperature point as in line 2 of Algorithm 2, and the configuration that was arrived at after the burn in period of 10^4 updates was saved. For the subsequent 7 runs, the very first lattice configuration for temperature point T_i was chosen to be equivalent to the saved lattice configuration from the very first run at temperature point T_i. After doing the first run separately to obtain the initial configurations, the 7 remaining simulations to gather the relevant lattice configurations were run for 8 days on 41 computing nodes in parallel in a computer cluster.

3.2. Methodology for Analyzing the Elementary Cellular Automata

The average information distribution was estimated in all 88 inequivalent elementary cellular automata. (While there are 256 different rules in total, some of them are computationally equivalent. In particular, exchanging the roles of black and white in the rule table and reflecting the rule through a vertical axis does not change the computational capabilities of the automaton. Not considering rules that are equivalent under these transformations yields 88 rules that are of interest). To gather the probability distributions for the PID estimator, 88 automata with 10^4 cells were simulated for 10^3 time steps using periodic boundary conditions. For each automaton, a random initial configuration was generated, such that each cell at time step $t = 0$ was associated with a value taken uniformly at random from the set $\{0, 1\}$.

The input pair for the PID was taken to be the cell's 2 neighbors (considered as a single random vector) and the cell itself at time step t, while the output was the cell's value at the next time step $t + 1$. This is indeed a logical setup to use, as it ensures that the input set contains all the variables that the output is a function of. Using these random variables, a single global distribution was generated for each rule. Note that this differs from the methodology that was used in the case of the Ising model, where a subset of the sites was chosen for analysis, yielding 100 different local distributions and PID values, the latter of which were subsequently averaged to obtain estimates of the global measures.

Because the emergent dynamics of a cellular automaton depend on the initial configuration of the lattice, the above experiment (generating initial configurations for each of the 88 automata, simulating the dynamics and generating the distribution that is fed into the estimator) was repeated 5 times, after which the resulting 5 PIDs of each rule were averaged.

4. Results

Next, we provide the results of applying a PID estimator to the dynamics of neighboring units in two different complex systems. The focus of the first section is on the Ising model, while the second concentrates on elementary cellular automata.

4.1. Ising Model: Partial Information Decomposition as a Function of Temperature

First, we consider the case in which the partial information decomposition is evaluated for triplets of neighboring spins in the lattice. In Figure 3, the average mutual information and PID terms are given as a function of the temperature.

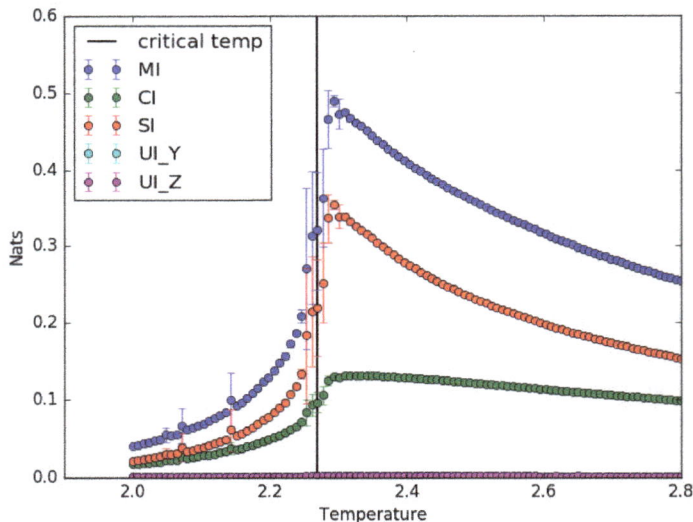

Figure 3. Average mutual information and PID terms (with two random neighbors of every "center" spin considered as inputs) of a 128×128 lattice Ising model evaluated at 102 temperature points spaced evenly over the interval [2.0, 2.8]. All information functionals are given in nats. Error bars represent the standard deviation over eight runs.

As seen from the figure, mutual information peaks around the phase transition (more precisely, at $T \approx 2.293$)—a phenomenon that agrees with previous theoretical and numerical work [26]. In addition, since, in the experiment under discussion, the mutual information was measured between a site and two of its neighbors, as opposed to measuring it between two neighboring sites only, it would be reasonable to expect the resulting mutual information to be higher in the current experiment. Indeed, two neighbors should have more information about their center site than a single neighbor has. Barnett et al. [26] observed that the mutual information between two neighboring sites (the quantity I_{pw} in the paper) achieves a maximum value of less than 0.3. In agreement with intuition, the blue graph representing mutual information in Figure 3 achieves a peak value of just under 0.5.

Observing the partial information decomposition terms of the Ising model in Figure 3, one can see that the non-zero terms seem to peak around the phase transition, just as mutual information itself does. Shared information is the most dominant of the partial information decomposition near the phase transition (before and after) and it reaches its maximum at the critical point. Indeed, shared information follows a curve similar to the mutual information, with the exception of being shifted downwards about 0.15 nats for temperatures near the phase transition. The synergistic information as a function of temperature follows a different graph with noteworthy differences. First, numerically, it peaks slightly before mutual information does at $T \approx 2.333$. In addition, its overall behaviour also deviates from that of mutual information, with the graph being quite a bit flatter, not exhibiting a sharp peak.

The unique information terms are always near 0, no matter which neighbor is considered. First, it is reasonable that both of the unique information terms are identical, as the neighbors are chosen randomly. Second, the fact that there is no unique information in the system is also intuitively plausible, as each neighbor interacts with the center site in an identical fashion. Indeed, from corollary 8 in [8], a symmetry in the probability distributions $p(X, Y) = p(X, Z)$ between the two inputs Y and Z ensures that both unique information terms should be identically zero. This symmetry between two random neighbors in a 2D Ising model is expected to be maintained across all temperatures unless the neighboring sites would belong to different frozen clusters, which is a negligible event. Moreover, all computations of PID are averaged over many different sites.

In Figure 4, the results of measuring information-theoretic functionals between the center sites and all of their neighbors are illustrated. As expected, the mutual information term increases in value (about 0.1 nat at the critical point) compared to Figure 3 because, considering all four of the sites that interact with the center site, as opposed to just two, should reduce the amount of uncertainty one has about the center. Further inspection reveals that the PID term most responsible for the increased mutual information is shared information. The complementary and unique information terms have roughly the same values in both experiments. Specifically, at all temperature points, unique information terms are 0 and synergistic information varies around 0.1 nats in the disorder regime.

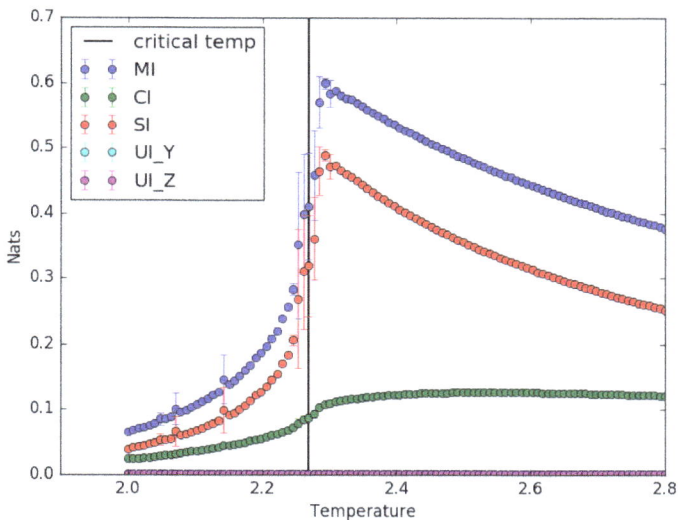

Figure 4. Average mutual information and PID terms (with all random neighbors considered as inputs) of a 128×128 lattice Ising model evaluated at 102 temperature points spaced evenly over the interval [2.0, 2.8]. Error bars represent the standard deviation over eight runs.

An unanticipated difference between the first (two neighbors) and second (four neighbors) experiment is that, when all neighbors are considered, the synergistic information term is flatter than before and peaks even deeper in the disorder phase, at temperature $T \approx 2.554$, while shared information does not change its maximum point across the two experiments.

As for the behaviour of synergistic information, we do not have an analytical explanation for the observed phenomenon. That said, it is possible that it is related to the peak of global transfer entropy (a form of conditional mutual information) in the disorder phase of the Ising model, as demonstrated by Barnett et al. [26]. According to Equation (3), when unique information vanishes, synergy becomes equal to conditional mutual information as well. However, the exact relationship between the synergy and transfer entropy in the Ising model remains unclear, as the random variables considered as arguments to the conditional mutual information functional in this paper do not correspond to the ones used by Barnett et al.

To confirm that the observed phenomena are not specific to a lattice of size 128×128, but are general characteristics of the computational properties of the Ising model, the simulations were repeated with a smaller, 64×64 lattice. The experimental setup was analogous to the one used in the previous experiments, with the exception that the measurements were averaged over six different runs (instead of eight) and, for each run, 50 different random sites were chosen for PID analysis (instead of 100). The simulations were run on 102 temperature points spaced evenly over the interval [2.0, 2.8].

Figure 5 depicts the results when only two random immediate neighbors are considered as input to the center site in the PID framework. Although the mutual, shared and synergistic information graphs are more shaky at the phase transition due to random fluctuations, in general, the graphs are almost identical to the corresponding graphs in Figure 3. The mutual and shared information quantities peak at $T \approx 2.277$, while synergistic information peaks at $T \approx 2.327$.

The results of measuring PID terms when all neighboring sites are considered as inputs to the center site are illustrated in Figure 6. Both mutual and shared information again peak at $T \approx 2.277$. Complementary information peaks at $T \approx 2.515$, a little closer to the phase transition than was the case when the lattice size was twice the size (Figure 4). This observation validates that the peak in synergy does not gradually move closer to the phase transition with increasing lattice sizes, suggesting that it could be a general property of the model.

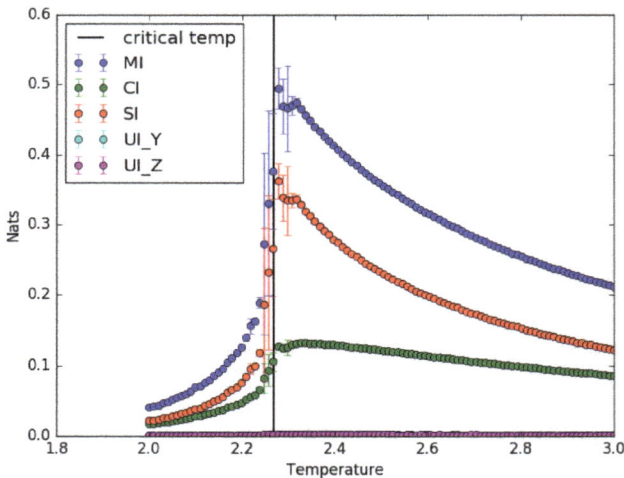

Figure 5. Average mutual information and PID terms (with two random neighbors considered as inputs) of a 64×64 lattice Ising model evaluated at 102 temperature points spaced evenly over the interval [2.0, 3.0]. Error bars represent the standard deviation over eight runs.

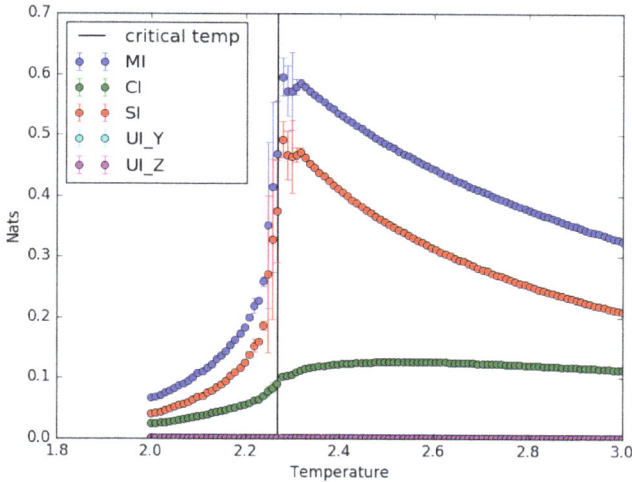

Figure 6. Average mutual information and PID terms (with all random neighbors considered as inputs) of a 64 × 64 lattice Ising model evaluated at 102 temperature points spaced evenly over the interval [2.0, 3.0]. Error bars represent the standard deviation over eight runs.

4.2. PID of Elementary Cellular Automata

In Figure 7, all 88 inequivalent elementary cellular automata have been depicted based on their PID terms. Each point represents a single rule, and the points are colored according to their Wolfram's class. Because there are four terms in PID, principal component analysis was used to project the four-dimensional PID vectors into three-dimensional space. It is important to explicitly mention that some "points" in the plot are actually clusters of several rules, but, due to their almost identical PID terms, they overlap with each other, yielding a single visual mark on the plot. For example, the cluster numbered as 1 appears to be a single point, but there are actually five different rules present at this location.

From the figure, it can be seen that the rules corresponding to Wolfram's class I are all clustered together in a single location separate from the rest of the automata. This is natural, as these class I rules quickly converge to a homogeneous all-white state, such that there is no uncertainty left in the system. In an all-white state, the entropy of the system is 0 implying that mutual information, and, accordingly, all of the PID terms to be 0 as well. While various other clusters appear, they do not correspond well to Wolfram's three other classes, meaning that there is no straightforward relationship between Wolfram's classification and the information distribution in elementary cellular automata.

To further investigate this claim, we show the distribution of PID values across Wolfram's classes in Figure 8. From Figure 8a, it can be seen that, in general, the synergy goes up when the complexity of the automata in terms of Wolfram's classification increases. However, there are many outliers in the second class and the variance of the third class is extremely high, making it hard to further draw any specific conclusions. On average, shared information seems to be higher in class 2 automata, while it is almost 0 for the majority of class 3 and 4 automata. Focusing on the last two panels (Figure 8c,d), it shows that, for these rules, two neighbors at the previous time step have usually more information about a cell's value at current time step than its own previous value does.

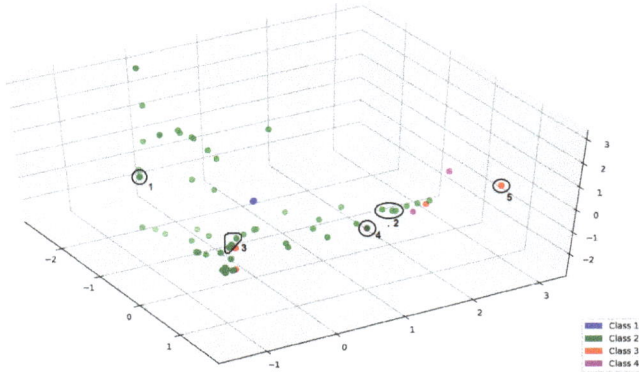

Figure 7. All 88 inequivalent cellular automata positioned on a three-dimensional space according to their information distribution. The automata are coloured based on their Wolfram's class. Some of the clusters of rules are highlighted and numbered, so that they can be referred to in the text.

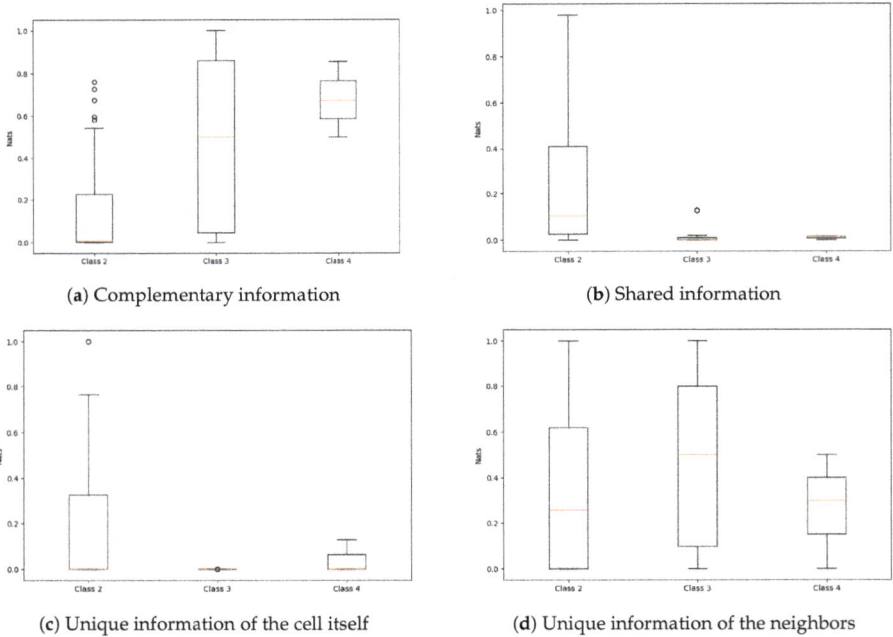

(a) Complementary information

(b) Shared information

(c) Unique information of the cell itself

(d) Unique information of the neighbors

Figure 8. Boxplots representing the distributions of specific PID terms of cellular automata belonging to Wolfram's classes II, III and IV.

In Figure 9, the top panels show the space-time diagrams of two different rules, where the dynamics were generated using random initial states. The two considered automata belong to Wolfram's second class because they quickly converge into a repetitive state. The diagrams look very alike visually as well, containing densely populated diagonal lines. It would not be unreasonable to expect these rules to be clustered together in Figure 7. Interestingly, however, these rules are partitioned into two different clusters that are spaced far apart from each other. In particular, rule 6 (and similarly rules 38 and

134) appears in cluster 2, while the automaton 130 (and similarly rules 24 and 152) belong to cluster 3. At first glance, this partitioning might be rather confusing, but the solution to the conundrum becomes apparent when one zooms in on the space-time diagrams. As can be seen from the bottom panels in Figure 9, the intricate structure of the diagonal lines is different between rules 6 and 130. It turns out that rules such as 6, 38 and 134 all have diagonal lines that are composed of small "inverted L" type blocks, while the diagonals of rules 130, 24, and 152 are much simpler, having a thickness of just a single cell. More generally, the PID terms seem to depend heavily on the specific local details of the emergent repeating, ubiquituous patterns in the space-time diagrams of cellular automata.

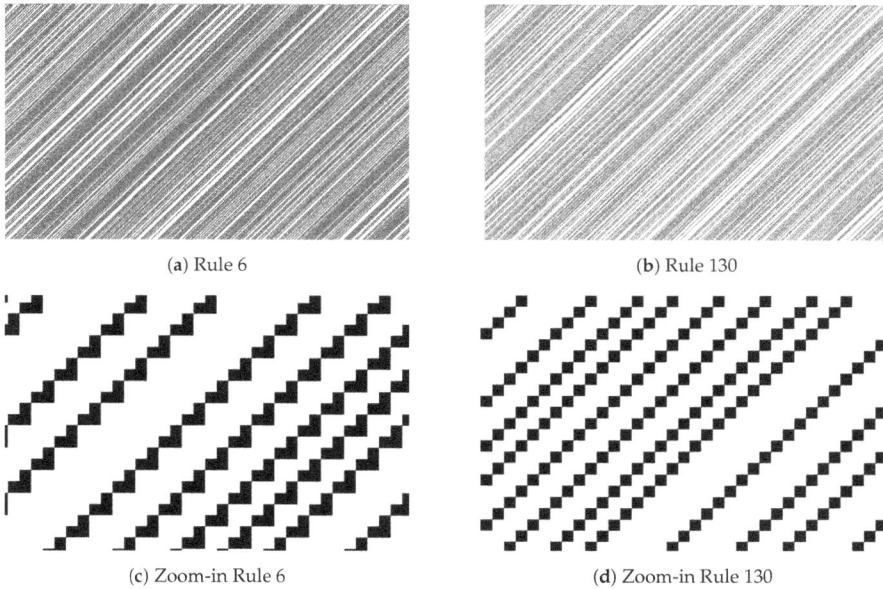

(a) Rule 6

(b) Rule 130

(c) Zoom-in Rule 6

(d) Zoom-in Rule 130

Figure 9. Top panels: space-time diagrams of elementary cellular automata belonging to Wolfram's class II. Rule 6 automaton belongs to cluster 2 in Figure 7, while Rule 130 belongs to cluster 3. Bottom panels: zoomed space-time diagrams for rules 6 and 130.

To better understand why the specific details of the diagonals yield a radical change in the PID terms, a closer quantitative look at the PID of the rules under discussion is in order. The mutual information of all of the six rules is almost exclusively divided between synergy and the unique information provided by the neighbors, leaving the remaining PID terms close to 0. The first three rules each have roughly about 0.55 nats of synergy and 0.25 nats of unique information. In contrast, the last three rules have no complementary information, but their neighbors have about twice as much unique information about the cell's next state, approximately 0.62 nats each. Thus, almost all of the information in the systems with simpler diagonals is provided uniquely by the neighbors of a site.

The former numeric observations are not surprising because, looking at the dynamics of rule 130 from Figure 9d, the new states are almost always uniquely determined by the neighbors alone. Indeed, the ubiquitous white background arises mainly because, if the right neighbor of a cell is white, this cell's next value will also be white. If, however, the left neighbor is white and the right is black, the cell's next state will be black. The latter relationship produces the diagonals. In the case of rule 6, there is a lot more synergy in the system because neither the cell's previous state or the neighbors are able to produce the complex "reversed L" shaped diagonals alone. The rather high unique information comes from the fact that the left neighbor being black completely determines that the cell's value will be white in the next step.

Some other clusters are not as straightforward to analyze, but, nevertheless, in many cases, it is still possible to give some intuitive justifications of the characterization that the PID has produced. For example, Figure 10 depicts the rules in cluster 4, which all have exactly 0.5 nats of synergy and 0.5 nats of unique information from the neighbors. While the automata look rather different from the distance, zooming into the lattices again reveals the similarities. Looking at the zoomed space-time diagrams in Figure 11, it can be seen that what the automata under observation have in common is that they all contain rather complex stairway-like structures traveling from the upper right to the lower left.

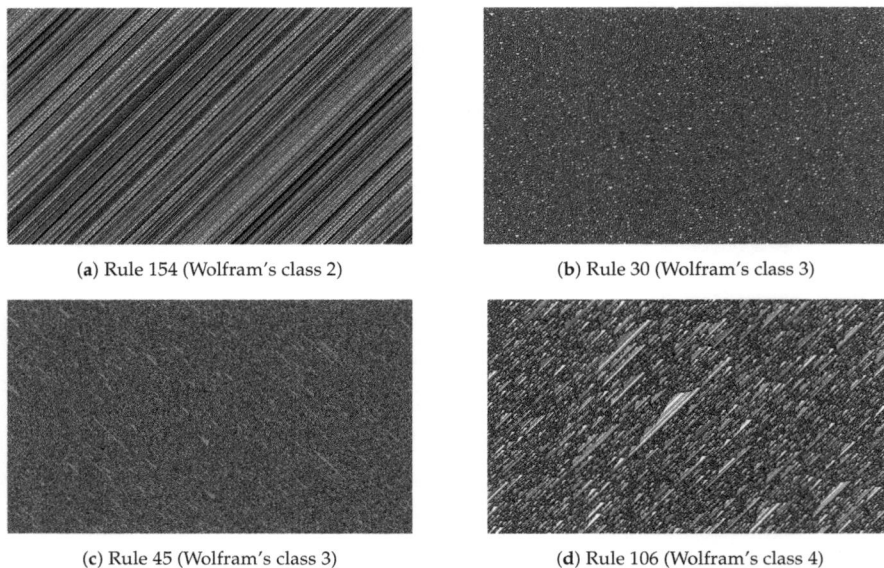

(**a**) Rule 154 (Wolfram's class 2)

(**b**) Rule 30 (Wolfram's class 3)

(**c**) Rule 45 (Wolfram's class 3)

(**d**) Rule 106 (Wolfram's class 4)

Figure 10. Space-time diagrams of elementary cellular automata belonging to cluster 4 in Figure 7.

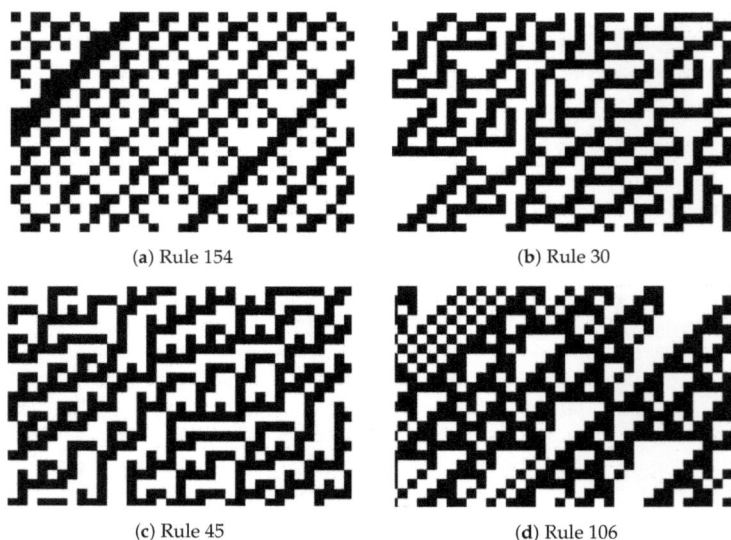

(**a**) Rule 154

(**b**) Rule 30

(**c**) Rule 45

(**d**) Rule 106

Figure 11. Zoomed space-time diagrams of the automata plotted in Figure 10.

Another noteworthy collection of rules is cluster 5, which consists of three automata that Wolfram has classified as chaotic. All the automata belonging to this cluster have 1 nat of mutual information, which is all exclusively provided by complementary information. The cluster is interesting because it shows that, at least for some subset of automata, their qualitative characterization coincides with the quantitative one provided by the PID.

5. Discussion

This section starts by putting the results obtained in the two complex systems into a larger context and by discussing their implications. The possibility of analyzing other dynamical complex systems with the information-theoretic tools used in this article is critically examined in the second section. Finally, we will conclude with several suggestions for further work.

5.1. Implications of the Results

In the paper "Information flow in a kinetic Ising model peaks in the disordered phase" [26], it is shown that global transfer entropy peaks in the disorder phase in the Ising model, just before the phase transition. This result might suggest the possibility that global transfer entropy might be used as an indicator of an impending phase transition before it actually takes place. In a subsequent commentary discussing this work [27], Lionel Barnett, one of the authors of the paper, argues that this result might also generalize to other real-world dynamical complex systems that undergo phase transitions. The practical importance of this could be high as a predictor of imminent phase transitions, but it needs to be tested in practice with real data.

In this article, it was found that one of the PID terms, complementary information or synergy also obtains a maximum in the disorder regime in the Ising system. Taking the commentary by Barnett into account, it would be worthwhile to study various real-world systems near phase transitions in terms of partial information decomposition. In particular, it would be interesting to measure the synergy between various components with the hope of predicting the arising phase transition in advance.

As for elementary cellular automata, the obtained characterization of the rules based on the PID can be a complementary perspective to Wolfram's classification. Wolfram's classification relies largely on human intuition and was developed by qualitatively analysing the space-time diagrams of all elementary cellular automata. In contrast, the characterization based on partial information decomposition is automatic and more grounded theoretically, not relying on qualitative observations. While Wolfram's classification is able to differentiate between different automata based on the global behaviour of the emergent structures, it is agnostic to the subtle details in the structures themselves. As for the characterization based on the PID terms, the opposite seems to be true.

5.2. Related Work

There is a large body of previous work in applying information theory to analyze dynamical complex systems that undergo phase transitions. Specifically, it has been shown that mutual information and other related information-theoretic measures peak at the critical point where the systems undergo an order–disorder transition. Such is the case for several mathematical models like random Boolean networks [28] and Vicsek's self-propelled particle model [29].

As for real-world systems, Harre and Bossomaier [30] measured mutual information between pairs of selected stocks and found that the peaks in information take place around known market crashes. In another paper [31], to better understand phase transitions in cognitive behaviours, the same authors analyzed mutual information between successive moves in the game of Go as a function of players' skill level. They found that information peaks around the transition from amateur to professional, "agreeing with other evidence that a radical shift in strategic thinking occurs at this juncture" [32].

Particularly relevant to the work at hand is the above-mentioned information-theoretic analysis of the Ising model. It has been analytically shown that, in a two-dimensional Ising model, the mutual information between joint states of two spin systems peaks at the critical temperature [33]. Barnett et al. [26] show empirically that mutual information measured between pairs of neighboring spins peaks at the phase transition. In the current paper, this result is replicated and extended by also measuring the decomposition terms of this mutual information. They further discovered that another related quantity called global transfer entropy peaks strictly in the disorder phase *before* the phase transition.

Not directly related to this article, but contextually rather relevant, are various works that have made use of information theory to quantitatively validate long-held hypotheses about information storage and transfer in elementary cellular automata. In the article "Local measures of information storage in complex distributed computation" [1], Lizier et al. found quantitative evidence that specific structures in elementary cellular automata called blinkers and background domains are "dominant information storage processes in these systems." In another closely related paper [34], the same authors conclude that "local transfer entropy provides the first quantitative evidence for the long-held conjecture that the emergent traveling coherent structures known as particles ... are the dominant information transfer agents in cellular automata."

Of particular interest to this paper is the work done by Chliamovitch et al. [35], in which the behaviour of multi-information, a generalization of mutual information to multiple variables, in elementary cellular automata was studied. It was found that, while it could be possible to establish a classification of cellular automata rules based on this measure, it would not correspond with Wolfram's four classes. This is because multi-information failed to discriminate between all pairs of Wolfram's classes except between classes I and IV.

5.3. Limitations

The two complex systems analyzed in this paper have an important property in common that makes their investigation with PID estimators very convenient, not to say possible. First, they are both binary, meaning that the individual elements of the systems can only be in two different states. Second, in both systems, each local part of the model is directly influenced by only a handful of other agents. Indeed, in the Ising model, the energy of a single site depends only on the spins of its four immediate neighbors, while the next value of a cell in elementary cellular automata is determined by the three cells in its local neighborhood. What follows is a discussion of why both of these characteristics are paramount to successful analysis of information distribution in complex systems.

First, the systems being binary, or more generally, discrete with relatively few possible states, ensures that the number of rows in the probability distribution that the PID numerical estimator takes as input is relatively small. The number of rows of the distribution increases polynomially in the number of states of the random variables that it contains. For example, a distribution with three random variables with 20 possible states would have 8000 rows. Such a large distribution is challenging for the numerical estimators we used, and the version at the moment we conducted this research was able to handle distributions with roughly 2500 rows. This challenge also can arise when the analyzed system has continuous elements, since a naive discretization strategy, or, in other words, dividing the continuous signal into a finite number of different states, will result in a large number of number of states. To analyze the performance of the estimator on discretized versions of continuous signals, a multivariate Gaussian probability distribution was generated, discretized, and fed into the estimator. The convergence of this discretization approach together with the study of the optimization challenges in the numerical estimators of PID are presented in [14].

Second, the systems having few directly interdependent components again ensures that the number of rows in the distributions is relatively small, the latter increasing polynomially in the number of random variables that the three random vectors contain. There is, however, an even more fundamental problem that has nothing to do with the numerical estimator, but rather with the fact that the PID mathematical framework has currently been developed for two logical input sets only. In particular, if the number of inputs in the system grows, and they are not naturally divisible into two distinct sets, it becomes increasingly hard to reasonably choose the two subsets of input channels. Even if the input space is composed of two logical sets, taking only a small subset of components from each might not yield desirable results. This is because there is exponentially many ways to choose the subsets with respect to each other, and there is often no straightforward way to know which configuration is the "right" one.

To better understand the argument put forth in the last paragraph, it is instructive to look at the results of another preliminary experiment that was carried out as part of this research. In particular, the average information distribution between the nodes in a feed-forward neural network was analyzed while it was trained on a classification task. The model consisted of two hidden layers, each containing 300 neurons. While such models usually have continuous activation functions, it is not feasible to discretize these continuous signals with fine enough granularity without making their analysis with the estimator unfeasible. Thus, binary activations were used in the hidden layers of the network, as introduced by Courbariaux et al. [36]. The output layer of the network consisted of softmax units. The network was trained on the MNIST handwritten digit database [37] for 150 epochs. The training and validation learning curves of this classifier exhibited smooth decaying graphs saturating at certain base levels.

To estimate the information distribution in the system, 200 triplets were taken for analysis. For each triplet, the two inputs were taken to be two random nodes from the last hidden layer of the network, and the output was taken to be the true target decimal value. The 200 probability distributions were subsequently fed into the PID numerical estimator and the results averaged. This procedure was repeated for each epoch, but the 200 triplets remained the same throughout the experiment.

We observed that the mutual information behaves similarly to the reflection of the training loss over the horizontal axis. This agrees with the observation made in Bard Sorngard's master's thesis "Information Theory for Analyzing Neural Networks" [38], in which the mutual information between the neurons in a toy neural network was measured during training. We also observed that the unique information terms follow the mutual information curve almost exactly, and that complementary and redundant information terms are both essentially 0. It is the authors' belief that the PID terms are rather uninteresting largely because the inputs do not come from two logically distinct subsystems (especially given the limitation that we used a PID framework so far restricted to characterize the information relations between one output and two input variables). Every neuron in the last layer has 299 neighbors, and there is no fundamental reason to prefer one neighbor over the other. This illustrates some challenges in finding a natural partition of a complex system in meaningful triplets of random variables to which one could apply most of the current versions of partial information decomposition.

5.4. Future Work

There are various promising research directions in the domain of partial information decomposition itself. First, the mathematical framework of partial information decomposition used here has so far been developed for the bivariate input case. The general decomposition of multi-variate information remains to be further developed, and it is expected to open the door to a refined characterization of information distribution in many classes of complex systems not considered in this article.

In the case of the Ising model, it might be of interest to study more theoretically how information is distributed between the different parts of the model. This would provide some further insight as to why the PID functionals behave as they do in this specific model. In addition, the results obtained in the Ising system should inspire further research into real-world complex systems in which it would be of importance to predict the occurrence of a phase transition in advance.

A system of major importance in which Ising models have been shown to be a good fit is the dynamics of ganglion cells in the vertebrate retina [39]. This system has been extensively researched as an excellent model to study neuronal population codes. In particular, a pressing question is to what degree these neurons code visual information in a redundant or independent manner, a question that can be directly addressed by the framework analyzed here.

In this paper, *elementary* cellular automata were studied, in which, by definition, each cell is directly influenced by only three cells in its local neighborhood. However, these relatively simple systems are just a special case of a larger class of models, called *one-dimensional cellular automata*, where cells can depend on an arbitrary fixed number of nearby cells. It is up to further work to study the information distribution in cellular automata that are not elementary. Das et al. [40] used genetic algorithms to discover different rules that are able to perform specific computational tasks, like classifying whether the majority of cells in the initial configuration have a value of 1. It could be worthwhile to study the information distribution in different automata that solve common tasks.

Finally, there is more work to be done in analyzing the information distribution in artificial neural networks. The PID measurements obtained from analyzing feed-forward neural networks in this work were uninteresting largely because there was no natural partitioning of nodes belonging to the same layer in this model. However, such a partitioning does exist in recurrent neural networks, where each neuron has both bottom-up inputs from the previous layer and lateral contextual inputs from the same layer at the previous time step. Applying the current numerical estimator to recurrent networks can prove to be difficult, however, as for the authors' knowledge, there is no existing work validating that binarizing the activations of a recurrent network yields a reasonable model.

More generally, we consider that, provided a meaningful partition of nodes in the network and armed with multivariate approaches [41,42], the concepts and tools from information theory can play an important role in characterizing and bringing a novel perspective on the training of neural networks.

6. Conclusions

Most of this paper is devoted to applying PID to empirically analyze the distribution of information in two well-known dynamical complex systems.

First, it was observed that complementary or synergistic information peaks in the disorder regime of the Ising model. If such phenomenon is to be generalizable to other phase transitions, this result could be of practical value. Second, a novel quantitative characterization of elementary cellular automata based on information distribution was obtained. The proposed characterization is complementary, and orthogonal, to the popular qualitative classification proposed by S. Wolfram. Third, feedforward neural networks were found to be difficult to characterize in information distribution terms within the current bivariate PID framework. Some more promising research directions in the study of neural networks and information dynamics include recurrent neural networks and multivariate formulations of PID.

Acknowledgments: The authors would like to thank M. Wibral and V. Priesemann for introducing us to the problem of PID and many insightful discussions. The authors also thank J. Lizier, P. Martinez Mediano, and L. Barnett for insightful discussions about PID in Ising and neural networks. R.V. also thanks the financial support from the Estonian Research Council through the personal research grant PUT1476. This work was supported by the Estonian Centre of Excellence in IT (EXCITE), funded by the European Regional Development Fund.

Author Contributions: Sten Sootla was responsible for implementing the relevant models in code and carrying out all of the subsequent analyses. Dirk Oliver Theis and Raul Vicente conceptualized and supervised the research, and developed the PID estimator used in this work.

Conflicts of Interest: The authors declare no conflict of interest.

References

1. Lizier, J.T.; Prokopenko, M.; Zomaya, A.Y. Local measures of information storage in complex distributed computation. *Inf. Sci.* **2012**, *208*, 39–54.
2. Schreiber, T. Measuring Information Transfer. *Phys. Rev. Lett.* **2000**, *85*, 461–464.

3. Vicente, R.; Wibral, M.; Lindner, M.; Pipa, G. Transfer entropy—A model-free measure of effective connectivity for the neurosciences. *J. Comput. Neurosci.* **2011**, *30*, 45–67.
4. Wibral, M.; Vicente, R.; Lindner, M. *Transfer Entropy in Neuroscience*; Springer: Berlin, Germany, 2014.
5. Wibral, M.; Vicente, R.; Lizier, J.T. *Directed Information Measures in Neuroscience*; Springer: Berlin, Germany, 2014.
6. Lizier, J.T.; Prokopenko, M.; Zomaya, A.Y. Information modification and particle collisions in distributed computation. *Chaos* **2010**, *20*, 037109.
7. Williams, P.L.; Beer, R.D. Nonnegative Decomposition of Multivariate Information. *arXiv* **2010**, arXiv:1004.2515.
8. Bertschinger, N.; Rauh, J.; Olbrich, E.; Jost, J.; Ay, N. Quantifying Unique Information. *Entropy* **2014**, *16*, 2161–2183.
9. Harder, M.; Salge, C.; Polani, D. Bivariate measure of redundant information. *Phys. Rev. E* **2013**, *87*, 012130.
10. Griffith, V.; Koch, C. Quantifying Synergistic Mutual Information. In *Guided Self-Organization: Inception*; Prokopenko, M., Ed.; Springer: Berlin/Heidelberg, Germany, 2014; pp. 159–190.
11. Ince, R.A. The Partial Entropy Decomposition: Decomposing Multivariate Entropy and Mutual Information via Pointwise Common Surprisal. *arXiv* **2017**, arXiv:1702.01591.
12. Wibral, M.; Lizier, J.T.; Priesemann, V. Bits from brains for biologically inspired computing. *Front. Robot. AI* **2015**, *2*, 5.
13. Wibral, M.; Priesemann, V.; Kay, J.W.; Lizier, J.T.; Phillips, W.A. Partial Information Decomposition as a Unified Approach to the Specification of Neural Goal Functions. *arXiv* **2015**, arXiv:1510.00831.
14. Makkeh, A.; Theis, D.O.; Vicente, R. Bivariate Partial Information Decomposition: The Optimization Perspective. *Entropy* **2017**, *19*, 530.
15. Boyd, S.; Vandenberghe, L. *Convex Optimization*; Cambridge University Press: New York, NY, USA, 2004.
16. Andersen, M.S.; Dahl, J.; Vandenberghe, L. CVXOPT: A Python Package for Convex Optimization. Available online: http://cvxopt.org/ (accessed on 2 November 2017).
17. Niss, M. History of the Lenz-Ising Model 1920-1950: From Ferromagnetic to Cooperative Phenomena. *Arch. Hist. Exact Sci.* **2005**, *59*, 267–318.
18. Huang, K. *Statistical Mechanics*, 2nd ed.; John Wiley & Sons: Hoboken, NJ, USA, 1987.
19. Wolfram, S. Random Sequence Generation by Cellular Automata. *Adv. Appl. Math.* **1986**, *7*, 123–169.
20. David, A.; Rosenblueth, C.G. A Model of City Traffic Based on Elementary Cellular Automata. *Complex Syst.* **2011**, *19*, 305.
21. Cook, M. Universality in Elementary Cellular Automata. *Complex Syst.* **2004**, *15*, 1–40.
22. Weisstein, E.W. Elementary Cellular Automaton. From MathWorld—A Wolfram Web Resource. Available online: http://mathworld.wolfram.com/ElementaryCellularAutomaton.html (accessed on 4 May 2017).
23. Wolfram, S. Universality and Complexity in Cellular Automata. *Phys. D Nonlinear Phenom.* **1984**, *10D*, 1–35.
24. Wolfram, S. *A New Kind of Science*; Wolfram Media Inc.: Champaign, IL, USA, 2002.
25. Glauber, R.J. Time-dependent statistics of the Ising model. *J. Math. Phys.* **1963**, *4*, 294–307.
26. Barnett, L.; Lizier, J.T.; Harré, M.; Seth, A.K.; Bossomaier, T. Information flow in a kinetic Ising model peaks in the disordered phase. *Phys. Rev. Lett.* **2013**, *111*, 177203.
27. Barnett, L. A Commentary on Information Flow in a Kinetic Ising Model Peaks in the Disordered Phase. Available online: http://users.sussex.ac.uk/~lionelb/Ising_TE_commentary.html (accessed on 6 April 2017).
28. Lizier, J.T.; Prokopenko, M.; Zomaya, A.Y. The information dynamics of phase transitions in random boolean networks. In Proceedings of the Eleventh International Conference on the Simulation and Synthesis of Living Systems (ALife XI), Winchester, UK, 5–8 August 2008; pp. 374–381.
29. Wicks, R.T.; Chapman, S.C.; Dendy, R.O. Mutual information as a tool for identifying phase transitions in dynamical complex systems with limited data. *Phys. Rev. E* **2007**, *75*, 051125.
30. Harré, M.; Bossomaier, T. Phase-transition-like behaviour of information measures in financial markets. *EPL* **2009**, *87*, 18009.
31. Harré, M.S.; Bossomaier, T.; Gillett, A.; Snyder, A. The aggregate complexity of decisions in the game of Go. *Eur. Phys. J. B* **2011**, *80*, 555–563.
32. Bossomaier, T.; Barnett, L.; Harré, M. Information and phase transitions in socio-economic systems. *Complex Adapt. Syst. Model.* **2013**, *1*, 9.

33. Matsuda, H.; Kudo, K.; Nakamura, R.; Yamakawa, O.; Murata, T. Mutual information of Ising systems. *Int. J. Theor. Phys.* **1996**, *35*, 839–845.

34. Lizier, J.T.; Prokopenko, M.; Zomaya, A.Y. Local information transfer as a spatiotemporal filter for complex systems. *Phys. Rev. E* **2008**, *77*, 026110.

35. Chliamovitch, G.; Chopard, B.; Dupuis, A. On the Dynamics of Multi-information in Cellular Automata. In Proceedings of the Cellular Automata—11th International Conference on Cellular Automata for Research and Industry (ACRI) 2014, Krakow, Poland, 22–25 September 2014; pp. 87–95.

36. Courbariaux, M.; Bengio, Y. BinaryNet: Training Deep Neural Networks with Weights and Activations Constrained to +1 or −1. *arXiv* **2016**, arXiv:1602.02830.

37. Lecun, Y.; Cortes, C.; Burges, C.J. The MNIST Database of Handwritten Digits. Available online: http://yann.lecun.com/exdb/mnist/ (accessed on 4 May 2017).

38. Sorngard, B. Information Theory for Analyzing Neural Networks. Master's Thesis, Norwegian University of Science and Technology, Trondheim, Norway, 2014.

39. Tkačik, G.; Mora, T.; Marre, O.; Amodei, D.; Palmer, S.E.; Berry, M.J.; Bialek, W. Thermodynamics and signatures of criticality in a network of neurons. *Proc. Natl. Acad. Sci. USA* **2015**, *112*, 11508–11513.

40. Das, R.; Mitchell, M.; Crutchfield, J.P. A genetic Algorithm discovers particle-based computation in cellular automata. In *Parallel Problem Solving from Nature—PPSN III: International Conference on Evolutionary Computation, Proceedings of the Third Conference on Parallel Problem Solving from Nature, Jerusalem, Israel, 9–14 October 1994*; Davidor, Y., Schwefel, H.P., Männer, R., Eds.; Springer: Berlin/Heidelberg, Germany, 1994; pp. 344–353.

41. Shwartz-Ziv, R.; Tishby, N. Opening the Black Box of Deep Neural Networks via Information. *arXiv* **2017**, arXiv:1703.00810.

42. Tax, T.; Mediano, P.A.; Shanahan, M. The Partial Information Decomposition of Generative Neural Network Models. *Entropy* **2017**, *19*, 474.

MDPI

St. Alban-Anlage 66

4052 Basel

Switzerland

Tel. +41 61 683 77 34

Fax +41 61 302 89 18

www.mdpi.com

Entropy Editorial Office

E-mail: entropy@mdpi.com

www.mdpi.com/journal/entropy

www.ingramcontent.com/pod-product-compliance
Lightning Source LLC
Chambersburg PA
CBHW051713210326
41597CB00032B/5463